$\mathcal{L}_1$ Adaptive Control Theory

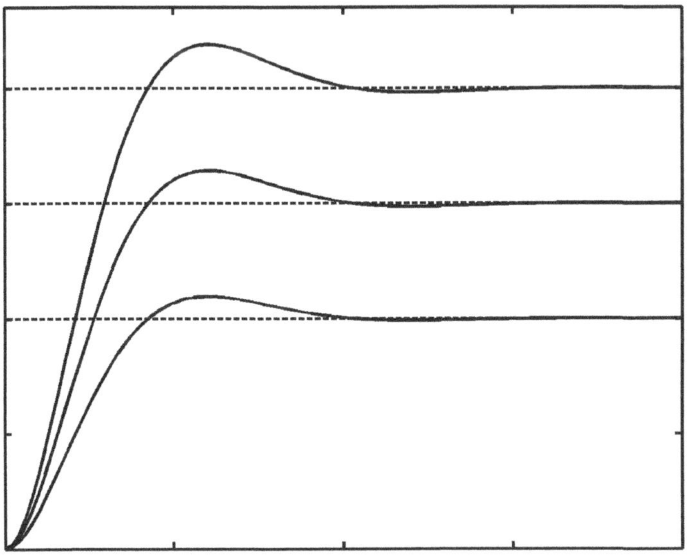

Advances in Design and Control

SIAM's Advances in Design and Control series consists of texts and monographs dealing with all areas of design and control and their applications. Topics of interest include shape optimization, multidisciplinary design, trajectory optimization, feedback, and optimal control. The series focuses on the mathematical and computational aspects of engineering design and control that are usable in a wide variety of scientific and engineering disciplines.

Series Volumes

Khalil, Hassan K., *High-Gain Observers in Nonlinear Feedback Control*
Bauso, Dario, *Game Theory with Engineering Applications*
Corless, M., King, C., Shorten, R., and Wirth, F., *AIMD Dynamics and Distributed Resource Allocation*
Walker, Shawn W., *The Shapes of Things: A Practical Guide to Differential Geometry and the Shape Derivative*
Michiels, Wim and Niculescu, Silviu-Iulian, *Stability, Control, and Computation for Time-Delay Systems: An Eigenvalue-Based Approach, Second Edition*
Narang-Siddarth, Anshu and Valasek, John, *Nonlinear Time Scale Systems in Standard and Nonstandard Forms: Analysis and Control*
Bekiaris-Liberis, Nikolaos and Krstic, Miroslav, *Nonlinear Control Under Nonconstant Delays*
Osmolovskii, Nikolai P. and Maurer, Helmut, *Applications to Regular and Bang-Bang Control: Second-Order Necessary and Sufficient Optimality Conditions in Calculus of Variations and Optimal Control*
Biegler, Lorenz T., Campbell, Stephen L., and Mehrmann, Volker, eds., *Control and Optimization with Differential-Algebraic Constraints*
Delfour, M. C. and Zolésio, J.-P., *Shapes and Geometries: Metrics, Analysis, Differential Calculus, and Optimization, Second Edition*
Hovakimyan, Naira and Cao, Chengyu, *$\mathcal{L}_1$ Adaptive Control Theory: Guaranteed Robustness with Fast Adaptation*
Speyer, Jason L. and Jacobson, David H., *Primer on Optimal Control Theory*
Betts, John T., *Practical Methods for Optimal Control and Estimation Using Nonlinear Programming, Second Edition*
Shima, Tal and Rasmussen, Steven, eds., *UAV Cooperative Decision and Control: Challenges and Practical Approaches*
Speyer, Jason L. and Chung, Walter H., *Stochastic Processes, Estimation, and Control*
Krstic, Miroslav and Smyshlyaev, Andrey, *Boundary Control of PDEs: A Course on Backstepping Designs*
Ito, Kazufumi and Kunisch, Karl, *Lagrange Multiplier Approach to Variational Problems and Applications*
Xue, Dingyü, Chen, YangQuan, and Atherton, Derek P., *Linear Feedback Control: Analysis and Design with MATLAB*
Hanson, Floyd B., *Applied Stochastic Processes and Control for Jump-Diffusions: Modeling, Analysis, and Computation*
Michiels, Wim and Niculescu, Silviu-Iulian, *Stability and Stabilization of Time-Delay Systems: An Eigenvalue-Based Approach*
Ioannou, Petros and Fidan, Barış, *Adaptive Control Tutorial*
Bhaya, Amit and Kaszkurewicz, Eugenius, *Control Perspectives on Numerical Algorithms and Matrix Problems*
Robinett III, Rush D., Wilson, David G., Eisler, G. Richard, and Hurtado, John E., *Applied Dynamic Programming for Optimization of Dynamical Systems*
Huang, J., *Nonlinear Output Regulation: Theory and Applications*
Haslinger, J. and Mäkinen, R. A. E., *Introduction to Shape Optimization: Theory, Approximation, and Computation*
Antoulas, Athanasios C., *Approximation of Large-Scale Dynamical Systems*
Gunzburger, Max D., *Perspectives in Flow Control and Optimization*
Delfour, M. C. and Zolésio, J.-P., *Shapes and Geometries: Analysis, Differential Calculus, and Optimization*
Betts, John T., *Practical Methods for Optimal Control Using Nonlinear Programming*
El Ghaoui, Laurent and Niculescu, Silviu-Iulian, eds., *Advances in Linear Matrix Inequality Methods in Control*
Helton, J. William and James, Matthew R., *Extending H^∞ Control to Nonlinear Systems: Control of Nonlinear Systems to Achieve Performance Objectives*

$\mathcal{L}_1$ Adaptive Control Theory

Guaranteed Robustness with Fast Adaptation

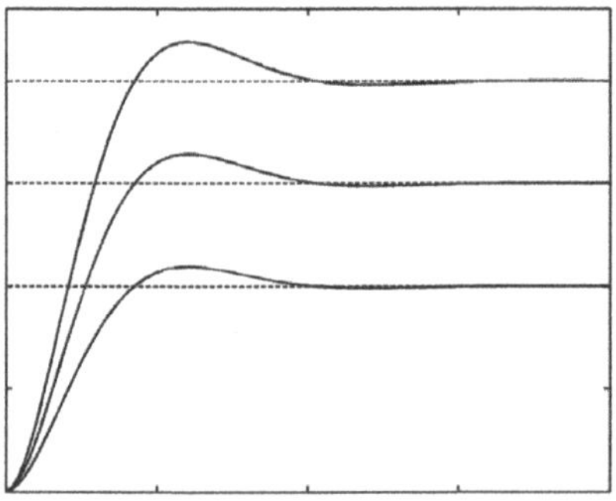

Naira Hovakimyan
University of Illinois
Urbana, Illinois

Chengyu Cao
University of Connecticut
Storrs, Connecticut

siam.

Society for Industrial and Applied Mathematics
Philadelphia

10 9 8 7 6 5 4 3 2

Figure 1 in the preface appears courtesy of the United States Army.

Figure 2 in the preface appears courtesy of the Naval Postgraduate School.

Figures 3 & 4 in the preface appear courtesy of NASA.

Figure 6.1 used with permission from NASA.

Library of Congress Cataloging-in-Publication Data

Hovakimyan, Naira.
$\mathcal{L}_1$ adaptive control theory : guaranteed robustness with fast adaptation / Naira Hovakimyan, Chengyu Cao.
p. cm. – (Advances in design and control ; 21)
Includes bibliographical references and index.
ISBN 978-0-898717-04-4
1. Adaptive control systems. 2. Robust control. I. Cao, Chengyu. II. Title.
TJ217.H68 2010
629.8'36–dc22

2010013646

To my parents Emma and Viktor, and to my sister Anna with love and gratitude
NH

To my wife Xingwei and our son Lucas Bochao, as well as to our parents
Jinrong, Guangju, Runkuan, and Yageng with love and gratitude
CC

Contents

Foreword			**xi**
Preface			**xiii**
1	**Introduction**		**1**
	1.1	Historical Overview	1
	1.2	Two Different Architectures of Adaptive Control	4
		1.2.1 Direct MRAC	4
		1.2.2 Direct MRAC with State Predictor	6
		1.2.3 Tuning Challenges	7
	1.3	Saving the Time-Delay Margin	8
	1.4	Uniformly Bounded Control Signal	12
2	**State Feedback in the Presence of Matched Uncertainties**		**17**
	2.1	Systems with Unknown Constant Parameters	17
		2.1.1 Problem Formulation	18
		2.1.2 $\mathcal{L}_1$ Adaptive Control Architecture	18
		2.1.3 Analysis of the $\mathcal{L}_1$ Adaptive Controller: Scaling	19
		2.1.4 Design of the $\mathcal{L}_1$ Adaptive Controller: Robustness and Performance	25
		2.1.5 Simulation Example	29
		2.1.6 Loop Shaping via State-Predictor Design	31
	2.2	Systems with Uncertain System Input Gain	35
		2.2.1 Problem Formulation	36
		2.2.2 $\mathcal{L}_1$ Adaptive Control Architecture	37
		2.2.3 Performance Bounds of the $\mathcal{L}_1$ Adaptive Controller	39
		2.2.4 Performance in the Presence of Nonzero Trajectory Initialization Error	44
		2.2.5 Time-Delay Margin Analysis	47
		2.2.6 Gain-Margin Analysis	59
		2.2.7 Simulation Example: Robotic Arm	59
	2.3	Extension to Systems with Unmodeled Actuator Dynamics	68
		2.3.1 Problem Formulation	68
		2.3.2 $\mathcal{L}_1$ Adaptive Control Architecture	69
		2.3.3 Analysis of the $\mathcal{L}_1$ Adaptive Controller	71

2.3.4 Simulation Example: Rohrs' Example 77
2.4 $\mathcal{L}_1$ Adaptive Controller for Nonlinear Systems 81
2.4.1 Problem Formulation 81
2.4.2 $\mathcal{L}_1$ Adaptive Control Architecture 82
2.4.3 Analysis of the $\mathcal{L}_1$ Adaptive Controller 84
2.4.4 Simulation Example: Wing Rock 91
2.5 $\mathcal{L}_1$ Adaptive Controller in the Presence of Nonlinear Unmodeled Dynamics 95
2.5.1 Problem Formulation 96
2.5.2 $\mathcal{L}_1$ Adaptive Control Architecture 97
2.5.3 Analysis of the $\mathcal{L}_1$ Adaptive Controller 99
2.5.4 Simulation Example 107
2.6 Filter Design for Performance and Robustness Trade-Off 113
2.6.1 Overview of Stochastic Optimization Algorithms 115
2.6.2 LMI-Based Filter Design 116

3 State Feedback in the Presence of Unmatched Uncertainties 123
3.1 $\mathcal{L}_1$ Adaptive Controller for Nonlinear Strict-Feedback Systems 123
3.1.1 Problem Formulation 123
3.1.2 $\mathcal{L}_1$ Adaptive Control Architecture 124
3.1.3 Analysis of the $\mathcal{L}_1$ Adaptive Controller 127
3.1.4 Simulation Example 138
3.2 $\mathcal{L}_1$ Adaptive Controller in the Presence of Unmatched Uncertainties . 142
3.2.1 Problem Formulation 143
3.2.2 $\mathcal{L}_1$ Adaptive Control Architecture 144
3.2.3 Analysis of the $\mathcal{L}_1$ Adaptive Controller 147
3.2.4 Simulation Example 156
3.3 Piecewise-Constant Adaptive Laws for Systems with Unmatched Uncertainties 161
3.3.1 Problem Formulation 161
3.3.2 $\mathcal{L}_1$ Adaptive Control Architecture 162
3.3.3 Analysis of the $\mathcal{L}_1$ Adaptive Controller 166
3.3.4 Simulation Example 174

4 Output Feedback 179
4.1 $\mathcal{L}_1$ Adaptive Output Feedback Controller for First-Order Reference Systems 179
4.1.1 Problem Formulation 180
4.1.2 $\mathcal{L}_1$ Adaptive Control Architecture 180
4.1.3 Analysis of the $\mathcal{L}_1$ Adaptive Output Feedback Controller 183
4.1.4 Design for the $\mathcal{L}_1$-Norm Condition 189
4.1.5 Simulation Example 190
4.2 $\mathcal{L}_1$ Adaptive Output Feedback Controller for Non-SPR Reference Systems 192
4.2.1 Problem Formulation 192
4.2.2 $\mathcal{L}_1$ Adaptive Control Architecture 193
4.2.3 Analysis of the $\mathcal{L}_1$ Adaptive Controller 198

4.2.4 Simulation Example: Two-Cart Benchmark Problem . . . 207

5 $\mathcal{L}_1$ Adaptive Controller for Time-Varying Reference Systems 211

5.1 $\mathcal{L}_1$ Adaptive Controller for Linear Time-Varying Systems 211
5.1.1 Problem Formulation 211
5.1.2 $\mathcal{L}_1$ Adaptive Control Architecture 212
5.1.3 Analysis of $\mathcal{L}_1$ Adaptive Controller 214
5.1.4 Simulation Example 218
5.2 $\mathcal{L}_1$ Adaptive Controller for Nonlinear Systems with Unmodeled Dynamics 223
5.2.1 Problem Formulation 223
5.2.2 $\mathcal{L}_1$ Adaptive Control Architecture 224
5.2.3 Analysis of the $\mathcal{L}_1$ Adaptive Controller 226
5.2.4 Simulation Example 234

6 Applications, Conclusions, and Open Problems 241

6.1 $\mathcal{L}_1$ Adaptive Control in Flight 241
6.1.1 Flight Validation of $\mathcal{L}_1$ Adaptive Control at Naval Postgraduate School 243
6.1.2 $\mathcal{L}_1$ Adaptive Control Design for the NASA AirSTAR Flight Test Vehicle 254
6.1.3 Other Applications 259
6.2 Key Features, Extensions, and Open Problems 261
6.2.1 Main Features of the $\mathcal{L}_1$ Adaptive Control Theory 261
6.2.2 Extensions Not Covered in the Book 261
6.2.3 Open Problems 262

A Systems Theory 265

A.1 Vector and Matrix Norms 265
A.1.1 Vector Norms 265
A.1.2 Induced Norms of Matrices 266
A.2 Symmetric and Positive Definite Matrices 267
A.3 $\mathcal{L}$-spaces and $\mathcal{L}$-norms 267
A.4 Impulse Response of Linear Time-Invariant Systems 269
A.5 Impulse Response of Linear Time-Varying Systems 270
A.6 Lyapunov Stability 271
A.6.1 Autonomous Systems 271
A.6.2 Time-Varying Systems 272
A.7 $\mathcal{L}$-Stability 274
A.7.1 BIBO Stability of LTI Systems 275
A.7.2 BIBO Stability for LTV Systems 278
A.8 Linear Parametrization of Nonlinear Systems 280
A.9 Linear Time-Varying Representation of Systems with Linear Unmodeled Dynamics 283
A.10 Linear Time-Varying Representation of Systems with Linear Unmodeled Actuator Dynamics 285
A.11 Properties of Controllable Systems 287

A.11.1 Linear Time-Invariant Systems 287
A.11.2 Linear Time-Varying Systems 288
A.12 Special Case of State-to-Input Stability 289
A.12.1 Linear Time-Invariant Systems 289
A.12.2 Linear Time-Varying Systems 290

B Projection Operator for Adaptation Laws **293**

C Basic Facts on Linear Matrix Inequalities **297**
C.1 Linear Matrix Inequalities and Convex Optimization 297
C.2 LMIs for Computation of $\mathcal{L}_1$-Norm of LTI Systems 298
C.3 LMIs for Stability Analysis of Systems with Time Delay 299
C.4 LMIs in the Presence of Uncertain Parameters 299

Bibliography **301**

Index **317**

Foreword

This book has been inspired by the problems of adaptive flight control. Research in this field started with attempts to develop adaptive autopilots for supersonic aircraft in the mid 1950s. An interesting perspective on the early development is given in the proceedings of the self-adaptive flight control symposium at Wright Development Center in March 1959 [62]. The papers describing Honeywell's self-oscillating adaptive controller [153] and the model reference adaptive controller (MRAC) based on the MIT rule [174] are particularly noteworthy. At the symposium it was announced that flight tests would be performed on the F-101A (the MIT system) and on an X-15 test vehicle (Honeywell's system) [61]. Results of the tests appeared after a few years [22, 113]. The interest in adaptive flight control generated a lot of research on adaptive control in industry and academia. However, a crash of the X-15 [164], which was partially attributed to the adaptive system, gave adaptive flight control a bad reputation.

The early development of adaptive control was dominated by experiments, and there was very little support from theory [14]. Advances made in stability theory and system identification inspired development of the theory for adaptive systems. Understanding of MRAC improved significantly when the Lyapunov theory was applied to stability analysis [25, 139]. The paper [56] by Goodwin, Ramage, and Caines, which gave conditions for global stability, was a landmark. There was also an analogous development of the self-tuning regulator (STR). Stability proofs were given in [48, 114]. A key assumption was that the parameter estimates must be bounded. Projection methods were introduced to ensure this. The assumptions required for global stability, however, were restrictive, as illustrated by the example of Rohrs [147], which showed lack of robustness in the presence of unmodeled dynamics. Various fixes, like filtering, projection, and normalization, were introduced to cope with the difficulties. The relation between the STR and the MRAS was also clarified [48].

Significant advances in applications of adaptive control in specific areas resulted in commercial products. The ship-steering autopilot Steermaster [80], developed in the 1970s, is still on the market (`http://www.es.northropgrumman.com/solutions/steermaster/index.html`). The company First Control (`http://www.firstcontrol.se/`) developed systems with STR as the primary control algorithm. Thousands of systems have been installed in the process industry since 1985. Typical applications are rolling mills and continuous casting. Use of adaptive feedforward turned out to be particularly useful.

In the late 1980s, there was a reasonable understanding of adaptive systems, many books appeared, and many applications were presented [13]. Adaptive controllers worked well in specific applications, but they did not become the widely used universal controllers that many of us dreamed about. Interest in adaptive control in academia declined in the 1990s because the easy problems had been solved and ideas for tackling the difficult problems were lacking. Backstepping [100] was the last major development.

Interest in adaptive control for aerospace applications reemerged at the turn of the century. The driving forces were requirements for reconfiguration and damage control and a strong desire to simplify extensive and costly verification and validation procedures. The Air Force, the Navy, and NASA worked with industry and academia to develop adaptive techniques for air vehicles and munition. Major flight tests were also performed [161, 175].

The first results on $\mathcal{L}_1$ adaptive control were presented by Cao and Hovakimyan at the 2006 American Control Conference [27], and the journal paper appears 2 years later [29]. This book, written by the creators of $\mathcal{L}_1$ adaptive control, gives a comprehensive account of the ideas and the theory and glimpses of flight control applications. $\mathcal{L}_1$ adaptive control can be viewed as a modified model reference adaptive control scheme where the basic architecture is based on the internal model principle.

The key theoretical results are bounds on the $\mathcal{L}_\infty$-norms of the errors in model states and control signals. The theory gives a nice way to explore adaptation rates, a long-standing open problem in adaptive control. A main result is that the error norms are (uniformly) inversely proportional to the square root of the adaptation gains. High values of the adaptation gains are thus advantageous. Another feature is that the control signal is filtered to avoid high frequencies in the control signals. The filter is also used to shape the nominal response. The conditions are given in terms of $\mathcal{L}_1$-norms of certain transfer functions that involve the filter and the largest values of the unknown parameters. The performance bounds are the key to establishing performance guarantees for adaptive control. Because of the internal model structure of the controller, it admits a good delay margin even for high adaptation gains [34]. In practice, however, the largest adaptation gains are limited by the computational power and the high-frequency sensor noise. Design of the filter is crucial; it can, however, be handled with linear theory. Even if $\mathcal{L}_1$ adaptive control was inspired by flight control, the concept of course can also be applied to other systems which require adaptation. $\mathcal{L}_1$ adaptive control has been tested in several applications, most notably flight control for aircraft, missiles, and spacecraft. The flight tests cover control surface and sensor failures and other sources of unmodeled dynamics. Current flight tests are performed on a subscale commercial jet at NASA.

Günter Stein, who gave a balanced account of adaptive flight control in the 1980s, summarized the state of the art as follows: *The main point made is that for conventional flight control problems, adaptive control is the losing alternative in a historical competition with explicit airdata scheduling*. The flight tests of $\mathcal{L}_1$ adaptive control show great promise, but only time will tell if it will be a viable alternative to gain scheduling. The material presented in this book is a good start for those who want to explore an alternative to gain scheduling.

Lund, Sweden *Karl Johan Åström*

February 2010

Preface

This book gives a comprehensive overview of the recently developed $\mathcal{L}_1$ adaptive control theory with detailed proofs of the fundamental results. The key feature of $\mathcal{L}_1$ adaptive control architectures is the guaranteed robustness in the presence of fast adaptation. This is possible to achieve by appropriate formulation of the control objective with the understanding that the uncertainties in any feedback loop can be compensated for *only* within the bandwidth of the control channel. By explicitly building the robustness specification into the problem formulation, it is possible to decouple adaptation from robustness via continuous feedback and to increase the speed of adaptation, subject only to hardware limitations. With $\mathcal{L}_1$ adaptive control architectures, fast adaptation appears to be beneficial both for performance and robustness, while the trade-off between the two is resolved via the selection of the underlying filtering structure. The latter can be addressed via conventional methods from classical and robust control. Moreover, the performance bounds of $\mathcal{L}_1$ adaptive control architectures can be analyzed to determine the extent of the modeling of the system that is required for the given set of hardware.

The book is organized into six chapters and has an appendix that summarizes the main mathematical results, used to develop the proofs.

Chapter 1 starts with a brief historical overview of adaptive control theory. It proceeds with an introduction to the main ideas of the $\mathcal{L}_1$ adaptive controller. Two equivalent architectures of model reference adaptive controllers (MRAC) are considered next, and the challenges of tuning these schemes are discussed. The chapter proceeds with analysis of a stable scalar system with constant disturbance and introduces the main idea of the $\mathcal{L}_1$ adaptive controller. Two key features are analyzed in detail: the closed-loop system's guaranteed phase margin and the uniform bound for its control signal.

Chapter 2 presents the $\mathcal{L}_1$ adaptive controller for systems in the presence of matched uncertainties. It starts from linear systems with constant unknown parameters and develops the proofs of stability and the performance bounds. The results in this section prove that the $\mathcal{L}_1$ adaptive controller leads to guaranteed, uniform, and decoupled performance bounds for both the system's input and output signals. First, it is proved that with fast adaptation the state and the control signal of the closed-loop nonlinear $\mathcal{L}_1$ adaptive system follow the same signals of a reference linear time-invariant (LTI) system for all $t \geq 0$. As compared to the original reference system of MRAC, which assumes perfect cancelation of uncertainties, the reference LTI system in $\mathcal{L}_1$ adaptive control theory assumes only partial cancelation of uncertainties, namely, those that are within the bandwidth of the control channel. Despite this, and because it is an LTI system, its response scales uniformly with changes in the initial conditions, reference inputs, and uncertain parameters. Therefore, the response of the closed-loop nonlinear $\mathcal{L}_1$ adaptive system also scales with all the changes in initial conditions, reference inputs, and uncertainties. Next, it is proved that this LTI reference system can be designed to achieve the desired control specifications. This step is the key to the trade-off between performance and robustness and is reduced to tuning the structure and the bandwidth of a stable strictly proper bandwidth-limited linear filter. Thus, the complete performance bounds of the nonlinear $\mathcal{L}_1$ adaptive controller are presented via two terms: the first is inversely proportional to the rate of adaptation, while the second depends upon the bandwidth of a linear filter. This decoupling between adaptation and robustness is the key feature of the $\mathcal{L}_1$ adaptive controller. The chapter proceeds by extending the class of systems to accommodate an uncertain system input gain, time-varying parameters, and disturbances. A rigorous proof for a lower bound of the time-delay margin of the closed-loop $\mathcal{L}_1$ adaptive system is provided in the case of open-loop linear systems with unknown constant parameters. This lower bound is computed from an LTI system, using its phase margin and its gain crossover frequency. The loop transfer function of this LTI system has a decoupled structure, which allows for tuning its phase margin or time-delay margin via the selection of the bandwidth-limited filter. The chapter proceeds by considering unmodeled actuator dynamics and nonlinear systems in the presence of unmodeled dynamics and uses the well-known Rohrs' example to provide further insights into the $\mathcal{L}_1$ adaptive controller. Other benchmark applications are discussed. An overview of tuning methods for the design of this filter for a performance–robustness trade-off is presented toward the end, and as an example, an LMI-based solution is described with certain (conservative) guarantees.

Chapter 3 extends the $\mathcal{L}_1$ adaptive controller to accommodate unmatched uncertainties. It starts with nonlinear strict-feedback systems, for which the $\mathcal{L}_1$ adaptive backstepping scheme is presented. The chapter proceeds with an extension to multi-input multi-output (MIMO) nonlinear systems in the presence of general unmatched uncertainties and unmodeled dynamics or, alternately, unknown time- and state-dependent nonlinear cross-coupling, which cannot be controlled by recursive (backstepping-type) design methods. Two different adaptive laws are introduced, one of which, being piecewise constant, is directly related to the sampling parameter of the CPU. There are certain advantages to this new type of adaptive law. It updates the parametric estimate based on the hardware (CPU) provided specification. At the sampling times, the adaptive law reduces one of the components of the identification error to zero, with the residual being proportional to the sampling interval of integration. This implies that by increasing the rate of sampling, one can reduce the influence of the residual term on the performance bounds. The uniform performance bounds are derived for the control signal and the system state as compared to the corresponding signals of a bounded closed-loop reference system, which assumes partial cancelation of uncertainties within the bandwidth of the control channel. This MIMO architecture has been applied to NASA's Generic Transport Model (GTM), which is part of the AirSTAR system, and to Boeing's X-48B blended wing body aircraft. Appropriate references are provided.

Chapter 4 presents the output feedback solution. It starts by considering first-order reference systems for performance specifications. Next, it considers more general reference systems that do not verify the SPR property for their input-output transfer function. In the second case, the piecewise-constant adaptive law is invoked for compensation of the effect of uncertainties on the system's regulated output. Similar to state-feedback architectures, a closed-loop reference system is considered, which assumes partial cancelation of uncertainties within the bandwidth of the control channel. However, unlike the state-feedback case, the sufficient condition for stability in this case couples the system uncertainty with the desired reference-system behavior and the filter design. The two-cart benchmark example is discussed as an illustration of this extension. Also, the flight tests at the Naval Postgraduate School are based on the solutions from this chapter.

Chapter 5 presents an extension to accommodate linear time-varying (LTV) reference systems. This extension is critical for practical applications. For example, in flight control, quite often the performance specifications across the flight envelope are different at different operational conditions. This leads to a time-varying reference system, the analysis of which cannot be captured by the tools developed in previous chapters. Appropriate mathematical tools for addressing this class of systems are presented in the appendices. The chapter also presents a complete solution for nonlinear systems in the presence of unmodeled dynamics. The uniform performance bounds of the system state and the control signal are computed with respect to the corresponding signals of an LTV reference system, which meets different transient specifications at different points of the operational envelope.

Chapter 6 summarizes some of the further extensions not captured within this book, gives an overview of the applications and the flight tests that have used this theory, and states the open problems and the challenges for future thinking. Appropriate references are provided.

The book concludes with appendices, where basic mathematical facts are collected to support the main proofs.

The book can be used for teaching a graduate-level special-topics course in robust adaptive control. Supplemental files for the book are posted at

`www.siam.org/books/dc21.`

Notations

The book interchangeably uses time-domain and frequency-domain language for representation of signals and systems. For example, $\xi(t)$ and $\xi(s)$ denote the function of time and its Laplace transform, respectively. However, this should not confuse the reader, as all the equations in the book are written using only one argument, either t or s. There are no mixed notations in any of the equations of the book. By smoothly moving from one form of representation to another, we streamlined the analysis and proofs. Whenever needed, thorough explanations are provided. Unless otherwise noted, $||\cdot||$ will be used for the 2-norm of a vector. Finally, $\mathfrak{L}(\xi(t))$ is used to denote the Laplace transform of the time signal $\xi(t)$.

Acknowledgments

The most valuable performance metric for every feedback solution is the experimental validation of it. We are especially grateful to Isaac Kaminer and Vladimir Dobrokhodov from the Naval Postgraduate School for sharing their challenging problems with us and for their insights into the performance of the solutions. Their numerous successful flight tests in the context of very challenging multiconstraint applications in Camp Roberts, CA, since 2006 have built strong credibility in this theory. The invaluable insights shared with us by Kevin Wise and Eugene Lavretsky from Boeing and by Brett Ridgely and Richard Hindman from Raytheon lie behind the assumptions, solutions, and techniques in this book.

Kevin Wise and Eugene Lavretsky motivated many of the problems and posed a number of *good questions* to us. Kevin provided tremendous encouragement, and if not for his efforts of transitioning adaptive control into industry products and his feedback from that experience, this theory would not have been developed. We are grateful to Randy Beard from Brigham Young University, Jonathan How from MIT, and Ralph Smith from North Carolina State University for their experimental results in different applications.

We would like to acknowledge a number of research scientists, postdoctoral fellows, and graduate students who have worked with us at different times and whose Ph.D. dissertations and technical papers contribute to the chapters in this book. Among these, we are especially thankful to Vijay Patel of the Indian Ministry of Defense for his help and support with flight control applications during his sabbatical stay at Virginia Tech in the initial stages of the theory's development. Graduate students Jiang Wang, Dapeng Li, Enric Xargay, Zhiyuan Li, Evgeny Kharisov, Hui Sun, and Kwang-Ki Kim contributed to the development of the results in this book while working on their dissertations. We are profoundly grateful to Evgeny Kharisov for his support with the editing of the book and his careful check of all the proofs and examples and for generating a large number of new simulations to support the theoretical claims. We would like to acknowledge Tyler Leman for his efforts in transitioning the theory to the X-48B platform and Alan Gostin for developing the MATLAB-based toolbox in support of this theory. Enric Xargay's contributions to the flight tests of NASA's GTM (AirSTAR) model deserve special mention. His feedback from that experience significantly enhanced the understanding of the theory. Enric's help with the last proofreading of the book was extremely useful and constructive.

We had invaluable comments from Karl J. Åström regarding the structure and the layout of the book. His recommendations regarding the additional simulations and examples helped to provide deeper insights for the reader. He also helped us with the editing of the historical overview. We are grateful to Roberto Tempo for numerous useful discussions on the design and the optimization problems of the methods in this book. Geir Dullerud from the University of Illinois at Urbana-Champaign provided a number of useful comments. Several discussions with Allen Tannenbaum and Tryphon Georgiou have influenced our thinking over the years. We are thankful to Xiaofeng Wang for extending the theory to event-triggered networked systems and to Tamer Başar for his interest in quantization of the $\mathcal{L}_1$ adaptive controller.

Irene Gregory from the NASA Langley Research Center (LaRC), Dave Doman and Michael Bolender from the Wright-Patterson Air Force Research Laboratory (AFRL), and John Burken from the NASA Dryden Flight Research Center have shared different mid- to high-fidelity models with us, explaining the particular challenges and objectives of those applications. We would like to thank Irene with a special mention for her efforts in transitioning the theory to the flight tests on the GTM at NASA LaRC. We also would like to acknowledge Kevin Cunningham, Austin Murch, and the rest of the staff of the AirSTAR flight test facility for sharing their insights into flight dynamics and for their support with implementation of the control law on the GTM. Our interactions with Johnny Evers from the Eglin AFRL, Siva Banda from the Wright-Patterson AFRL, and the research groups they lead were always full of inspiration. The financial support for this research was initially provided by the Air Force Office of Scientific Research, and it was leveraged into various programs across the country, including *Certification of Advanced Flight Critical Systems* at Wright-Patterson AFRL and NASA's *Integrated Resilient Aircraft Control.*

Last, but not least, we would like to thank our families for their unconditional dedication, love, and support, and to whom—with our humble gratitude—we dedicate this book.

Champaign, Illinois
Storrs, Connecticut

Naira Hovakimyan
Chengyu Cao

March 2010

Figure 1: SIG Rascal 110 research aircraft
(Camp Roberts, CA).

Figure 2: Flight tests with Naval Postgraduate School
(Fort Hunter Liggett, CA).

Figure 3: AirSTAR T1 and T2 research aircraft
(NASA Wallops Flight Facility, VA).

Figure 4: AirSTAR Mobile Operations Station
(NASA Wallops Flight Facility, VA).

Chapter 1

Introduction

1.1 Historical Overview

Research in adaptive control was motivated by the design of autopilots for highly agile aircraft that need to operate at a wide range of speeds and altitudes, experiencing large parametric variations. In the early 1950s adaptive control was conceived and proposed as a technology for automatically adjusting the controller parameters in the face of changing aircraft dynamics [61, 126]. In [14], that period is called the *brave era* because "there was a very short path from idea to flight test with very little analysis in between." The tragic flight test of the X-15 was the first trial of an adaptive flight control system [164]. It clearly indicated a lack of depth in understanding the robustness properties of adaptive feedback loops.

The initial results in adaptive control were inspired by system identification [115], which led to an architecture consisting of an online parameter estimator combined with automatic control design [16, 81]. Two architectures of adaptive control emerged: the direct method, where only controller parameters were estimated, and the indirect method, where process parameters were estimated and the controller parameters were obtained using some design procedure. To achieve identifiability, it was necessary to introduce a condition of persistency of excitation [15] in order to guarantee that the parameter estimates converge. The relationships between the architectures were clarified in [48].

The progress in systems theory led to fundamental theory for development of stable adaptive control architectures (see [18, 20, 49, 57, 102, 103, 127, 132, 150, 151, 159] and references therein). This was accompanied by several examples, including Rohrs' example, challenging the robustness of adaptive controllers in the presence of unmodeled dynamics, [147]. Although [147] included a rigorous proof of the existence of two infinite-gain operators in the closed-loop adaptive system, the explanation given for the phenomena observed in the simulations, which was based on qualitative considerations, was not complete. A thorough explanation was provided in later papers by Åström [12] and Anderson [5]. Nevertheless, with his example, Rohrs brought up an important point: the available adaptive control algorithms to that date were unable to adjust the bandwidth of the closed-loop system and guarantee its robustness. The results and conclusions of this paper led to an ideological controversy, and other authors started to investigate the robustness and convergence of adaptive controllers. The works of Ioannou and Kokotović [72–74], Peterson and Narendra [142], Kresselmeier and Narendra [96], and Narendra and Annaswamy [130] deserve special mention. In these papers, the authors analyzed the causes of instability and proposed damping-type modifications of adaptive laws to prevent them. The basic idea of all the modifications was to limit the gain of the adaptation loop and to eliminate its integral action. Examples of these modifications are the σ-modification [74] and the *e*-modification [130]. All these modifications attempted to provide a solution to the problem of *parameter drift*; however, they did not directly address the *architectural problem* identified by Rohrs. We notice that lack of robustness of adaptive controllers has been analyzed in robust control literature [55]. An incomplete overview of robustness and stability issues of adaptive controllers can be found in [5].

On the other hand, an example presented in [185] demonstrated that the system output can have overly poor transient tracking behavior before ideal asymptotic convergence takes place. In [182], the author proved that it may not be possible to optimize $\mathcal{L}_2$ and $\mathcal{L}_\infty$ performance simultaneously by using a constant adaptation rate. Following these results, modifications of adaptive controllers were proposed in [43, 163] that render the tracking error arbitrarily small in terms of both mean-square and $\mathcal{L}_\infty$-bounds. Further, it was shown in [42] that the modifications proposed in [43, 163] could be derived as a linear feedback of the tracking error, and the improved performance was obtained only due to a nonadaptive high-gain feedback. In [159], a composite adaptive controller was proposed, which suggests a new adaptation law using both tracking error and prediction error and leads to less oscillatory behavior in the presence of high adaptation gains as compared to model reference adaptive control (MRAC). In [125], a high-gain switching MRAC technique was introduced to achieve arbitrary good transient tracking performance under a relaxed set of assumptions as compared to MRAC, and the results were shown to be of existence type only. In [131], a multiple model switching scheme was proposed to improve the transient performance of adaptive controllers. In [10], it was shown that an arbitrarily close transient bound can be achieved by enforcing a parameter-dependent persistent excitation condition. In [101], computable $\mathcal{L}_2$- and $\mathcal{L}_\infty$-bounds for the output tracking error signals were obtained for a special class of adaptive controllers using backstepping. The underlying linear nonadaptive controller possesses a parametric robustness property. However, for a large parametric uncertainty it requires high-gain feedback. In [136], dynamic certainty equivalent controllers with unnormalized estimators were used for adaptation, which permit derivation of a uniform upper bound for the $\mathcal{L}_2$-norm of the tracking error in terms of the initial parameter estimation error. In the presence of sufficiently small initial conditions, the author proved that the $\mathcal{L}_\infty$-norm of the tracking error is upper bounded by the $\mathcal{L}_\infty$-norm of the reference input. In [9, 50, 137], a differential game theoretic type $\mathcal{H}_\infty$ approach was investigated for achieving arbitrarily close disturbance attenuation for tracking performance, albeit at the price of increased control effort. In [187], a new certainty-equivalence-based adaptive controller was presented using a backstepping-type controller with a normalized adaptive law to achieve asymptotic stability and guarantee performance bounds comparable with the tuning functions scheme, without the use of higher-order nonlinearities. References [128, 129] developed the supervisory control approach that defines a fast switching scheme between candidate controllers leading to guaranteed performance bounds. However, robustness of these schemes to unmodeled dynamics appears to be limited by the frequency of switching [6, 66].

As compared to the linear systems theory, several important aspects of the transient performance analysis seem to be missing in these efforts. First, the bounds are computed for tracking errors only, not for control signals. Although the latter can be deduced from the former, it is straightforward to verify that the ability to adjust the former may not extend to the latter in case of nonlinear control laws. Second, since the purpose of adaptive control is to ensure stable performance in the presence of modeling uncertainties, one needs to ensure that (admissible) changes in reference commands and system dynamics due to possible faults or unexpected uncertainties do not lead to unacceptable transient deviations or oscillatory control signals, implying that a *retuning of adaptation parameters* is required. Finally, one needs to ensure that the modifications or solutions, suggested for performance improvement of adaptive controllers, are not achieved via high-gain feedback.

In brief summary, the development of the theory of adaptive control over the years has taken rather the trend of defining a *larger and larger class* of systems, for which a Lyapunov proof can be done for asymptotic stability. *At which location should the uncertainty appear, what should be the degree of mismatch, how should the adaptive law be modified, etc., to get a negative definite (semidefinite) derivative of the associated candidate Lyapunov function for* a new class *of systems?* These questions or one of them is present in almost every paper addressing the next stage of development in the theory of adaptive control. Significant efforts have been reported on relaxation of the matching conditions by extending the backstepping-design approach to a broader class of systems, including strict-parametric feedback and feedforward systems [9, 45, 97, 100, 137, 138], analysis of robustness of these schemes to unmodeled dynamics [4, 8, 70, 71, 77, 134], extensions to output feedback with an objective to achieve global or semiglobal output feedback stabilization [76, 84, 98, 99, 119, 120], extension to systems with time-varying parameters [68, 122, 135, 186], relaxation of the relative degree [strictly positive real (SPR)] requirement via input-filtered transformations [118, 121], extension to nonminimum phase systems [69, 75], etc.

These fundamental results provide sufficient conditions on the bounds of uncertainties and initial conditions, which would guarantee that with the given adaptive feedback architecture, the signals in the feedback loop remain bounded. Though very important, when dealing with practical applications, boundedness, ultimate boundedness, or even asymptotic convergence are weak properties for nonlinear (adaptive) feedback systems. On one hand, unmodeled dynamics, latencies, and noise require precise quantification of the robustness and the stability margins of the underlying feedback loop. On the other hand, performance requirements in real applications necessitate a *predictable* response for the closed-loop system, dependent upon the changes in system dynamics. In adaptive control, the nature of the adaptation process plays a central role in both robustness and performance. Ideally, one would like *adaptation* to *correctly* respond to all the changes in initial conditions, reference inputs, and uncertainties by *quickly identifying* a set of control parameters that would provide a satisfactory system response. This, of course, demands *fast estimation schemes with high adaptation rates* and, as a consequence, leads to the fundamental question of determining the upper bound on the adaptation rate that would *not* result in poor robustness characteristics. We notice that the results available in the literature consistently limited the rate of variation of uncertainties, by providing examples of destabilization due to fast adaptation [75, p. 549], while the transient performance analysis was continually reduced to persistency of excitation-type assumptions, which, besides being a highly undesirable phenomenon, cannot be verified a priori. The lack of analytical quantification of the relationship between the rate of adaptation, the transient response, and the robustness margins led to *gain-scheduled designs of adaptive controllers*, examples of which are the successful flight tests of the late 1990s by the Air Force and Boeing [175, 176]. The flight tests relied on intensive Monte Carlo analysis for determination of the best rate of adaptation for various flight conditions. It was apparent that fast adaptation was leading to high frequencies in control signals and increased sensitivity to time delays. The fundamental question was thus reduced to determining an *architecture*, which would allow for *fast adaptation without losing robustness*. It was clearly understood that such an architecture can reduce the amount of gain scheduling, and possibly eliminate it, as *fast adaptation—in the presence of guaranteed robustness*—should be able to compensate for the negative effects of rapid variation of uncertainties on the system response.

The $\mathcal{L}_1$ adaptive control theory addressed precisely this question by setting an architecture in place for which adaptation is *decoupled* from robustness. The speed of adaptation in these architectures is limited only by the available hardware, while robustness is resolved via conventional methods from classical and robust control. The architectures of $\mathcal{L}_1$ adaptive control theory have *guaranteed transient performance* and *guaranteed robustness* in the presence of *fast adaptation*, without introducing or enforcing persistence of excitation, without any gain scheduling in the controller parameters, and without resorting to high-gain feedback. With $\mathcal{L}_1$ adaptive controller in the feedback loop, the response of the closed-loop system can be predicted a priori, thus significantly reducing the amount of Monte Carlo analysis required for verification and validation of such systems. These features of $\mathcal{L}_1$ adaptive control theory were verified—*consistently with the theory*—in a large number of flight tests and in mid- to high-fidelity simulation environments [19, 35, 36, 46, 51, 58, 59, 67, 82, 83, 88, 94, 104–106, 110, 124, 140, 141, 170, 172].

To facilitate the development of $\mathcal{L}_1$ adaptive control theory, in the next section we introduce two equivalent architectures of MRAC, which lead to the same error dynamics from the same initial conditions. We later use one of these structures as a basis for development of the main results in this book.

1.2 Two Different Architectures of Adaptive Control

In this section we present two different, but equivalent, architectures of adaptive control. Although their implementation is different, they both lead to the same error dynamics from the same initial conditions. The difference in their implementation principle is the key to the development of $\mathcal{L}_1$ adaptive control architectures in this book.

1.2.1 Direct MRAC

Let the system dynamics propagate according to the following differential equation:

$$\begin{aligned} \dot{x}(t) &= A_m x(t) + b\left(u(t) + k_x^\top x(t)\right), \quad x(0) = x_0, \\ y(t) &= c^\top x(t), \end{aligned} \tag{1.1}$$

where $x(t) \in \mathbb{R}^n$ is the state of the system (measured), $A_m \in \mathbb{R}^{n\times n}$ is a known Hurwitz matrix that defines the desired dynamics for the closed-loop system, $b, c \in \mathbb{R}^n$ are known constant vectors, $k_x \in \mathbb{R}^n$ is a vector of unknown constant parameters, $u(t) \in \mathbb{R}$ is the control input, and $y(t) \in \mathbb{R}$ is the regulated output. Given a uniformly bounded piecewise-continuous reference input $r(t) \in \mathbb{R}$, the objective is to define an adaptive feedback signal $u(t)$ such that $y(t)$ tracks $r(t)$ with desired specifications, while all the signals remain bounded.

The MRAC architecture proceeds by considering the nominal controller

$$u_{\text{nom}}(t) = -k_x^\top x(t) + k_g r(t), \tag{1.2}$$

where

$$k_g \triangleq \frac{-1}{c^\top A_m^{-1} b}. \tag{1.3}$$

This nominal controller assumes perfect cancelation of the uncertainties in (1.1) and leads to the desired (ideal) reference system

$$\begin{aligned} \dot{x}_m(t) &= A_m x_m(t) + b k_g r(t), \quad x_m(0) = x_0, \\ y_m(t) &= c^\top x_m(t), \end{aligned} \tag{1.4}$$

where $x_m(t) \in \mathbb{R}^n$ is the state of the reference model. The choice of k_g according to (1.3) ensures that $y_m(t)$ tracks step reference inputs with zero steady-state error.

The direct model reference adaptive controller is given by

$$u(t) = -\hat{k}_x^\top(t) x(t) + k_g r(t), \tag{1.5}$$

where $\hat{k}_x(t) \in \mathbb{R}^n$ is the estimate of k_x. Substituting (1.5) into (1.1) yields the closed-loop system dynamics

$$\begin{aligned} \dot{x}(t) &= (A_m - b\tilde{k}_x^\top(t)) x(t) + b k_g r(t), \quad x(0) = x_0, \\ y(t) &= c^\top x(t), \end{aligned}$$

where $\tilde{k}_x(t) \triangleq \hat{k}_x(t) - k_x$ denotes the parametric estimation error.

Letting $e(t) \triangleq x_m(t) - x(t)$ be the tracking error signal, the tracking error dynamics can be written as

$$\dot{e}(t) = A_m e(t) + b\tilde{k}_x^\top(t) x(t), \quad e(0) = 0. \tag{1.6}$$

The update law for the parametric estimate is given by

$$\dot{\hat{k}}_x(t) = -\Gamma x(t) e^\top(t) P b, \quad \hat{k}_x(0) = k_{x0}, \tag{1.7}$$

where $\Gamma \in \mathbb{R}^+$ is the adaptation gain and $P = P^\top > 0$ solves the algebraic Lyapunov equation

$$A_m^\top P + P A_m = -Q$$

for arbitrary $Q = Q^\top > 0$. The block diagram of the closed-loop system is given in Figure 1.1.

Consider the following Lyapunov function candidate:

$$V(e(t), \tilde{k}_x(t)) = e^\top(t) P e(t) + \frac{1}{\Gamma} \tilde{k}_x^\top(t) \tilde{k}_x(t). \tag{1.8}$$

Its time derivative along the system trajectories (1.6)–(1.7) is given by

$$\begin{aligned}\dot{V}(t) &= -e^\top(t)Qe(t) + 2e^\top(t)Pb\tilde{k}_x^\top(t)x(t) + \frac{2}{\Gamma}\tilde{k}_x^\top(t)\dot{\hat{k}}_x(t)\\ &= -e^\top(t)Qe(t) + 2\tilde{k}_x^\top(t)\left(\frac{1}{\Gamma}\dot{\hat{k}}_x(t) + x(t)e^\top(t)Pb\right)\\ &= -e^\top(t)Qe(t) \le 0.\end{aligned}$$

Hence, the equilibrium of (1.6)–(1.7) is Lyapunov stable, i.e., the signals $e(t)$, $\tilde{k}_x(t)$ are bounded. Since $x(t) = x_m(t) - e(t)$, and $x_m(t)$ is the state of a stable reference model, then $x(t)$ is bounded. To show that the tracking error converges asymptotically to zero, we compute the second derivative of $V(e(t), \tilde{k}_x(t))$ as

$$\ddot{V}(t) = -2e^\top(t)Q\dot{e}(t).$$

It follows from (1.6) that $\dot{e}(t)$ is uniformly bounded, and hence $\ddot{V}(t)$ is bounded, implying that $\dot{V}(t)$ is uniformly continuous. Application of Barbalat's lemma (see A.6.1) yields

$$\lim_{t\to\infty} \dot{V}(t) = 0,$$

which consequently proves that $e(t) \to 0$ as $t \to \infty$. Thus, $x(t)$ asymptotically converges to $x_m(t)$. This in turn implies that $y(t) = c^\top x(t)$ asymptotically converges to $y_m(t) = c^\top x_m(t)$, which follows $r(t)$ with desired specifications.

Notice that asymptotic convergence of parametric estimation errors to zero is not guaranteed. The parametric estimation errors are guaranteed only to stay bounded.

1.2.2 Direct MRAC with State Predictor

Next, we consider a reparameterization of the above architecture using a state predictor (or identifier), given by

$$\begin{aligned}\dot{\hat{x}}(t) &= A_m\hat{x}(t) + b(u(t) + \hat{k}_x^\top(t)x(t)), \quad \hat{x}(0) = x_0,\\ \hat{y}(t) &= c^\top\hat{x}(t),\end{aligned} \tag{1.9}$$

where $\hat{x}(t) \in \mathbb{R}^n$ is the state of the predictor. The system in (1.9) replicates the system structure from (1.1) with the unknown parameter k_x replaced by its estimate $\hat{k}_x(t)$. By subtracting (1.1) from (1.9), we obtain the *prediction error dynamics* (or identification error dynamics), independent of the control choice,

$$\dot{\tilde{x}}(t) = A_m\tilde{x}(t) + b\tilde{k}_x^\top(t)x(t), \quad \tilde{x}(0) = 0,$$

where $\tilde{x}(t) \triangleq \hat{x}(t) - x(t)$ and $\tilde{k}_x(t) \triangleq \hat{k}_x(t) - k_x$. Notice that these error dynamics are identical to the error dynamics in (1.6).

Next, let the adaptive law for $\hat{k}_x(t)$ be given as

$$\dot{\hat{k}}_x(t) = -\Gamma x(t)\tilde{x}^\top(t)Pb, \quad \hat{k}_x(0) = k_{x0}, \tag{1.10}$$

where $\Gamma \in \mathbb{R}^+$ is the adaptation rate and $A_m^\top P + PA_m = -Q$, $Q = Q^\top > 0$. This adaptive law is similar to (1.7) in its structure, except that the tracking error $e(t)$ is replaced by the prediction error $\tilde{x}(t)$. The choice of the Lyapunov function candidate

$$V(\tilde{x}(t), \tilde{k}_x(t)) = \tilde{x}^\top(t)P\tilde{x}(t) + \frac{1}{\Gamma}\tilde{k}_x^\top(t)\tilde{k}_x(t)$$

leads to

$$\dot{V}(t) = -\tilde{x}^\top(t)Q\tilde{x}(t) \leq 0,$$

implying that the errors $\tilde{x}(t)$ and $\tilde{k}_x(t)$ are uniformly bounded. Notice, however, that without introducing the feedback signal $u(t)$ one cannot apply Barbalat's lemma to conclude asymptotic convergence of $\tilde{x}(t)$ to zero. Both $x(t)$ and $\hat{x}(t)$ can diverge at the same rate, keeping $\tilde{x}(t)$ uniformly bounded.

If we use (1.5) in (1.9), with account of (1.10), the closed-loop state predictor replicates the bounded reference system of (1.4):

$$\dot{\hat{x}}(t) = A_m\hat{x}(t) + bk_g r(t), \quad \hat{x}(0) = x_0,$$
$$\hat{y}(t) = c^\top \hat{x}(t).$$

Hence, Barbalat's lemma can be invoked to conclude that $\tilde{x}(t) \to 0$ as $t \to \infty$. The block diagram of the closed-loop system with the predictor is given in Figure 1.2.

Figures 1.1 and 1.2 illustrate the fundamental difference between the direct MRAC and the predictor-based adaptation. In Figure 1.2, the control signal is provided as input to both systems, the system and the predictor, while in Figure 1.1 the control signal serves only as input to the system. This feature is the key to the development of $\mathcal{L}_1$ adaptive control architectures with quantifiable performance bounds.

1.2.3 Tuning Challenges

From the above Lyapunov analysis, we notice that the tracking error can be upper bounded in the following way:

$$\|e(t)\|(= \|\tilde{x}(t)\|) \leq \sqrt{\frac{V(t)}{\lambda_{\min}(P)}} \leq \sqrt{\frac{V(0)}{\lambda_{\min}(P)}} = \frac{\|\tilde{k}_x(0)\|}{\sqrt{\lambda_{\min}(P)\Gamma}}, \quad \forall\, t \geq 0.$$

This bound shows that the tracking error can be arbitrarily reduced for all $t \geq 0$ (including the transient phase) by increasing the adaptation gain Γ [100]. However, from the control law in (1.5) and the adaptive laws in (1.7) and (1.10), it follows that large adaptive gains result in high-gain feedback control, which manifests itself in high-frequency oscillations in the control signal and reduced tolerance to time delays. Moreover, applications requiring identification schemes with time scales comparable with those of the closed-loop dynamics appear to be extremely challenging due to undesirable interactions of the two processes [5]. Due to the lack of systematic design guidelines to select an adequate adaptation gain, tuning of such applications is being commonly resolved by either computationally expensive Monte Carlo simulations or trial-and-error methods following some empirical guidelines or engineering intuition. As a consequence, proper tuning of MRAC architectures (or their equivalent state-predictor-based reparameterizations) represents a major challenge and has largely remained an open question in the literature.

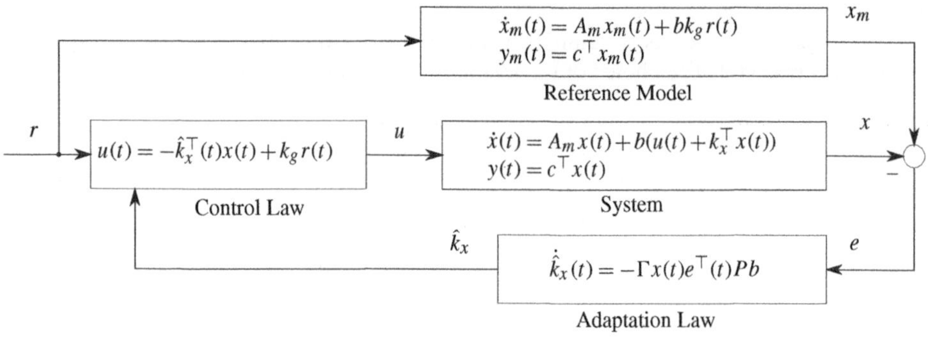

Figure 1.1: Closed-loop direct MRAC architecture.

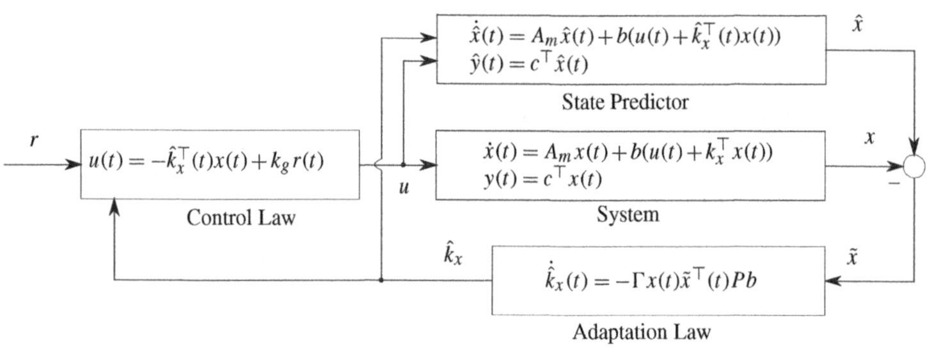

Figure 1.2: Closed-loop MRAC architecture with state predictor.

1.3 Saving the Time-Delay Margin

Next we will introduce the key ideas of the $\mathcal{L}_1$ adaptive controller, which enables *fast adaptation* with *guaranteed robustness*. We will start with a simple stable scalar system with constant disturbance, which can be analyzed by resorting to tools from classical control. We notice that in this case, MRAC reduces to a linear (model-following) integral controller. Since the closed-loop system remains linear, we use the Nyquist criterion to analyze stability and robustness of this system. Taking advantage of its linear structure, we present (i) some of the benefits of $\mathcal{L}_1$ adaptive control architectures and (ii) different concepts and tools that will be used throughout the book. In particular, we will show that *fast adaptation of $\mathcal{L}_1$ adaptive control architectures is beneficial for robustness*. We will also derive the *uniform* performance bounds of the $\mathcal{L}_1$ adaptive controller, for both the state and the control signal, and show the role of the bandwidth-limited filter of the $\mathcal{L}_1$ architecture in obtaining these uniform bounds.

Toward that end, consider the scalar system

$$\dot{x}(t) = -x(t) + \theta + u(t), \quad x(0) = x_0, \tag{1.11}$$

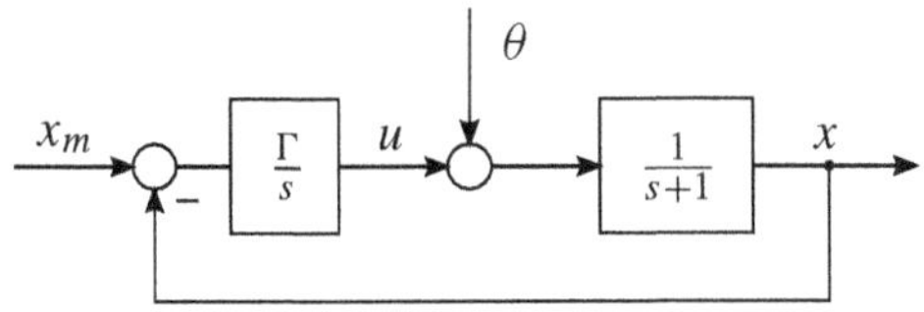

Figure 1.3: Closed-loop system with MRAC-type integral controller.

where θ is the unknown constant to be rejected by the control input $u(t)$. Let the objective be stabilization of the origin. For this system, the MRAC architecture described in (1.4) and (1.5) reduces to an integral controller of the structure

$$u(t) = -\hat{\theta}(t), \tag{1.12}$$

where $\hat{\theta}(t)$ is the estimate of θ, given by

$$\dot{\hat{\theta}}(t) = -\Gamma(x_m(t) - x(t)), \quad \hat{\theta}(0) = \theta_0, \quad \Gamma > 0, \tag{1.13}$$

and $x_m(t)$ is the reference signal, generated by the system

$$\dot{x}_m(t) = -x_m(t), \quad x_m(0) = x_0.$$

We notice that this reference system is obtained from the original system (1.11) by substitution of the ideal nominal controller $u_{\mathrm{nom}}(t) = -\theta$ into it, thus assuming perfect cancelation of the uncertain parameter θ in the system (1.11). The block diagram of the closed-loop system is shown in Figure 1.3.

The loop transfer function of this system (with negative feedback) is

$$L_1(s) = \frac{\Gamma}{s(s+1)}. \tag{1.14}$$

Because the closed-loop system remains linear time-invariant (LTI), one can use standard tools from classical control to analyze the stability margins of this system. The two most commonly used stability margins are the gain and the phase margin. From Figure 1.4(a), it is obvious that the Nyquist plot of $L_1(s)$ never crosses the negative part of the real line; therefore, the closed-loop system has infinite gain margin ($g_m = \infty$). The gain crossover frequency ω_{gc} can be computed from

$$|L_1(j\omega_{gc})| = \frac{\Gamma}{\omega_{gc}\sqrt{\omega_{gc}^2 + 1}} = 1,$$

which leads to the phase margin

$$\phi_m = \pi + \angle L_1(j\omega_{gc}) = \arctan\left(\frac{1}{\omega_{gc}}\right).$$

Careful analysis indicates that increasing Γ leads to higher gain crossover frequency and consequently reduces the phase margin. The reduction of phase margin with large Γ can also be observed in Figure 1.4(a). So, if increasing Γ improves the tracking performance for all $t \geq 0$, including the *transient phase*, then it obviously hurts the robustness (or relative stability) of the closed-loop system. Thus, the adaptation rate Γ is the key to the trade-off between performance and robustness. Since tracking and robustness cannot be achieved simultaneously, there is nothing surprising about this, but we would like to explore if the architecture can be modified so that the trade-off between tracking and robustness is resolved differently and the adaptation gain Γ can be safely increased for transient performance improvement without hurting the robustness of the closed-loop system (see Section 1.2.3).

To obtain the $\mathcal{L}_1$ adaptive controller for this system, the controller in (1.12)–(1.13) will be modified in two ways. First, we introduce the state predictor,

$$\dot{\hat{x}}(t) = -\hat{x}(t) + \hat{\theta}(t) + u(t), \quad \hat{x}(0) = x_0,$$

which leads to the following prediction error dynamics, independent of the control choice:

$$\dot{\tilde{x}}(t) = -\tilde{x}(t) + \tilde{\theta}(t), \quad \tilde{x}(0) = 0, \tag{1.15}$$

where $\tilde{x}(t) \triangleq \hat{x}(t) - x(t)$ and $\tilde{\theta}(t) \triangleq \hat{\theta}(t) - \theta$. The parametric estimate, given by (1.13), is thus replaced by

$$\dot{\hat{\theta}}(t) = -\Gamma\tilde{x}(t), \quad \theta(0) = \theta_0, \quad \Gamma > 0.$$

Next, instead of choosing the adaptive controller as $u(t) = -\hat{\theta}(t)$, we use a low-pass filtered version of $\hat{\theta}(t)$,

$$u(s) = -C(s)\hat{\theta}(s), \tag{1.16}$$

where $u(s)$ and $\hat{\theta}(s)$ are the Laplace transforms of $u(t)$ and $\hat{\theta}(t)$, respectively, and $C(s)$ is a bounded-input bounded-output (BIBO) stable strictly proper transfer function subject to $C(0) = 1$ with zero initialization for its state-space realization. The block diagram of this system is given in Figure 1.5. In the foregoing analysis, we further consider a first-order low-pass filter

$$C(s) = \frac{\omega_c}{s + \omega_c}; \tag{1.17}$$

however, similar results can be obtained using more complex filters. The loop transfer function of this system (with negative feedback) is

$$L_2(s) = \frac{\Gamma C(s)}{s(s+1) + \Gamma(1 - C(s))}. \tag{1.18}$$

Notice that in the absence of the filter, i.e., with $C(s) = 1$, the controller in (1.16) reduces to the MRAC-type integral controller introduced earlier, and (1.18) reduces to (1.14), that is, $L_2(s) = L_1(s)$.

Although (1.18) has a more complex structure than (1.14), the Nyquist plot in Figure 1.4(b) shows that the phase and the gain margins of the $\mathcal{L}_1$ controller are not significantly affected by large values of Γ. The effect of the adaptive gain on the robustness margins of the two closed-loop systems is clearly presented in Figure 1.6. The figure shows that, while the phase margin of the MRAC-type integral controller vanishes as one increases the adaptation gain Γ, the $\mathcal{L}_1$ adaptive controller has a guaranteed bounded-away-from-zero phase and gain margins in the presence of *fast adaptation*.

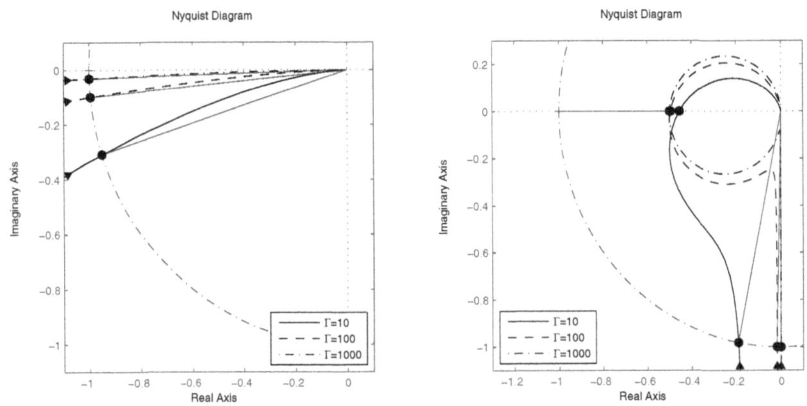

(a)
(a) Integral controller

(b) $\mathcal{L}_1$ controller

Figure 1.4: Nyquist plots for the loop transfer functions.

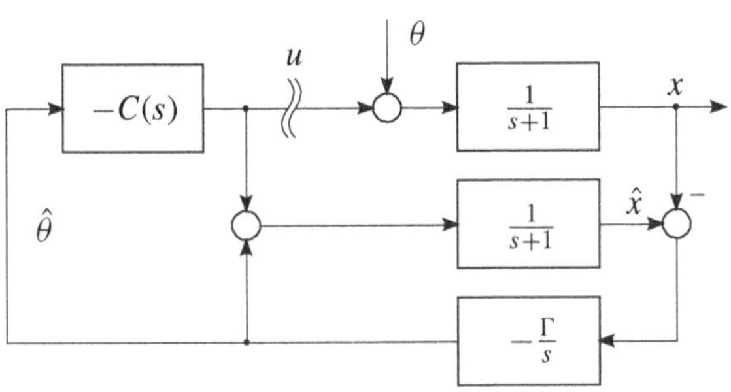

Figure 1.5: Closed-loop system with $\mathcal{L}_1$ adaptive controller.

Further, notice that as $\Gamma \to \infty$, the expression in (1.18) leads to the following limiting loop transfer function:

$$L_{2l}(s) = \frac{C(s)}{1 - C(s)} = \frac{\omega_c}{s} . \tag{1.19}$$

This loop transfer function has an infinite gain margin ($g_m = \infty$) and a phase margin of $\phi_m = \pi/2$. However, from Figure 1.6(a), we notice that the gain margin is always finite and actually converges to $g_m = 6.02$ dB with the increase of Γ. We note that the (high-frequency) dynamics of the adaptation loop do not appear in the limiting loop transfer function in (1.19). Then, since the phase crossover frequency tends to infinity as the adaptation gain Γ increases, this limiting loop transfer function cannot be used to analyze the gain margin of the closed-loop system with the $\mathcal{L}_1$ adaptive controller. However, the gain crossover frequency stays in the low-frequency range, where the limiting loop transfer function in (1.19) is a good approximation of the actual loop transfer function in (1.18). Consequently, the limiting loop transfer function can be used to analyze the phase margin of the closed-loop adaptive system.

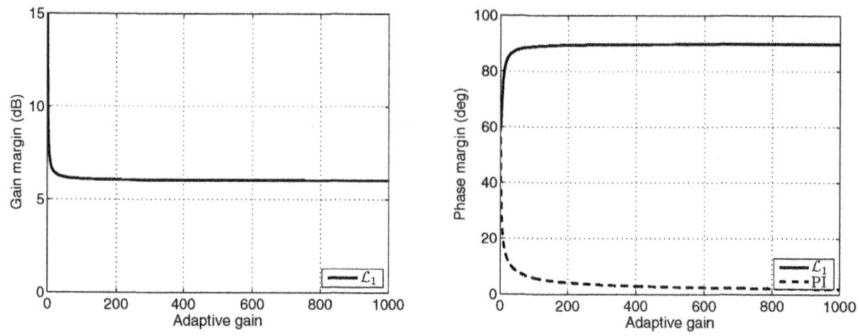

(a) gain margin (b) phase margin

Figure 1.6: Effect of high adaptation gain on the stability margins.

One can equivalently measure the robustness of the system by computing its *time-delay margin*, which is defined as the amount of time delay that brings the system to the verge of instability. The additional phase lag in the system due to a time delay τ is given by $\phi_\tau(\omega) = \angle\left(e^{-\tau j\omega}\right) = -\tau\omega$. Recalling the definition of the phase margin, one can compute the time-delay margin $\mathcal{T}$ as the amount of delay introduced in the system that reduces the phase margin to zero:

$$\phi_m = \mathcal{T}\omega_{gc} \quad \Rightarrow \quad \mathcal{T} = \frac{\phi_m}{\omega_{gc}}.$$

From (1.19), it follows that $\omega_{gc} = \omega_c$, which further implies that the $\mathcal{L}_1$ adaptive controller has the following time-delay margin as $\Gamma \to \infty$:

$$\mathcal{T} = \frac{\phi_m}{\omega_{gc}} = \frac{\pi}{2\omega_c}.$$

Hence, we observe that the $\mathcal{L}_1$ adaptive controller, defined by (1.16), retains guaranteed robustness in the presence of large values of Γ, while the MRAC-type integral controller obviously loses its phase margin in the presence of fast adaptation.

1.4 Uniformly Bounded Control Signal

Next, we analyze a key property of the $\mathcal{L}_1$ adaptive controller, which is inherently related to the robustness features discussed above. We start by considering the following closed-loop structure:

$$\begin{aligned} x_{\text{ref}}(s) &= \frac{1}{s+1}\left(\frac{\theta}{s} + u_{\text{ref}}(s)\right) + \frac{x_0}{s+1}, \\ u_{\text{ref}}(s) &= -C(s)\frac{\theta}{s}. \end{aligned} \tag{1.20}$$

This system is constructed from (1.11) and (1.16) by using $\theta/s = \mathfrak{L}(\theta)$ instead of $\hat{\theta}(s)$ in (1.16) and, hence, represents a closed-loop architecture using the *ideal* nonadaptive version of the $\mathcal{L}_1$ controller. We will refer to this system as a *(closed-loop) reference system*, as it is with respect to this system that we are able to compute uniform performance bounds. Notice that the reference controller $u_{\text{ref}}(s) = -C(s)\frac{\theta}{s}$, as compared to the nominal controller $u_{\text{nom}}(s) = -\frac{\theta}{s}$ of MRAC, assumes only partial cancelation of uncertainties, i.e., it compensates only for the uncertainties within the bandwidth of $C(s)$. This reference system defines the *best achievable performance* with the $\mathcal{L}_1$ adaptive architecture. The response of this closed-loop reference system can be written as

$$x_{\text{ref}}(s) = \frac{1}{s(s+1)}(1 - C(s))\theta + \frac{x_0}{s+1}.$$

Similarly, the response of the system in (1.11) with the $\mathcal{L}_1$ controller in (1.16) takes the form (in the frequency domain)

$$x(s) = \frac{1}{s+1}\left(\frac{\theta}{s} - C(s)\hat{\theta}(s)\right) + \frac{x_0}{s+1} = \frac{1}{s+1}\left((1 - C(s))\frac{\theta}{s} - C(s)\tilde{\theta}(s)\right) + \frac{x_0}{s+1},$$

where $\hat{\theta}(s)$ and $\tilde{\theta}(s)$ are the Laplace transforms of $\hat{\theta}(t)$ and $\tilde{\theta}(t)$, respectively. Notice that

$$x_{\text{ref}}(s) - x(s) = \frac{1}{s+1}C(s)\tilde{\theta}(s). \tag{1.21}$$

Also, it follows from (1.15) that

$$\tilde{x}(s) = \frac{1}{s+1}\tilde{\theta}(s), \tag{1.22}$$

which allows for rewriting (1.21) as

$$x_{\text{ref}}(s) - x(s) = C(s)\tilde{x}(s). \tag{1.23}$$

Moreover, notice that

$$\tilde{\theta}(s) = \hat{\theta}(s) - \theta/s = -\Gamma\tilde{x}(s)/s + \hat{\theta}_0/s - \theta/s.$$

Substituting the above expression into (1.22) and solving for $\tilde{x}(s)$ leads to

$$\tilde{x}(s) = -\frac{1}{s^2 + s + \Gamma}(\theta - \hat{\theta}_0).$$

We can now take the inverse Laplace transform of $\tilde{x}(s)$ for $\Gamma > 1/4$ to obtain

$$\tilde{x}(t) = -\frac{\theta - \hat{\theta}_0}{\sqrt{\Gamma - 1/4}} e^{-\frac{1}{2}t} \sin(\sqrt{\Gamma - 1/4}\,t). \tag{1.24}$$

This expression yields the following uniform upper bound on the prediction error:

$$|\tilde{x}(t)| \leq \frac{|\theta - \hat{\theta}_0|}{\sqrt{\Gamma - 1/4}}, \quad \forall\, t \geq 0.$$

Letting $\gamma_0 \triangleq |\theta - \hat{\theta}_0|/\sqrt{\Gamma - 1/4}$, we can write

$$|\tilde{x}(t)| \leq \gamma_0, \qquad \forall\, t \geq 0.$$

Notice that $\lim_{\Gamma \to \infty} \gamma_0 = 0$. Also notice from (1.24) that $\lim_{t \to \infty} \tilde{x}(t) = 0$.

From (1.23) we can also derive the following uniform upper bound:

$$|x_{\mathrm{ref}}(t)-x(t)| = \left|\int_0^t h_c(\tau)\tilde{x}(t-\tau)d\tau\right| \leq \int_0^t |h_c(\tau)\tilde{x}(t-\tau)|d\tau$$
$$\leq \gamma_0 \int_0^t |h_c(\tau)|d\tau \leq \gamma_0 \int_0^\infty |h_c(\tau)|d\tau, \quad \forall\, t \geq 0,$$

where $h_c(t)$ is the impulse response of $C(s)$. In particular, for the $C(s)$ in (1.17), the impulse response can be explicitly computed, leading to the following uniform upper bound:

$$|x_{\mathrm{ref}}(t)-x(t)| \leq \gamma_0, \quad \forall\, t \geq 0.$$

This implies that the error between the closed-loop system with the $\mathcal{L}_1$ adaptive controller and the closed-loop reference system, which uses the reference controller, can be uniformly bounded by a constant inverse proportional to the square root of the rate of adaptation.

Similarly, using (1.16), (1.20), and (1.22), we can derive

$$u_{\mathrm{ref}}(s)-u(s) = C(s)\tilde{\theta}(s) = C(s)(s+1)\tilde{x}(s). \tag{1.25}$$

Denoting $H_u(s) \triangleq C(s)(s+1)$ and letting $h_u(t)$ be the impulse response for $H_u(s)$, we obtain the following upper bound:

$$|u_{\mathrm{ref}}(t)-u(t)| \leq \gamma_0 \int_0^\infty |h_u(\tau)|d\tau, \quad \forall\, t \geq 0. \tag{1.26}$$

Because $C(s)$ is strictly proper and BIBO stable, $H_u(s) \triangleq C(s)(s+1)$ is proper and BIBO stable, and hence it has uniformly bounded impulse response, that is $\int_0^\infty |h_u(\tau)|d\tau < \infty$. Further, since $\lim_{\Gamma\to\infty}\gamma_0 = 0$, we can conclude from (1.26) that the time history of the $\mathcal{L}_1$ adaptive controller can be rendered arbitrarily close to the one of the reference controller for all $t \geq 0$ by increasing the rate of adaptation Γ.

Notice that without the low-pass filter, i.e., with $C(s) = 1$, equation (1.25) reduces to

$$u_{\mathrm{ref}}(s)-u(s) = (s+1)\tilde{x}(s).$$

From this expression, it is obvious that the transfer function from $\tilde{x}(t)$ to $u_{\mathrm{ref}}(t)-u(t)$ is improper, and hence, in the absence of the filter $C(s)$, one cannot *uniformly* upper bound $|u_{\mathrm{ref}}(t)-u(t)|$ as we did in (1.26).

This simple analysis illustrates the role of $C(s)$ toward obtaining *a uniform performance bound* for the control signal of the $\mathcal{L}_1$ adaptive control architecture, as compared to its nonadaptive version (which is uniformly bounded by definition). We further notice that this uniform bound is inverse proportional to the square root of the rate of adaptation, similar to the tracking error. Thus, both performance bounds can be systematically improved by increasing the rate of adaptation. The remaining issue is the design of the low-pass linear filter $C(s)$ to ensure that the reference system in (1.20) achieves desired performance specifications in the presence of unknown θ.

In the remainder of this book, we will show that, similar to this simple example and with appropriate extension of the above described concepts to nonlinear closed-loop adaptive systems, the $\mathcal{L}_1$ adaptive control theory shifts the tuning issue from determining the rate of the *nonlinear* gradient minimization scheme to the design of a *linear* strictly proper and stable filter, implying that the trade-off between performance and robustness of the closed-loop adaptive system can be systematically addressed using well-established tools from classical and robust control.

Finally, we note that the uniform bounds for the system's state and control signals are expressed in terms of the impulse response of proper BIBO-stable transfer functions, which correspond to the $\mathcal{L}_1$-norms of the underlying systems. Consequently, the corresponding control architectures are referred to as $\mathcal{L}_1$ *adaptive controllers*.

Chapter 2

State Feedback in the Presence of Matched Uncertainties

In this chapter we present the full state-feedback solution for several different classes of systems in the presence of *matched* uncertainties. We start with linear systems with constant unknown parameters and present the $\mathcal{L}_1$ adaptive controller for this class of systems. We derive the performance bounds of this controller and show that these can be clearly decoupled into two distinct components, the adaptation and the robustness bounds. The *adaptation bounds* can be improved by increasing the rate of adaptation, while the *robustness bounds* can be appropriately addressed via known methods from linear-systems theory. We proceed by considering linear time-varying systems in the presence of unknown system input gain and present a new *architecture* for this class of systems. We analyze the time-delay margin for the case of constant unknown parameters and provide a guaranteed lower bound for it via the phase margin of an auxiliary LTI system. Then, we extend the $\mathcal{L}_1$ adaptive controller to nonlinear systems in the presence of unmodeled dynamics and analyze several well-known benchmark examples from the literature. We further discuss various methods for the design of the underlying filter toward achieving the desired performance-robustness trade-off for the closed-loop adaptive system with the $\mathcal{L}_1$ adaptive controller and provide a conservative, but guaranteed, solution via linear matrix inequalities (LMIs).

2.1 Systems with Unknown Constant Parameters

This section considers LTI systems in the presence of unknown constant parameters. The $\mathcal{L}_1$ adaptive controller ensures uniformly bounded transient and steady-state tracking for both of the system's signals, input and output, as compared to the same signals of a bounded reference LTI system, which assumes partial cancelation of uncertainties within the bandwidth of the control channel. The time histories of the signals of this reference LTI system can be made arbitrarily close to the signals of a different LTI system, called *design system*, the output of which can be used for control specifications [29]. This *decoupling* of the performance bounds between adaptation and robustness is further illustrated in simulations.

2.1.1 Problem Formulation

Consider the class of systems

$$\begin{aligned} \dot{x}(t) &= Ax(t) + b(u(t) + \theta^\top x(t)), \quad x(0) = x_0, \\ y(t) &= c^\top x(t), \end{aligned} \tag{2.1}$$

where $x(t) \in \mathbb{R}^n$ is the system state vector (measured); $u(t) \in \mathbb{R}$ is the control signal; $b, c \in \mathbb{R}^n$ are known constant vectors; A is the known $n \times n$ matrix, with (A, b) controllable; θ is the unknown parameter, which belongs to a given compact convex set $\theta \in \Theta \subset \mathbb{R}^n$; and $y(t) \in \mathbb{R}$ is the regulated output. In this section we present an adaptive control solution, which ensures that the system output $y(t)$ follows a given piecewise-continuous bounded reference signal $r(t)$ with quantifiable transient and steady-state performance bounds.

2.1.2 $\mathcal{L}_1$ Adaptive Control Architecture

Consider the control structure

$$u(t) = u_m(t) + u_{\mathrm{ad}}(t), \quad u_m(t) = -k_m^\top x(t), \tag{2.2}$$

where $k_m \in \mathbb{R}^n$ renders $A_m \triangleq A - bk_m^\top$ Hurwitz, while $u_{\mathrm{ad}}(t)$ is the adaptive component, to be defined shortly. The static feedback gain k_m leads to the following partially closed-loop system:

$$\begin{aligned} \dot{x}(t) &= A_m x(t) + b(\theta^\top x(t) + u_{\mathrm{ad}}(t)), \quad x(0) = x_0, \\ y(t) &= c^\top x(t). \end{aligned} \tag{2.3}$$

For the linearly parameterized system in (2.3), we consider the state predictor

$$\begin{aligned} \dot{\hat{x}}(t) &= A_m \hat{x}(t) + b(\hat{\theta}^\top(t) x(t) + u_{\mathrm{ad}}(t)), \quad \hat{x}(0) = x_0, \\ \hat{y}(t) &= c^\top \hat{x}(t), \end{aligned} \tag{2.4}$$

where $\hat{x}(t) \in \mathbb{R}^n$ is the state of the predictor and $\hat{\theta}(t) \in \mathbb{R}^n$ is the estimate of the parameter θ, governed by the following projection-type adaptive law:

$$\dot{\hat{\theta}}(t) = \Gamma \operatorname{Proj}(\hat{\theta}(t), -\tilde{x}^\top(t) P b x(t)), \quad \hat{\theta}(0) = \hat{\theta}_0 \in \Theta, \tag{2.5}$$

where $\tilde{x}(t) \triangleq \hat{x}(t) - x(t)$ is the prediction error, $\Gamma \in \mathbb{R}^+$ is the adaptation gain, and $P = P^\top > 0$ solves the algebraic Lyapunov equation $A_m^\top P + P A_m = -Q$ for arbitrary symmetric $Q = Q^\top > 0$. The projection is confined to the set Θ (see Definition B.3). The Laplace transform of the adaptive control signal is defined as

$$u_{\mathrm{ad}}(s) = -C(s)\left(\hat{\eta}(s) - k_g r(s)\right), \tag{2.6}$$

where $r(s)$ and $\hat{\eta}(s)$ are the Laplace transforms of $r(t)$ and $\hat{\eta}(t) \triangleq \hat{\theta}^\top(t) x(t)$, respectively, $k_g \triangleq -1/(c^\top A_m^{-1} b)$, and $C(s)$ is a BIBO-stable and strictly proper transfer function with DC gain $C(0) = 1$, and its state-space realization assumes zero initialization.

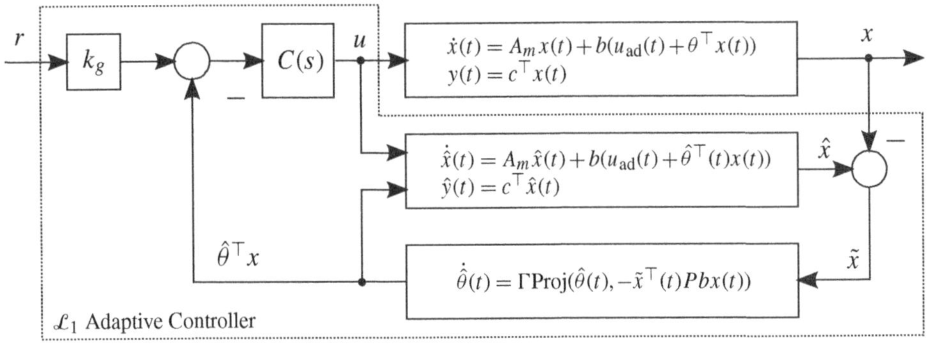

Figure 2.1: Closed-loop $\mathcal{L}_1$ adaptive system.

The $\mathcal{L}_1$ adaptive controller is defined via the relationships in (2.2), (2.4)–(2.6), with k_m and $C(s)$ verifying the following $\mathcal{L}_1$-norm condition:

$$\lambda \triangleq \|G(s)\|_{\mathcal{L}_1} L < 1, \tag{2.7}$$

where

$$G(s) \triangleq H(s)(1 - C(s)), \quad H(s) \triangleq (s\mathbb{I} - A_m)^{-1} b, \quad L \triangleq \max_{\theta \in \Theta} \|\theta\|_1 . \tag{2.8}$$

The $\mathcal{L}_1$ adaptive control architecture with its main elements is represented in Figure 2.1.

2.1.3 Analysis of the $\mathcal{L}_1$ Adaptive Controller: Scaling

Closed-Loop Reference System

Consider the following nonadaptive version of the adaptive control system in (2.1), (2.2), (2.6), which defines the *closed-loop reference system* for the class of systems in (2.1):

$$\begin{aligned} \dot{x}_{\mathrm{ref}}(t) &= A x_{\mathrm{ref}}(t) + b(\theta^\top x_{\mathrm{ref}}(t) + u_{\mathrm{ref}}(t)), \qquad x_{\mathrm{ref}}(0) = x_0, \\ u_{\mathrm{ref}}(s) &= -C(s)(\theta^\top x_{\mathrm{ref}}(s) - k_g r(s)) - k_m^\top x_{\mathrm{ref}}(s), \\ y_{\mathrm{ref}}(s) &= c^\top x_{\mathrm{ref}}(s). \end{aligned} \tag{2.9}$$

As compared to the nominal controller of MRAC in (1.2), the controller in (2.9) attempts to compensate only for uncertainties in the system that are within the bandwidth of $C(s)$. The block diagram of this system is given in Figure 2.2. Notice that $C(s) = 1$ leads to the reference model of MRAC, which was introduced in (1.4).

Lemma 2.1.1 If $\|G(s)\|_{\mathcal{L}_1} L < 1$, then the system in (2.9) is bounded-input bounded-state (BIBS) stable with respect to $r(t)$ and x_0.

Proof. From the definition of the closed-loop reference system in (2.9), it follows that

$$x_{\mathrm{ref}}(s) = H(s) k_g C(s) r(s) + G(s) \theta^\top x_{\mathrm{ref}}(s) + x_{\mathrm{in}}(s),$$

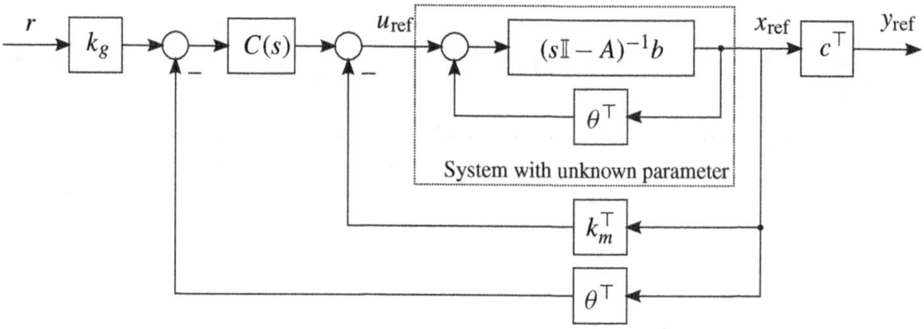

Figure 2.2: Closed-loop reference system.

where $x_{\rm in}(s) \triangleq (s\mathbb{I}-A_m)^{-1}x_0$. Recalling the fact that $H(s)$, $C(s)$, and $G(s)$ are proper BIBO-stable transfer functions, it follows from (2.9) that for all $\tau \in [0,\infty)$ the following bound holds:

$$\|x_{\rm ref\tau}\|_{\mathcal{L}_\infty} \le \|H(s)k_g C(s)\|_{\mathcal{L}_1}\|r_\tau\|_{\mathcal{L}_\infty} + \|G(s)\theta^\top\|_{\mathcal{L}_1}\|x_{\rm ref\tau}\|_{\mathcal{L}_\infty} + \|x_{\rm in\tau}\|_{\mathcal{L}_\infty}\,.$$

Since A_m is Hurwitz, $x_{\rm in}(t)$ is uniformly bounded. Definition A.7.4 and the relationships in (2.7) and (2.8) imply that

$$\|G(s)\theta^\top\|_{\mathcal{L}_1} = \max_{i=1,\ldots,n}\|G_i(s)\|_{\mathcal{L}_1}\sum_{j=1}^{n}|\theta_j| \le \|G(s)\|_{\mathcal{L}_1}L < 1\,. \tag{2.10}$$

Consequently,

$$\|x_{\rm ref\tau}\|_{\mathcal{L}_\infty} \le \frac{\|H(s)k_g C(s)\|_{\mathcal{L}_1}}{1-\|G(s)\theta^\top\|_{\mathcal{L}_1}}\|r_\tau\|_{\mathcal{L}_\infty} + \frac{\|x_{\rm in\tau}\|_{\mathcal{L}_\infty}}{1-\|G(s)\theta^\top\|_{\mathcal{L}_1}}\,. \tag{2.11}$$

Since $r(t)$ and $x_{\rm in}(t)$ are uniformly bounded, and (2.11) holds uniformly for all $\tau \in [0,\infty)$, $x_{\rm ref}(t)$ is uniformly bounded. Boundedness of $y_{\rm ref}(t)$ follows from its definition. This completes the proof. □

Transient and Steady-State Performance

The following error dynamics can be derived from (2.3) and (2.4):

$$\dot{\tilde{x}}(t) = A_m\tilde{x}(t) + b\tilde{\theta}^\top(t)x(t)\,, \quad \tilde{x}(0)=0\,, \tag{2.12}$$

where $\tilde{\theta}(t) \triangleq \hat{\theta}(t)-\theta$. Letting $\tilde{\eta}(t) \triangleq \tilde{\theta}^\top(t)x(t)$, with $\tilde{\eta}(s)$ being its Laplace transform, the error dynamics in (2.12) can be written in the frequency domain as

$$\tilde{x}(s) = H(s)\tilde{\eta}(s)\,. \tag{2.13}$$

Lemma 2.1.2 The prediction error in (2.12) is uniformly bounded:

$$\|\tilde{x}\|_{\mathcal{L}_\infty} \le \sqrt{\frac{\theta_{\max}}{\lambda_{\min}(P)\Gamma}}\,, \quad \theta_{\max} \triangleq 4\max_{\theta\in\Theta}\|\theta\|^2\,, \tag{2.14}$$

where $\lambda_{\min}(P)$ is the minimum eigenvalue of P.

Proof. Consider the following Lyapunov function candidate:

$$V(\tilde{x}(t),\tilde{\theta}(t)) = \tilde{x}^\top(t)P\tilde{x}(t) + \frac{1}{\Gamma}\tilde{\theta}^\top(t)\tilde{\theta}(t).$$

Using Property B.2 of the projection operator, we can upper bound the derivative of the Lyapunov function along the trajectories of the system as

$$\begin{aligned}
\dot{V}(t) &= \dot{\tilde{x}}^\top(t)P\tilde{x}(t) + \tilde{x}^\top(t)P\dot{\tilde{x}}(t) + \frac{1}{\Gamma}\left(\dot{\tilde{\theta}}^\top(t)\tilde{\theta}(t) + \tilde{\theta}^\top(t)\dot{\tilde{\theta}}(t)\right)\\
&= \tilde{x}^\top(t)(A_m^\top P + PA_m)\tilde{x}(t) + 2\tilde{x}^\top(t)Pb\tilde{\theta}^\top(t)x(t) + \frac{2}{\Gamma}\tilde{\theta}^\top(t)\dot{\hat{\theta}}(t)\\
&= -\tilde{x}^\top(t)Q\tilde{x}(t) + 2\tilde{x}^\top(t)Pb\tilde{\theta}^\top(t)x(t) + 2\tilde{\theta}^\top(t)\mathrm{Proj}(\hat{\theta}(t), -x(t)\tilde{x}^\top(t)Pb)\\
&= -\tilde{x}^\top(t)Q\tilde{x}(t) + 2\tilde{\theta}^\top(t)\left(x(t)\tilde{x}^\top(t)Pb + \mathrm{Proj}(\hat{\theta}(t), -x(t)\tilde{x}^\top(t)Pb)\right)\\
&\le -\tilde{x}^\top(t)Q\tilde{x}(t),
\end{aligned}$$

which implies that $\tilde{x}(t)$ and $\tilde{\theta}(t)$ are uniformly bounded. Next, since $\tilde{x}(0) = 0$, it follows that

$$\lambda_{\min}(P)\|\tilde{x}(t)\|^2 \le V(t) \le V(0) = \frac{\tilde{\theta}^\top(0)\tilde{\theta}(0)}{\Gamma}.$$

The projection operator ensures that $\hat{\theta}(t) \in \Theta$, and therefore

$$\frac{\tilde{\theta}^\top(0)\tilde{\theta}(0)}{\Gamma} \le \frac{4\max_{\theta\in\Theta}\|\theta\|^2}{\Gamma},$$

which leads to the following upper bound:

$$\|\tilde{x}(t)\|^2 \le \frac{\theta_{\max}}{\lambda_{\min}(P)\Gamma}.$$

Since $\|\cdot\|_\infty \le \|\cdot\|$, and this bound is uniform, the bound above yields

$$\|\tilde{x}_\tau\|_{\mathcal{L}_\infty} \le \sqrt{\frac{\theta_{\max}}{\lambda_{\min}(P)\Gamma}},$$

which holds for every $\tau \ge 0$. The result in (2.14) immediately follows from this inequality, as it holds uniformly in τ. □

We notice that the bound in (2.14) is derived independently of $u_{\mathrm{ad}}(t)$. This implies that both $x(t)$ and $\hat{x}(t)$ can diverge at the same rate, maintaining a uniformly bounded error between the two. Next, we prove that, with the adaptive feedback given by (2.6), the state of the predictor remains bounded and consequently leads to asymptotic convergence of the tracking error $\tilde{x}(t)$ to zero.

Lemma 2.1.3 If $u_{\mathrm{ad}}(t)$ is defined according to (2.6), and the condition in (2.7) holds, then we have the following asymptotic result:

$$\lim_{t\to\infty}\tilde{x}(t) = 0. \tag{2.15}$$

Proof. To prove asymptotic convergence of $\tilde{x}(t)$ to zero, one needs to ensure that $\hat{x}(t)$ in (2.4), with $u_{\rm ad}(t)$ given by (2.6), is uniformly bounded.

First, we notice that

$$\hat{x}(s) = G(s)\hat{\eta}(s) + k_g H(s)C(s)r(s) + x_{\rm in}(s),$$

which leads to the following upper bound:

$$\|\hat{x}_\tau\|_{\mathcal{L}_\infty} \leq \|G(s)\|_{\mathcal{L}_1}\|\hat{\eta}_\tau\|_{\mathcal{L}_\infty} + \|k_g H(s)C(s)\|_{\mathcal{L}_1}\|r_\tau\|_{\mathcal{L}_\infty} + \|x_{\rm in\tau}\|_{\mathcal{L}_\infty}. \tag{2.16}$$

Next, applying the triangular relationship for norms to the bound in (2.14), we have

$$\big|\|\hat{x}_\tau\|_{\mathcal{L}_\infty} - \|x_\tau\|_{\mathcal{L}_\infty}\big| \leq \sqrt{\frac{\theta_{\max}}{\lambda_{\min}(P)\Gamma}}.$$

The projection in (2.5) ensures that $\hat{\theta}(t) \in \Theta$, and hence $\|\hat{\eta}_\tau\|_{\mathcal{L}_\infty} \leq L\|x_\tau\|_{\mathcal{L}_\infty}$. Substituting for $\|x_\tau\|_{\mathcal{L}_\infty}$ yields

$$\|\hat{\eta}_\tau\|_{\mathcal{L}_\infty} \leq L\left(\|\hat{x}_\tau\|_{\mathcal{L}_\infty} + \sqrt{\frac{\theta_{\max}}{\lambda_{\min}(P)\Gamma}}\right). \tag{2.17}$$

Then, the bounds on $\|\hat{x}_\tau\|_{\mathcal{L}_\infty}$ and $\|\hat{\eta}_\tau\|_{\mathcal{L}_\infty}$ in (2.16) and (2.17), with account of the stability condition in (2.7), lead to

$$\|\hat{x}_\tau\|_{\mathcal{L}_\infty} \leq \frac{\lambda\sqrt{\frac{\theta_{\max}}{\lambda_{\min}(P)\Gamma}} + \|k_g H(s)C(s)\|_{\mathcal{L}_1}\|r_\tau\|_{\mathcal{L}_\infty} + \|x_{\rm in\tau}\|_{\mathcal{L}_\infty}}{1-\lambda}.$$

Since the bound on the right-hand side is uniform, $\hat{x}(t)$ is uniformly bounded. Application of Barbalat's lemma leads to the asymptotic result in (2.15). □

Remark 2.1.1 The above presented proof can be straightforwardly extended to accommodate faster prediction error dynamics by considering a more general structure for the predictor as compared to (2.4):

$$\dot{\hat{x}}(t) = A_m\hat{x}(t) + b(\hat{\theta}^\top(t)x(t) + u_{\rm ad}(t)) - K_{\rm sp}\tilde{x}(t), \quad \hat{x}(0) = x_0,$$
$$\hat{y}(t) = c^\top\hat{x}(t),$$

where $K_{\rm sp} \in \mathbb{R}^{n\times n}$ can be used to assign faster poles for $(A_m - K_{\rm sp})$. The idea of having different poles for the prediction error dynamics as compared to the original $H(s) = (s\mathbb{I} - A_m)^{-1}b$ was first introduced in [27].

To streamline the subsequent derivation of the performance bounds for the class of systems considered in this section, we first note that (A_m, b) is the state-space realization of $H(s)$, and since the pair (A, b) is controllable and $A_m = A - bk_m^\top$, then (A_m, b) is also controllable. Hence, Lemma A.12.1 implies that there exists c_o such that

$$H_1(s) \triangleq C(s)\frac{1}{c_o^\top H(s)}c_o^\top \tag{2.18}$$

is proper and BIBO stable.

Theorem 2.1.1 For the system in (2.1) and the controller defined via (2.2) and (2.4)–(2.6), subject to the $\mathcal{L}_1$-norm condition in (2.7), we have

$$\|x_{\rm ref} - x\|_{\mathcal{L}_\infty} \le \frac{\gamma_1}{\sqrt{\Gamma}}, \qquad \|u_{\rm ref} - u\|_{\mathcal{L}_\infty} \le \frac{\gamma_2}{\sqrt{\Gamma}}, \tag{2.19}$$

$$\lim_{t\to\infty} \|x_{\rm ref}(t) - x(t)\| = 0, \qquad \lim_{t\to\infty} \|u_{\rm ref}(t) - u(t)\| = 0, \tag{2.20}$$

where

$$\gamma_1 \triangleq \frac{\|C(s)\|_{\mathcal{L}_1}}{1 - \|G(s)\|_{\mathcal{L}_1} L} \sqrt{\frac{\theta_{\max}}{\lambda_{\min}(P)}},$$

$$\gamma_2 \triangleq \|H_1(s)\|_{\mathcal{L}_1} \sqrt{\frac{\theta_{\max}}{\lambda_{\min}(P)}} + \|C(s)\theta^\top + k_m^\top\|_{\mathcal{L}_1} \gamma_1 .$$

Proof. The response of the closed-loop system in (2.3) with the $\mathcal{L}_1$ adaptive controller in (2.6) can be written (in the frequency domain) as

$$x(s) = H(s) k_g C(s) r(s) + G(s)\theta^\top x(s) - H(s)C(s)\tilde{\eta}(s) + x_{\rm in}(s).$$

Also, from the definition of the closed-loop reference system in (2.9), it follows that

$$x_{\rm ref}(s) = H(s) k_g C(s) r(s) + G(s)\theta^\top x_{\rm ref}(s) + x_{\rm in}(s).$$

The two expressions above and the prediction error dynamics in (2.13) lead to

$$x_{\rm ref}(s) - x(s) = G(s)\theta^\top (x_{\rm ref}(s) - x(s)) + C(s)\tilde{x}(s), \tag{2.21}$$

which, along with Lemma A.7.1, implies that

$$\|(x_{\rm ref} - x)_\tau\|_{\mathcal{L}_\infty} \le \|G(s)\theta^\top\|_{\mathcal{L}_1} \|(x_{\rm ref} - x)_\tau\|_{\mathcal{L}_\infty} + \|C(s)\|_{\mathcal{L}_1} \|\tilde{x}_\tau\|_{\mathcal{L}_\infty} .$$

Then, the bounds in (2.10) and (2.14) lead to the uniform upper bound

$$\|(x_{\rm ref} - x)_\tau\|_{\mathcal{L}_\infty} \le \frac{\|C(s)\|_{\mathcal{L}_1}}{1 - \|G(s)\|_{\mathcal{L}_1} L} \|\tilde{x}_\tau\|_{\mathcal{L}_\infty} \le \frac{\|C(s)\|_{\mathcal{L}_1}}{1 - \|G(s)\|_{\mathcal{L}_1} L} \sqrt{\frac{\theta_{\max}}{\lambda_{\min}(P)\Gamma}}, \tag{2.22}$$

which proves the first bound in (2.19). Moreover, it follows from (2.21) that

$$x_{\rm ref}(s) - x(s) = (\mathbb{I} - G(s)\theta^\top)^{-1} C(s)\tilde{x}(s).$$

From the Final Value Theorem and Lemma 2.1.3, we know that

$$\lim_{s\to 0} s\tilde{x}(s) = \lim_{t\to\infty} \tilde{x}(t) = 0.$$

Then, since the reference system is BIBS stable (see Lemma 2.1.1), $(\mathbb{I} - G(s)\theta^\top)^{-1}$ is stable, and hence, the Final Value Theorem and the limit for $\tilde{x}(t)$ above lead to

$$\begin{aligned}
\lim_{t\to\infty} (x_{\rm ref}(t) - x(t)) &= \lim_{s\to 0} s\,(x_{\rm ref}(s) - x(s)) \\
&= \lim_{s\to 0} s(\mathbb{I} - G(s)\theta^\top)^{-1} C(s)\tilde{x}(s) \\
&= \lim_{s\to 0} (\mathbb{I} - G(s)\theta^\top)^{-1} C(s)\,(s\tilde{x}(s)) \\
&= 0,
\end{aligned}$$

which leads to the first limit in (2.20).

To derive the second bound in (2.19), we first notice that the expressions in (2.2), (2.6), and (2.9) lead to the following relationship:

$$u_{\mathrm{ref}}(s) - u(s) = C(s)\tilde{\eta}(s) - (C(s)\theta^\top + k_m^\top)(x_{\mathrm{ref}}(s) - x(s)) . \tag{2.23}$$

Then, it follows from the error dynamics in (2.13) and the definition of $H_1(s)$ in (2.18) that

$$C(s)\tilde{\eta}(s) = H_1(s)\tilde{x}(s) ,$$

which implies that

$$u_{\mathrm{ref}}(s) - u(s) = H_1(s)\tilde{x}(s) - (C(s)\theta^\top + k_m^\top)(x_{\mathrm{ref}}(s) - x(s)) ,$$

and, consequently, the following bound holds:

$$\|(u_{\mathrm{ref}} - u)_\tau\|_{\mathcal{L}_\infty} \leq \|H_1(s)\|_{\mathcal{L}_1} \|\tilde{x}_\tau\|_{\mathcal{L}_\infty} + \|C(s)\theta^\top + k_m^\top\|_{\mathcal{L}_1} \|(x_{\mathrm{ref}} - x)_\tau\|_{\mathcal{L}_\infty} .$$

Hence, the bounds in (2.14) and (2.22) lead to the second upper bound in (2.19). To show the second limit in (2.20), we notice that the boundedness and the uniform continuity of $\tilde{x}(t)$ and $\tilde{\eta}(t)$ imply that $\dot{\tilde{x}}(t)$ is bounded and uniformly continuous, and application of Barbalat's lemma leads thus to $\lim_{t\to\infty} \dot{\tilde{x}}(t) = 0$. Consequently, $\lim_{t\to\infty} \tilde{\eta}(t) = 0$. Then, the second limit in (2.20) follows immediately from the expression in (2.23) and the first limit in (2.20). □

Remark 2.1.2 Since $C(0) = 1$, application of the Final Value Theorem to the closed-loop reference system in (2.9) in the case of constant $r(t) \equiv r$ leads to

$$\lim_{t\to\infty} y_{\mathrm{ref}}(t) = c^\top H(0)C(0)k_g r = r .$$

Remark 2.1.3 Notice that if we set $C(s) = 1$, which corresponds to the MRAC architecture, the norm of $H_1(s)$ is reduced to

$$\|H_1(s)\|_{\mathcal{L}_1} = \left\| \frac{1}{c_o^\top H(s)} c_o^\top \right\|_{\mathcal{L}_1} ,$$

which is not bounded, since $c_o^\top H(s)$ is strictly proper. Therefore, one cannot conclude a uniform performance bound for the control signal of MRAC similar to the one in (2.19).

Remark 2.1.4 Theorem 2.1.1 implies that, by increasing the adaptive gain Γ, the time histories of $x(t)$ and $u(t)$ can be made arbitrarily close to $x_{\mathrm{ref}}(t)$ and $u_{\mathrm{ref}}(t)$ for all $t \geq 0$. Because $x_{\mathrm{ref}}(t)$ and $u_{\mathrm{ref}}(t)$ are signals of an LTI system, then all the changes in initial conditions, reference inputs, and parametric uncertainties will lead to *uniform scaled changes* in the time histories of these signals and consequently also in the time histories of the corresponding signals $x(t)$ and $u(t)$ of the $\mathcal{L}_1$ *adaptive nonlinear closed-loop system*. Thus, the control objective is reduced to selection of k_m and $C(s)$ to ensure that the LTI reference system has the desired response.

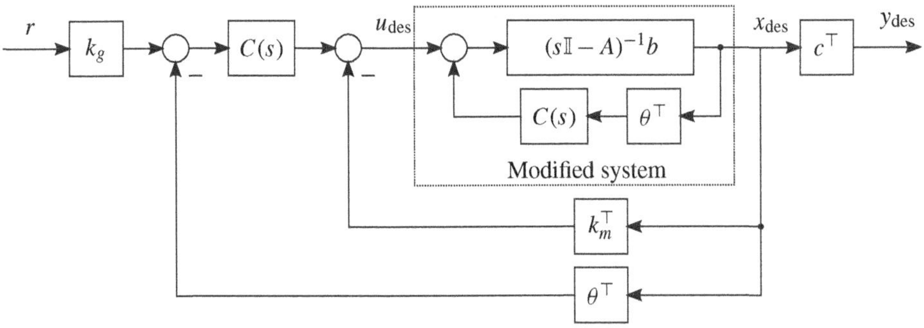

Figure 2.3: Design system for control specifications.

2.1.4 Design of the $\mathcal{L}_1$ Adaptive Controller: Robustness and Performance

Notice that the closed-loop reference system in (2.9) depends upon the vector θ of unknown parameters, and hence it cannot be used for introducing the transient specifications. Next, we consider the following LTI system, which will be referred to as a *design system*, with its output free of uncertainties:

$$\begin{aligned} x_{\text{des}}(s) &= C(s)k_g H(s)r(s) + x_{\text{in}}(s), \\ u_{\text{des}}(s) &= k_g C(s)r(s) - C(s)\theta^\top x_{\text{des}}(s) - k_m^\top x_{\text{des}}(s), && (2.24) \\ y_{\text{des}}(s) &= c^\top x_{\text{des}}(s). && (2.25) \end{aligned}$$

The block diagram of this system is shown in Figure 2.3. As compared to the reference system in Figure 2.2, we note that the filter $C(s)$ is also a part of the system definition.

Lemma 2.1.4 Subject to (2.7), the following upper bounds hold:

$$\|y_{\text{des}} - y_{\text{ref}}\|_{\mathcal{L}_\infty} \leq \frac{\lambda}{1-\lambda}\|c^\top\|_1 \left(\|k_g H(s)C(s)\|_{\mathcal{L}_1}\|r\|_{\mathcal{L}_\infty} + \|x_{\text{in}}\|_{\mathcal{L}_\infty}\right), \quad (2.26)$$

$$\|x_{\text{des}} - x_{\text{ref}}\|_{\mathcal{L}_\infty} \leq \frac{\lambda}{1-\lambda}\left(\|k_g H(s)C(s)\|_{\mathcal{L}_1}\|r\|_{\mathcal{L}_\infty} + \|x_{\text{in}}\|_{\mathcal{L}_\infty}\right),$$

$$\begin{aligned} \|u_{\text{des}} - u_{\text{ref}}\|_{\mathcal{L}_\infty} \leq{}& \frac{\lambda}{1-\lambda}\|C(s)\theta^\top + k_m^\top\|_{\mathcal{L}_1} \\ &\cdot\left(\|k_g H(s)C(s)\|_{\mathcal{L}_1}\|r\|_{\mathcal{L}_\infty} + \|x_{\text{in}}\|_{\mathcal{L}_\infty}\right). \quad (2.27) \end{aligned}$$

Proof. It follows from (2.9) that

$$x_{\text{ref}}(s) = (\mathbb{I} - G(s)\theta^\top)^{-1}(H(s)k_g C(s)r(s) + x_{\text{in}}(s)).$$

Since this reference system is BIBS stable, $(\mathbb{I} - G(s)\theta^\top)^{-1}$ is stable, and hence it can be expanded into convergent series:

$$(\mathbb{I} - G(s)\theta^\top)^{-1} = \mathbb{I} + \sum_{i=1}^{\infty}(G(s)\theta^\top)^i.$$

Therefore, the reference system can be rewritten as

$$x_{\text{ref}}(s) = \left(\mathbb{I} + \sum_{i=1}^{\infty}(G(s)\theta^\top)^i\right)(k_g H(s)C(s)r(s) + x_{\text{in}}(s))$$
$$= x_{\text{des}}(s) + \sum_{i=1}^{\infty}(G(s)\theta^\top)^i(k_g H(s)C(s)r(s) + x_{\text{in}}(s)).$$

Since $\|G(s)\theta^\top\|_{\mathcal{L}_1} \le \|G(s)\|_{\mathcal{L}_1} L \triangleq \lambda < 1$, then

$$\begin{aligned}\|x_{\text{des}} - x_{\text{ref}}\|_{\mathcal{L}_\infty} &\le \sum_{i=1}^{\infty}\lambda^i(\|k_g H(s)C(s)\|_{\mathcal{L}_1}\|r\|_{\mathcal{L}_\infty} + \|x_{\text{in}}\|_{\mathcal{L}_\infty}) \\ &= \frac{\lambda}{1-\lambda}(\|k_g H(s)C(s)\|_{\mathcal{L}_1}\|r\|_{\mathcal{L}_\infty} + \|x_{\text{in}}\|_{\mathcal{L}_\infty}).\end{aligned} \tag{2.28}$$

Recalling the definitions of $y_{\text{ref}}(t) = c^\top x_{\text{ref}}(t)$ and $y_{\text{des}}(t) = c^\top x_{\text{des}}(t)$, the bound in (2.28) leads to the bound in (2.26).

From (2.9) and (2.24), one can derive

$$u_{\text{des}}(s) - u_{\text{ref}}(s) = -(C(s)\theta^\top + k_m^\top)(x_{\text{des}}(s) - x_{\text{ref}}(s))$$

and further use the bound in (2.28) to obtain (2.27). □

Taking into consideration that $x_{\text{in}}(t)$ is exponentially decaying, the control objective can be achieved via proper selection of the static feedback gain k_m and the low-pass filter $C(s)$. In particular, the design of k_m and $C(s)$ needs to ensure that $C(s)c^\top H(s)$ (which does not depend on the unknown parameters) has the desired transient and steady-state performance characteristics, while simultaneously guaranteeing a small value of λ (or $\|G(s)\|_{\mathcal{L}_1}$). In general, k_m is chosen so that the state matrix A_m specifies desired closed-loop dynamics, while the bandwidth-limited filter $C(s)$ is designed to track reference signals and compensate for the undesirable effects of the uncertainties within a prespecified range of frequencies.

In the case when $C(s)$ is a low-pass filter, the system $G(s)$, which was defined in (2.8) as $G(s) = H(s)(1 - C(s))$, can be seen as the cascade of a low-pass system $H(s)$ and a high-pass system $(1 - C(s))$. Then, if the bandwidth of $C(s)$, which approximately corresponds to the cut-off frequency of $(1 - C(s))$, is designed to be larger than the bandwidth of $H(s)$, the resulting $G(s)$ is a "no-pass filter" with small $\mathcal{L}_1$-norm. The illustration is given in Figure 2.4. Hence, it follows that $\|G(s)\|_{\mathcal{L}_1}$ can be rendered arbitrarily small by

(i) increasing the bandwidth of the low-pass filter $C(s)$ for a given set of closed-loop performance specifications (in terms of the state matrix A_m). This solution leads to small design bounds and, therefore, yields a closed-loop adaptive systems with desired behavior. However, low-pass filters with high bandwidths may result in high-gain feedback and thus lead to closed-loop systems with overly small robustness margins and susceptible to measurement noise.

(ii) reducing the bandwidth of $H(s)$ by slowing down the eigenvalues of the matrix A_m for a given filter design. With this solution, a certain amount of performance is sacrificed to maintain a desired level of robustness.

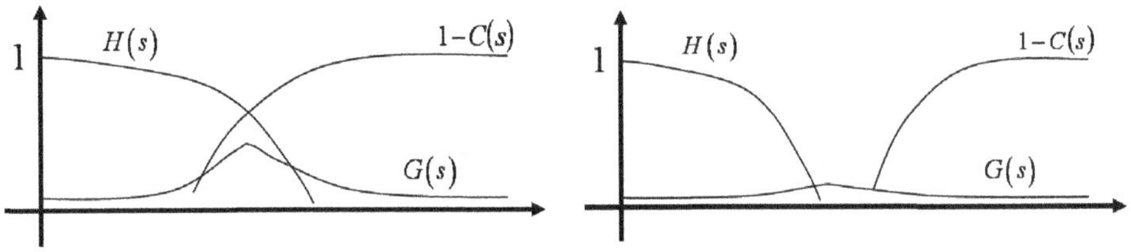

Figure 2.4: Cascaded systems.

It is important to emphasize that the use of a static feedback gain k_m in the control signal is not necessary. In fact, in $\mathcal{L}_1$ adaptive control architectures, the desired closed-loop dynamics are specified through the state matrix A_m of the predictor. However, if the desired closed-loop dynamics are far from the actual open-loop system dynamics, the upper bound on the uncertain parameter vector θ might be large, which may lead to high-gain feedback solutions as one tries to satisfy the $\mathcal{L}_1$-norm condition. Under these circumstances, the use of a static feedback gain can be useful for achieving designs with the desired performance and satisfactory robustness margins.

The lemma below tries to illustrate some of the points discussed above by showing that in the case of a first-order low-pass filter, the $\mathcal{L}_1$-norm of the system $G(s)$ can be rendered arbitrarily small by increasing the bandwidth of this filter.

Lemma 2.1.5 Let $C(s) = \omega_c/(s+\omega_c)$. For the arbitrary strictly proper BIBO-stable system $H(s)$, the following is true:

$$\lim_{\omega_c \to \infty} \|(1 - C(s))H(s)\|_{\mathcal{L}_1} = 0.$$

Proof. Notice that

$$(1 - C(s))H(s) = \frac{sH(s)}{s+\omega_c}.$$

Therefore

$$\|(1 - C(s))H(s)\|_{\mathcal{L}_1} \leq \left\| \frac{1}{s+\omega_c} \right\|_{\mathcal{L}_1} \|sH(s)\|_{\mathcal{L}_1} = \frac{1}{\omega_c} \|sH(s)\|_{\mathcal{L}_1}.$$

From the fact that $H(s)$ is strictly proper and stable, it follows that $sH(s)$ is proper and stable, which implies that $\|sH(s)\|_{\mathcal{L}_1}$ exists and is bounded. This leads to

$$\lim_{\omega_c \to \infty} \|(1 - C(s))H(s)\|_{\mathcal{L}_1} \leq \lim_{\omega_c \to \infty} \frac{1}{\omega_c} \|sH(s)\|_{\mathcal{L}_1} = 0.$$

The proof is complete. □

Remark 2.1.5 Theorem 2.1.1 and Lemma 2.1.4 imply that the $\mathcal{L}_1$ adaptive controller can generate a system response to track (2.24) and (2.25) both in transient and steady-state if we select the adaptive gain large and minimize λ. Notice that $u_{\text{des}}(t)$ in (2.24) depends upon the unknown parameter θ, while $y_{\text{des}}(t)$ in (2.25) does not. This implies that for different values of θ, the $\mathcal{L}_1$ adaptive controller will generate different control signals (dependent on θ) to ensure uniform system response (independent of θ). This is exactly what one would expect from an adaptive controller in terms of adapting to unknown parameters: the control signal changes dependent upon the uncertainties exactly so that the system output retains uniform (scaled) performance during the transient and steady-state phases. This basically states that the $\mathcal{L}_1$ adaptive architecture controls an unknown system as an LTI feedback controller would have done if the parameters were known.

Remark 2.1.6 It follows from Theorem 2.1.1 that in the presence of a large adaptive gain, the $\mathcal{L}_1$ adaptive controller and the system state approximate $u_{\text{ref}}(t)$ and $x_{\text{ref}}(t)$, respectively. Therefore, $y(t)$ approximates the output response of the LTI system $c^\top(\mathbb{I}-G(s)\theta^\top)^{-1}k_g H(s)C(s)$ to the input $r(t)$, and hence its transient performance specifications, such as overshoot and settling time, can be derived for every value of θ. If we further minimize λ, it follows from Lemma 2.1.4 that $y(t)$ approximates the output response of the LTI system $C(s)c^\top H(s)$ to the input signal $r(t)$. In this case, the $\mathcal{L}_1$ adaptive controller leads to uniform transient performance of $y(t)$ independent of the value of the unknown parameter θ. For the resulting $\mathcal{L}_1$ adaptive control signal one can characterize the transient specifications such as its amplitude and rate of change for every $\theta \in \Theta$, using $u_{\text{des}}(t)$.

Example 2.1.2 Next, we compare the performance of the $\mathcal{L}_1$ adaptive controller to that of a linear high-gain feedback. Consider the scalar system dynamics

$$\dot{x}(t) = \theta x(t) + u(t), \quad x(0) = 0,$$

where $\theta \in [\theta_{\min}, \theta_{\max}]$, and let the desired system have a pole at -2 and unity DC gain. Since for this example we have $b = 1$, it follows from the definition of $H(s)$ that

$$H(s) = \frac{1}{s+2}.$$

Then, in order to ensure that the desired system has unity DC gain and the closed-loop system tracks step-reference signals with zero steady-state error, a feedforward gain k_g is needed. So we set

$$k_g = 2,$$

which leads to the desired system

$$k_g H(s) = \frac{2}{s+2}.$$

First, let the high-gain feedback controller be given by

$$u(t) = -kx(t) + kr(t),$$

leading to the following closed-loop dynamics:

$$\dot{x}(t) = (\theta - k)x(t) + kr(t).$$

One needs to choose $k > \theta_{\max}$ to guarantee stability. We notice that both the steady-state error and the transient performance depend on the unknown parameter θ. By further introducing a proportional-integral (PI) controller, one can achieve zero steady-state error. If one chooses

$$k \gg \max\{|\theta_{\max}|, |\theta_{\min}|\},$$

the response of the closed-loop system (in the frequency domain) is given by

$$x(s) = \frac{k}{s-(\theta-k)} r(s) \approx \frac{k}{s+k} r(s),$$

which is obviously different from the performance specified by $k_g H(s)$. Next, we apply the $\mathcal{L}_1$ adaptive controller. Letting $u_m = -2x$, the partially closed-loop dynamics can be written as

$$\dot{x}(t) = -2x + (u_{\text{ad}}(t) + \theta x(t)).$$

Selecting $k_g = 2$ and $C(s) = \frac{\omega_c}{s+\omega_c}$ with large ω_c, and setting the adaptive gain Γ large, it follows from (2.19) that

$$x(s) \approx x_{\text{ref}}(s) \approx x_{\text{des}}(s) = C(s) k_g H(s) r(s) = \frac{\omega_c}{s+\omega_c} \frac{2}{s+2} r(s) \approx \frac{2}{s+2} r(s),$$

$$u(s) \approx u_{\text{ref}}(s) = (-2 - C(s)\theta) x_{\text{ref}}(s) + C(s) 2 r(s) \approx (-2-\theta) x_{\text{ref}}(s) + 2r(s).$$

The first of these relationships implies that the control objective is met, while the second states that the $\mathcal{L}_1$ adaptive controller approximates $u_{\text{ref}}(t)$, which partially cancels θ. ∎

2.1.5 Simulation Example

Consider the system in (2.1) and let

$$A = \begin{bmatrix} 0 & 1 \\ -1 & -1.4 \end{bmatrix}, \quad b = \begin{bmatrix} 0 \\ 1 \end{bmatrix}, \quad c = \begin{bmatrix} 1 \\ 0 \end{bmatrix}, \quad \theta = \begin{bmatrix} -4 \\ 4.5 \end{bmatrix}.$$

Further, let $\Theta = \{\vartheta = [\vartheta_1, \vartheta_2]^\top \in \mathbb{R}^2 : \vartheta_i \in [-10, 10], \text{ for all } i = 1,2\}$, which leads to $L = 20$. Letting $k_m = 0$, we implement the $\mathcal{L}_1$ adaptive controller following (2.2), (2.4)–(2.6). Let

$$\Gamma = 10000, \quad C_1(s) = \frac{\omega_c}{s+\omega_c}.$$

The $\mathcal{L}_1$-norm $\|G_1(s)\|_{\mathcal{L}_1} = \|H(s)(1-C_1(s))\|_{\mathcal{L}_1}$ can be calculated numerically. In particular, Figure 2.5(a) shows $\lambda_1 = \|G_1(s)\|_{\mathcal{L}_1} L$ with respect to the bandwidth of the low-pass filter ω_c. Notice that for $\omega_c > 50$, we have $\lambda_1 < 1$. Choosing $\omega_c = 160\ \text{s}^{-1}$ leads to $\lambda_1 = \|G_1(s)\|_{\mathcal{L}_1} L \approx 0.3 < 1$, which yields improved performance bounds in (2.26)–(2.27).

The simulation results for the $\mathcal{L}_1$ adaptive controller are shown in Figures 2.6(a) and 2.6(b) for step reference inputs $r = 25,\ 100,\ 400$. We notice that the $\mathcal{L}_1$ adaptive controller leads to scaled control inputs and scaled system outputs for scaled reference inputs. Figures 2.7(a) and 2.7(b) show the performance for the time-varying reference signal $r(t) = 100\cos(0.2t)$ without any retuning of the controller.

Next, consider the following design:

$$\Gamma = 400, \quad C_2(s) = \frac{3\omega_c^2 s + \omega_c^3}{(s+\omega_c)^3}.$$

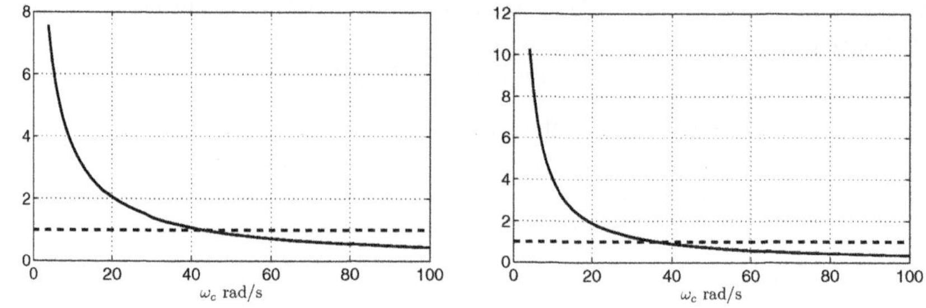

(a) $\lambda_1(\omega_c)$ (b) $\lambda_2(\omega_c)$

Figure 2.5: λ_1 and λ_2 with respect to ω_c.

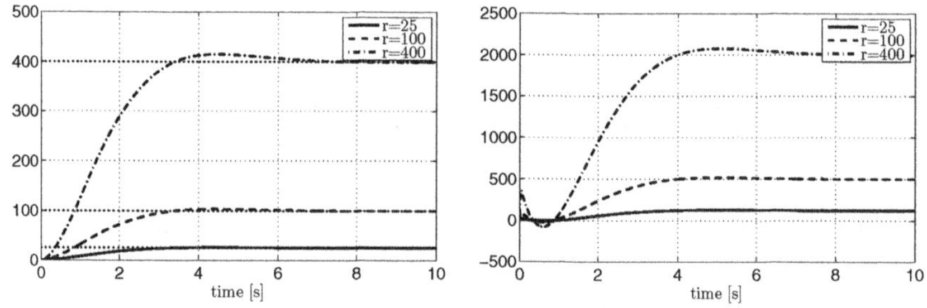

(a) Time history of $y(t)$ (b) Time history of $u(t)$

Figure 2.6: Performance of $\mathcal{L}_1$ adaptive controller with $C_1(s) = \frac{160}{s+160}$ for step-reference inputs.

Figure 2.5(b) shows $\lambda_2 = \|G_2(s)\|_{\mathcal{L}_1} L = \|H(s)(1 - C_2(s))\|_{\mathcal{L}_1} L$ as a function of ω_c. Notice that for $\omega_c > 40$, we have $\lambda_2 < 1$. Setting $\omega_c = 50\ \text{s}^{-1}$ leads to $\lambda_2 = 0.71$. The simulation results are shown in Figures 2.8(a) and 2.8(b) for constant reference inputs $r = 25,\ 100,\ 400$, which are again scaled accordingly.

Remark 2.1.7 The example above illustrates that high-order filters $C(s)$ may give the opportunity to use relatively small adaptive gains. While a rigorous relationship between the choice of the adaptive gain and the order of the filter cannot be derived, an insight into this can be gained from the following analysis. It follows from (2.1), (2.2), and (2.6) that

$$x(s) = k_g H(s)C(s)r(s) + H(s)\theta^\top x(s) - H(s)C(s)\hat{\eta}(s) + x_{\text{in}}(s),$$

while the state predictor can be rewritten as

$$\hat{x}(s) = k_g H(s)C(s)r(s) + H(s)(1 - C(s))\hat{\eta}(s) + x_{\text{in}}(s).$$

We note that the low-frequency component of the parameter estimate, $C(s)\hat{\eta}(s)$, is the input to the actual system, while the complementary high-frequency component, $(1 - C(s))\hat{\eta}(s)$,

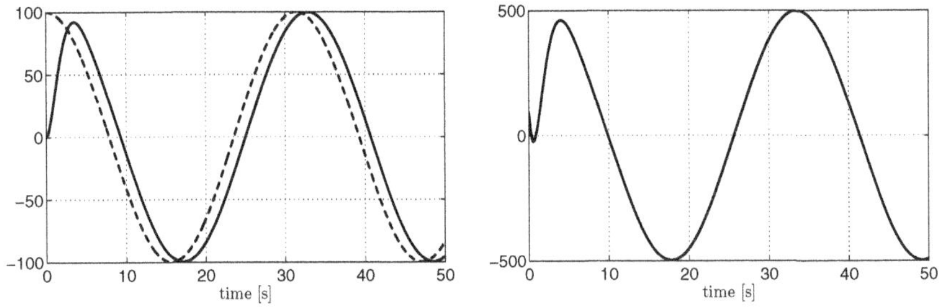

(a) $y(t)$ (solid) and $r(t)$ (dashed) (b) Time history of $u(t)$

Figure 2.7: Performance of $\mathcal{L}_1$ adaptive controller with $C_1(s) = \frac{160}{s+160}$ for $r = 100\cos(0.2t)$.

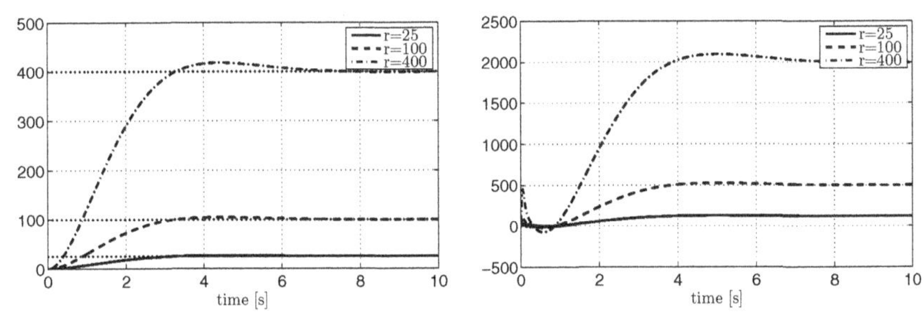

(a) Time history of $y(t)$ (b) Time history of $u(t)$

Figure 2.8: Performance of $\mathcal{L}_1$ adaptive controller with $C_2(s) = \frac{7500s+50^3}{(s+50)^3}$ for step-reference inputs.

goes into the state predictor. Since a low-pass filter $C(s)$ can only attenuate frequency content above its bandwidth, $\mathcal{L}_1$ adaptive designs using filters with high bandwidths will require large adaptive gains in order to generate frequencies beyond the bandwidths of the filters. A properly designed high-order $C(s)$ can be more effective to serve the purpose of filtering with reduced tailing effects and, hence, can generate similar λ with a smaller bandwidth. This further implies that a similar level of performance can be achieved with a smaller adaptive gain.

2.1.6 Loop Shaping via State-Predictor Design

To get further insights into the $\mathcal{L}_1$ adaptive controller, we consider the following first-order system, corrupted by input and output disturbances:

$$
\begin{aligned}
\dot{x}(t) &= -x(t) + u(t) + \sigma(t), \quad x(0) = x_0, \\
y(t) &= x(t) + d(t),
\end{aligned}
$$

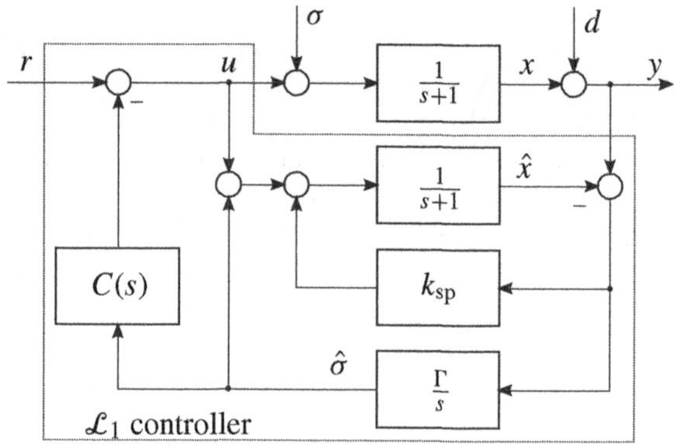

Figure 2.9: Closed-loop $\mathcal{L}_1$ system with modified predictor.

where $x(t) \in \mathbb{R}$ is the system state, $y(t) \in \mathbb{R}$ is the measured output, $u(t) \in \mathbb{R}$ is the control input, and $\sigma(t),\ d(t) \in \mathbb{R}$ are unknown bounded signals, representing input and output disturbances, respectively.

For the design of the $\mathcal{L}_1$ adaptive controller, following (2.18), we consider the following state predictor:

$$\begin{aligned}\dot{\hat{x}}(t) &= -\hat{x}(t) + u(t) + \hat{\sigma}(t) - k_{\mathrm{sp}}\tilde{x}(t), \quad \hat{x}(0) = x_0,\\ \hat{y}(t) &= \hat{x}(t),\end{aligned} \tag{2.29}$$

where $\hat{x}(t) \in \mathbb{R}$ is the predictor state, $k_{\mathrm{sp}} \geq 0$ is a constant that can be tuned to shape the frequency response of the closed-loop system, $\tilde{x}(t) \triangleq \hat{x}(t) - y(t)$, and $\hat{\sigma}(t) \in \mathbb{R}$ is the disturbance estimation, governed by the following adaptation law:

$$\dot{\hat{\sigma}}(t) = -\Gamma\tilde{x}(t), \quad \Gamma > 0.$$

Finally, let $C(s)$ be a strictly proper stable transfer function with DC gain $C(0) = 1$, and assume zero initialization for its state-space realization. The $\mathcal{L}_1$ controller for this system can be defined as follows:

$$u(s) = -C(s)\hat{\sigma}(s) + r(s),$$

where $u(s)$, $\hat{\sigma}(s)$, and $r(s)$ are the Laplace transforms of $u(t)$, $\hat{\sigma}(t)$, $r(t)$, respectively. The block diagram of this system is given in Figure 2.9.

The modification of the state predictor in (2.29) affects the prediction error dynamics,

$$\dot{\tilde{x}}(t) = -(1 + k_{\mathrm{sp}})\tilde{x}(t) + \tilde{\sigma}(t), \quad \tilde{x}(0) = 0,$$

where $\tilde{\sigma}(t) \triangleq \hat{\sigma}(t) - \sigma(t)$. In the frequency domain, these error dynamics can be rewritten as

$$\tilde{x}(s) = H_m(s)\tilde{\sigma}(s), \quad H_m(s) \triangleq \frac{1}{s + 1 + k_{\mathrm{sp}}}.$$

Table 2.1: Closed-loop transfer functions for the system in Figure 2.9.

	$r(s)$	$\sigma(s)$	$d(s)$
$x(s)$	$\frac{1}{s+1}$	$\frac{s^2+(k_{sp}+1)s+\Gamma(1-C(s))}{(s+1)P(s)}$	$-\Gamma\frac{C(s)}{P(s)}$
$y(s)$	$\frac{1}{s+1}$	$\frac{s^2+(k_{sp}+1)s+\Gamma(1-C(s))}{(s+1)P(s)}$	$\frac{s^2+(k_{sp}+1)s+\Gamma(1-C(s))}{P(s)}$
$u(s)$	1	$-\Gamma\frac{C(s)}{P(s)}$	$-\Gamma\frac{(s+1)C(s)}{P(s)}$

$P(s) \triangleq s^2 + (k_{sp} + 1)s + \Gamma$

Notice that the closed-loop system is LTI, and therefore we can use classical control tools to investigate its properties [17]. We compute the transfer functions from the signals $r(t)$, $\sigma(t)$, and $d(t)$ to each of the outputs $x(t)$, $y(t)$, and the control $u(t)$. These transfer functions are summarized in Table 2.1. One can see that there are only six different transfer functions. Moreover, the transfer functions $H_{xr}(s) = H_{yr}(s) = 1/(s+1)$ and $H_{ur}(s) = 1$ do not depend upon the controller parameters. Therefore, we will consider only the following four transfer functions:

$$\begin{aligned} H_{x\sigma}(s) = H_{y\sigma}(s) &= \frac{s^2+(k_{sp}+1)s+\Gamma(1-C(s))}{(s+1)P(s)}, & H_{yd}(s) &= \frac{s^2+(k_{sp}+1)s+\Gamma(1-C(s))}{P(s)}, \\ H_{u\sigma}(s) = H_{xd}(s) &= \frac{-\Gamma C(s)}{P(s)}, & H_{ud}(s) &= \frac{-\Gamma(s+1)C(s)}{P(s)}, \end{aligned} \tag{2.30}$$

where $P(s) \triangleq s^2 + (k_{sp} + 1)s + \Gamma$ is the characteristic polynomial of the adaptation loop. For the purpose of analysis, let

$$C(s) = \frac{1}{s+1}.$$

Figure 2.10 shows the Bode plots for the transfer functions in (2.30) for the $\mathcal{L}_1$ adaptive controller (with $k_{sp} = 0$) for different adaptation gains. One can see that the Bode plots of the transfer functions $H_{yd}(s)$, $H_{u\sigma}(s)$, and $H_{ud}(s)$ have a peak. The frequency location of the peak depends upon Γ, and it moves to the right toward higher frequencies as one increases Γ. For this simple LTI closed-loop system, we can analytically explain this phenomenon by considering the common part of the denominator of the transfer functions, which is given by $P(s) = s^2 + s + \Gamma = s^2 + 2\zeta\omega_n s + \omega_n^2$, where $\omega_n \triangleq \sqrt{\Gamma}$ and $\zeta \triangleq 1/(2\sqrt{\Gamma})$. One can see that the damping ζ decreases with the increase of the adaptation gain Γ. This peak points to a sensitivity to noise and disturbances, which is typical as well for other adaptive controllers in the presence of high adaptation rates, including MRAC. However, from the Bode plots of $H_{u\sigma}$ and $H_{x\sigma}$, one can see that with the growth of Γ, while moving to the right, the peak never crosses the horizontal axis, which is consistent with the theoretical results proved in Section 2.1.3. Moreover, with the $\mathcal{L}_1$ adaptive controller, this issue can be addressed by appropriate tuning of the state predictor, as will be illustrated shortly.

In fact, in the presence of the term $-k_{sp}\tilde{x}(t)$ in the predictor dynamics, the new characteristic polynomial for the adaptation loop is given by $P(s) \triangleq s^2 + (k_{sp} + 1)s + \Gamma$. By

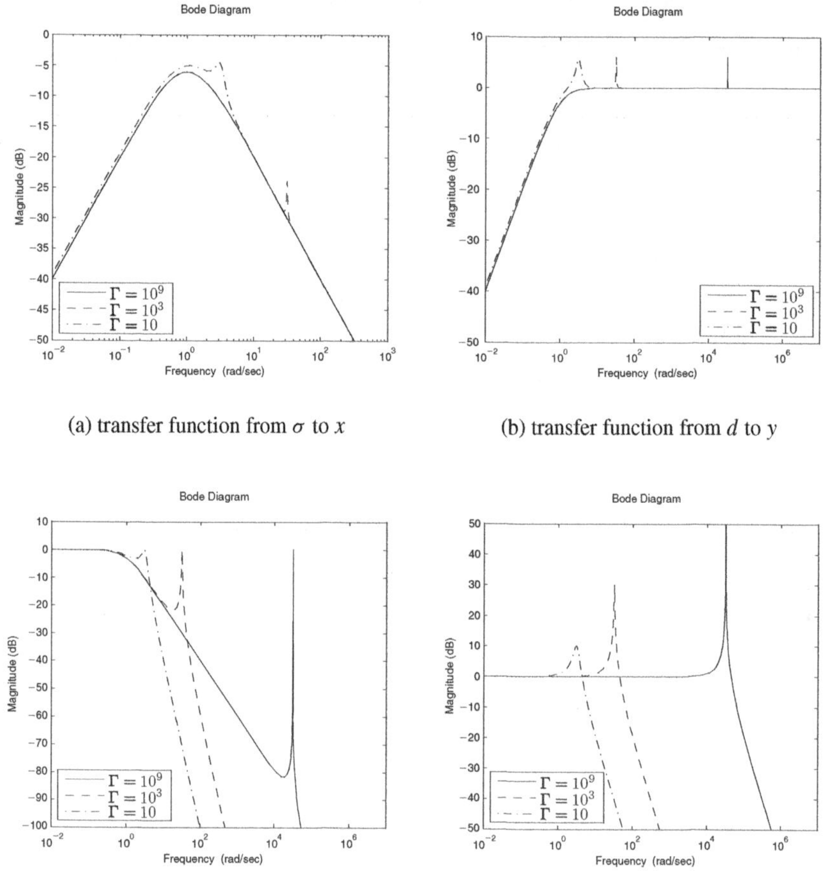

(a) transfer function from σ to x

(b) transfer function from d to y

(c) transfer function from σ to u

(d) transfer function from d to u

Figure 2.10: Bode plots for the closed-loop transfer functions for different adaptation gains Γ and for $k_{sp} = 0$.

appropriate tuning of k_{sp}, we can adjust the damping of the adaptation loop $\zeta = (k_{sp} + 1)/(2\sqrt{\Gamma})$ to an arbitrary desired value. Figure 2.11 shows the effect of the coefficient k_{sp} on the Bode plots for a fixed value of the adaptive gain $\Gamma = 10000$. One can see that by setting the value of k_{sp} appropriately, we can eliminate the peak described earlier in all the transfer functions. Therefore, in the following discussion we consider $k_{sp} = 1.4\sqrt{\Gamma} - 1$, which yields $\zeta = 0.7$.

Figure 2.12 shows the Bode plots for the closed-loop system with the modified (faster) state predictor for different values of the adaptation gain Γ. One can see that the adaptive control system provides good disturbance rejection within the bandwidth of the low-pass filter, and the control channel is not affected by high-frequency disturbance content. A detailed discussion can be found in [90]. Notice that similar results in noise attenuation can also be achieved via a higher order $C(s)$.

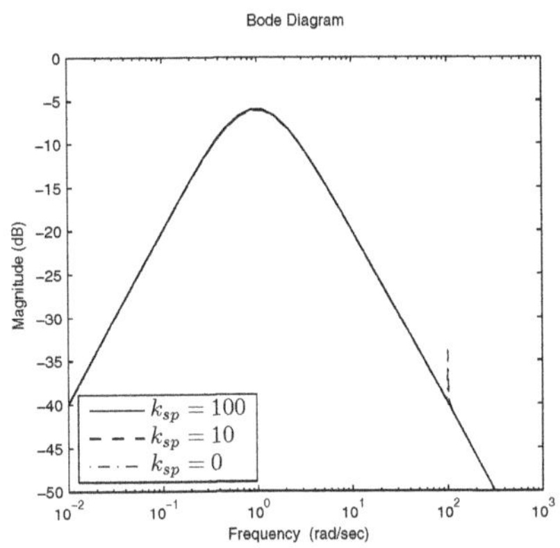

(a) transfer function from σ to x

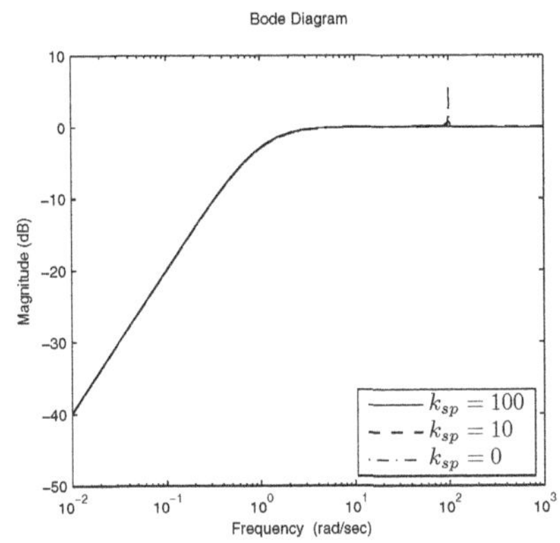

(b) transfer function from d to y

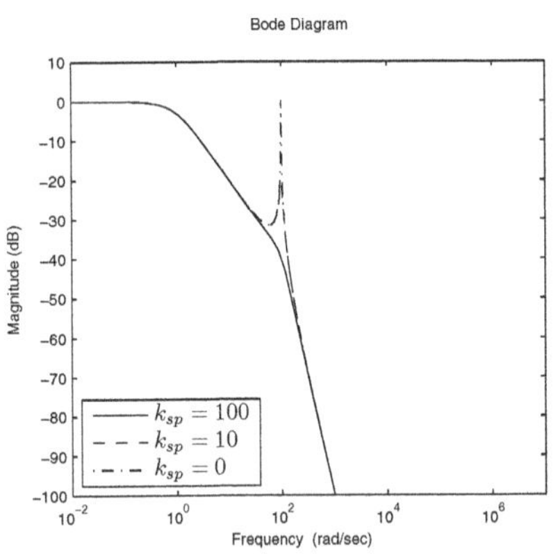

(c) transfer function from σ to u

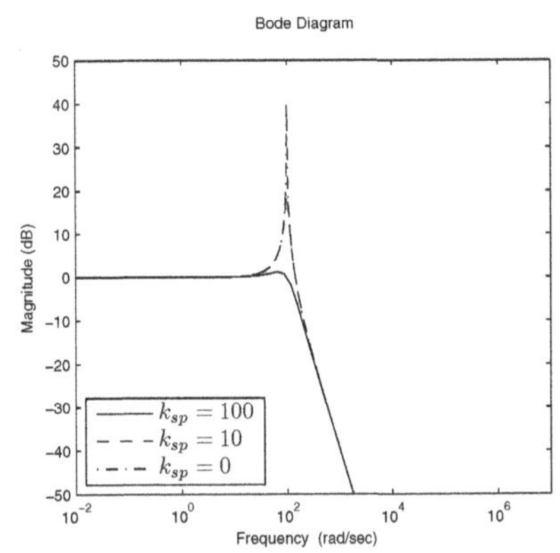

(d) transfer function from d to u

Figure 2.11: Bode plots for the closed-loop transfer functions for different $k_{\rm sp}$ for $\Gamma = 10000$.

2.2 Systems with Uncertain System Input Gain

The results in Section 2.1 imply that *fast adaptation* ensures uniform performance bounds for the system's signals, both input and output, as compared to the corresponding signals of a bounded-reference LTI system. This gives the opportunity to extend the class of systems beyond the one with constant unknown parameters and consider systems with time-varying parameters and disturbances. In this section, we also incorporate uncertainty in the system input gain. We introduce a *new* control architecture, which gives an opportunity to compensate for the effect of the unknown system input gain on the output performance. We also derive the time-delay margin of this closed-loop architecture and analyze the effect of nonzero initialization errors on the system's performance [34].

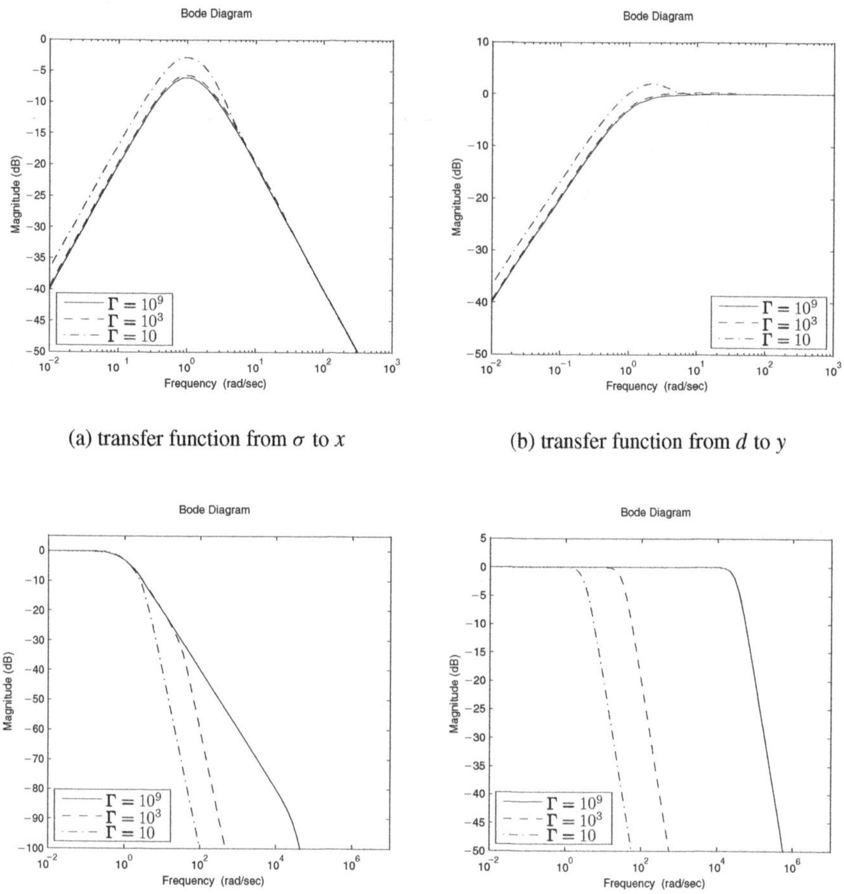

(a) transfer function from σ to x

(b) transfer function from d to y

(c) transfer function from σ to u

(d) transfer function from d to u

Figure 2.12: Bode plots for the closed-loop transfer functions for different adaptation gains Γ and for $k_{sp} = 1.4\sqrt{\Gamma} - 1$.

2.2.1 Problem Formulation

Consider the following class of systems:

$$\begin{aligned} \dot{x}(t) &= A_m x(t) + b\left(\omega u(t) + \theta^\top(t) x(t) + \sigma(t)\right), \quad x(0) = x_0, \\ y(t) &= c^\top x(t), \end{aligned} \tag{2.31}$$

where $x(t) \in \mathbb{R}^n$ is the system state vector (measured); $u(t) \in \mathbb{R}$ is the control input; $y(t) \in \mathbb{R}$ is the regulated output; $b, c \in \mathbb{R}^n$ are known constant vectors; A_m is a known Hurwitz $n \times n$ matrix specifying the desired closed-loop dynamics; $\omega \in \mathbb{R}$ is an unknown constant with known sign; $\theta(t) \in \mathbb{R}^n$ is a vector of time-varying unknown parameters; and $\sigma(t) \in \mathbb{R}$ models input disturbances.

Assumption 2.2.1 (Uniform boundedness of unknown parameters) Let

$$\theta(t) \in \Theta, \quad |\sigma(t)| \leq \Delta_0, \quad \forall\, t \geq 0,$$

where Θ is a known convex compact set and $\Delta_0 \in \mathbb{R}^+$ is a known (conservative) bound of $\sigma(t)$.

Assumption 2.2.2 (Uniform boundedness of the rate of variation of parameters) Let $\theta(t)$ and $\sigma(t)$ be continuously differentiable with uniformly bounded derivatives:

$$\|\dot{\theta}(t)\| \leq d_\theta < \infty, \quad |\dot{\sigma}(t)| \leq d_\sigma < \infty, \quad \forall\, t \geq 0.$$

Assumption 2.2.3 (Partial knowledge of uncertain system input gain) Let

$$\omega \in \Omega_0 \triangleq [\omega_{l_0},\, \omega_{u_0}],$$

where $0 < \omega_{l_0} < \omega_{u_0}$ are given known upper and lower bounds on ω.

The control objective is to design a full-state feedback adaptive controller to ensure that $y(t)$ tracks a given bounded piecewise-continuous reference signal $r(t)$ with quantifiable performance bounds.

2.2.2 $\mathcal{L}_1$ Adaptive Control Architecture

State Predictor

We consider the following state predictor:

$$\begin{aligned} \dot{\hat{x}}(t) &= A_m \hat{x}(t) + b\left(\hat{\omega}(t)u(t) + \hat{\theta}^\top(t)x(t) + \hat{\sigma}(t)\right), \quad \hat{x}(0) = x_0, \\ \hat{y}(t) &= c^\top \hat{x}(t), \end{aligned} \tag{2.32}$$

which has the same structure as the system in (2.31); the only difference is that the unknown parameters ω, $\theta(t)$, and $\sigma(t)$ are replaced by their adaptive estimates $\hat{\omega}(t)$, $\hat{\theta}(t)$, and $\hat{\sigma}(t)$.

Adaptation Laws

The adaptive process is governed by the following projection-based adaptation laws:

$$\begin{aligned} \dot{\hat{\theta}}(t) &= \Gamma \mathrm{Proj}(\hat{\theta}(t), -\tilde{x}^\top(t)Pbx(t)), \quad \hat{\theta}(0) = \hat{\theta}_0, \\ \dot{\hat{\sigma}}(t) &= \Gamma \mathrm{Proj}(\hat{\sigma}(t), -\tilde{x}^\top(t)Pb), \quad \hat{\sigma}(0) = \hat{\sigma}_0, \\ \dot{\hat{\omega}}(t) &= \Gamma \mathrm{Proj}(\hat{\omega}(t), -\tilde{x}^\top(t)Pbu(t)), \quad \hat{\omega}(0) = \hat{\omega}_0, \end{aligned} \tag{2.33}$$

where $\tilde{x}(t) \triangleq \hat{x}(t) - x(t)$, $\Gamma \in \mathbb{R}^+$ is the adaptation rate, and $P = P^\top > 0$ is the solution of the algebraic Lyapunov equation $A_m^\top P + PA_m = -Q$ for arbitrary $Q = Q^\top > 0$. In the implementation of the projection operator we use the compact set Θ as given in Assumption 2.2.1, while we replace Δ_0 and Ω_0 by Δ and $\Omega \triangleq [\omega_l,\, \omega_u]$ such that

$$\Delta_0 < \Delta, \quad 0 < \omega_l < \omega_{l_0} < \omega_{u_0} < \omega_u\,. \tag{2.34}$$

The purpose of this definition for the projection bounds will be clarified in the analysis of the time-delay and the gain margins.

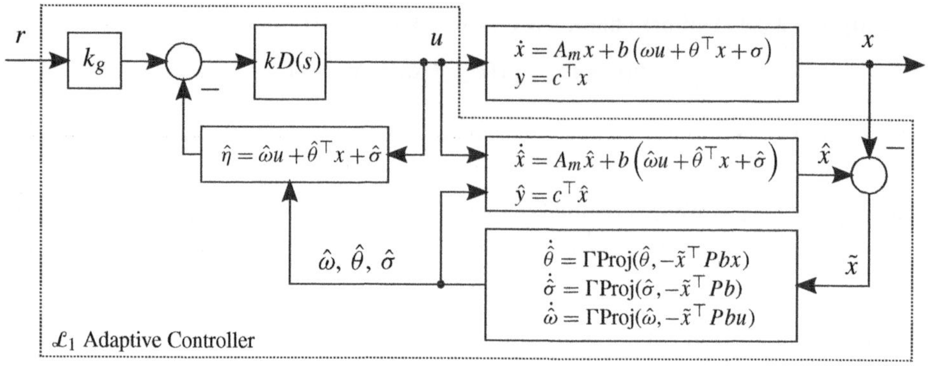

Figure 2.13: Closed-loop adaptive system.

Control Law

The control signal is generated as the output of the following (feedback) system:

$$u(s) = -kD(s)(\hat{\eta}(s) - k_g r(s)), \tag{2.35}$$

where $r(s)$ and $\hat{\eta}(s)$ are the Laplace transforms of $r(t)$ and $\hat{\eta}(t) \triangleq \hat{\omega}(t)u(t) + \hat{\theta}^\top(t)x(t) + \hat{\sigma}(t)$, respectively; $k_g \triangleq -1/(c^\top A_m^{-1} b)$; and $k > 0$ and $D(s)$ are a feedback gain and a strictly proper transfer function leading to a strictly proper stable

$$C(s) \triangleq \frac{\omega k D(s)}{1 + \omega k D(s)} \qquad \forall\, \omega \in \Omega_0 \tag{2.36}$$

with DC gain $C(0) = 1$. One simple choice is $D(s) = 1/s$, which yields a first-order strictly proper $C(s)$ of the form

$$C(s) = \frac{\omega k}{s + \omega k}.$$

As before, let

$$L \triangleq \max_{\theta \in \Theta} \|\theta\|_1, \quad H(s) \triangleq (s\mathbb{I} - A_m)^{-1} b, \quad G(s) \triangleq H(s)(1 - C(s)). \tag{2.37}$$

The $\mathcal{L}_1$ adaptive controller is defined via (2.32), (2.33), (2.35), subject to the following $\mathcal{L}_1$-norm condition:

$$\|G(s)\|_{\mathcal{L}_1} L < 1. \tag{2.38}$$

The $\mathcal{L}_1$ adaptive control architecture with its main elements is represented in Figure 2.13.

In the case of constant $\theta(t)$, the $\mathcal{L}_1$-norm condition can be simplified. For the specific choice of $D(s) = 1/s$, it is reduced to

$$A_g \triangleq \begin{bmatrix} A_m + b\theta^\top & b\omega \\ -k\theta^\top & -k\omega \end{bmatrix}, \tag{2.39}$$

being Hurwitz for all $\theta \in \Theta$ and $\omega \in \Omega_0$.

2.2.3 Performance Bounds of the $\mathcal{L}_1$ Adaptive Controller

Closed-Loop Reference System

Consider the following closed-loop reference system, which again corresponds to the non-adaptive version of the $\mathcal{L}_1$ adaptive controller:

$$\dot{x}_{\text{ref}}(t) = A_m x_{\text{ref}}(t) + b\left(\omega u_{\text{ref}}(t) + \theta^\top(t) x_{\text{ref}}(t) + \sigma(t)\right),\; x_{\text{ref}}(0) = x_0, \tag{2.40}$$

$$u_{\text{ref}}(s) = \frac{C(s)}{\omega}(k_g r(s) - \eta_{\text{ref}}(s)), \tag{2.41}$$

$$y_{\text{ref}}(t) = c^\top x_{\text{ref}}(t), \tag{2.42}$$

where $r(s)$ and $\eta_{\text{ref}}(s)$ are the Laplace transforms of $r(t)$ and $\eta_{\text{ref}}(t) \triangleq \theta^\top(t) x_{\text{ref}}(t) + \sigma(t)$, respectively. The next lemma establishes stability of the closed-loop system in (2.40)–(2.42).

Lemma 2.2.1 If k and $D(s)$ verify the $\mathcal{L}_1$-norm condition in (2.38), the closed-loop reference system in (2.40)–(2.42) is BIBS stable with respect to $r(t)$ and x_0.

Proof. It follows from (2.40)–(2.42) that

$$x_{\text{ref}}(s) = G(s)\eta_{\text{ref}}(s) + H(s)C(s)k_g r(s) + x_{\text{in}}(s),$$

where, similar to previous analysis, we have set $x_{\text{in}}(s) \triangleq (s\mathbb{I} - A_m)^{-1} x_0$. Since A_m is Hurwitz, $x_{\text{in}}(t)$ is uniformly bounded. Next, it follows from Lemma A.7.1 that

$$\|x_{\text{ref}\tau}\|_{\mathcal{L}_\infty} \leq \|G(s)\|_{\mathcal{L}_1} \|\eta_{\text{ref}\tau}\|_{\mathcal{L}_\infty} + \|H(s)C(s)k_g\|_{\mathcal{L}_1} \|r\|_{\mathcal{L}_\infty} + \|x_{\text{in}}\|_{\mathcal{L}_\infty}.$$

Using the definition in (2.37), we have the following upper bound for $\eta_{\text{ref}}(t)$:

$$\|\eta_{\text{ref}\tau}\|_{\mathcal{L}_\infty} \leq L\|x_{\text{ref}\tau}\|_{\mathcal{L}_\infty} + \|\sigma_\tau\|_{\mathcal{L}_\infty}.$$

Substituting and solving for $\|x_{\text{ref}\tau}\|_{\mathcal{L}_\infty}$, one obtains

$$\|x_{\text{ref}\tau}\|_{\mathcal{L}_\infty} \leq \frac{\|H(s)C(s)k_g\|_{\mathcal{L}_1}\|r\|_{\mathcal{L}_\infty} + \|G(s)\|_{\mathcal{L}_1}\Delta + \|x_{\text{in}}\|_{\mathcal{L}_\infty}}{1 - \|G(s)\|_{\mathcal{L}_1} L}.$$

Then, because k and $D(s)$ verify the condition in (2.38), $\|x_{\text{ref}\tau}\|_{\mathcal{L}_\infty}$ is uniformly bounded for all $\tau \geq 0$. Hence, the closed-loop reference system in (2.40)–(2.42) is BIBS stable. □

Lemma 2.2.2 If $\theta(t) \equiv \theta$ is constant, and $D(s) = 1/s$, then the closed-loop reference system in (2.40)–(2.42) is BIBS stable with respect to $r(t)$ and x_0 if and only if the matrix A_g in (2.39) is Hurwitz for all $\theta \in \Theta$ and $\omega \in \Omega_0$.

Proof. In the case of constant $\theta(t)$, the state-space realization of (2.40)–(2.42) is given by

$$\begin{bmatrix} \dot{x}_{\text{ref}}(t) \\ \dot{u}_{\text{ref}}(t) \end{bmatrix} = A_g \begin{bmatrix} x_{\text{ref}}(t) \\ u_{\text{ref}}(t) \end{bmatrix} + \begin{bmatrix} b\sigma(t) \\ kk_g r(t) - k\sigma(t) \end{bmatrix}, \quad \begin{array}{rcl} x_{\text{ref}}(0) & = & x_0, \\ u_{\text{ref}}(0) & = & 0, \end{array}$$

which is stable if and only if A_g is Hurwitz for all $\theta \in \Theta$ and $\omega \in \Omega_0$. □

Transient and Steady-State Performance

The system dynamics in (2.31) and the state predictor in (2.32) lead to the following prediction-error dynamics:

$$\dot{\tilde{x}}(t) = A_m\tilde{x}(t) + b(\tilde{\omega}(t)u(t) + \tilde{\theta}^\top(t)x(t) + \tilde{\sigma}(t)), \quad \tilde{x}(0) = 0, \tag{2.43}$$

where $\tilde{\theta}(t) \triangleq \hat{\theta}(t) - \theta(t)$, $\tilde{\sigma}(t) \triangleq \hat{\sigma}(t) - \sigma(t)$, and $\tilde{\omega}(t) \triangleq \hat{\omega}(t) - \omega$. Let $\tilde{\eta}(t) \triangleq \tilde{\omega}(t)u(t) + \tilde{\theta}^\top(t)x(t) + \tilde{\sigma}(t)$, and let $\tilde{\eta}(s)$ be the Laplace transform of $\tilde{\eta}(t)$. Then, the error dynamics in (2.43) can be rewritten in frequency domain as

$$\tilde{x}(s) = H(s)\tilde{\eta}(s). \tag{2.44}$$

Lemma 2.2.3 The prediction error $\tilde{x}(t)$ is uniformly bounded,

$$\|\tilde{x}\|_{\mathcal{L}_\infty} \leq \sqrt{\frac{\theta_m}{\lambda_{\min}(P)\Gamma}}, \tag{2.45}$$

where

$$\theta_m \triangleq 4\max_{\theta\in\Theta}\|\theta\|^2 + 4\Delta^2 + (\omega_u - \omega_l)^2 + 4\frac{\lambda_{\max}(P)}{\lambda_{\min}(Q)}\left(d_\theta \max_{\theta\in\Theta}\|\theta\| + d_\sigma\Delta\right). \tag{2.46}$$

Proof. Consider the Lyapunov function candidate:

$$V(\tilde{x}(t),\tilde{\theta}(t),\tilde{\omega}(t),\tilde{\sigma}(t)) = \tilde{x}^\top(t)P\tilde{x}(t) + \frac{1}{\Gamma}\left(\tilde{\theta}^\top(t)\tilde{\theta}(t) + \tilde{\omega}^2(t) + \tilde{\sigma}^2(t)\right). \tag{2.47}$$

First, we prove that

$$V(t) \leq \frac{\theta_m}{\Gamma}.$$

Since $\hat{x}(0) = x(0)$, we can easily verify that

$$V(0) \leq \frac{1}{\Gamma}\left(4\max_{\theta\in\Theta}\|\theta\|^2 + 4\Delta^2 + (\omega_u - \omega_l)^2\right) < \frac{\theta_m}{\Gamma}.$$

Using the projection-based adaptive laws in (2.33), one can derive the following upper bound:

$$\begin{aligned}
\dot{V}(t) &= -\tilde{x}^\top(t)Q\tilde{x}(t) + 2\tilde{x}^\top(t)Pb\tilde{\omega}(t)u(t) + 2\tilde{x}^\top(t)Pb\tilde{\theta}^\top(t)x(t) \\
&\quad + 2\tilde{x}^\top(t)Pb\tilde{\sigma}(t) + \frac{2}{\Gamma}\left(\tilde{\theta}^\top(t)\dot{\hat{\theta}}(t) + \tilde{\omega}(t)\dot{\hat{\omega}}(t) + \tilde{\sigma}(t)\dot{\hat{\sigma}}(t)\right) \\
&\quad - \frac{2}{\Gamma}\left(\tilde{\theta}^\top(t)\dot{\theta}(t) + \tilde{\sigma}(t)\dot{\sigma}(t)\right) \\
&= -\tilde{x}^\top(t)Q\tilde{x}(t) + 2\tilde{\omega}(t)\left(\tilde{x}^\top(t)Pbu(t) + \mathrm{Proj}(\hat{\omega}(t), -\tilde{x}^\top(t)Pbu(t))\right) \\
&\quad + 2\tilde{\theta}^\top(t)\left(x(t)\tilde{x}^\top(t)Pb + \mathrm{Proj}(\hat{\theta}(t), -x(t)\tilde{x}^\top(t)Pb)\right) \\
&\quad + 2\tilde{\sigma}(t)\left(x^\top(t)Pb + \mathrm{Proj}(\hat{\sigma}(t), -\tilde{x}^\top(t)Pb)\right) \\
&\quad - \frac{2}{\Gamma}\left(\tilde{\theta}^\top(t)\dot{\theta}(t) + \tilde{\sigma}(t)\dot{\sigma}(t)\right) \\
&\leq -\tilde{x}^\top(t)Q\tilde{x}(t) + \frac{2}{\Gamma}\left(|\tilde{\theta}^\top(t)\dot{\theta}(t)| + |\tilde{\sigma}(t)\dot{\sigma}(t)|\right).
\end{aligned} \tag{2.48}$$

The projection operator ensures that $\hat{\theta}(t) \in \Theta$, $\hat{\omega}(t) \in \Omega$, $|\hat{\sigma}(t)| \leq \Delta$ for all $t \geq 0$, and therefore, the upper bounds in (2.2.2) lead to the following upper bound:

$$|\tilde{\theta}^\top(t)\dot{\theta}(t)| + |\tilde{\sigma}(t)\dot{\sigma}(t)| \leq 2(\max_{\theta\in\Theta} \|\theta\| d_\theta + d_\sigma \Delta).$$

Moreover, the projection operator also ensures that

$$\max_{t\geq 0}\left(\frac{1}{\Gamma}\left(\tilde{\theta}^\top(t)\tilde{\theta}(t) + \tilde{\omega}^2(t) + \tilde{\sigma}^2(t)\right)\right) \leq \frac{1}{\Gamma}\left(4\max_{\theta\in\Theta}\|\theta\|^2 + 4\Delta^2 + (\omega_u - \omega_l)^2\right),$$

which holds for all $t \geq 0$.

If at any time $t_1 > 0$, one has $V(t_1) > \theta_m/\Gamma$, then it follows from (2.46) and (2.47) that

$$\tilde{x}^\top(t_1)P\tilde{x}(t_1) > 4\frac{\lambda_{\max}(P)}{\Gamma\lambda_{\min}(Q)}\left(d_\theta \max_{\theta\in\Theta}\|\theta\| + d_\sigma\Delta\right),$$

and thus

$$\tilde{x}^\top(t_1)Q\tilde{x}(t_1) \geq \frac{\lambda_{\min}(Q)\tilde{x}^\top(t_1)P\tilde{x}(t_1)}{\lambda_{\max}(P)} > 4\frac{d_\theta \max_{\theta\in\Theta}\|\theta\| + d_\sigma\Delta}{\Gamma}. \tag{2.49}$$

Hence, if $V(t_1) > \theta_m/\Gamma$, then from (2.48) and (2.49) we have

$$\dot{V}(t_1) < 0, \tag{2.50}$$

and it follows from (2.50) that for all $t \geq 0$,

$$V(t) \leq \frac{\theta_m}{\Gamma}.$$

Since $\lambda_{\min}(P)\|\tilde{x}(t)\|^2 \leq \tilde{x}^\top(t)P\tilde{x}(t) \leq V(t)$, then

$$\|\tilde{x}(t)\|^2 \leq \frac{\theta_m}{\lambda_{\min}(P)\Gamma}, \qquad \forall\, t \geq 0,$$

which leads to (2.45). □

We further notice that the bound in (2.45) is proportional to the square root of the rate of variation of uncertainties and is inverse proportional to the square root of the adaptation gain.

Theorem 2.2.1 Given the system in (2.31) and the $\mathcal{L}_1$ adaptive controller defined via (2.32), (2.33), and (2.35), subject to the $\mathcal{L}_1$-norm condition in (2.38), we have

$$\|x_{\text{ref}} - x\|_{\mathcal{L}_\infty} \leq \frac{\gamma_1}{\sqrt{\Gamma}}, \qquad \|u_{\text{ref}} - u\|_{\mathcal{L}_\infty} \leq \frac{\gamma_2}{\sqrt{\Gamma}}, \tag{2.51}$$

where

$$\gamma_1 \triangleq \frac{\|C(s)\|_{\mathcal{L}_1}}{1 - \|G(s)\|_{\mathcal{L}_1}L}\sqrt{\frac{\theta_m}{\lambda_{\min}(P)}},$$

$$\gamma_2 \triangleq \left\|\frac{C(s)}{\omega}\right\|_{\mathcal{L}_1} L\gamma_1 + \left\|\frac{H_1(s)}{\omega}\right\|_{\mathcal{L}_1}\sqrt{\frac{\theta_m}{\lambda_{\min}(P)}},$$

and $H_1(s) = C(s)\frac{1}{c_o^\top H(s)}c_o^\top$ was introduced in (2.18).

Proof. Let $\eta(t) \triangleq \theta^\top(t)x(t) + \sigma(t)$. It follows from (2.35) that

$$u(s) = -kD(s)(\omega u(s) + \eta(s) - k_g r(s) + \tilde{\eta}(s)).$$

Consequently

$$u(s) = -\frac{kD(s)}{1+\omega kD(s)}(\eta(s) - k_g r(s) + \tilde{\eta}(s)).$$

Using the definition of $C(s)$ in (2.36), we can write

$$u(s) = -\frac{C(s)}{\omega}(\eta(s) - k_g r(s) + \tilde{\eta}(s)), \tag{2.52}$$

and the system in (2.31) takes the form

$$x(s) = G(s)\eta(s) + H(s)C(s)k_g r(s) - H(s)C(s)\tilde{\eta}(s) + x_{\text{in}}(s).$$

Similarly, it follows from (2.40)–(2.42) that

$$x_{\text{ref}}(s) = G(s)\eta_{\text{ref}}(s) + H(s)C(s)k_g r(s) + x_{\text{in}}(s).$$

Then

$$x_{\text{ref}}(s) - x(s) = G(s)\eta_e(s) + H(s)C(s)\tilde{\eta}(s), \quad \eta_e(t) \triangleq \theta^\top(t)(x_{\text{ref}}(t) - x(t)). \tag{2.53}$$

Using (2.44), we can rewrite

$$x_{\text{ref}}(s) - x(s) = G(s)\eta_e(s) + C(s)\tilde{x}(s).$$

Lemma A.7.1 gives the following upper bound:

$$\|(x_{\text{ref}} - x)_\tau\|_{\mathcal{L}_\infty} \leq \|G(s)\|_{\mathcal{L}_1}\|\eta_{e\tau}\|_{\mathcal{L}_\infty} + \|C(s)\|_{\mathcal{L}_1}\|\tilde{x}_\tau\|_{\mathcal{L}_\infty}. \tag{2.54}$$

From the definition of L in (2.37) it follows that $\|\eta_{e\tau}\|_{\mathcal{L}_\infty} \leq L\|(x_{\text{ref}} - x)_\tau\|_{\mathcal{L}_\infty}$. Substituting this back into (2.54), and solving for $\|(x_{\text{ref}} - x)_\tau\|_{\mathcal{L}_\infty}$, with account of the upper bound from Lemma 2.2.3 and the condition in (2.38), one gets

$$\|(x_{\text{ref}} - x)_\tau\|_{\mathcal{L}_\infty} \leq \frac{\|C(s)\|_{\mathcal{L}_1}}{1 - \|G(s)\|_{\mathcal{L}_1}L}\sqrt{\frac{\theta_m}{\lambda_{\min}(P)\Gamma}},$$

which holds uniformly for all $\tau \geq 0$, leading to the first bound in (2.51).

To prove the second bound in (2.51), we notice that from (2.41) and (2.52) one can derive

$$u_{\text{ref}}(s) - u(s) = -\frac{C(s)}{\omega}\eta_e(s) + \frac{C(s)}{\omega}\tilde{\eta}(s). \tag{2.55}$$

Similar to the proof of (2.19) in Theorem 2.1.1, we refer to Lemma A.12.1 and rewrite

$$\frac{C(s)}{\omega}\tilde{\eta}(s) = \frac{1}{\omega}H_1(s)\tilde{x}(s).$$

Since $C(s)$ is strictly proper and stable, the system $H_1(s)$ is proper and stable. Then it follows from Lemma A.7.1 that the difference in (2.55) can be upper bounded as

$$\|(u_{\text{ref}}-u)_\tau\|_{\mathcal{L}_\infty} \leq \left\|\frac{C(s)}{\omega}\right\|_{\mathcal{L}_1} L\|(x_{\text{ref}}-x)_\tau\|_{\mathcal{L}_\infty} + \left\|\frac{H_1(s)}{\omega}\right\|_{\mathcal{L}_1} \|\tilde{x}_\tau\|_{\mathcal{L}_\infty}.$$

Using the first bound in (2.51) and the upper bound from Lemma 2.2.3 in the above expression leads to the second bound in (2.51). □

Remark 2.2.1 Notice that letting $k \to \infty$ leads to $C(s) \to 1$, and thus the reference controller in (2.41), in the limit, leads to perfect cancelation of uncertainties and recovers the performance of the ideal desired system. In this case, the uniform bound for the control signal is lost, as $C(s)=1$ corresponds to an $H_1(s)$, which is improper and, hence, its $\mathcal{L}_1$-norm does not exist.

Theorem 2.2.2 For the closed-loop system in (2.31) with the $\mathcal{L}_1$ adaptive controller defined via (2.32), (2.33), and (2.35), if A_g in (2.39) is Hurwitz, $\theta(t) \equiv \theta$ is constant and $D(s)=1/s$, we have

$$\|x_{\text{ref}}-x\|_{\mathcal{L}_\infty} \leq \frac{\gamma_3}{\sqrt{\Gamma}}, \quad \|u_{\text{ref}}-u\|_{\mathcal{L}_\infty} \leq \frac{\gamma_4}{\sqrt{\Gamma}}, \tag{2.56}$$

where

$$\gamma_3 \triangleq \left\|H_g(s)H_1(s)\right\|_{\mathcal{L}_1}\sqrt{\frac{\theta_m}{\lambda_{\min}(P)}}, \quad H_g(s) \triangleq (s\mathbb{I}-A_g)^{-1}\begin{bmatrix} b \\ 0 \end{bmatrix},$$

$$\gamma_4 \triangleq \left\|\frac{C(s)}{\omega}\theta^\top\right\|_{\mathcal{L}_1}\gamma_3 + \left\|\frac{H_1(s)}{\omega}\right\|_{\mathcal{L}_1}\sqrt{\frac{\theta_m}{\lambda_{\min}(P)}}.$$

Proof. Let $e(t) \triangleq x_{\text{ref}}(t)-x(t)$ and $\varsigma(s) \triangleq -\frac{C(s)}{\omega}\theta^\top e(s)$. With this notation, (2.53) can be written as

$$e(s) = H(s)(\theta^\top e(s) + \omega\varsigma(s) + C(s)\tilde{\eta}(s))$$

and further expressed in state-space form as

$$\begin{bmatrix} \dot{e}(t) \\ \dot{\varsigma}(t) \end{bmatrix} = A_g \begin{bmatrix} e(t) \\ \varsigma(t) \end{bmatrix} + \begin{bmatrix} b \\ 0 \end{bmatrix}\tilde{\eta}_C(t), \quad \begin{matrix} e(0)=0, \\ \varsigma(0)=0, \end{matrix} \quad \tilde{\eta}_C(s) \triangleq C(s)\tilde{\eta}(s).$$

Let $x_\varsigma(t) \triangleq [e^\top(t)\ \varsigma(t)]^\top$. Given a Hurwitz matrix A_g, the system $H_g(s)$ is BIBO stable and strictly proper. Since

$$x_\varsigma(s) = H_g(s)\tilde{\eta}_C(s) = H_g(s)H_1(s)\tilde{x}(s),$$

we can follow similar arguments as in the proof of Theorem (2.2.1) and derive the two bounds in (2.56). □

Thus, the tracking errors between $x_{\text{ref}}(t)$ and $x(t)$ and between $u_{\text{ref}}(t)$ and $u(t)$ are uniformly bounded by a constant inverse proportional to $\sqrt{\Gamma}$, implying that one can arbitrarily improve the tracking performance for both signals simultaneously by increasing Γ. In the case of constant θ and σ, one can prove in addition the following asymptotic result.

Lemma 2.2.4 If in the system (2.31) the parameters $\theta(t) \equiv \theta$ and $\sigma(t) \equiv \sigma$ are constant, A_g in (2.39) is Hurwitz, and $D(s) = 1/s$, the $\mathcal{L}_1$ adaptive controller, defined via (2.32), (2.33), and (2.35), leads to the following asymptotic result:

$$\lim_{t\to\infty} \tilde{x}(t) = 0.$$

Proof. In the case of constant $\theta(t)$ and $\sigma(t)$, the derivative of the Lyapunov function in (2.48) takes the form $\dot{V}(t) = -\tilde{x}^\top(t)Q\tilde{x}(t) \leq 0$, implying boundedness of $\tilde{x}(t)$ and $\tilde{\theta}(t)$, $\tilde{\omega}(t)$, $\tilde{\sigma}(t)$. It follows from Theorem 2.2.2 that $x(t)$ is bounded, and consequently $\hat{x}(t)$ is bounded. Also, the adaptive laws in (2.33) ensure that $\hat{\theta}(t)$, $\hat{\omega}(t)$, $\hat{\sigma}(t)$ are bounded. Hence, it can be checked straightforwardly that $\dot{\tilde{x}}(t)$ is bounded, which leads to uniform boundedness of $\ddot{V}(t)$, and hence uniform continuity of $\dot{V}(t)$. It then follows from Barbalat's lemma that $\dot{V}(t) \to 0$ as $t \to \infty$, leading to $\lim_{t\to\infty} \tilde{x}(t) = 0$. □

2.2.4 Performance in the Presence of Nonzero Trajectory Initialization Error

In this section we prove that in the case of constant θ, arbitrary nonzero trajectory initialization errors lead to exponentially decaying transient errors in the input and output signals of the system. Thus, we consider the system in (2.31) with constant unknown θ while retaining the time-varying disturbance $\sigma(t)$:

$$\begin{aligned} \dot{x}(t) &= A_m x(t) + b(\omega u(t) + \theta^\top x(t) + \sigma(t)), \quad x(0) = x_0, \\ y(t) &= c^\top x(t). \end{aligned} \tag{2.57}$$

We consider the same state predictor as in the previous section,

$$\begin{aligned} \dot{\hat{x}}(t) &= A_m \hat{x}(t) + b\left(\hat{\omega}(t)u(t) + \hat{\theta}^\top(t)x(t) + \hat{\sigma}(t)\right), \quad \hat{x}(0) = \hat{x}_0, \\ \hat{y}(t) &= c^\top \hat{x}(t), \end{aligned} \tag{2.58}$$

where, however, $\hat{x}_0$ might not be equal to x_0 in general. Since $\hat{x}_0 \neq x_0$, then the prediction-error dynamics take the form

$$\dot{\tilde{x}}(t) = A_m \tilde{x}(t) + b(\tilde{\omega}(t)u(t) + \tilde{\theta}^\top(t)x(t) + \tilde{\sigma}(t)), \quad \tilde{x}(0) = \hat{x}_0 - x_0. \tag{2.59}$$

Lemma 2.2.5 For the prediction-error dynamics in (2.59), the following upper bound holds:

$$\|\tilde{x}(t)\| \leq \rho(t), \quad \forall\, t \geq 0,$$

where

$$\begin{aligned} \rho(t) &\triangleq \sqrt{\frac{\left(V(0) - \frac{\theta_n}{\Gamma}\right)e^{-\alpha t}}{\lambda_{\min}(P)} + \frac{\theta_n}{\lambda_{\min}(P)\Gamma}}, \quad \alpha \triangleq \frac{\lambda_{\min}(Q)}{\lambda_{\max}(P)}, \\ \theta_n &\triangleq 4\max_{\theta\in\Theta} \|\theta\|^2 + 4\Delta^2 + (\omega_u - \omega_l)^2 + \frac{4\Delta d_\sigma}{\alpha} \end{aligned} \tag{2.60}$$

with $V(t)$ being the Lyapunov function in (2.47).

Proof. Consider the Lyapunov function candidate in (2.47). Since ω and θ are constant, it follows from (2.48) that

$$\dot{V}(t) \leq -\tilde{x}^\top(t)Q\tilde{x}(t) + 2\Gamma^{-1}|\tilde{\sigma}(t)\dot{\sigma}(t)|. \tag{2.61}$$

Projection ensures that $\hat{\theta}(t) \in \Theta$, $|\hat{\sigma}(t)| \leq \Delta$, $\hat{\omega}(t) \in \Omega$, and therefore

$$\frac{1}{\Gamma}\max_{t\geq 0}(\tilde{\omega}^2(t) + \tilde{\theta}^\top(t)\tilde{\theta}(t) + \tilde{\sigma}^2(t)) \leq \frac{1}{\Gamma}\left(4\max_{\theta\in\Theta}\|\theta\|^2 + 4\Delta^2 + (\omega_u - \omega_l)^2\right).$$

Hence

$$V(t) \leq \tilde{x}^\top(t)P\tilde{x}(t) + \frac{\theta_n}{\Gamma} - \frac{4\Delta d_\sigma \lambda_{\max}(P)}{\lambda_{\min}(Q)\Gamma},$$

where θ_n is given in (2.60). Further, since

$$\tilde{x}^\top(t)Q\tilde{x}(t) \geq \frac{\lambda_{\min}(Q)}{\lambda_{\max}(P)}\tilde{x}^\top(t)P\tilde{x}(t) \geq \frac{\lambda_{\min}(Q)}{\lambda_{\max}(P)}\left(V(t) - \frac{\theta_n}{\Gamma}\right) + \frac{4\Delta d_\sigma}{\Gamma},$$

the upper bound in (2.61) can be used to obtain

$$\dot{V}(t) \leq -\frac{\lambda_{\min}(Q)}{\lambda_{\max}(P)}V(t) + \frac{\lambda_{\min}(Q)}{\lambda_{\max}(P)}\frac{\theta_n}{\Gamma},$$

which leads to

$$V(t) \leq \left(V(0) - \frac{\theta_n}{\Gamma}\right)e^{-\alpha t} + \frac{\theta_n}{\Gamma}$$

with α being given in (2.60). Substituting this expression into

$$\|\tilde{x}(t)\|^2 \leq \frac{\tilde{x}^\top(t)P\tilde{x}(t)}{\lambda_{\min}(P)} \leq \frac{V(t)}{\lambda_{\min}(P)}$$

completes the proof. □

To derive the performance bounds for the closed-loop adaptive system with the $\mathcal{L}_1$ adaptive controller, we first need to introduce the following definition and prove a preliminary result (Lemma 2.2.6). For an m-input n-output stable proper transfer function $F(s)$ with impulse response $f(t)$, let

$$\Psi_F(t) = \max_{i=1,\ldots,n}\sqrt{\sum_{j=1}^{m} f_{ij}^2(t)}, \tag{2.62}$$

where $f_{ij}(t)$ is the ith row, jth column of the impulse response matrix of $F(s)$.

Lemma 2.2.6 Consider an m-input n-output stable proper transfer function $F(s)$ and let $p(s) = F(s)q(s)$. If $\|q(t)\| \leq \mu(t)$, then $\|p(t)\|_\infty \leq \Psi_F(t) * \mu(t)$, where $*$ is the convolution operation.

Proof. Since $p(s) = F(s)q(s)$, then

$$p_i(t) = \int_0^t f_i(t-\tau)q(\tau)d\tau, \quad \forall\, i = 1,\ldots,n,$$

where f_i corresponds to the ith row of the impulse response matrix for $F(s)$. Upper bounding $p_i(t)$ gives

$$|p_i(t)| \leq \int_0^t \|f_i(t-\tau)\|\|q(\tau)\|d\tau \leq \int_0^t \Psi_F(t-\tau)\mu(\tau)d\tau, \quad \forall\, i = 1,\ldots,n,$$

which completes the proof. □

Let

$$H_2(s) \triangleq (\mathbb{I} - G(s)\theta^\top)^{-1}C(s), \quad H_3(s) \triangleq -\frac{C(s)\theta^\top}{\omega}.$$

Theorem 2.2.3 Given the system in (2.57) and the $\mathcal{L}_1$ adaptive controller, defined via (2.58), (2.33), and (2.35), subject to the $\mathcal{L}_1$-norm condition in (2.38), the following bounds hold for all $t \geq 0$:

$$\|x_{\text{ref}}(t) - x(t)\|_\infty \leq \gamma(t), \tag{2.63}$$

$$\|u_{\text{ref}}(t) - u(t)\|_\infty \leq \frac{1}{\omega}\Psi_{H_1}(t) * (\rho(t) + \|\tilde{x}_{\text{in}}(t)\|) + \Psi_{H_3}(t) * \gamma(t), \tag{2.64}$$

where

$$\gamma(t) \triangleq \Psi_{H_2}(t) * (\rho(t) + \|\tilde{x}_{\text{in}}(t)\|), \tag{2.65}$$

and $\tilde{x}_{\text{in}}(t)$ is the inverse Laplace transform of $\tilde{x}_{\text{in}}(s) \triangleq (s\mathbb{I} - A_m)^{-1}(\hat{x}_0 - x_0)$.

Proof. It follows from (2.52) that

$$u(s) = \frac{C(s)}{\omega}(-\theta^\top x(s) - \sigma(s) + k_g r(s) - \tilde{\eta}(s)), \tag{2.66}$$

and the system in (2.57) consequently takes the form

$$x(s) = H(s)\Big(C(s)k_g r(s) + (1 - C(s))(\theta^\top x(s) + \sigma(s)) - C(s)\tilde{\eta}(s)\Big) + x_{\text{in}}(s).$$

In the case of constant θ, the reference system in (2.40)–(2.42) is given by

$$\begin{aligned} x_{\text{ref}}(s) &= H(s)\Big(C(s)k_g r(s) + (1 - C(s))(\theta^\top x_{\text{ref}}(s) + \sigma(s))\Big) + x_{\text{in}}(s), \\ u_{\text{ref}}(s) &= \frac{C(s)}{\omega}\Big(-\theta^\top x_{\text{ref}}(s) - \sigma(s) + k_g r(s)\Big). \end{aligned} \tag{2.67}$$

Thus

$$x_{\text{ref}}(s) - x(s) = H(s)\Big((1 - C(s))\theta^\top(x_{\text{ref}}(s) - x(s)) + C(s)\tilde{\eta}(s)\Big) = H_2(s)H(s)\tilde{\eta}(s).$$

It follows from (2.57) and (2.58) that

$$\tilde{x}(s) = H(s)\tilde{\eta}(s) + \tilde{x}_{\text{in}}(s),$$

and consequently

$$x_{\text{ref}}(s) - x(s) = H_2(s)(\tilde{x}(s) - \tilde{x}_{\text{in}}(s)).$$

Using the upper bound in Lemma 2.2.5, we have

$$\|\tilde{x}(t) - \tilde{x}_{\text{in}}(t)\| \leq \|\tilde{x}(t)\| + \|\tilde{x}_{\text{in}}(t)\| \leq \rho(t) + \|\tilde{x}_{\text{in}}(t)\|,$$

and, hence, Lemma 2.2.6 leads to the upper bound in (2.63).

To prove the bound in (2.64), we notice that from (2.66) and (2.67) one has

$$\begin{aligned} u_{\text{ref}}(s) - u(s) &= -\frac{C(s)\theta^\top}{\omega}(x_{\text{ref}}(s) - x(s)) + \frac{C(s)}{\omega}\tilde{\eta}(s) \\ &= -\frac{C(s)\theta^\top}{\omega}(x_{\text{ref}}(s) - x(s)) + \frac{C(s)}{\omega}\frac{1}{c_o^\top H(s)}c_o^\top H(s)\tilde{\eta}(s) \\ &= H_3(s)(x_{\text{ref}}(s) - x(s)) + \frac{H_1(s)}{\omega}(\tilde{x}(s) - \tilde{x}_{\text{in}}(s)), \end{aligned}$$

and, therefore, application of Lemma 2.2.6 leads to the bound in (2.64). □

Remark 2.2.2 We notice that the above bounds are derived using a conservative estimation $\|\tilde{x}(t) - \tilde{x}_{\text{in}}(t)\| \leq \|\tilde{x}(t)\| + \|\tilde{x}_{\text{in}}(t)\|$, which leads to a conservative upper bound for $\|\tilde{x}(t) - \tilde{x}_{\text{in}}(t)\|$. In fact, $\tilde{x}(t)$ and $\tilde{x}_{\text{in}}(t)$ tend to cancel each other in certain cases, leading to very small transient deviation, as observed in simulations.

2.2.5 Time-Delay Margin Analysis

$\mathcal{L}_1$ Adaptive Controller in the Presence of Time Delay

In this section, we derive the time-delay margin of the $\mathcal{L}_1$ adaptive controller for the system in (2.57) with the predictor given in (2.58), subject to $\hat{x}_0 \neq x_0$. For simplicity, we choose $D(s) = 1/s$, although arbitrary $D(s)$ satisfying (2.38) can be accommodated in the analysis below.

We proceed by considering the following three systems.

System 1. Let $\tau > 0$ denote the unknown constant time delay in the control channel. The system in (2.57), when closed with the delayed $\mathcal{L}_1$ adaptive controller, takes the form

$$\dot{x}(t) = A_m x(t) + b(\omega u_d(t) + \theta^\top x(t) + \sigma(t)), \quad x(0) = x_0, \tag{2.68}$$

where

$$u_d(t) = \begin{cases} 0, & 0 \leq t < \tau, \\ u(t-\tau), & t \geq \tau, \end{cases} \tag{2.69}$$

with $u(t)$ being computed via (2.33), (2.35), (2.58) and $u_d(t)$ being the delayed control signal. We further notice that this closed-loop system has a unique solution. It is the stability of this closed-loop system that we are trying to determine dependent upon τ. We further notice that due to the delay in the control channel, one could not derive the dynamics of the error between the system and the predictor, the boundedness of which is stated in Lemmas 2.2.3 and 2.2.5. Theorems 2.2.1, 2.2.2, and 2.2.3 also do not hold.

System 2. Next, we consider the following closed-loop system without delay but with additional disturbance:

$$\dot{x}_q(t) = A_m x_q(t) + b\left(\omega u_q(t) + \theta^\top x_q(t) + \sigma(t) + \nu(t)\right), \quad x_q(0) = x_0, \tag{2.70}$$

and let $u_q(t)$ be given via (2.33), (2.35), and (2.58) with $x(t)$ being replaced by $x_q(t)$. We notice that compared to the system in (2.57), the system in (2.70) has an additional disturbance signal $\nu(t)$. If $\nu(t)$ is continuously differentiable with uniformly bounded derivative and

$$|\sigma(t) + \nu(t)| \leq \Delta, \quad \forall t \geq 0, \tag{2.71}$$

where Δ was introduced in (2.34), then application of the $\mathcal{L}_1$ adaptive controller to the system in (2.70) is well defined, and hence the results of Lemma 2.2.5 and Theorem 2.2.3 are valid.

System 3. Finally, we consider the open-loop system in (2.57) and apply $u_{q_d}(t)$ to it with $u_{q_d}(t)$ being the τ-delayed signal of $u_q(t)$:

$$u_{q_d}(t) = \begin{cases} 0, & 0 \leq t < \tau, \\ u_q(t-\tau), & t \geq \tau. \end{cases} \tag{2.72}$$

We denote the system's response to this signal by $x_o(t)$, where the subindex o indicates the open-loop nature of this system. It is important to notice that at this point we view $u_{q_d}(t)$ as a time-varying input signal for (2.57), not as a feedback signal, so that (2.57) remains an open-loop system in this context. These last two systems are illustrated in Figure 2.14.

In preparation for the proof of the main result in Theorem 2.2.4, we now derive some preliminary results that will be used in future derivations.

Lemma 2.2.7 If the output of the open-loop System 3 has the same time history as the closed-loop output of System 2, i.e.,

$$x_o(t) \equiv x_q(t), \quad \forall t \geq 0, \tag{2.73}$$

then $u(t) \equiv u_q(t)$, $x(t) \equiv x_q(t)$, for all $t \geq 0$.

Proof. Equation (2.73) implies that the open-loop time-delayed System 3, given by (2.57), (2.72), generates $x_q(t)$ in response to the input $u_q(t)$. When applied to (2.70), $u_q(t)$ leads to $x_q(t)$. Hence, $u_q(t)$ and $x_q(t)$ are also solutions of the closed-loop adaptive System 1, defined via (2.33), (2.35), (2.58), (2.68), and (2.69). □

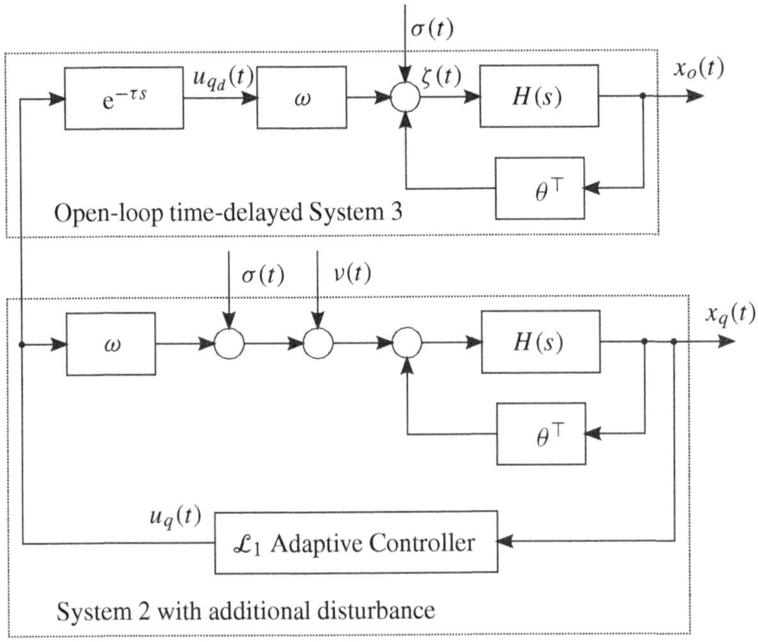

Figure 2.14: Block diagram of Systems 2 and 3.

Lemma 2.2.7 consequently implies that to ensure the stability of System 1 in the presence of a given time delay τ, it is sufficient to prove the existence of $\nu(t)$ in System 2, satisfying (2.71) and verifying (2.73). We notice, however, that the closed-loop System 2 is a nonlinear system due to the nonlinear adaptive laws, so the proof of existence of such $\nu(t)$ for this system and the explicit construction of the bound Δ is not straightforward. Moreover, we note that the condition in (2.73) relates the time delay τ of System 1 (or System 3) to the signal $\nu(t)$ implicitly. In the next section we introduce an LTI system that helps to prove the existence of such $\nu(t)$ and leads to explicit computation of Δ. The definition of this LTI system is the key step in the overall time-delay margin analysis. The LTI system has an exogenous input that lumps the time trajectories of the nonlinear elements of the closed-loop System 2. For this LTI system, the time-delay margin can be computed via its loop transfer function, which consequently defines a conservative, but guaranteed, lower bound for the time-delay margin of the closed-loop adaptive system.

LTI System in the Presence of the Same Time Delay

We consider the following closed-loop LTI system in the presence of the same time delay τ:

$$\dot{x}_l(t) = (A_m + b\theta^\top)x_l(t) + b\zeta_l(t), \quad x_l(0) = x_0, \tag{2.74}$$

$$\zeta_l(s) = \omega u_{l_d}(s) + \sigma(s), \tag{2.75}$$

$$u_{l_d}(t) = \begin{cases} 0, & 0 \le t < \tau, \\ u_l(t-\tau), & t \ge \tau, \end{cases} \tag{2.76}$$

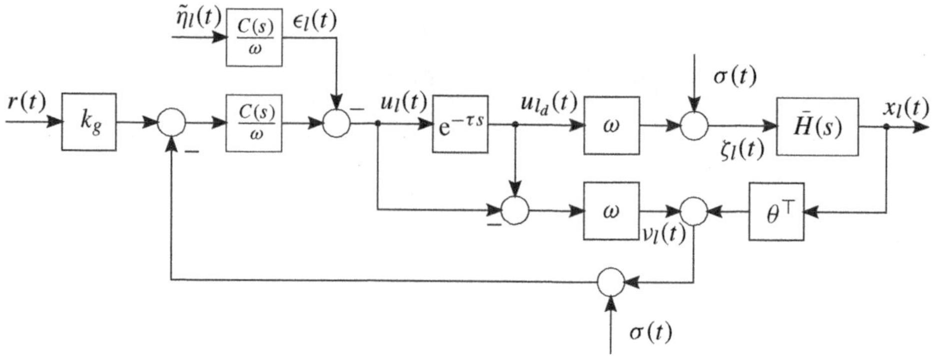

Figure 2.15: Equivalent LTI system in (2.74)–(2.79).

$$u_l(s) = \frac{C(s)}{\omega}\Big(k_g r(s) - \theta^\top x_l(s) - \sigma(s) - v_l(s)\Big) - \epsilon_l(s), \tag{2.77}$$

$$\epsilon_l(s) = \frac{C(s)}{\omega}\tilde{\eta}_l(s), \tag{2.78}$$

$$v_l(s) = \omega u_{l_d}(s) - \omega u_l(s), \tag{2.79}$$

where $x_l(t)$, $u_l(t)$, and $\epsilon_l(t)$ are the states of this LTI system; $\tilde{\eta}_l(t)$ is an exogenous signal; $r(t)$ and $\sigma(t)$ are the same bounded reference and disturbance input as in (2.57) with $|\sigma(t)| \leq \Delta_0$; and $C(s)$ is the same low-pass filter used in the definition of the reference system in (2.40)–(2.42). Further, let

$$\bar{H}(s) \triangleq (s\mathbb{I} - A_m - b\theta^\top)^{-1} b.$$

The system (2.74)–(2.79) is highly coupled, as shown in Figure 2.15.

Remark 2.2.3 In the absence of time delay, i.e., when $\tau = 0$, we have $u_{l_d}(t) = u_l(t)$, and the system in (2.74)–(2.79) is reduced to

$$\dot{x}_l(t) = (A_m + b\theta^\top)x_l(t) + b(\omega u_l(t) + \sigma(t)), \quad x_l(0) = x_0,$$
$$u_l(s) = \frac{1}{\omega}C(s)(k_g r(s) - \theta^\top x_l(s) - \sigma(s)) - \epsilon_l(s),$$
$$\epsilon_l(s) = \frac{C(s)}{\omega}\tilde{\eta}_l(s).$$

We notice that the trajectories of this LTI system are uniquely defined, once $\tilde{\eta}_l(t)$ and $r(t)$ are specified. Omitting the nonzero initial condition x_0, we notice the following relationship between this LTI system and the original reference system (2.40)–(2.42) (with $\theta(t) \equiv \theta$):

$$\frac{x_l(s)}{r(s)} = k_g C(s)\big(\mathbb{I} + C(s)\bar{H}(s)\theta^\top\big)^{-1}\bar{H}(s) = \frac{x_{\text{ref}}(s)}{r(s)},$$
$$\frac{x_l(s)}{\sigma(s)} = \big(1 - C(s)\big)\big(\mathbb{I} + C(s)\bar{H}(s)\theta^\top\big)^{-1}\bar{H}(s) = \frac{x_{\text{ref}}(s)}{\sigma(s)}.$$

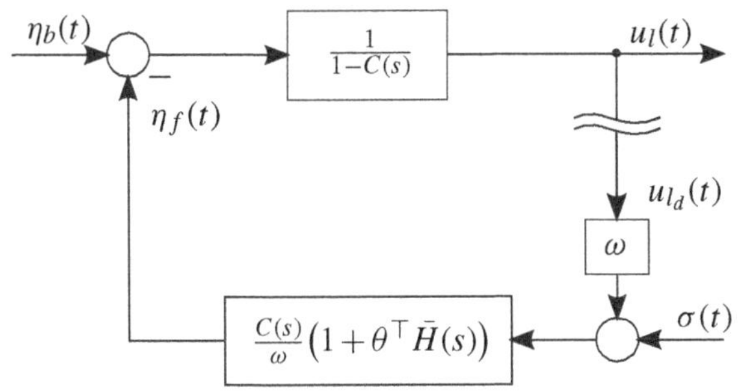

Figure 2.16: LTI system for time-delay margin prediction.

Time-Delay Margin of the LTI System

We notice that the phase margin of the LTI system in (2.74)–(2.79) can be determined from its open-loop transfer function from $u_{l_d}(t)$ to $u_l(t)$, and it does not depend upon its initial condition. Omitting the initial condition $x_l(0)$ for the time being, it follows from (2.74)–(2.79) that

$$u_l(s) = \frac{1}{1-C(s)}\left(\frac{C(s)}{\omega}k_g r(s) - \frac{C(s)}{\omega}(1+\theta^\top \bar{H}(s))\big(\omega u_{l_d}(s)+\sigma(s)\big) - \epsilon_l(s)\right).$$

Therefore, it can be equivalently presented as in Figure 2.16 with the following signals:

$$u_l(s) = \frac{\eta_b(s)-\eta_f(s)}{1-C(s)},$$
$$\eta_f(s) = \frac{C(s)}{\omega}(1+\theta^\top \bar{H}(s))\big(\omega u_{l_d}(s)+\sigma(s)\big),$$
$$\eta_b(s) = \frac{C(s)}{\omega}k_g r(s) - \epsilon_l(s).$$

Assume $\tilde{\eta}_l(t)$ is bounded. Then, the resulting $\epsilon_l(t)$ is bounded. Also, the loop transfer function of the system in Figure 2.16 is

$$L_o(s) = \frac{C(s)}{1-C(s)}(1+\theta^\top \bar{H}(s)). \tag{2.80}$$

Since $\sigma(t)$ and $r(t)$ are bounded, the phase margin ϕ_m of (2.80) can be derived from its Bode plot. Its time-delay margin is given by

$$\mathcal{T} = \frac{\phi_m}{\omega_{gc}}, \tag{2.81}$$

where ω_{gc} is the gain crossover frequency of $L_o(s)$. The next lemma states a sufficient condition for boundedness of all the states in the system (2.74)–(2.79), including the internal states.

Lemma 2.2.8 For arbitrary bounded signal $\epsilon_l(t)$, if

$$\tau < \mathcal{T}, \tag{2.82}$$

the signals $x_l(t)$, $u_l(t)$, $v_l(t)$ are bounded.

Proof. From the definition of the time-delay margin of LTI systems, it follows that for all $\tau < \mathcal{T}$, if $\epsilon_l(t)$ is bounded, $u_l(t)$ is bounded. Hence, it follows from (2.76) that $u_{l_d}(t)$ is bounded, and the relationship in (2.79) implies that $\nu_l(t)$ is bounded. Since $u_l(t)$, $\epsilon_l(t)$, $\nu_l(t)$, $r(t)$, and $\sigma(t)$ are bounded, it follows from (2.77) that $\theta^\top x_l(t)$ is bounded. From (2.74) we notice that

$$x_l(s) = H(s)(\theta^\top x_l(s) + \zeta_l(s)) + x_{\rm in}(s),$$

which leads to boundedness of $x_l(t)$. □

Lemma 2.2.9 Let $\epsilon_l(t)$ be bounded, and let τ comply with (2.82). If $\tilde{\eta}_l(t)$ is bounded, then $\dot{\nu}_l(t)$ is bounded.

Proof. Using Lemma 2.2.8, we immediately conclude that $x_l(t)$ and $u_l(t)$ are bounded. Since $C(s)$ is strictly proper and stable, bounded $\tilde{\eta}_l(t)$ ensures that $\epsilon_l(t)$ is continuously differentiable with uniformly bounded derivative. Using similar arguments, one can conclude that both $u_l(t)$ and $u_{l_d}(t)$ have bounded derivatives. Hence, it follows from (2.79) that $\dot{\nu}_l(t)$ is bounded. □

Time-Delay Margin of the Closed-Loop Adaptive System

Let

$$\epsilon_b(\epsilon_\alpha, t) \triangleq \Psi_{H_1}(t) * (\kappa_\alpha(\epsilon_\alpha, t) + \|\tilde{x}_{\rm in}(t)\|) + \epsilon_\beta,$$

$$\kappa_\alpha(\epsilon_\alpha, t) \triangleq \sqrt{\frac{(V(0) - \epsilon_\alpha)e^{-\alpha t} + \epsilon_\alpha}{\lambda_{\min}(P)}},$$

where $\Psi_{H_1}(t)$ is determined from (2.62) and (2.18), $\|\tilde{x}_{\rm in}(t)\|$ is introduced in (2.65), and $V(t)$ is the Lyapunov function in (2.47), while $\epsilon_\alpha > 0$ and $\epsilon_\beta > 0$ are arbitrary positive constants. We notice that $\epsilon_b(\epsilon_\alpha, t)$ is positive for all $t \geq 0$ and is uniformly bounded. Further assume that $\epsilon_l(t)$ verifies the upper bound

$$|\epsilon_l(t)| \leq \epsilon_b(\epsilon_\alpha, t), \quad \forall\, t \geq 0. \tag{2.83}$$

For all $\tau < \mathcal{T}$, where $\mathcal{T}$ is the time-delay margin of $L_o(s)$, Lemma 2.2.8 guarantees that the map $\Delta_n : \mathbb{R}^+ \times [0, \mathcal{T}) \to \mathbb{R}^+$, given by

$$\Delta_n(\epsilon_\alpha, \tau) \triangleq \max_{|\epsilon_l(t)| \leq \epsilon_b(\epsilon_\alpha, t)} \|\sigma + \nu_l\|_{\mathcal{L}_\infty}, \tag{2.84}$$

is well defined. Finally, let $\delta_1 > 0$ be an arbitrary constant, and let

$$\Delta \triangleq \Delta_n(\epsilon_\alpha, \tau) + \delta_1. \tag{2.85}$$

Remark 2.2.4 Notice that the computation of Δ_n in (2.84) depends upon $\sigma(t)$, the control signal $u_l(t)$, and the delay τ, involved in definition of $\nu_l(t)$. While Δ_0 is independent of $\epsilon_l(t)$, maximization of $\nu_l(t)$ depends upon $u_l(t)$ and $u_{l_d}(t)$ and hence upon $\epsilon_l(t)$. If the external signals in the LTI system (2.74)–(2.79) are well defined, then one can compute a conservative estimate for Δ_n from (2.84). We also notice that the definition of $\epsilon_b(t)$ depends upon the nonzero trajectory initialization error.

Similarly, we define

$$\Delta_u(\epsilon_\alpha,\tau) \triangleq \max_{|\epsilon_l(t)|\leq \epsilon_b(\epsilon_\alpha,t)} \|u_l\|_{\mathcal{L}_\infty}, \quad \Delta_x(\epsilon_\alpha,\tau) \triangleq \max_{|\epsilon_l(t)|\leq \epsilon_b(\epsilon_\alpha,t)} \|x_l\|_{\mathcal{L}_\infty}.$$

If $\tilde{\eta}_l(t)$ verifies the upper bound

$$\|\tilde{\eta}_l\|_{\mathcal{L}_\infty} \leq (\omega_u - \omega_l)\Delta_u(\epsilon_\alpha,\tau) + 2L\Delta_x(\epsilon_\alpha,\tau) + 2\Delta, \tag{2.86}$$

and the resulting $\epsilon_l(t)$ complies with (2.83), Lemma 2.2.9 guarantees that for all $\tau < \mathcal{T}$

$$\Delta_d \triangleq \max_{\tilde{\eta}_l(t)} \|\dot{\sigma} + \dot{\nu}_l\|_{\mathcal{L}_\infty} \tag{2.87}$$

is a computable (possibly unknown) constant. Further, let

$$\begin{aligned}
\epsilon_c(t) &\triangleq \Psi_{H_1}(t) * (\kappa_\beta(\theta_q,t) + \|\tilde{x}_{\text{in}}(t)\|),\\
\kappa_\beta(\theta_q,t) &\triangleq \sqrt{\frac{\left(V(0) - \frac{\theta_q}{\Gamma}\right)\mathrm{e}^{-\alpha t} + \frac{\theta_q}{\Gamma}}{\lambda_{\min}(P)}},\\
\theta_q &\triangleq 4\max_{\theta\in\Theta}\|\theta\|^2 + 4\Delta^2 + (\omega_u - \omega_l)^2 + 4\frac{\lambda_{\max}(P)\Delta_d\Delta}{\lambda_{\min}(Q)},
\end{aligned} \tag{2.88}$$

where $V(t)$ is the Lyapunov function in (2.47). We notice that for arbitrary $\epsilon_\alpha > 0$ and arbitrary $\tau > 0$ verifying (2.82), if $\tilde{\eta}_l(t)$ complies with (2.86) and the resulting $\epsilon_l(t)$ complies with (2.83), one has bounded Δ and Δ_d and bounded positive $\epsilon_c(t)$.

Theorem 2.2.4 Consider System 1 and the LTI system in (2.74)–(2.79) in the presence of the same time delay τ. For arbitrary $\epsilon_\alpha \in \mathbb{R}^+$, choose Δ according to (2.85) and restrict the adaptive gain to the following lower bound:

$$\Gamma > \frac{\theta_q}{\epsilon_\alpha}. \tag{2.89}$$

Then, for every $\tau < \mathcal{T}$, there exists a bounded exogenous signal $\tilde{\eta}_l(t)$ over $[0,\infty)$ verifying (2.86) such that the resulting $\epsilon_l(t)$ complies with (2.83), and $x_l(t) = x(t)$, $u_l(t) = u(t)$ $\forall\, t \geq 0$.

Proof. To proceed with the proof, we introduce the following notations. Let $x_h(t)$ be the state variable of a general LTI system $H_x(s)$, and let $x_i(t)$ and $x_s(t)$ be the input and the output signals of it. We notice that for arbitrary time instant t_1 and arbitrary fixed time interval $[t_1, t_2]$, given a continuous input signal $x_i(t)$ over $[t_1, t_2)$, $x_s(t)$ is uniquely defined for $t \in [t_1, t_2]$. Let $\mathcal{S}$ be the map

$$x_s(t)|_{t\in[t_1,t_2]} = \mathcal{S}(H_x(s), x_h(t_1), x_i(t)|_{t\in[t_1,t_2)}).$$

If $x_i(t)$ is continuous, then $\mathcal{S}$ is a continuous map. We further notice that $x_s(t)$ is defined over a closed interval $[t_1, t_2]$, although $x_i(t)$ is defined over the corresponding open set $[t_1, t_2)$. It follows from the definition of $\mathcal{S}$ that given

$$\begin{aligned}
x_{s_1}(t)|_{t\in[t_1,t_2]} &= \mathcal{S}(H_x(s), x_{h_1}, x_{i_1}(t)|_{t\in[t_1,t_2)}),\\
x_{s_2}(t)|_{t\in[t_1,t_2]} &= \mathcal{S}(H_x(s), x_{h_2}, x_{i_2}(t)|_{t\in[t_1,t_2)}),
\end{aligned}$$

if $x_{h_1} = x_{h_2}$ and $x_{i_1}(t) = x_{i_2}(t)$ over $[t_1, t_2)$, then $x_{s_1}(t) = x_{s_2}(t)$ for arbitrary $t \in [t_1, t_2]$.

In (2.70), we notice that if $|\sigma(t)+\nu(t)| \le \Delta$, for all $t \in [0, t^*]$, for some $t^* > 0$, and $\sigma(t)$, $\nu(t)$ have bounded derivatives over $[0, t^*]$, then application of the $\mathcal{L}_1$ adaptive controller is well defined over $[0, t^*]$. Letting

$$d_{t^*} \triangleq \|(\dot{\sigma}+\dot{\nu})_{t^*}\|_{\mathcal{L}_\infty},$$

it follows from (2.58) and (2.70) that

$$\dot{\tilde{x}}_q(t) = A_m \tilde{x}_q(t) + b\tilde{\eta}_q(t), \quad \tilde{x}_q(0) = \hat{x}(0) - x(0),$$

where $\tilde{x}_q(t) \triangleq \hat{x}(t) - x_q(t)$, and

$$\tilde{\eta}_q(t) \triangleq \tilde{\omega}(t)u_q(t) + \tilde{\theta}^\top(t)x_q(t) + \tilde{\sigma}_q(t), \quad \tilde{\sigma}_q(t) \triangleq \hat{\sigma}(t) - (\sigma(t)+\nu(t)). \tag{2.90}$$

The choice of $D(s) = 1/s$ leads to a first-order low-pass filter,

$$C(s) = \frac{\omega k}{s+\omega k}.$$

The expression above, along with the definition of $u_q(t)$, implies that

$$\begin{aligned} u_q(t)|_{t\in[0,t^*]} &= \mathcal{S}\Big(C(s)/\omega, u_q(0), \big(k_g r(t) - \theta^\top x_q(t) - \sigma(t) - \nu(t) - \tilde{\eta}_q(t)\big)|_{t\in[0,t^*)}\Big), \\ \tilde{x}_q(t)|_{t\in[0,t^*]} &= \mathcal{S}(H(s), \tilde{x}_q(0), \tilde{\eta}_q(t)|_{t\in[0,t^*)}). \end{aligned}$$

We notice that $u_q(t)|_{t\in[0,t^*]}$ can be equivalently presented as

$$\begin{aligned} u_q(t)|_{t\in[0,t^*]} = {} & \mathcal{S}\Big(C(s)/\omega, u_q(0), \big(k_g r(t) - \theta^\top x_q(t) - \sigma(t) - \nu(t)\big)|_{t\in[0,t^*)}\Big) \\ & - \epsilon(t)|_{t\in[0,t^*]}, \end{aligned} \tag{2.91}$$

where

$$\epsilon(t)|_{t\in[0,t^*]} = \mathcal{S}(C(s)/\omega, 0, \tilde{\eta}_q(t)|_{t\in[0,t^*)}). \tag{2.92}$$

Next, let

$$\begin{aligned} \kappa(\theta_{t^*}, t) &\triangleq \sqrt{\frac{\left(V(0) - \frac{\theta_{t^*}}{\Gamma}\right)e^{-\alpha t} + \frac{\theta_{t^*}}{\Gamma}}{\lambda_{\min}(P)}}, \quad t \in [0, t^*], \quad \alpha = \frac{\lambda_{\min}(Q)}{\lambda_{\max}(P)}, \\ \theta_{t^*} &\triangleq 4\max_{\theta\in\Theta}\|\theta\|^2 + 4\Delta^2 + (\omega_u - \omega_l)^2 + \frac{4\Delta d_{t^*}}{\alpha}. \end{aligned} \tag{2.93}$$

From Lemma 2.2.5, it follows that $\|\tilde{x}_q(t)\| \le \kappa(\theta_{t^*}, t)$, for all $t \in [0, t^*]$. Since

$$\begin{aligned} \epsilon(t)|_{t\in[0,t^*]} &= \mathcal{S}\left(\frac{C(s)}{\omega c_o^\top H(s)} c_o^\top H(s), 0, \tilde{\eta}_q(t)|_{t\in[0,t^*)}\right) \\ &= \mathcal{S}\left(\frac{C(s)}{\omega c_o^\top H(s)} c_o^\top, 0, (\tilde{x}_q(t) - \tilde{x}_{\text{in}}(t))|_{t\in[0,t^*)}\right), \end{aligned}$$

then $\epsilon(t)$ can be upper bounded as

$$|\epsilon(t)| \leq \Psi_{H_1}(t) * (\kappa(\theta_{t^*}, t) + \|\tilde{x}_{\text{in}}(t)\|), \quad \forall\, t \in [0, t^*]. \tag{2.94}$$

Next, we prove the existence of a continuously differentiable $\nu(t)$, $t \geq 0$, in the closed-loop adaptive System 2 and the existence of bounded $\tilde{\eta}_l(t)$ in the time-delayed LTI system such that for all $t \geq 0$,

$$|\sigma(t) + \nu(t)| < \Delta, \quad |\dot{\sigma}(t) + \dot{\nu}(t)| < \Delta_d, \quad x_o(t) = x_q(t), \tag{2.95}$$
$$u_l(t) = u_q(t), \quad x_l(t) = x_q(t), \quad \epsilon_l(t) = \epsilon(t), \quad |\epsilon_l(t)| < \epsilon_b(\epsilon_\alpha, t). \tag{2.96}$$

We notice that these two systems are well defined if the external inputs $\nu(t)$ and $\tilde{\eta}_l(t)$ are specified. With (2.95), Lemma 2.2.7 implies that $x(t) = x_q(t)$ and $u(t) = u_q(t)$ for all $t \geq 0$, while (2.96) proves Theorem 2.2.4.

Let $\zeta(t) \triangleq \omega u_{q_d}(t) + \sigma(t)$, $t \geq 0$. It follows from the definition of the map $\mathcal{S}$ and the definition of the time-delayed open-loop System 3 that, for all $i \geq 0$, one can write

$$x_o(t)|_{t\in[i\tau,(i+1)\tau]} = \mathcal{S}\left(\bar{H}(s), x_o(i\tau), \zeta(t)|_{t\in[i\tau,(i+1)\tau)}\right). \tag{2.97}$$

In the three steps below, we prove by iteration that for all $t \in [0, i\tau)$:

$$\begin{array}{lll} u_q(t) = u_l(t), & \epsilon(t) = \epsilon_l(t), & x_o(t) = x_q(t) = x_l(t), \\ u_{q_d}(t) = u_{l_d}(t), & |\epsilon(t)| < \epsilon_b(\epsilon_\alpha, t), & \end{array} \tag{2.98}$$

for all $i \geq 0$. In addition, we prove that over $t \in [0, i\tau)$ there exist bounded $\tilde{\eta}_l(t)$ and continuously differentiable $\nu(t)$ such that for all $i \geq 0$

$$\nu(t) = \nu_l(t), \quad |\sigma(t) + \nu(t)| < \Delta, \quad |\dot{\sigma}(t) + \dot{\nu}(t)| < \Delta_d. \tag{2.99}$$

Step 1: In this step, we prove that (2.98)–(2.99) hold for $i = 0$. It follows from (2.92) that $\epsilon(0) = 0$. Due to zero initialization of $C(s)$, one has $u_q(0) = u_l(0) = 0$. Recall that $x_o(0) = x_l(0) = x_0$. The expressions in (2.72) and (2.76) imply that $u_{q_d}(t) = u_{l_d}(t) = 0$, $t \in [0, \tau)$. Hence, it follows that for $i = 0$, we have

$$\begin{array}{l} u_q(i\tau) = u_l(i\tau), \quad \epsilon(i\tau) = \epsilon_l(i\tau), \quad x_o(i\tau) = x_q(i\tau) = x_l(i\tau), \\ u_{q_d}(t) = u_{l_d}(t), \quad \forall\, t \in [i\tau, (i+1)\tau), \\ |\epsilon(i\tau)| < \epsilon_b(\epsilon_\alpha, i\tau). \end{array}$$

For $i = 0$, the existence of $\nu(t)$ satisfying (2.99) is trivial.

Step 2: Assume that, for some $i \geq 0$, the following conditions hold:

$$u_q(t) = u_l(t), \quad \forall\, t \in [0, i\tau], \tag{2.100}$$
$$\epsilon(t) = \epsilon_l(t), \quad \forall\, t \in [0, i\tau], \tag{2.101}$$
$$x_o(t) = x_q(t) = x_l(t), \quad \forall\, t \in [0, i\tau], \tag{2.102}$$
$$u_{q_d}(t) = u_{l_d}(t), \quad \forall\, t \in [i\tau, (i+1)\tau), \tag{2.103}$$
$$|\epsilon(t)| < \epsilon_b(\epsilon_\alpha, t), \quad \forall\, t \in [0, i\tau], \tag{2.104}$$

and there exist bounded $\tilde{\eta}_l(t)$ and continuously differentiable $\nu(t)$ such that

$$\nu(t) = \nu_l(t), \quad |\sigma(t)+\nu(t)| < \Delta, \quad |\dot{\sigma}(t)+\dot{\nu}(t)| < \Delta_d, \quad t \in [0, i\tau). \tag{2.105}$$

We prove below that there exist bounded $\tilde{\eta}_l(t)$ and continuously differentiable $\nu(t)$ with bounded derivative such that (2.100)–(2.105) hold for $(i+1)$.

We notice that the relationship in (2.74) implies

$$x_l(t)|_{t\in[i\tau,(i+1)\tau]} = \mathcal{S}\left(\bar{H}(s), x_l(i\tau), \zeta_l(t)|_{t\in[i\tau,(i+1)\tau)}\right). \tag{2.106}$$

It follows from (2.103) that $\zeta_l(t) = \zeta(t)$, for all $t \in [i\tau, (i+1)\tau)$. Using (2.102), it follows from (2.97) and (2.106) that

$$x_o(t) = x_l(t), \quad \forall\, t \in [0, (i+1)\tau]. \tag{2.107}$$

We now define $\nu(t)$ over $[i\tau, (i+1)\tau)$ as

$$\nu(t) = \omega u_{q_d}(t) - \omega u_q(t), \quad t \in [i\tau, (i+1)\tau). \tag{2.108}$$

Since (2.70) implies that

$$x_q(t)|_{t\in[i\tau,(i+1)\tau]} = \mathcal{S}(\bar{H}(s), x_q(i\tau), (\omega u_q(t)+\sigma(t)+\nu(t))|_{t\in[i\tau,(i+1)\tau)}),$$

it follows from (2.108) that

$$x_q(t)|_{t\in[i\tau,(i+1)\tau]} = \mathcal{S}(\bar{H}(s), x_q(i\tau), \zeta(t)_{t\in[i\tau,(i+1)\tau)}).$$

Along with (2.97) and (2.102), the expression above ensures that

$$x_q(t) = x_o(t), \quad \forall\, t \in [0, (i+1)\tau]. \tag{2.109}$$

We need to prove that the definition in (2.108) guarantees

$$|\sigma(t)+\nu(t)| < \Delta, \quad t \in [i\tau, (i+1)\tau), \tag{2.110}$$

which is required for application of the $\mathcal{L}_1$ adaptive controller. Assume that it is not true. Since (2.105) holds for all $t \in [0, i\tau)$ and $\nu(t)$ is continuous over $[i\tau, (i+1)\tau)$, there must exist $t' \in [i\tau, (i+1)\tau)$ such that $|\sigma(t)+\nu(t)| < \Delta$ for all $t < t'$ and

$$|\sigma(t')+\nu(t')| = \Delta. \tag{2.111}$$

It follows from (2.97) and (2.108) that

$$x_o(t)_{t\in[i\tau,t']} = \mathcal{S}\Big(\bar{H}(s), x_o(i\tau), (\omega u_q(t)+\sigma(t)+\nu(t))|_{t\in[i\tau,t')}\Big),$$

while the relationships in (2.91) and (2.92) imply that

$$u_q(t)|_{t\in[i\tau,t']} = \mathcal{S}\left(\frac{C(s)}{\omega}, u_q(i\tau)+\epsilon(i\tau), (k_g r(t) - \theta^\top x_q(t) - \sigma(t) - \nu(t))|_{t\in[i\tau,t')}\right) - \epsilon(t)|_{t\in[i\tau,t']}, \tag{2.112}$$

where

$$\epsilon(t)|_{t\in[i\tau,t']} = \mathcal{S}\Big(C(s)/\omega, \epsilon(i\tau), \tilde{\eta}_q(t)|_{t\in[i\tau,t')}\Big). \tag{2.113}$$

Let

$$\tilde{\eta}_l(t) = \tilde{\eta}_q(t), \quad t \in [i\tau, t'). \tag{2.114}$$

Then, we have

$$\epsilon_l(t)|_{t\in[i\tau,t']} = \mathcal{S}\Big(C(s)/\omega, \epsilon_l(i\tau), \tilde{\eta}_q(t)|_{t\in[i\tau,t')}\Big),$$

which along with (2.101) and (2.113) implies

$$\epsilon_l(t) = \epsilon(t), \quad \forall\, t \in [i\tau, t']. \tag{2.115}$$

Substituting (2.100), (2.107), (2.109), and (2.115) into (2.112) yields

$$u_q(t)|_{t\in[i\tau,t']} = \mathcal{S}\Big(C(s)/\omega, u_l(i\tau)+\epsilon_l(i\tau), (k_g r(t) - \theta^\top x_l(t) - \sigma(t) - \nu(t))|_{t\in[i\tau,t')}\Big)$$
$$- \epsilon_l(t)|_{t\in[i\tau,t']}. \tag{2.116}$$

The relationships in (2.79) and (2.103) imply that

$$\nu_l(t) = \omega u_{q_d}(t) - \omega u_l(t), \quad t \in [i\tau, t']. \tag{2.117}$$

Since (2.77) implies

$$u_l(t)|_{t\in[i\tau,t']} = \mathcal{S}\Big(C(s)/\omega, u_l(i\tau)+\epsilon_l(i\tau), (k_g r(t) - \theta^\top x_l(t) - \sigma(t) - \nu_l(t))|_{t\in[i\tau,t')}\Big)$$
$$- \epsilon_l(t)|_{t\in[i\tau,t']},$$

it follows from (2.108), (2.116), and (2.117) that

$$u_q(t) = u_l(t), \quad \forall\, t \in [i\tau, t'], \tag{2.118}$$

$$\nu(t) = \nu_l(t), \quad \forall\, t \in [i\tau, t'). \tag{2.119}$$

It follows from (2.105) and (2.119) that

$$\nu(t) = \nu_l(t), \quad \forall\, t \in [0, t'). \tag{2.120}$$

We now prove by contradiction that

$$|\epsilon(t)| < \epsilon_b(\epsilon_\alpha, t), \quad \forall\, t \in [i\tau, t']. \tag{2.121}$$

If (2.121) is not true, then since $\epsilon(t)$ is continuous, there exists $\bar{t} \in [i\tau, t']$ such that $|\epsilon(t)| < \epsilon_b(\epsilon_\alpha, t)$, for all $t \in [i\tau, \bar{t})$, and

$$|\epsilon(\bar{t})| = \epsilon_b(\epsilon_\alpha, t). \tag{2.122}$$

It follows from (2.104) that

$$|\epsilon(t)| \le \epsilon_b(\epsilon_\alpha, t), \quad \forall\, t \in [0, \bar{t}]. \tag{2.123}$$

The relationships in (2.100), (2.102), (2.107), (2.109), and (2.118) imply that $u_q(t) = u_l(t)$, $x_q(t) = x_l(t)$ for all $t \in [0, \bar{t}]$. Therefore, (2.90) and (2.114) imply that

$$\tilde{\eta}_l(t) = \tilde{\omega}(t) u_l(t) + \tilde{\theta}^\top(t) x_l(t) + \tilde{\sigma}_q(t),$$

and consequently

$$\|\tilde{\eta}_{l_{\bar{t}}}\|_{\mathcal{L}_\infty} \leq (\omega_u - \omega_l)\|u_{l_{\bar{t}}}\|_{\mathcal{L}_\infty} + 2L\|x_{l_{\bar{t}}}\|_{\mathcal{L}_\infty} + 2\Delta\,. \tag{2.124}$$

It follows from (2.101) and (2.115) that $\epsilon_l(t) = \epsilon(t)$, for all $t \in [0, t']$, and hence (2.123) implies

$$|\epsilon_l(t)| \leq \epsilon_b(\epsilon_\alpha, t), \quad \forall\, t \in [0, \bar{t}]\,. \tag{2.125}$$

Using (2.125), Lemma 2.2.8 implies that $\nu_l(t)$, $u_l(t)$, $x_l(t)$ are bounded subject to

$$|\sigma(t) + \nu_l(t)| \leq \Delta_n(\epsilon_\alpha, \tau) < \Delta, \quad \|u_{l_{\bar{t}}}\|_{\mathcal{L}_\infty} \leq \Delta_u(\epsilon_\alpha, \tau), \quad \|x_{l_{\bar{t}}}\|_{\mathcal{L}_\infty} < \Delta_x(\epsilon_\alpha, \tau)\,. \tag{2.126}$$

It follows from (2.124) and (2.126) that

$$\|\tilde{\eta}_{l_{\bar{t}}}\|_{\mathcal{L}_\infty} \leq (\omega_u - \omega_l)\Delta_u(\epsilon_\alpha, \tau) + 2L\Delta_x(\epsilon_\alpha, \tau) + 2\Delta\,,$$

which along with (2.125) verifies (2.83) and (2.86). Hence, it follows from (2.87) that $|\dot{\sigma}(t) + \dot{\nu}_l(t)| \leq \Delta_d$ for all $t \in [0, \bar{t}]$. Since (2.120) holds for all $t \in [0, t')$, $\nu(t)$ is bounded and differentiable such that

$$|\sigma(t) + \nu(t)| \leq \Delta_n(\epsilon_\alpha, \tau) < \Delta, \quad |\dot{\sigma}(t) + \dot{\nu}(t)| \leq \Delta_d, \quad \forall\, t \in [0, \bar{t}]\,. \tag{2.127}$$

It follows from (2.94) that

$$|\epsilon(t)| \leq \Psi_{H_1}(t) * (\kappa_\beta(\theta_{\bar{t}}, t) + \|\tilde{x}_{\text{in}}(t)\|)$$

for all $t \in [0, \bar{t}]$. The relationships in (2.88), (2.93), and (2.127) imply that $\theta_{\bar{t}} \leq \theta_q$, and using the upper bound from (2.94) we have

$$|\epsilon(t)| \leq \Psi_{H_3}(t) * (\kappa(\theta_q, t) + \|\tilde{x}_{\text{in}}(t)\|) = \epsilon_c(t)\,.$$

From (2.89) we have $\frac{\theta_q}{\Gamma} < \epsilon_\alpha$ and hence $\kappa_\beta(\theta_q, t) \leq \kappa_\alpha(\epsilon_\alpha, t)$, for all $t \geq 0$. From the definition of $\Psi_{H_1}(t)$ and the properties of convolution, it follows that $\epsilon_c(t) < \epsilon_b(\epsilon_\alpha, t)$ and hence $|\epsilon(t)| < \epsilon_b(\epsilon_\alpha, t)$ for all $t \in [0, \bar{t}]$, which contradicts (2.122). Therefore, (2.121) holds.

Since (2.121) is true, it follows from (2.127) that $|\sigma(t) + \nu(t)| \leq \Delta_n < \Delta$, for all $t \in [0, t']$, which contradicts (2.111). Thus, the upper bound in (2.110) holds. Therefore, the relationships in (2.107), (2.109), (2.110), (2.115), (2.118), (2.119), (2.121), and (2.127) imply that there exist bounded $\tilde{\eta}_l(t)$ and continuously differentiable $\nu(t)$ in $[0, (i+1)\tau)$ such that

$$\begin{aligned} x_o(t) = x_q(t) = x_l(t), \qquad u_q(t) = u_l(t), \qquad \epsilon(t) = \epsilon_l(t), \\ |\epsilon(t)| < \epsilon_b(\epsilon_\alpha, t), \quad \forall\, t \in [0, (i+1)\tau], \\ \nu(t) = \nu_l(t), \quad |\sigma(t) + \nu(t)| < \Delta, \\ |\dot{\sigma}(t) + \dot{\nu}(t)| \leq \Delta_d, \quad \forall\, t \in [0, (i+1)\tau). \end{aligned} \tag{2.128}$$

It follows from (2.72), (2.76), and the fact that $u_q(t) = u_l(t)$ for all $t \in [i\tau, (i+1)\tau]$ that $u_{q_d}(t) = u_{l_d}(t)$ for all $t \in [(i+1)\tau, (i+2)\tau)$, which along with (2.128) proves (2.100)–(2.104) for $i+1$.

Step 3: By iterating the results from Step 2, we prove (2.98)–(2.99) for arbitrary $i \geq 0$.

We note that the relationships in (2.98)–(2.99) lead to (2.95)–(2.96) directly, which completes the proof. □

Theorem 2.2.4 establishes the equivalence of state and control trajectories of the closed-loop adaptive system and the LTI system in (2.74)–(2.79) in the presence of the same time delay. Therefore the time-delay margin of the system in (2.74)–(2.79) can be used as a conservative lower bound for the time-delay margin of the closed-loop adaptive system. The proof of the following result follows from Lemma 2.2.8 and Theorem 2.2.4 directly.

Corollary 2.2.1 Subject to (2.38) and (2.82), if Γ and Δ are selected appropriately large, the closed-loop system in (2.68) with the τ-delayed controller, given by (2.32), (2.33), (2.35), and (2.69), is stable.

Remark 2.2.5 Notice that the bound Δ used in the projection-based adaptive law for $\hat{\sigma}(t)$ depends upon the initial condition x_0 via $\eta_l(t)$. This implies that the result is semiglobal, in a sense that larger x_0 may require larger Δ to ensure the same time-delay margin given by (2.81).

2.2.6 Gain-Margin Analysis

We now analyze the gain margin of the system in (2.57) with the $\mathcal{L}_1$ adaptive controller. Consider the system

$$\dot{x}(t) = A_m x(t) + b\left(\omega_g u(t) + \theta^\top(t)x(t) + \sigma(t)\right), \quad x(0) = x_0,$$

where $\omega_g \triangleq g\omega$, with g being a positive constant. We note that this transformation implies that the set Ω in the application of the projection operator for adaptive laws needs to increase accordingly. However, increasing Ω will not violate the condition in (2.38) required for stability, as the latter depends only on the bounds of $\theta(t)$. Thus, it follows from (2.34) that the gain margin of the $\mathcal{L}_1$ adaptive controller is determined by

$$g_m = [\omega_l/\omega_{l_0}, \omega_u/\omega_{u_0}].$$

We note that the lower bound of g_m is greater than zero, while its definition implies that arbitrary gain margin can be obtained through appropriate choice of Ω.

2.2.7 Simulation Example: Robotic Arm

System Dynamics and Assumptions

Consider the dynamics of a single-link robot arm rotating on a vertical plane:

$$I\ddot{q}(t) = u(t) + \frac{\mathrm{Mgl}\cos(q(t))}{2} + \bar{\sigma}(t) + F_1(t)q(t) + F(t)\dot{q}(t),$$

where $q(t)$ and $\dot{q}(t)$ are the angular position and velocity, respectively, $u(t)$ is the input torque, I is the unknown moment of inertia, M is the unknown mass, l is the unknown length, $F(t)$ is an unknown time-varying friction coefficient, $F_1(t)$ is the position-dependent external torque coefficient, and $\bar{\sigma}(t)$ is an unknown uniformly bounded disturbance. The equations of motion can be cast into the following normal form:

$$\begin{aligned}\dot{x}(t) &= A_m x(t) + b(\omega u(t) + \theta^\top(t)x(t) + \sigma(t)),\\ y(t) &= c^\top x(t),\end{aligned}$$

where $\omega = 1/I$ is the unknown control effectiveness, and

$$\begin{aligned}A_m &= \begin{bmatrix} 0 & 1 \\ -1 & -1.4 \end{bmatrix}, \quad b = \begin{bmatrix} 0 \\ 1 \end{bmatrix}, \quad \theta(t) = \left[1 + \frac{F_1(t)}{I},\ 1.4 + \frac{F(t)}{I}\right]^\top,\\ \sigma(t) &= \frac{\mathrm{Mgl}\cos(x_1(t))}{2I} + \frac{\bar{\sigma}(t)}{I}.\end{aligned}$$

Consider three cases of parametric uncertainties,

$$\begin{aligned}\omega_1 &= 1, & \theta_1(t) &= [2+\cos(\pi t),\ 2+0.3\sin(\pi t)+0.2\cos(2t)]^\top,\\ \omega_2 &= 1.5, & \theta_2(t) &= [\sin(0.5\pi t)+\cos(\pi t),\ -1+0.1\sin(3\pi t)]^\top,\\ \omega_3 &= 0.8, & \theta_3(t) &= [4.5,\ 3-\sin(t)]^\top,\end{aligned}$$

and the following three cases of disturbances:

$$\begin{aligned}\sigma_1(t) &= \sin\left(\frac{\pi}{2}t\right),\\ \sigma_2(t) &= \cos(x_1(t)) + 2\sin(\pi t) + \cos\left(\frac{7\pi}{5}t\right),\\ \sigma_3(t) &= \cos(x_1(t)) + 2\sin(2\pi t) + \cos\left(\frac{16\pi}{5}t\right).\end{aligned}$$

The compact sets can be conservatively set to $\Omega_0 = [0.5,\ 1.8]$, $\Theta = \{\vartheta = [\vartheta_1,\ \vartheta_2]^\top \in \mathbb{R}^2 : \vartheta_i \in [-5,\ 5],\ i = 1, 2\}$, and $\Delta_0 = 10$.

Design of the $\mathcal{L}_1$ Adaptive Controller

We implement the $\mathcal{L}_1$ adaptive controller according to (2.32), (2.33), and (2.35) subject to the $\mathcal{L}_1$-norm condition in (2.38). Letting $D(s) = 1/s$, we have

$$G(s) = \frac{s}{s+\omega k} H(s), \quad H(s) = \left[\frac{1}{s^2+1.4s+1},\ \frac{s}{s^2+1.4s+1}\right]^\top.$$

It is straightforward to verify numerically that for $\omega k > 30$, one has $\|G(s)\|_{\mathcal{L}_1} L < 1$. Since $\omega > 0.5$, we set $k = 60$. For implementation of the adaptation laws we select the following larger projection bounds $\Omega = [0.1,\ 2]$, $\Delta = 50$ for ω and $\sigma(t)$, and retain the original Θ for $\theta(t)$. Finally, let $\Gamma = 100000$.

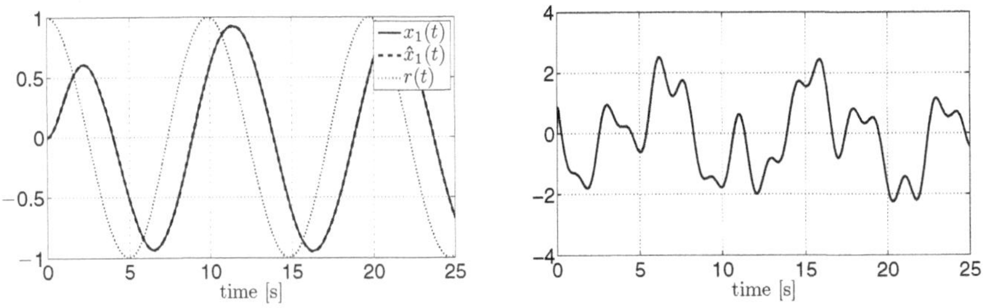

(a) $x_1(t)$, $\hat{x}_1(t)$, and $r(t)$ (b) Time history of $u(t)$

Figure 2.17: Performance of the $\mathcal{L}_1$ adaptive controller for ω_1, $\theta_1(t)$ and $\sigma_1(t)$.

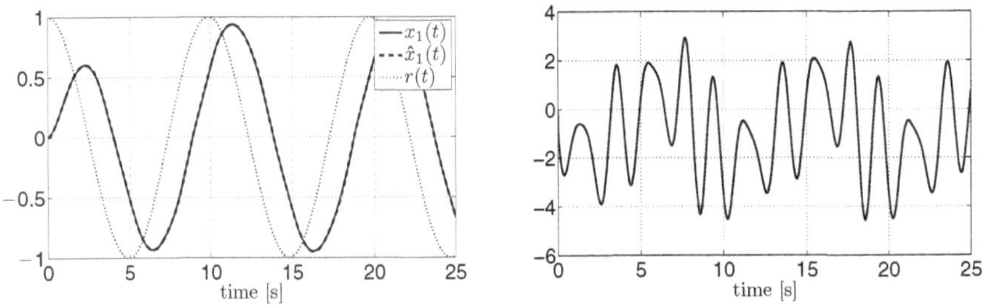

(a) $x_1(t)$, $\hat{x}_1(t)$, and $r(t)$ (b) Time history of $u(t)$

Figure 2.18: Performance of the $\mathcal{L}_1$ adaptive controller for ω_1, $\theta_1(t)$ and $\sigma_2(t)$.

Performance Verification for Different Uncertainties and Disturbances

Let $r(t) = \cos(2t/\pi)$ be the reference trajectory, and let the uncertainties be given by ω_1 and $\theta_1(t)$. The simulation results obtained with the $\mathcal{L}_1$ adaptive controller for different disturbances $\sigma_i(t)$, $i = 1,2,3$, without any retuning, are shown in Figures 2.17 through 2.19. We note that the fast adaptation of the $\mathcal{L}_1$ adaptive controller guarantees smooth and uniform transient performance in the presence of different unknown time-varying disturbances. The frequencies in the control signal match the frequencies of the disturbance that the controller is supposed to compensate for. Notice that $x_1(t)$ and $\hat{x}_1(t)$ are almost the same in all the figures.

Further, let the disturbance be given by $\sigma_1(t)$. The simulation results with the $\mathcal{L}_1$ adaptive controller (without any retuning) for different uncertainties ω_i and $\theta_i(t)$, $i = 1,2,3$ are shown in Figures 2.17, 2.20, and 2.21. One can see that the $\mathcal{L}_1$ adaptive controller retains its uniform performance.

Next, we simulate the response of the closed-loop adaptive system with the $\mathcal{L}_1$ adaptive system for various step reference inputs. For this example, let the disturbance be given by $\sigma_1(t)$ and let the uncertainties be given by ω_1 and $\theta_1(t)$. The results are given in Figure 2.22. One can see from the plot that the system response scales uniformly. This scaling property is typical of linear systems and is consistent with the claims in Section 2.2.3, which state that in the presence of fast adaptation, the input-output signals of the closed-loop $\mathcal{L}_1$ adaptive system remain close to the same signals of a bounded linear reference

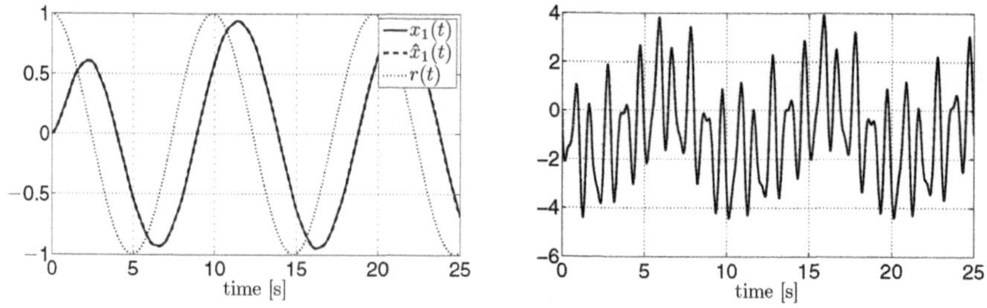

(a) $x_1(t)$, $\hat{x}_1(t)$, and $r(t)$ (b) Time history of $u(t)$

Figure 2.19: Performance of the $\mathcal{L}_1$ adaptive controller for ω_1, $\theta_1(t)$, and $\sigma_3(t)$.

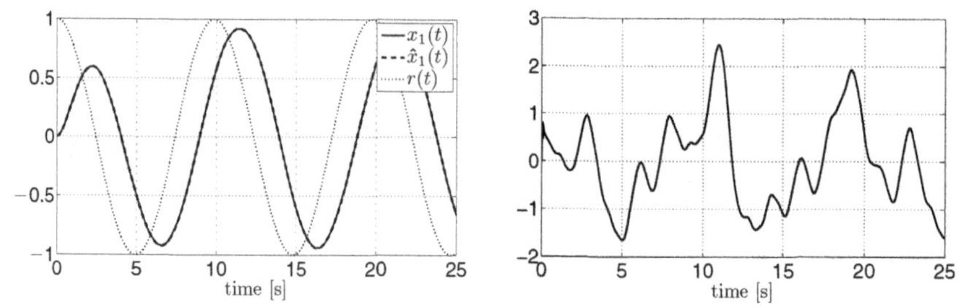

(a) $x_1(t)$, $\hat{x}_1(t)$, and $r(t)$ (b) Time history of $u(t)$

Figure 2.20: Performance of the $\mathcal{L}_1$ adaptive controller for ω_2, $\theta_2(t)$, and $\sigma_1(t)$.

system.

Performance Verification for Nonzero Trajectory Initialization Error

To check the performance in the presence of nonzero initialization errors, we set $\omega = 1$, $\theta(t) = \theta = [2,\ 2]^\top$, and let $\sigma(t) = \sigma_1(t)$. We use (2.58), along with (2.33), and (2.35) subject to (2.38), setting $\hat{x}(0) = [1,\ 1]^\top$ and $x(0) = [0,\ 0]^\top$. The simulation results in Figure 2.23 verify the performance of the $\mathcal{L}_1$ adaptive controller in the presence of nonzero initialization errors. We emphasize that there is no retuning of the $\mathcal{L}_1$ adaptive controller from the previous case.

Next we test the performance of $\mathcal{L}_1$ adaptive controller in the case of time-varying uncertainties. We set $\omega = 1$, $\theta(t) = \theta_1(t)$, $\sigma(t) = \sigma_1(t)$, and use the same $\mathcal{L}_1$ adaptive controller without retuning. Figure 2.24 demonstrates no degradation in performance.

Stability Margins

Next, we verify the stability margins of the $\mathcal{L}_1$ adaptive controller for this system. For this purpose, we use the same constants $\omega = 1$, $\theta = [2,\ 2]^\top$, and let $\sigma(t) = \sigma_1(t)$. We derive $L_o(s)$ in (2.80) and compute the worst-case time-delay margin for all $\omega \in \Omega_0$ and $\theta \in \Theta$ from the open-loop system's Bode plot. The worst-case values for the unknown parameters

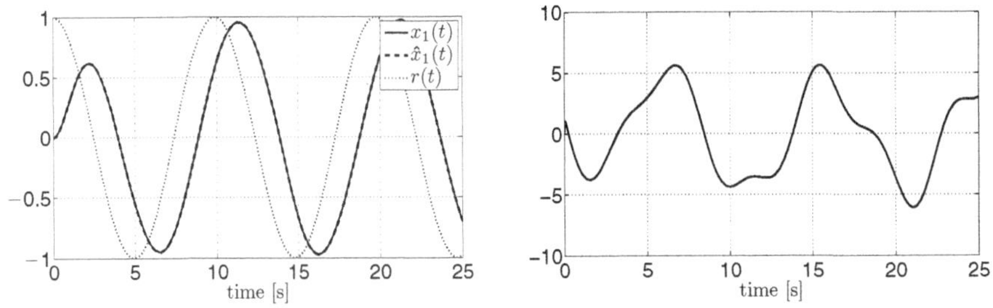

(a) $x_1(t)$, $\hat{x}_1(t)$, and $r(t)$

(b) Time history of $u(t)$

Figure 2.21: Performance of the $\mathcal{L}_1$ adaptive controller for ω_3, $\theta_3(t)$, and $\sigma_1(t)$.

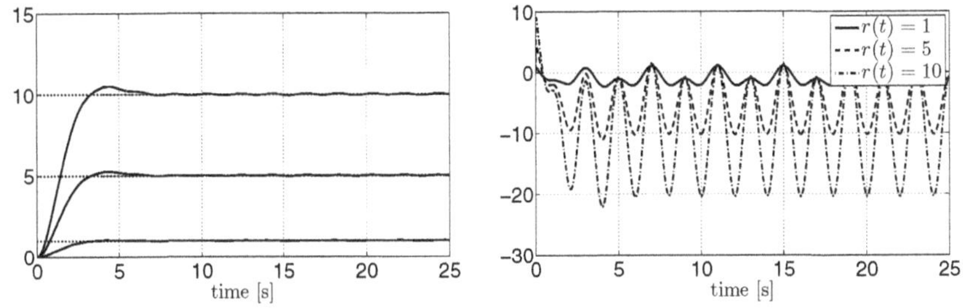

(a) $x_1(t)$ and $r(t)$

(b) Time history of $u(t)$

Figure 2.22: Performance of the $\mathcal{L}_1$ adaptive controller for step reference inputs.

are $\omega = 1.625$ and $\theta = [-5,\ 2.5]^\top$, which lead to the phase margin $\phi_m = 88.53$ deg and the time-delay margin $\mathcal{T} = 0.0158$ s.

From (2.80) it follows that in the structure of $L_o(s)$ the system $\bar{H}(s)$ is decoupled from $C(s)$, and therefore the time-delay margin can be tuned by selection of $C(s)$. Thus, choosing $k = 13$ and

$$D_1(s) = \frac{s+0.1}{s(s+0.9)}$$

leads to $C_1(s) = (13s+1.3)/(s^2+13.9s+1.3)$, which has a lower bandwidth compared to the previous filter and consequently leads to improved time-delay margins (see Table 2.2). The Bode plots for both $C(s)$ and $C_1(s)$ are given in Figure 2.25. Notice that the simulation results for the closed-loop adaptive system without time delay in the loop, given in Figure 2.26, show that there is some degradation in the tracking performance, as expected. Thus, improving the time-delay margin hurts the transient performance, which is consistent with the conventional claims in classical and robust control.

The example above clearly illustrates the main advantage of the $\mathcal{L}_1$ adaptive controller, the tuning of which is limited to the design of a linear low-pass filter, as opposed to the selection of the rate of the nonlinear gradient minimization scheme in the adaptation laws. From this perspective, the $\mathcal{L}_1$ adaptive control paradigm achieves clear *separation between adaptation and robustness*: the adaptation can be as fast as the CPU permits, while

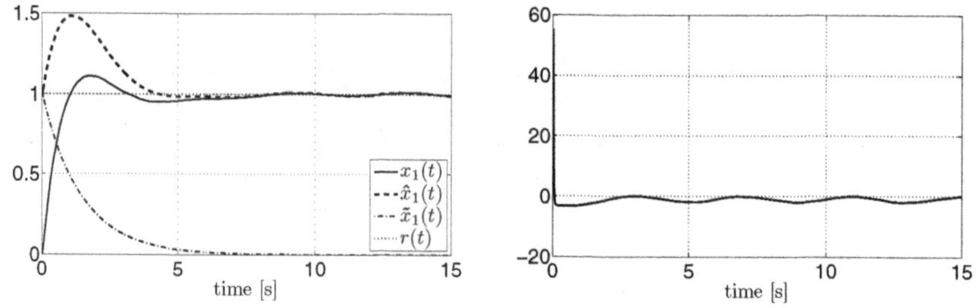

(a) $x_1(t)$, $\hat{x}_1(t)$, $\tilde{x}(t)$ and $r(t)$ (b) Time history of $u(t)$

Figure 2.23: Performance of the $\mathcal{L}_1$ adaptive controller in the presence of nonzero initialization error for constant parametric uncertainties.

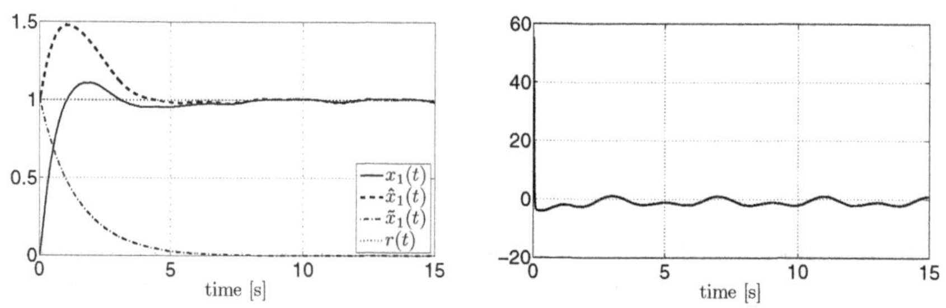

(a) $x_1(t)$, $\hat{x}_1(t)$, $\tilde{x}(t)$ and $r(t)$ (b) Time history of $u(t)$

Figure 2.24: Performance of the $\mathcal{L}_1$ adaptive controller in the presence of nonzero initialization error for time-varying uncertainties.

robustness can be resolved via conventional methods from linear feedback control.

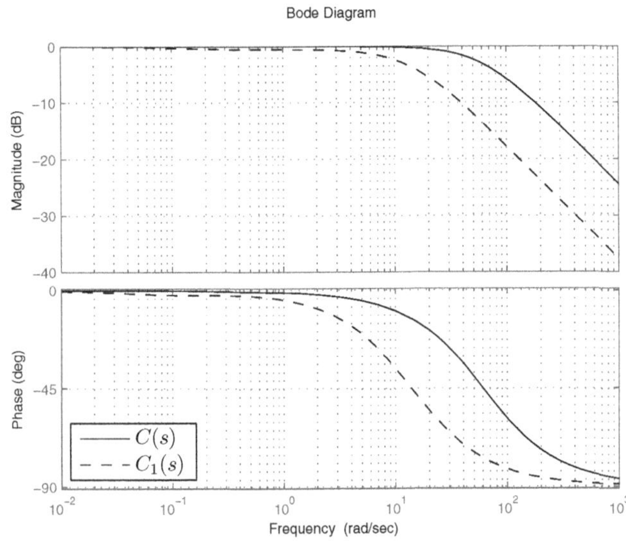

Figure 2.25: Bode plot for the low-pass filters.

Table 2.2: Worst-case phase and time-delay margins (TDM) of $L_o(s)$ and the actual closed-loop adaptive system.

	$D(s)$	$D_1(s)$
Phase margin of $L_o(s)$	88.53 deg	85.40 deg
Gain crossover frequency of $L_o(s)$	97.6 rad/s	21.4 rad/s
Predicted time-delay margin	0.0158 s	0.0698 s
Numerically verified TDM ($\Gamma = 10^8$)	0.0258 s	0.115 s
Numerically verified TDM ($\Gamma = 10^5$)	0.0236 s	0.114 s

As stated in Corollary 2.2.1, the time-delay margin of the LTI system in (2.80), computed for the worst-case parameters, provides a conservative lower bound for the time-delay margin of the closed-loop adaptive system. So, we set $\Delta = 10^9$, $\Omega = [0.1,\ 10^9]$, $\Gamma = 10^8$, and simulate the $\mathcal{L}_1$ adaptive controller and obtain the values for the time-delay margin for both filters numerically. These are given in Table 2.2. Notice that one can reduce the level of conservatism in the estimate of the lower bound of the time-delay margin if more accurate information about the unknown parameters is available, which can be used for setting tighter bounds in the implementation of the projection operator. For example, considering $D(s)$ and letting $\Omega_0 = [0.5,\ 1.5]$, and $\Theta = \{\vartheta = [\vartheta_1,\ \vartheta_2]^\top \in \mathbb{R}^2 : \vartheta_i \in [0,\ 3], \text{for all } i = 1, 2\}$ leads to $\mathcal{T} = 0.021$ s. This value is much closer to the real time-delay margin observed in

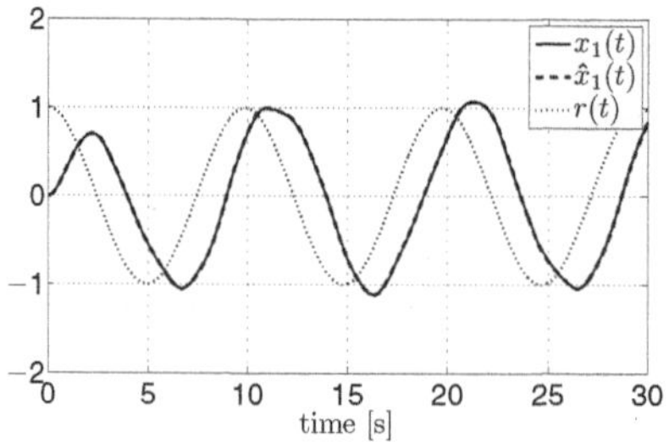

(a) $x_1(t)$, $\hat{x}_1(t)$, and $r(t)$

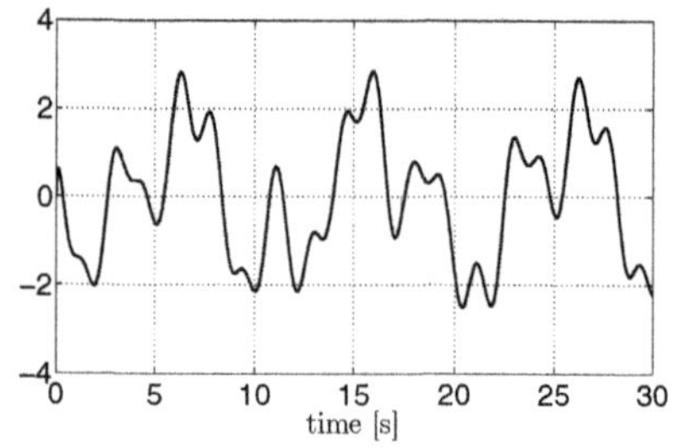

(b) Time history of $u(t)$

Figure 2.26: Performance of the $\mathcal{L}_1$ adaptive controller with filter $D_1(s)$ for $\sigma_1(t)$.

simulations. This implies that our conservative knowledge of uncertainty hurts our ability to predict the margin accurately, but not the margin itself.

Further, notice that a smaller value of Γ is preferable from an implementation point of view. Simulations show that for the closed-loop adaptive system with $D(s)$, if Γ decreases from 10^8 to 10^5, the time-delay margin decreases from 0.0258 s to 0.0236 s (i.e., about 9%). However, for the case with $D_1(s)$ it decreases from 0.115 s to 0.114 s (i.e., less than 1%), which is related to the fact that $D_1(s)$ results in $C_1(s)$ with smaller bandwidth as compared to $C(s)$ (see Figure 2.25). Thus, with smaller adaptation gain Γ, the design with $D_1(s)$ is more suitable in terms of robustness as compared to $D(s)$. A similar observation was made in Section 2.1.5, where increasing the order of the low-pass filter helped to achieve a similar performance level with a smaller adaptive gain Γ. Table 2.2 summarizes this results.

Next, keeping the same projection bounds ($\Delta = 10^9$, $\Omega = [0.1,\ 10^9]$) and setting the adaptation gain $\Gamma = 10^6$, in Figure 2.27 we give the simulation results for the closed-loop system with $D(s)$ in the presence of time delay of 0.015 s. The simulations verify Corollary 2.2.1. Moreover, notice that a moderate level of the time delay does not influence the transient tracking significantly.

Finally, we test the performance of the same $\mathcal{L}_1$ adaptive controller for the system with time-varying uncertainties without any retuning. Figure 2.28 shows the simulation

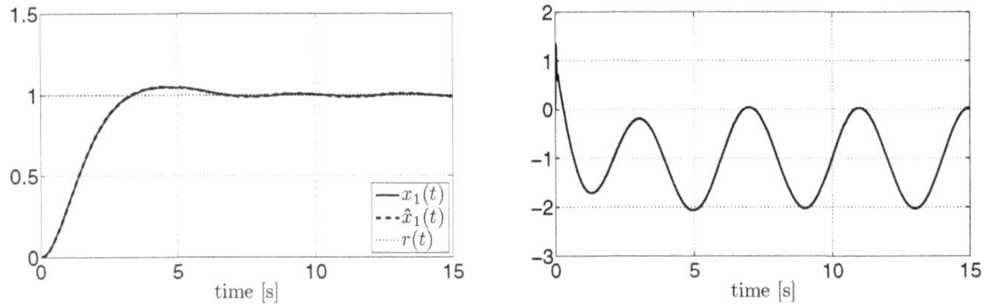

(a) $x_1(t)$, $\hat{x}_1(t)$, and $r(t)$ (b) Time history of $u(t)$

Figure 2.27: Performance of the $\mathcal{L}_1$ adaptive controller with time delay of 15 ms for constant parametric uncertainties.

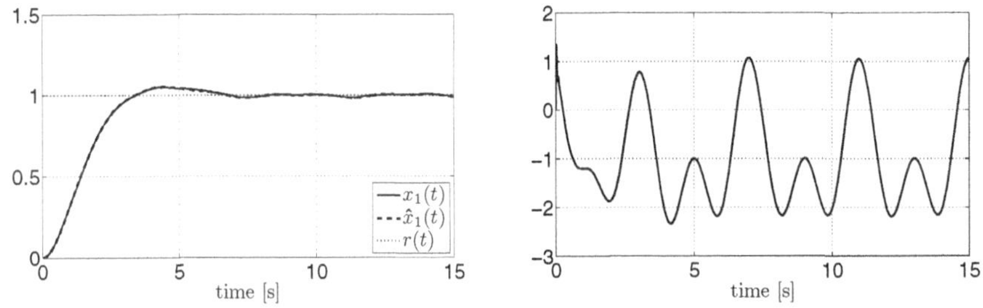

(a) $x_1(t)$, $\hat{x}_1(t)$, and $r(t)$ (b) Time history of $u(t)$

Figure 2.28: Performance of the $\mathcal{L}_1$ adaptive controller with time delay of 15 ms for time-varying uncertainties.

results for $\omega = 1$, $\theta(t) = \theta_1(t)$, $\sigma(t) = \sigma_1(t)$. We notice that the simulation results are very similar to the case of constant parametric uncertainties, given in Figure 2.27.

Remark 2.2.6 Recall the scalar system with the $\mathcal{L}_1$ controller considered in Section 1.3:

$$\begin{aligned}
\dot{x}(t) &= -x(t) + \theta + u(t), \quad x(0) = 0, \\
\dot{\hat{x}}(t) &= -\hat{x}(t) + \hat{\theta}(t) + u(t), \quad \hat{x}(0) = 0, \\
u(s) &= -C(s)\hat{\theta}(s), \quad C(s) = \frac{\omega_c}{s + \omega_c}, \\
\dot{\hat{\theta}}(t) &= -\Gamma \tilde{x}(t), \quad \hat{\theta}(0) = 0, \\
\tilde{x}(t) &= \hat{x}(t) - x(t).
\end{aligned}$$

The block diagram of this linear system is shown in Figure 1.5. The key feature of this system is that the system $L_o(s)$ from (2.80), giving the lower bound for the time-delay margin, depends only upon $C(s)$ and is free of uncertainties:

$$L_o(s) = \frac{C(s)}{1 - C(s)} = \frac{\omega_c}{s}.$$

Recall that the loop transfer function of the system in Figure 1.5 for its phase-margin analysis is given by

$$L_2(s) = \frac{\Gamma C(s)}{s(s+1)+\Gamma(1-C(s))}.$$

Notice that as $\Gamma \to \infty$ we obtain exactly $L_o(s)$ in the limit:

$$L_{2l}(s) \triangleq \lim_{\Gamma\to\infty} L_{2l}(s) = \frac{C(s)}{1-C(s)}.$$

This verifies that the guaranteed lower bound for the time-delay margin of the $\mathcal{L}_1$ adaptive controller, provided by Theorem 2.2.4 and Corollary 2.2.1, is achievable for at least one LTI system; i.e., the result is not conservative.

2.3 Extension to Systems with Unmodeled Actuator Dynamics

This section presents the $\mathcal{L}_1$ adaptive control architecture for the class of uncertain linear time-varying systems with unmodeled actuator dynamics. We prove that subject to a set of mild assumptions, the control architecture from Section 2.2 can be used for compensation of uncertainties within the bandwidth of the control channel, provided the selected adaptation gain is sufficiently large [28].

2.3.1 Problem Formulation

Consider the following class of systems:

$$\begin{aligned} \dot{x}(t) &= A_m x(t) + b\big(\mu(t) + \theta^\top(t)x(t) + \sigma_0(t)\big), \quad x(0) = x_0, \\ y(t) &= c^\top x(t), \end{aligned} \tag{2.129}$$

where $x(t) \in \mathbb{R}^n$ is the system state vector (measured); $y(t) \in \mathbb{R}$ is the system regulated output; $A_m \in \mathbb{R}^{n\times n}$ is a known Hurwitz matrix specifying the desired closed-loop dynamics; $b, c \in \mathbb{R}^n$ are known constant vectors; $\theta(t) \in \mathbb{R}^n$ is a vector of time-varying unknown parameters; $\sigma_0(t) \in \mathbb{R}$ is a time-varying disturbance; and $\mu(t) \in \mathbb{R}$ is the output of the system

$$\mu(s) = F(s)u(s),$$

where $u(t) \in \mathbb{R}$ is the control input and $F(s)$ is an unknown BIBO-stable transfer function with known sign of its DC gain.

We repeat the set of assumptions from Section 2.2 and impose one additional assumption on the stability of the unmodeled actuator dynamics.

Assumption 2.3.1 (Uniform boundedness of unknown parameters) Let

$$\theta(t) \in \Theta, \quad |\sigma_0(t)| \le \Delta_0, \quad \forall\, t \ge 0,$$

where Θ is a known convex compact set and $\Delta_0 \in \mathbb{R}^+$ is a known (conservative) bound of $\sigma_0(t)$.

Assumption 2.3.2 (Uniform boundedness of the rate of variation of parameters) Let $\theta(t)$ and $\sigma_0(t)$ be continuously differentiable with uniformly bounded derivatives:

$$\|\dot{\theta}(t)\| \leq d_\theta < \infty\,, \quad |\dot{\sigma}_0(t)| \leq d_{\sigma_0} < \infty\,, \quad \forall\, t \geq 0\,.$$

Assumption 2.3.3 (Partial knowledge of actuator dynamics) There exists $L_F > 0$ verifying $\|F(s)\|_{\mathcal{L}_1} \leq L_F$. Also, we assume that there exist known constants $\omega_l,\ \omega_u \in \mathbb{R}$ satisfying

$$0 < \omega_l \leq F(0) \leq \omega_u\,,$$

where, without loss of generality, we have assumed $F(0) > 0$. Finally, we assume (for design purposes) that we know a set $\mathbb{F}_\Delta$ of all admissible actuator dynamics.

The control objective is to design a full-state feedback-adaptive controller to ensure that $y(t)$ tracks a given bounded piecewise-continuous reference signal $r(t)$ with quantifiable performance bounds.

2.3.2 $\mathcal{L}_1$ Adaptive Control Architecture

Definitions and $\mathcal{L}_1$-Norm Sufficient Condition for Stability

As in Section 2.2.2, the design of the $\mathcal{L}_1$ adaptive controller proceeds by considering a positive-feedback gain $k > 0$ and a strictly proper stable transfer function $D(s)$, which imply that

$$C(s) \triangleq \frac{kF(s)D(s)}{1+kF(s)D(s)} \tag{2.130}$$

is a strictly proper stable transfer function with DC gain $C(0) = 1$ for all $F(s) \in \mathbb{F}_\Delta$. Next, similar to the previous development, let $x_{\text{in}}(s) \triangleq (s\mathbb{I} - A_m)^{-1}x_0$. Notice that from the fact that A_m is Hurwitz, it follows that $\|x_{\text{in}}\|_{\mathcal{L}_\infty}$ is bounded, and $x_{\text{in}}(t)$ is exponentially decaying.

For the proofs of stability and performance bounds, the choice of k and $D(s)$ needs to ensure that the following $\mathcal{L}_1$-norm condition holds:

$$\|G(s)\|_{\mathcal{L}_1} L < 1\,, \tag{2.131}$$

where

$$G(s) \triangleq (1 - C(s))H(s)\,, \quad H(s) \triangleq (s\mathbb{I} - A_m)^{-1}b\,, \quad L \triangleq \max_{\theta\in\Theta} \|\theta\|_1\,.$$

To streamline the subsequent analysis of stability and performance bounds, we introduce the following notation. Let $C_u(s) \triangleq C(s)/F(s)$. Notice that from the definition of $C(s)$ given in (2.130) and the fact that $D(s)$ is strictly proper and stable, while $F(s)$ is proper and stable, it follows that $C_u(s)$ is a strictly proper and stable transfer function. Next, let

$$H_1(s) \triangleq C_u(s)\frac{1}{c_o^\top H(s)}c_o^\top\,, \tag{2.132}$$

where $c_o \in \mathbb{R}^n$ is a vector that renders $H_1(s)$ BIBO stable. Existence of such c_o is proved in Lemma A.12.1. Further, let

$$\begin{aligned}
\rho_r &\triangleq \frac{\|G(s)\|_{\mathcal{L}_1}\Delta_0 + \|H(s)C(s)k_g\|_{\mathcal{L}_1}\|r\|_{\mathcal{L}_\infty} + \|x_{\text{in}}\|_{\mathcal{L}_\infty}}{1-\|G(s)\|_{\mathcal{L}_1}L},\\
\rho_{ur} &\triangleq \|C_u(s)\|_{\mathcal{L}_1}(|k_g|\|r\|_{\mathcal{L}_\infty} + L\rho_r + \Delta_0),\\
\rho &\triangleq \rho_r + \gamma_1,\\
\rho_u &\triangleq \rho_{ur} + \gamma_2,
\end{aligned}$$

where $k_g \triangleq -1/(c^\top A_m^{-1} b)$, and

$$\gamma_1 \triangleq \frac{\|C(s)\|_{\mathcal{L}_1}}{1-\|G(s)\|_{\mathcal{L}_1}L}\gamma_0 + \beta, \tag{2.133}$$

$$\gamma_2 \triangleq \|C_u(s)\|_{\mathcal{L}_1}L\gamma_1 + \|H_1(s)\|_{\mathcal{L}_1}\gamma_0, \tag{2.134}$$

with $\beta > 0$ and $\gamma_0 > 0$ being arbitrarily small positive constants.

Finally, using the conservative knowledge of $F(0)$, let

$$\Delta_\mu \triangleq \max_{F(s)\in\mathbb{F}_\Delta} \|F(s) - (\omega_l + \omega_u)/2\|_{\mathcal{L}_1}\rho_u, \tag{2.135}$$

$$\Delta \triangleq \Delta_0 + \Delta_\mu, \tag{2.136}$$

$$\rho_{\dot{u}} \triangleq \|ksD(s)\|_{\mathcal{L}_1}(\rho_u\omega_u + L\rho + \Delta + |k_g|\|r\|_{\mathcal{L}_\infty}).$$

Remark 2.3.1 Notice that if $F(s) = F$ is an unknown constant, the system in (2.129) degenerates into the system in (2.31). Consequently, the $\mathcal{L}_1$-norm condition in (2.131) reduces to the one in (2.38).

The elements of the $\mathcal{L}_1$ adaptive control architecture are introduced next.

State Predictor

We consider the following state predictor:

$$\begin{aligned}
\dot{\hat{x}}(t) &= A_m\hat{x}(t) + b\left(\hat{\omega}(t)u(t) + \hat{\theta}^\top(t)x(t) + \hat{\sigma}(t)\right), \quad \hat{x}(0) = x_0,\\
\hat{y}(t) &= c^\top\hat{x}(t),
\end{aligned} \tag{2.137}$$

where $\hat{x}(t) \in \mathbb{R}^n$ is the predictor state, while $\hat{\omega}(t), \hat{\sigma}(t) \in \mathbb{R}$, and $\hat{\theta}(t) \in \mathbb{R}^n$ are the adaptive estimates.

Adaptation Laws

The adaptive laws are defined using the projection operator:

$$\begin{aligned} \dot{\hat{\omega}}(t) &= \Gamma \operatorname{Proj}\left(\hat{\omega}(t), -\tilde{x}^\top(t) P b u(t)\right), \quad \hat{\omega}(0) = \hat{\omega}_0, \\ \dot{\hat{\theta}}(t) &= \Gamma \operatorname{Proj}\left(\hat{\theta}(t), -\tilde{x}^\top(t) P b x(t)\right), \quad \hat{\theta}(0) = \hat{\theta}_0, \\ \dot{\hat{\sigma}}(t) &= \Gamma \operatorname{Proj}\left(\hat{\sigma}(t), -\tilde{x}^\top(t) P b\right), \quad \hat{\sigma}(0) = \hat{\sigma}_0, \end{aligned} \tag{2.138}$$

where $\tilde{x}(t) \triangleq \hat{x}(t) - x(t)$, $\Gamma \in \mathbb{R}^+$ is the adaptation gain, $\operatorname{Proj}(\cdot,\cdot)$ denotes the projection operator, and the symmetric positive definite matrix $P = P^\top > 0$ solves the Lyapunov equation $A_m^\top P + P A_m = -Q$ for arbitrary $Q = Q^\top > 0$. The projection operator ensures that $\hat{\omega} \in \Omega = [\omega_l, \omega_u]$, $\hat{\theta} \in \Theta$, $|\hat{\sigma}| \leq \Delta$.

Control Law

The control signal is generated as the output of the following (feedback) system:

$$u(s) = -kD(s)(\hat{\eta}(s) - k_g r(s)), \tag{2.139}$$

where $r(s)$ and $\hat{\eta}(s)$ are the Laplace transforms of $r(t)$ and $\hat{\eta}(t) \triangleq \hat{\omega}(t)u(t) + \hat{\theta}^\top(t)x(t) + \hat{\sigma}(t)$, while k and $D(s)$ were introduced in (2.130).

The complete $\mathcal{L}_1$ adaptive controller is defined via (2.137), (2.138), and (2.139), subject to the $\mathcal{L}_1$-norm condition in (2.131).

2.3.3 Analysis of the $\mathcal{L}_1$ Adaptive Controller

Closed-Loop Reference System

Consider the following closed-loop reference system:

$$\begin{aligned} \dot{x}_{\text{ref}}(t) &= A_m x_{\text{ref}}(t) + b\big(\mu_{\text{ref}}(t) + \theta^\top(t) x_{\text{ref}}(t) + \sigma_0(t)\big), \quad x_{\text{ref}}(0) = x_0, \\ \mu_{\text{ref}}(s) &= F(s) u_{\text{ref}}(s), \\ u_{\text{ref}}(s) &= C_u(s)(k_g r(s) - \eta_{\text{ref}}(s)), \\ y_{\text{ref}}(t) &= c^\top x_{\text{ref}}(t), \end{aligned} \tag{2.140}$$

where $x_{\text{ref}}(t) \in \mathbb{R}^n$ is the reference system state vector and $\eta_{\text{ref}}(s)$ is the Laplace transform of $\eta_{\text{ref}}(t) \triangleq \theta^\top(t) x_{\text{ref}}(t) + \sigma_0(t)$. Notice that this reference system is not implementable, as it depends upon the unknown $\theta(t)$, $\sigma_0(t)$, and $F(s)$. Similar to previous sections, this reference system is used only for analysis purposes. The next lemma proves the stability of this closed-loop reference system.

Lemma 2.3.1 For the closed-loop reference system given in (2.140), subject to the $\mathcal{L}_1$-norm condition in (2.131), the following bounds hold:

$$\|x_{\text{ref}}\|_{\mathcal{L}_\infty} \leq \rho_r, \quad \|u_{\text{ref}}\|_{\mathcal{L}_\infty} \leq \rho_{ur}. \tag{2.141}$$

Proof. The closed-loop reference system in (2.140) can be rewritten in the frequency domain as follows:

$$x_{\text{ref}}(s) = G(s)\eta_{\text{ref}}(s) + H(s)C(s)k_g r(s) + x_{\text{in}}(s)\,. \qquad (2.142)$$

Lemma A.7.1, along with the fact that for bounded signals $\|(\cdot)_\tau\|_{\mathcal{L}_\infty} \leq \|\cdot\|_{\mathcal{L}_\infty}$, implies

$$\|x_{\text{ref}\tau}\|_{\mathcal{L}_\infty} \leq \|G(s)\|_{\mathcal{L}_1}\|\eta_{\text{ref}\tau}\|_{\mathcal{L}_\infty} + \|H(s)C(s)k_g\|_{\mathcal{L}_1}\|r\|_{\mathcal{L}_\infty} + \|x_{\text{in}}\|_{\mathcal{L}_\infty}\,.$$

It follows from Assumption 2.3.1 that

$$\|\eta_{\text{ref}\tau}\|_{\mathcal{L}_\infty} \leq L\|x_{\text{ref}\tau}\|_{\mathcal{L}_\infty} + \Delta_0\,, \qquad (2.143)$$

which leads to

$$\|x_{\text{ref}\tau}\|_{\mathcal{L}_\infty} \leq \|G(s)\|_{\mathcal{L}_1}(L\|x_{\text{ref}\tau}\|_{\mathcal{L}_\infty} + \Delta_0) + \|H(s)C(s)k_g\|_{\mathcal{L}_1}\|r\|_{\mathcal{L}_\infty} + \|x_{\text{in}}\|_{\mathcal{L}_\infty}\,.$$

Keeping in mind the $\mathcal{L}_1$-norm condition in (2.131), we solve for $\|x_{\text{ref}\tau}\|_{\mathcal{L}_\infty}$ in the expression above to obtain the following upper bound:

$$\|x_{\text{ref}\tau}\|_{\mathcal{L}_\infty} \leq \frac{\|G(s)\|_{\mathcal{L}_1}\Delta_0 + \|H(s)C(s)k_g\|_{\mathcal{L}_1}\|r\|_{\mathcal{L}_\infty} + \|x_{\text{in}}\|_{\mathcal{L}_\infty}}{1 - \|G(s)\|_{\mathcal{L}_1}L} = \rho_r\,.$$

Notice that this upper bound holds uniformly for all $\tau \geq 0$ and therefore

$$\|x_{\text{ref}}\|_{\mathcal{L}_\infty} \leq \rho_r\,. \qquad (2.144)$$

To prove the second bound in (2.141), notice that from (2.143) and (2.144) we have

$$\|\eta_{\text{ref}\tau}\|_{\mathcal{L}_\infty} \leq L\rho_r + \Delta_0\,.$$

Using

$$u_{\text{ref}}(s) = C_u(s)(k_g r(s) - \eta_{\text{ref}}(s)),$$

along with Lemma A.7.1, gives

$$\begin{aligned}\|u_{\text{ref}\tau}\|_{\mathcal{L}_\infty} &\leq \|C_u(s)\|_{\mathcal{L}_1}(|k_g|\|r\|_{\mathcal{L}_\infty} + \|\eta_{\text{ref}\tau}\|_{\mathcal{L}_\infty})\\ &\leq \|C_u(s)\|_{\mathcal{L}_1}(|k_g|\|r\|_{\mathcal{L}_\infty} + L\rho_r + \Delta_0) = \rho_{ur}\,,\end{aligned}$$

which holds uniformly for all $\tau \geq 0$ and proves the second bound in (2.141). □

Equivalent Linear Time-Varying System

Following Lemma A.10.1, on every finite time interval the system with unmodeled multiplicative dynamics can be transformed into an equivalent linear time-varying system with uncertain system input gain and with an additional disturbance. Thus, we transform the original system with unmodeled dynamics in (2.129) into an equivalent linear time-varying system with unknown time-varying parameters. This transformation requires us to impose the following assumptions on the signals of the system: the control signal $u(t)$ is continuous, and moreover the following bounds hold:

$$\|u_\tau\|_{\mathcal{L}_\infty} \leq \rho_u\,, \quad \|\dot{u}_\tau\|_{\mathcal{L}_\infty} \leq \rho_{\dot{u}}\,, \quad \forall\, \tau \geq 0. \qquad (2.145)$$

These assumptions will be verified later in the proof of Theorem 2.3.1.

Consider the system in (2.129) with $u(t)$ subject to (2.145). Then, Lemma A.10.1 implies that $\mu(t)$ can be rewritten as

$$\mu(t) = \omega u(t) + \sigma_\mu(t),$$

where ω is an unknown constant, $\omega \in (\omega_l, \omega_u)$, and $\sigma_\mu(t)$ is a continuous signal with (piecewise)-continuous derivative defined over $t \in [0, \tau]$, such that

$$|\sigma_\mu(t)| \leq \Delta_\mu, \quad |\dot{\sigma}_\mu(t)| \leq d_{\sigma_\mu},$$

with Δ_μ being defined in (2.135), and $d_{\sigma_\mu} \triangleq \|F(s) - (\omega_l + \omega_u)/2\|_{\mathcal{L}_1} \rho_{\dot{u}}$. This implies that one can rewrite the system in (2.129) over $t \in [0, \tau]$ as follows:

$$\begin{aligned} \dot{x}(t) &= A_m x(t) + b\big(\omega u(t) + \theta^\top(t) x(t) + \sigma(t)\big), \quad x(0) = x_0, \\ y(t) &= c^\top x(t), \end{aligned} \tag{2.146}$$

where $\sigma(t) \triangleq \sigma_0(t) + \sigma_\mu(t)$ is an unknown continuous time-varying signal, satisfying $|\sigma(t)| < \Delta$, with Δ being introduced in (2.136), and $|\dot{\sigma}(t)| < d_\sigma$ with $d_\sigma \triangleq d_{\sigma_0} + d_{\sigma_\mu}$.

Transient and Steady-State Performance

Using (2.146), one can write the error dynamics over $t \in [0, \tau]$ as

$$\dot{\tilde{x}}(t) = A_m \tilde{x}(t) + b(\tilde{\omega}(t) u(t) + \tilde{\theta}^\top(t) x(t) + \tilde{\sigma}(t)), \quad \tilde{x}(0) = 0, \tag{2.147}$$

where $\tilde{\omega}(t) \triangleq \hat{\omega}(t) - \omega$, $\tilde{\theta}(t) \triangleq \hat{\theta}(t) - \theta(t)$, and $\tilde{\sigma}(t) \triangleq \hat{\sigma}(t) - \sigma(t)$.

Lemma 2.3.2 For the error dynamics in (2.147), if $u(t)$ is continuous, and moreover the following bounds hold:

$$\|u_\tau\|_{\mathcal{L}_\infty} \leq \rho_u, \quad \|\dot{u}_\tau\|_{\mathcal{L}_\infty} \leq \rho_{\dot{u}},$$

then

$$\|\tilde{x}_\tau\|_{\mathcal{L}_\infty} \leq \sqrt{\frac{\theta_m(\rho_u, \rho_{\dot{u}})}{\lambda_{\min}(P)\Gamma}}, \tag{2.148}$$

where

$$\theta_m(\rho_u, \rho_{\dot{u}}) \triangleq (\omega_u - \omega_l)^2 + 4 \max_{\theta \in \Theta} \|\theta\|^2 + 4\Delta^2 + 4 \frac{\lambda_{\max}(P)}{\lambda_{\min}(Q)} (\max_{\theta \in \Theta} \|\theta\| d_\theta + \Delta d_\sigma).$$

Proof. Consider the following Lyapunov function candidate:

$$V(\tilde{x}(t), \tilde{\omega}(t), \tilde{\theta}(t), \tilde{\sigma}(t)) = \tilde{x}^\top(t) P \tilde{x}(t) + \frac{1}{\Gamma}(\tilde{\omega}^2(t) + \tilde{\theta}^\top(t)\tilde{\theta}(t) + \tilde{\sigma}^2(t)).$$

Using the adaptation laws in (2.138) and Property B.2 of the projection operator, we compute the upper bound on the derivative of the Lyapunov function similar to the proof of Lemma 2.2.3:

$$\dot{V}(t) \leq -\tilde{x}^\top(t) Q \tilde{x}(t) + \frac{2}{\Gamma}\left(|\tilde{\theta}^\top(t)\dot{\theta}(t)| + |\tilde{\sigma}(t)\dot{\sigma}(t)|\right).$$

Further, following steps similar to those of the proof of Lemma 2.2.3, we obtain the following uniform upper bound on the prediction error, which leads to the bound in (2.148):

$$\|\tilde{x}(t)\|^2 \leq \frac{\theta_m(\rho_u, \rho_{\dot{u}})}{\lambda_{\min}(P)\Gamma}, \qquad \forall\, t \in [0, \tau]. \qquad \square$$

Theorem 2.3.1 If the adaptive gain satisfies the design constraint

$$\Gamma \geq \frac{\theta_m(\rho_u, \rho_{\dot{u}})}{\lambda_{\min}(P)\gamma_0^2}, \tag{2.149}$$

where $\gamma_0 > 0$ is an arbitrary constant introduced in (2.133), we have

$$\|\tilde{x}\|_{\mathcal{L}_\infty} \leq \gamma_0, \tag{2.150}$$

$$\|x_{\text{ref}} - x\|_{\mathcal{L}_\infty} \leq \gamma_1, \tag{2.151}$$

$$\|u_{\text{ref}} - u\|_{\mathcal{L}_\infty} \leq \gamma_2. \tag{2.152}$$

Proof. We prove the bounds in (2.151) and (2.152) following a contradicting argument. Assume that (2.151) and (2.152) do not hold. Then, since

$$\|x_{\text{ref}}(0) - x(0)\|_\infty = 0, \quad \|u_{\text{ref}}(0) - u(0)\|_\infty = 0,$$

continuity of $x_{\text{ref}}(t)$, $x(t)$, $u_{\text{ref}}(t)$, $u(t)$ implies that there exists time $\tau > 0$ for which

$$\|x_{\text{ref}}(t) - x(t)\|_\infty < \gamma_1, \quad \|u_{\text{ref}}(t) - u(t)\|_\infty < \gamma_2, \qquad \forall t \in [0, \tau)$$

and

$$\|x_{\text{ref}}(\tau) - x(\tau)\|_\infty = \gamma_1 \quad \text{or} \quad \|u_{\text{ref}}(\tau) - u(\tau)\|_\infty = \gamma_2.$$

This implies that at least one of the following equalities holds:

$$\|(x_{\text{ref}} - x)_\tau\|_{\mathcal{L}_\infty} = \gamma_1, \quad \|(u_{\text{ref}} - u)_\tau\|_{\mathcal{L}_\infty} = \gamma_2. \tag{2.153}$$

Moreover, Lemma 2.3.1 yields the bounds

$$\|x_{\text{ref}\tau}\|_{\mathcal{L}_\infty} \leq \rho_r, \quad \|u_{\text{ref}\tau}\|_{\mathcal{L}_\infty} \leq \rho_{ur},$$

which, together with the bounds in (2.153), lead to

$$\|x_\tau\|_{\mathcal{L}_\infty} \leq \rho_r + \gamma_1 = \rho, \quad \|u_\tau\|_{\mathcal{L}_\infty} \leq \rho_{ur} + \gamma_2 = \rho_u. \tag{2.154}$$

Further, consider the control law in (2.139), whose derivative can be written in the frequency domain as

$$su(s) = -ksD(s)(\hat{\eta}(s) - k_g r(s)).$$

From the properties of the projection operator and the upper bounds in (2.154), we have

$$\|\hat{\eta}_\tau\|_{\mathcal{L}_\infty} \leq \omega_u \rho_u + L\rho + \Delta,$$

and consequently

$$\|\dot{u}_\tau\| \leq \|ksD(s)\|_{\mathcal{L}_1} \left(\omega_u \rho_u + L\rho + \Delta + |k_g| \|r\|_{\mathcal{L}_\infty}\right) = \rho_{\dot{u}}.$$

These bounds imply that the conditions of Lemma 2.3.2 hold. Then, selecting the adaptive gain Γ according to the design constraint in (2.149), it follows that

$$\|\tilde{x}_\tau\|_{\mathcal{L}_\infty} \leq \gamma_0. \tag{2.155}$$

Consider next the system in (2.129). Notice that for $t \in [0, \tau]$, we have

$$\hat{\eta}(t) = \mu(t) + \tilde{\eta}(t) + \eta(t),$$

where $\eta(t) \triangleq \theta^\top(t)x(t) + \sigma_0(t)$, $\tilde{\eta}(t) \triangleq \tilde{\omega}(t)u(t) + \tilde{\theta}^\top(t)x(t) + \tilde{\sigma}(t)$. Substituting the Laplace transform of $\hat{\eta}(t)$ into (2.139) yields

$$u(s) = -C_u(s)(\tilde{\eta}(s) + \eta(s) - k_g r(s)), \tag{2.156}$$

which implies that the closed-loop response of the system in (2.129) with the $\mathcal{L}_1$ adaptive controller can be rewritten (in the frequency domain) as

$$x(s) = G(s)\eta(s) - H(s)C(s)(\tilde{\eta}(s) - k_g r(s)) + x_{\text{in}}(s).$$

Recall that, in (2.142), the response of the reference system was presented as

$$x_{\text{ref}}(s) = G(s)\eta_{\text{ref}}(s) + H(s)C(s)k_g r(s) + x_{\text{in}}(s).$$

Then, the two expressions above lead to

$$x_{\text{ref}}(s) - x(s) = G(s)(\eta_{\text{ref}}(s) - \eta(s)) + H(s)C(s)\tilde{\eta}(s). \tag{2.157}$$

Since

$$\eta_{\text{ref}}(t) - \eta(t) \triangleq \theta^\top(t)(x_{\text{ref}}(t) - x(t)),$$

the following upper bound holds:

$$\|(\eta_{\text{ref}} - \eta)_\tau\|_{\mathcal{L}_\infty} \leq L\|(x_{\text{ref}} - x)_\tau\|_{\mathcal{L}_\infty}. \tag{2.158}$$

Moreover, it follows from (2.147) that

$$\tilde{x}(s) = H(s)\tilde{\eta}(s),$$

which, along with (2.157) and (2.158), leads to

$$\|(x_{\text{ref}} - x)_\tau\|_{\mathcal{L}_\infty} \leq \|G(s)\|_{\mathcal{L}_1} L\|(x_{\text{ref}} - x)_\tau\|_{\mathcal{L}_\infty} + \|C(s)\|_{\mathcal{L}_1}\|\tilde{x}_\tau\|_{\mathcal{L}_\infty}.$$

The bound in (2.155) and the definition of γ_1 in (2.133) yield

$$\|(x_{\text{ref}} - x)_\tau\|_{\mathcal{L}_\infty} \leq \frac{\|C(s)\|_{\mathcal{L}_1}}{1 - \|G(s)\|_{\mathcal{L}_1} L} \|\tilde{x}_\tau\|_{\mathcal{L}_\infty} \leq \frac{\|C(s)\|_{\mathcal{L}_1}}{1 - \|G(s)\|_{\mathcal{L}_1} L} \gamma_0 < \gamma_1 , \tag{2.159}$$

and, thus, we obtain a contradiction to the first equality in (2.153).

To show that the second equation in (2.153) also cannot hold, consider (2.140) and (2.156), which lead to

$$u_{\text{ref}}(s) - u(s) = C_u(s)(\eta(s) - \eta_{\text{ref}}(s)) + C_u(s)\tilde{\eta}(s) .$$

Using Lemma A.12.1 and the definition of $H_1(s)$ from (2.132), we can write

$$C_u(s)\tilde{\eta}(s) = H_1(s)\tilde{x}(s) ,$$

which, along with the result of Lemma A.7.1, yields the following upper bound:

$$\|(u_{\text{ref}} - u)_\tau\|_{\mathcal{L}_\infty} \leq L\|C_u(s)\|_{\mathcal{L}_1} \|(x_{\text{ref}} - x)_\tau\|_{\mathcal{L}_\infty} + \|H_1(s)\|_{\mathcal{L}_1} \|\tilde{x}_\tau\|_{\mathcal{L}_\infty} .$$

Then, from the the upper bounds on $\|\tilde{x}_\tau\|_{\mathcal{L}_\infty}$ and $\|(x_{\text{ref}} - x)_\tau\|_{\mathcal{L}_\infty}$ in (2.155) and (2.159), and the definition of γ_2 in (2.134), it follows that

$$\|(u_{\text{ref}} - u)_\tau\|_{\mathcal{L}_\infty} \leq \|C_u(s)\|_{\mathcal{L}_1} L(\gamma_1 - \beta) + \|H_1(s)\|_{\mathcal{L}_1} \gamma_0 < \gamma_2 , \tag{2.160}$$

which contradicts the second equality in (2.153). Notice that the upper bounds in (2.159) and (2.160) hold uniformly, which proves (2.151)–(2.152). Then, the upper bound in (2.150) follows from these two bounds and (2.155) directly. □

Remark 2.3.2 It follows from (2.149) that one can prescribe arbitrarily small γ_0 by increasing the adaptive gain, which further implies that the performance bounds γ_1 and γ_2 for the system's signals, both input and output, can be rendered arbitrarily small simultaneously.

Remark 2.3.3 Notice that letting $k \to \infty$ leads to $C(s) \to 1$, and thus the reference controller in the definition of the closed-loop reference system in (2.140) leads, in the limit, to perfect cancelation of uncertainties and recovers the performance of the ideal desired system. Notice that if we set $C(s) = 1$, the transfer function

$$C_u(s) = \frac{C(s)}{F(s)} = \frac{1}{F(s)}$$

becomes improper. Moreover, the norm of $H_1(s)$, given by

$$\|H_1(s)\|_{\mathcal{L}_1} = \left\| C_u(s) \frac{1}{c_o^\top H(s)} c_o^\top \right\|_{\mathcal{L}_1} ,$$

is not bounded since $c_o^\top H(s)$ is strictly proper and $C_u(s)$ is improper. Therefore, in the absence of $C(s)$, one cannot obtain a uniform performance bound for the control signal similar to Theorem 2.3.1.

Remark 2.3.4 We notice that the ideal system input signal

$$\mu_{\text{id}}(t) = k_g r(t) - \theta^\top(t)x(t) - \sigma_0(t) \tag{2.161}$$

is the one that leads to desired system response,

$$\begin{aligned} \dot{x}_{\text{id}}(t) &= A_m x_{\text{id}}(t) + b k_g r(t), \quad x(0) = x_0, \\ y_{\text{id}}(t) &= c^\top x_{\text{id}}(t), \end{aligned} \tag{2.162}$$

by canceling the uncertainties exactly. In the closed-loop reference system (2.140), $\mu_{\text{id}}(t)$ is further low-pass filtered by $C(s)$ to have guaranteed low-frequency range. Thus, the reference system in (2.140) has a different response as compared to (2.162) achieved with (2.161). Similar to the previous sections, the response of $x_{\text{ref}}(t)$ and $u_{\text{ref}}(t)$ can be made arbitrarily close to (2.162) by reducing $\|G(s)\|_{\mathcal{L}_1}$. If $F(s)$ is a relative degree one and minimum phase system, then $\|G(s)\|_{\mathcal{L}_1}$ can be made arbitrarily small by appropriately choosing the design parameters k and $D(s)$. However, for the general case of unknown $F(s)$, the design of k and $D(s)$ which satisfy (2.131), is an open question.

2.3.4 Simulation Example: Rohrs' Example

In this section we analyze Rohrs' example from [146, 147], which was constructed with a particular objective of analyzing the robustness properties of MRAC architectures. The system under consideration in [146] is a first-order-stable system with unknown time constant and DC gain, and with two highly damped unmodeled poles:

$$\begin{aligned} y(s) &= \frac{2}{s+1}\mu(s), \\ \mu(s) &= \frac{229}{s^2+30s+229}u(s). \end{aligned}$$

The system has a gain crossover frequency of $\omega_{gc} = 1.70$ rad/s and a phase margin of $\phi_m = 107.67$ deg. Its phase crossover frequency and the gain margin are $\omega_{\phi c} = 16.09$ rad/s and $g_m = 24.62$ dB, respectively. The control objective in [146] is given via the following first-order-stable reference model:

$$y_m(s) = \frac{3}{s+3}r(s).$$

Model Reference Adaptive Control: Parameter Drift

The conventional MRAC controller for this system takes the form

$$\begin{aligned} u(t) &= \hat{k}_y(t)y(t) + \hat{k}_r(t)r(t), \\ \dot{\hat{k}}_y(t) &= -e(t)y(t), \quad \hat{k}_y(0) = \hat{k}_{y0}, \\ \dot{\hat{k}}_r(t) &= -e(t)r(t), \quad \hat{k}_r(0) = \hat{k}_{r0}, \end{aligned}$$

where $e(t) = y(t) - y_m(t)$. The corresponding feedback loop of the MRAC architecture is shown in Figure 2.29.

For simulations we consider the same reference inputs as in [146]. The first reference input has the exact phase crossover frequency

$$r_1(t) = 0.3 + 1.85\sin(16.1t),$$

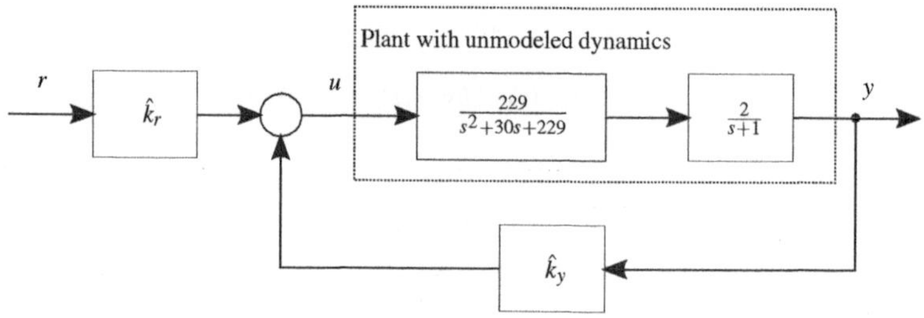

Figure 2.29: Rohrs' example. Closed-loop MRAC system.

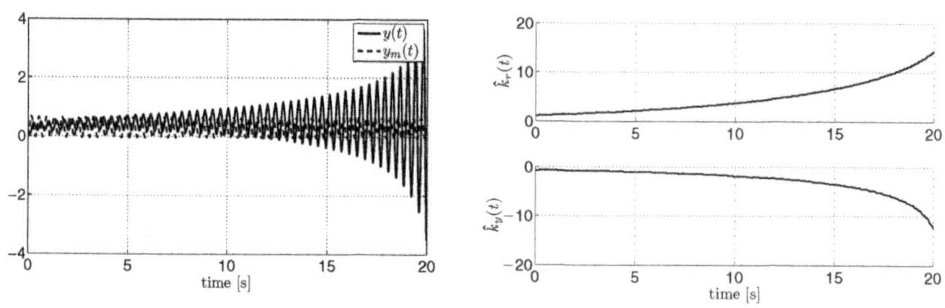

(a) System output $y(t)$, $y_m(t)$

(b) Controller parameters $\hat{k}_r(t)$, $\hat{k}_y(t)$

Figure 2.30: MRAC: Closed-loop system's response to $r_1(t)$.

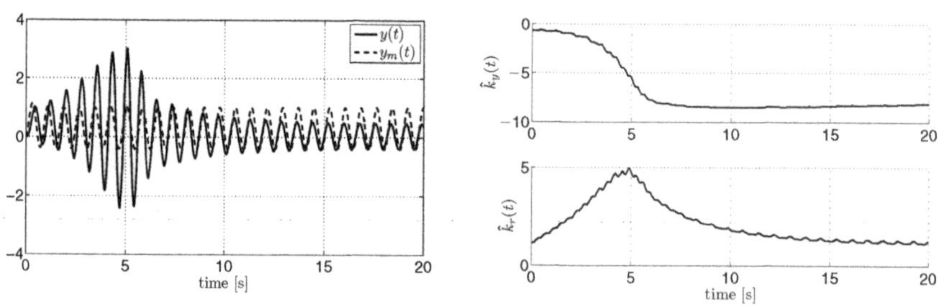

(a) System output $y(t)$, $y_m(t)$

(b) Controller parameters $\hat{k}_r(t)$, $\hat{k}_y(t)$

Figure 2.31: MRAC: Closed-loop system's response to $r_2(t)$.

while the second one is also a sinusoidal reference signal, but at a frequency which is approximately half the phase crossover frequency,

$$r_2(t) = 0.3 + 2\sin(8t).$$

We use the same initial conditions as in [146]: $y(0) = 0$, $\hat{k}_r(0) = 1.14$, and $\hat{k}_y(0) = -0.65$. The simulation results from [146] are reproduced here in Figures 2.30 and 2.31. In Figure 2.30, one can see that, while tracking $r_1(t)$, the closed-loop system is unstable due to parameter drift. In Figure 2.31, *bursting* takes place in the response of the closed-loop adaptive system to the reference signal $r_2(t)$.

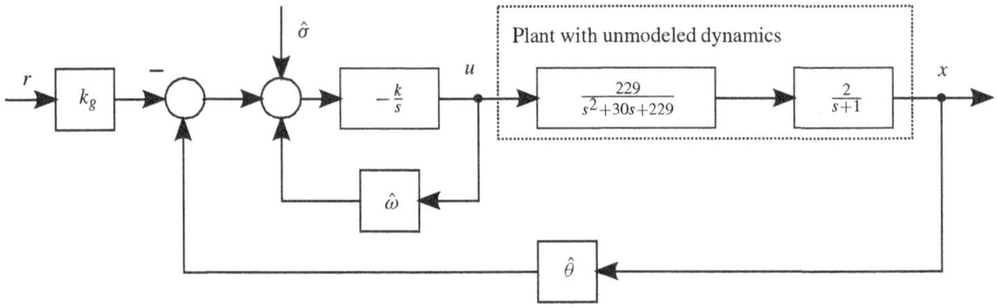

Figure 2.32: Rohrs' example. Closed-loop $\mathcal{L}_1$ control system.

$\mathcal{L}_1$ Adaptive Controller

It is straightforward to see that Rohrs' example can be cast into the framework of the $\mathcal{L}_1$ adaptive controller of this section. In fact, we can rewrite Rohrs' example in state-space form as

$$\dot{x}(t) = -3x(t) + 2(\mu(t) + x(t)), \quad x(0) = x_0,$$
$$y(t) = x(t),$$

where

$$\mu(s) = \frac{229}{s^2 + 30s + 229} u(s).$$

Then, the state predictor takes the form

$$\dot{\hat{x}}(t) = -3\hat{x}(t) + 2\left(\hat{\omega}(t)u(t) + \hat{\theta}(t)x(t) + \hat{\sigma}(t)\right), \quad \hat{x}(0) = x_0,$$
$$\hat{y}(t) = \hat{x}(t),$$

with $\hat{\omega}(t)$, $\hat{\theta}(t)$, and $\hat{\sigma}(t)$ being governed by

$$\dot{\hat{\omega}}(t) = \Gamma \text{Proj}\left(\hat{\omega}(t), -\tilde{x}(t)u(t)\right), \quad \hat{\omega}(0) = \hat{\omega}_0,$$
$$\dot{\hat{\theta}}(t) = \Gamma \text{Proj}\left(\hat{\theta}(t), -x(t)\tilde{x}(t)\right), \quad \hat{\theta}(0) = \hat{\theta}_0,$$
$$\dot{\hat{\sigma}}(t) = \Gamma \text{Proj}\left(\hat{\sigma}(t), -\tilde{x}(t)\right), \quad \hat{\sigma}(0) = \hat{\sigma}_0.$$

The control law is given by

$$u(s) = -kD(s)(\hat{\eta}(s) - k_g r(s)),$$

where $\hat{\eta}(t) \triangleq \hat{\omega}(t)u(t) + \hat{\theta}(t)x(t) + \hat{\sigma}(t)$ and $k_g = \frac{3}{2}$. The block diagram of the $\mathcal{L}_1$ adaptive control system is given in Figure 2.32.

The simulation plots for the inputs $r_1(t)$ and $r_2(t)$, using the $\mathcal{L}_1$ adaptive controller with $k = 5$, $D(s) = 1/s$, and $\Gamma = 1000$, are given in Figures 2.33 through 2.35. The initial conditions have been set to $x_0 = 0$, $\hat{\omega}_0 = 1.14$, $\hat{\theta}_0 = 0.65$, and $\hat{\sigma}_0 = 0$. The projection

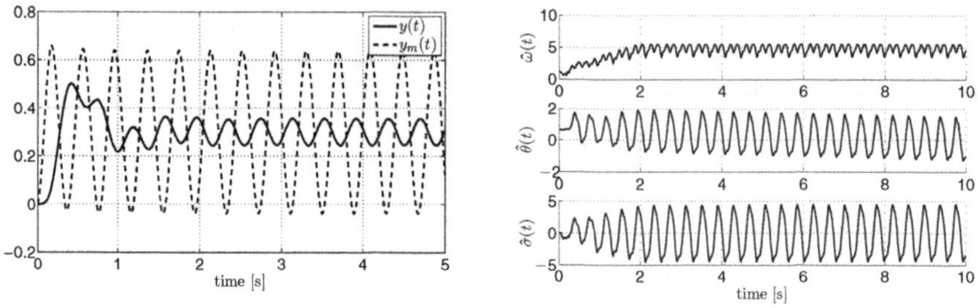

(a) System output $y(t)$, $y_m(t)$ (b) Controller parameters $\hat{\omega}(t)$, $\hat{\theta}(t)$, $\hat{\sigma}(t)$

Figure 2.33: $\mathcal{L}_1$ adaptive control: Closed-loop system response to $r_1(t)$.

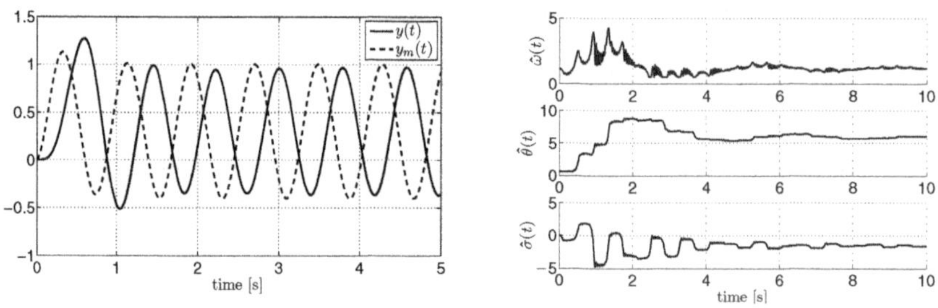

(a) System output $y(t)$, $y_m(t)$ (b) Controller parameters $\hat{\omega}(t)$, $\hat{\theta}(t)$, $\hat{\sigma}(t)$

Figure 2.34: $\mathcal{L}_1$ adaptive control: Closed-loop system response to $r_2(t)$.

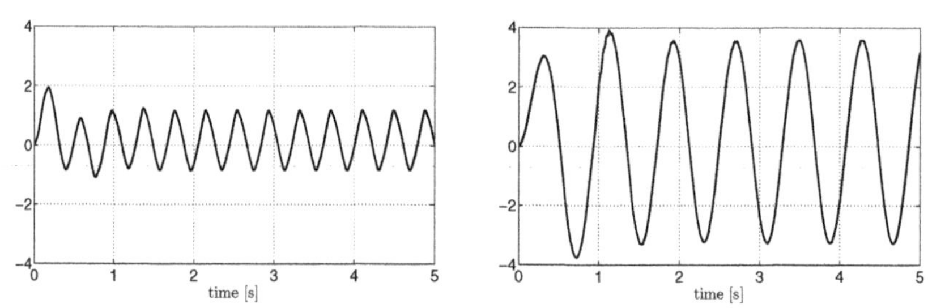

(a) $u(t)$ for $r_1(t)$ (b) $u(t)$ for $r_2(t)$

Figure 2.35: $\mathcal{L}_1$ adaptive control: Control signal time history.

bounds are set to $\Theta = [-10,\ 10]$, $\Delta = 10$, and $\Omega = [0.5\ 5.5]$. We see from the plots that the $\mathcal{L}_1$ adaptive controller guarantees that both the system output and the parameters remain bounded, while achieving an expected level of performance (as both reference commands are well beyond the bandwidth of the system).

We can get further insight into the $\mathcal{L}_1$ adaptive controller by analyzing the implementation block diagram in Figure 2.32 from a classical control perspective. We note that while the feedback gain $\hat{\theta}(t)$ plays a similar role as the feedback gain $\hat{k}_y(t)$ in MRAC, the adaptive parameter $\hat{\omega}(t)$ appears in the feedforward path as a feedback gain around the integrator $\frac{k}{s}$ and thus has the ability to adjust the bandwidth of the low-pass filter, the output of which is the feedback signal of the closed-loop adaptive system. Recall that in MRAC the feedforward gain $\hat{k}_r(t)$ did not play any role in the stabilization process and only scaled the reference input. It is important to note also that the various modifications of the adaptive laws, such as the σ-modification and the e-modification, are only means to ensure boundedness of the parameter estimates (and thus avoid the parameter drift) but by no means affect the *phase* in the system, as $\hat{\omega}(t)$ does in the $\mathcal{L}_1$ architecture. Instead, with the $\mathcal{L}_1$ adaptive controller, the open-loop system bandwidth changes as both $\hat{\omega}(t)$ and $\hat{\theta}(t)$ adapt, which leads to simultaneous *adaptation on the loop gain and the phase* of the closed-loop system.

2.4 $\mathcal{L}_1$ Adaptive Controller for Nonlinear Systems

In this section we present the $\mathcal{L}_1$ adaptive controller for systems with unknown state- and time-dependent nonlinearities. Under a mild set of assumptions, we prove that the $\mathcal{L}_1$ adaptive controller leads to uniform transient and steady-state performance bounds for the system's signals, both input and output, which can be systematically improved by increasing the rate of adaptation. Moreover, the semiglobal positively invariant sets, to which the state and the control signal are confined, are quantified dependent upon the filter parameters and the bounds on the partial derivatives of the unknown nonlinear function [30].

2.4.1 Problem Formulation

We consider the nonlinear system dynamics

$$\begin{aligned} \dot{x}(t) &= A_m x(t) + b(\omega u(t) + f(t,x(t))), \quad x(0) = x_0, \\ y(t) &= c^\top x(t), \end{aligned} \tag{2.163}$$

where $x(t) \in \mathbb{R}^n$ is the system state (measured); $A_m \in \mathbb{R}^{n\times n}$ is a known Hurwitz matrix specifying the desired closed-loop dynamics; $b, c \in \mathbb{R}^n$ are known constant vectors; $u(t) \in \mathbb{R}$ is the control input; $\omega \in \mathbb{R}$ is an unknown constant parameter with known sign, representing uncertainty in the system input gain; $f(t,x) : \mathbb{R} \times \mathbb{R}^n \to \mathbb{R}$ is an unknown nonlinear map continuous in its arguments; and $y(t) \in \mathbb{R}$ is the regulated output. The initial condition x_0 is assumed to be inside an arbitrarily large known set, i.e., $\|x_0\|_\infty \le \rho_0 < \infty$ with known $\rho_0 > 0$.

Assumption 2.4.1 (Partial knowledge of uncertain system input gain) Let

$$\omega \in \Omega \triangleq [\omega_l, \omega_u],$$

where $0 < \omega_l < \omega_u$ are known conservative bounds.

Assumption 2.4.2 (Uniform boundedness of $f(t,0)$) There exists $B > 0$ such that

$$|f(t,0)| \le B, \quad \forall t \ge 0.$$

Assumption 2.4.3 (Semiglobal uniform boundedness of partial derivatives) For arbitrary $\delta > 0$, there exist $d_{f_x}(\delta) > 0$ and $d_{f_t}(\delta) > 0$ independent of time, such that for arbitrary $\|x\|_\infty \leq \delta$, the partial derivatives of $f(t,x)$ are piecewise-continuous and bounded,

$$\left\|\frac{\partial f(t,x)}{\partial x}\right\|_1 \leq d_{f_x}(\delta), \quad \left|\frac{\partial f(t,x)}{\partial t}\right| \leq d_{f_t}(\delta).$$

The control objective is to design a full-state feedback adaptive controller to ensure that $y(t)$ tracks a given bounded piecewise-continuous reference signal $r(t)$ with quantifiable performance bounds.

2.4.2 $\mathcal{L}_1$ Adaptive Control Architecture

Definitions and $\mathcal{L}_1$–Norm Sufficient Condition for Stability

As in Section 2.2, the design of the $\mathcal{L}_1$ adaptive controller proceeds by considering a feedback gain $k > 0$ and a strictly proper transfer function $D(s)$, which lead, for all $\omega \in \Omega$, to a strictly proper stable transfer function

$$C(s) \triangleq \frac{\omega k D(s)}{1 + \omega k D(s)}$$

with DC gain $C(0) = 1$. As before, we let $x_{\text{in}}(t)$ be the signal with its Laplace transform being $x_{\text{in}}(s) \triangleq (s\mathbb{I} - A_m)^{-1}x_0$. Since A_m is Hurwitz, $\|x_{\text{in}}\|_{\mathcal{L}_\infty} \leq \rho_{\text{in}}$, where $\rho_{\text{in}} \triangleq \|s(s\mathbb{I} - A_m)^{-1}\|_{\mathcal{L}_1}\rho_0$. Further, let

$$L_\delta \triangleq \frac{\bar{\delta}(\delta)}{\delta} d_{f_x}(\bar{\delta}(\delta)), \quad \bar{\delta}(\delta) \triangleq \delta + \bar{\gamma}_1, \tag{2.164}$$

where $d_{f_x}(\cdot)$ was introduced in Assumption 2.4.3 and $\bar{\gamma}_1 > 0$ is an arbitrary positive constant.

For the proofs of stability and performance bounds, the choice of k and $D(s)$ needs to ensure that for a given ρ_0, there exists $\rho_r > \rho_{\text{in}}$, such that the following $\mathcal{L}_1$-norm condition can be verified:

$$\|G(s)\|_{\mathcal{L}_1} < \frac{\rho_r - \|H(s)C(s)k_g\|_{\mathcal{L}_1}\|r\|_{\mathcal{L}_\infty} - \rho_{\text{in}}}{L_{\rho_r}\rho_r + B}, \tag{2.165}$$

where $G(s) \triangleq H(s)(1 - C(s))$, $H(s) \triangleq (s\mathbb{I} - A_m)^{-1}b$, and $k_g \triangleq -1/(c^\top A_m^{-1}b)$ is the feedforward gain required for tracking of step reference command $r(t)$ with zero steady-state error.

To streamline the subsequent analysis of stability and performance bounds, we need to introduce some notation. Let

$$H_1(s) \triangleq C(s)\frac{1}{c_o^\top H(s)}c_o^\top,$$

where $c_o \in \mathbb{R}^n$ is a vector that renders $H_1(s)$ BIBO stable and proper. We refer to Lemma A.12.1 for the existence of such c_o.

We also define

$$\rho \triangleq \rho_r + \bar{\gamma}_1, \tag{2.166}$$

and let γ_1 be given by

$$\gamma_1 \triangleq \frac{\|C(s)\|_{\mathcal{L}_1}}{1 - \|G(s)\|_{\mathcal{L}_1} L_{\rho_r}} \gamma_0 + \beta, \tag{2.167}$$

where β and γ_0 are arbitrary small positive constants such that $\gamma_1 \leq \bar{\gamma}_1$. Moreover, let

$$\rho_u \triangleq \rho_{ur} + \gamma_2, \tag{2.168}$$

where ρ_{ur} and γ_2 are defined as

$$\rho_{ur} \triangleq \left\| \frac{C(s)}{\omega} \right\|_{\mathcal{L}_1} (|k_g| \|r\|_{\mathcal{L}_\infty} + L_{\rho_r} \rho_r + B), \tag{2.169}$$

$$\gamma_2 \triangleq \left\| \frac{C(s)}{\omega} \right\|_{\mathcal{L}_1} L_{\rho_r} \gamma_1 + \left\| \frac{H_1(s)}{\omega} \right\|_{\mathcal{L}_1} \gamma_0 . \tag{2.170}$$

Finally, let

$$\theta_b \triangleq d_{f_x}(\rho), \quad \Delta \triangleq B + \epsilon, \tag{2.171}$$

where $\epsilon > 0$ is an arbitrary constant.

Remark 2.4.1 In the following analysis we demonstrate that ρ_r and ρ characterize the positively invariant sets for the state of the closed-loop reference system (yet to be defined) and the state of the closed-loop adaptive system, respectively. We notice that, since $\bar{\gamma}_1$ can be set arbitrarily small, ρ can approximate ρ_r arbitrarily closely.

Remark 2.4.2 Notice that the $\mathcal{L}_1$-norm condition in (2.165) is a consequence of the *semiglobal* boundedness of the partial derivatives of $f(t,x)$, stated in Assumption 2.4.3. If $f(t,x)$ has a uniform bound for its derivative with respect to x, i.e., $\|\frac{\partial f}{\partial x}\| \leq d_{f_x} = L$ holds uniformly for all $x \in \mathbb{R}^n$, then

$$\lim_{\rho_r \to \infty} \frac{\rho_r - \|H(s)C(s)k_g\|_{\mathcal{L}_1} \|r\|_{\mathcal{L}_\infty} - \rho_{\text{in}}}{L\rho_r + B} = \frac{1}{L},$$

and the $\mathcal{L}_1$-norm condition in (2.165) degenerates into

$$\|G(s)\|_{\mathcal{L}_1} L < 1,$$

which is the same condition as the one in (2.7), derived in Chapter 2 for systems with constant unknown parameters. We will prove that (2.165) is a sufficient condition for stability of the closed-loop adaptive system.

The elements of $\mathcal{L}_1$ adaptive controller are introduced next.

State Predictor

We consider the following state predictor:

$$\begin{aligned} \dot{\hat{x}}(t) &= A_m \hat{x}(t) + b(\hat{\omega}(t)u(t) + \hat{\theta}(t)\|x(t)\|_\infty + \hat{\sigma}(t)), \quad \hat{x}(0) = x_0, \\ \hat{y}(t) &= c^\top \hat{x}(t), \end{aligned} \tag{2.172}$$

where $\hat{\omega}(t) \in \mathbb{R}$, $\hat{\theta}(t) \in \mathbb{R}$, and $\hat{\sigma}(t) \in \mathbb{R}$ are the adaptive estimates.

Adaptation Laws

The adaptive estimates $\hat{\omega}(t)$, $\hat{\theta}(t)$, and $\hat{\sigma}(t)$ are governed by the following adaptation laws:

$$\begin{aligned}\dot{\hat{\theta}}(t) &= \Gamma \text{Proj}(\hat{\theta}(t), -\tilde{x}^\top(t)Pb\|x(t)\|_\infty), \quad \hat{\theta}(0) = \hat{\theta}_0, \\ \dot{\hat{\sigma}}(t) &= \Gamma \text{Proj}(\hat{\sigma}(t), -\tilde{x}^\top(t)Pb), \quad \hat{\sigma}(0) = \hat{\sigma}_0, \\ \dot{\hat{\omega}}(t) &= \Gamma \text{Proj}(\hat{\omega}(t), -\tilde{x}^\top(t)Pbu(t)), \quad \hat{\omega}(0) = \hat{\omega}_0,\end{aligned} \tag{2.173}$$

where $\tilde{x}(t) \triangleq \hat{x}(t) - x(t)$, $\Gamma \in \mathbb{R}^+$ is the adaptation gain, while $P = P^\top > 0$ is the solution of the algebraic Lyapunov equation $A_m^\top P + PA_m = -Q$, for arbitrary symmetric $Q = Q^\top > 0$. The projection operator ensures that $\hat{\omega}(t) \in \Omega$, $\hat{\theta}(t) \in \Theta \triangleq [-\theta_b, \theta_b]$, $|\hat{\sigma}(t)| \leq \Delta$, with θ_b and Δ being defined in (2.171).

Control Law

The control signal is generated as the output of the following (feedback) system:

$$u(s) = -kD(s)(\hat{\eta}(s) - k_g r(s)), \tag{2.174}$$

where $\hat{\eta}(s)$ is the Laplace transform of $\hat{\eta}(t) \triangleq \hat{\omega}(t)u(t) + \hat{\theta}(t)\|x(t)\|_\infty + \hat{\sigma}(t)$.

The $\mathcal{L}_1$ adaptive controller is defined via (2.172), (2.173), and (2.174) subject to the $\mathcal{L}_1$-norm condition in (2.165).

2.4.3 Analysis of the $\mathcal{L}_1$ Adaptive Controller

Closed-Loop Reference System

We consider the following closed-loop reference system, in which the control signal attempts only to compensate for the uncertainties within the bandwidth of the low-pass filter $C(s)$:

$$\begin{aligned}\dot{x}_{\text{ref}}(t) &= A_m x_{\text{ref}}(t) + b(\omega u_{\text{ref}}(t) + f(t, x_{\text{ref}}(t))), \\ u_{\text{ref}}(s) &= \frac{C(s)}{\omega}(k_g r(s) - \eta_{\text{ref}}(s)), \\ y_{\text{ref}}(t) &= c^\top x_{\text{ref}}(t), \quad x_{\text{ref}}(0) = x_0,\end{aligned} \tag{2.175}$$

where $\eta_{\text{ref}}(s)$ is the Laplace transform of the signal $\eta_{\text{ref}}(t) \triangleq f(t, x_{\text{ref}}(t))$. The next lemma establishes the stability of the closed-loop reference system in (2.175).

Lemma 2.4.1 For the closed-loop reference system in (2.175), subject to the $\mathcal{L}_1$-norm condition in (2.165), if

$$\|x_0\|_\infty \leq \rho_0,$$

then

$$\|x_{\text{ref}}\|_{\mathcal{L}_\infty} < \rho_r, \tag{2.176}$$

$$\|u_{\text{ref}}\|_{\mathcal{L}_\infty} < \rho_{ur}, \tag{2.177}$$

where ρ_r and ρ_{ur} were introduced in (2.165) and (2.169), respectively.

Proof. The proof is done by contradiction. First, we note that from (2.175), it follows that

$$x_{\text{ref}}(s) = G(s)\eta_{\text{ref}}(s) + H(s)C(s)k_g r(s) + x_{\text{in}}(s), \tag{2.178}$$

and Lemma A.7.1 yields the following bound:

$$\|x_{\text{ref}\tau}\|_{\mathcal{L}_\infty} \leq \|G(s)\|_{\mathcal{L}_1}\|\eta_{\text{ref}\tau}\|_{\mathcal{L}_\infty} + \|H(s)C(s)k_g\|_{\mathcal{L}_1}\|r\|_{\mathcal{L}_\infty} + \|x_{\text{in}}\|_{\mathcal{L}_\infty}. \tag{2.179}$$

If (2.176) is not true, since $\|x_{\text{ref}}(0)\|_\infty = \|x_0\|_\infty \leq \rho_0 < \rho_r$ and $x_{\text{ref}}(t)$ is continuous, then there exists $\tau > 0$ such that

$$\|x_{\text{ref}}(t)\|_\infty < \rho_r, \quad \forall\, t \in [0,\tau),$$

and

$$\|x_{\text{ref}}(\tau)\|_\infty = \rho_r,$$

which implies that

$$\|x_{\text{ref}\tau}\|_{\mathcal{L}_\infty} = \rho_r. \tag{2.180}$$

Recalling the definition of $\bar{\delta}(\delta)$ in (2.164), we have $\rho_r < \bar{\rho}_r(\rho_r)$. Then, taking into consideration Assumptions 2.4.2 and 2.4.3, the equality in (2.180), together with the redefinition in (2.164), yields the following upper bound:

$$\|\eta_{\text{ref}\tau}\|_{\mathcal{L}_\infty} \leq L_{\rho_r}\rho_r + B. \tag{2.181}$$

Substituting this upper bound into (2.179), and noticing that for uniformly bounded signals $\|(\cdot)_\tau\|_{\mathcal{L}_\infty} \leq \|\cdot\|_{\mathcal{L}_\infty}$, we obtain

$$\|x_{\text{ref}\tau}\|_{\mathcal{L}_\infty} \leq \|G(s)\|_{\mathcal{L}_1}(L_{\rho_r}\rho_r + B) + \|H(s)C(s)k_g\|_{\mathcal{L}_1}\|r\|_{\mathcal{L}_\infty} + \rho_{\text{in}}.$$

The condition in (2.165) can be solved for ρ_r to obtain the upper bound

$$\|G(s)\|_{\mathcal{L}_1}(L_{\rho_r}\rho_r + B) + \|H(s)C(s)k_g\|_{\mathcal{L}_1}\|r\|_{\mathcal{L}_\infty} + \rho_{\text{in}} < \rho_r,$$

which implies that $\|x_{\text{ref}\tau}\|_{\mathcal{L}_\infty} < \rho_r$, thus contradicting (2.180). This proves the bound in (2.176).

Using (2.181), it follows from the definition of the reference control signal in (2.175) that

$$\|u_{\text{ref}}\|_{\mathcal{L}_\infty} \leq \left\|\frac{C(s)}{\omega}\right\|_{\mathcal{L}_1}(|k_g|\|r\|_{\mathcal{L}_\infty} + L_{\rho_r}\rho_r + B),$$

which proves the bound in (2.177). □

Equivalent (Semi-)Linear Time-Varying System

Next, we refer to Lemma A.8.1 to transform the nonlinear system in (2.163) into a (semi-)linear system with unknown time-varying parameters and disturbances.

Since

$$\|x_0\|_\infty \le \rho_0 < \rho\,, \quad u(0) = 0\,,$$

and $x(t)$, $u(t)$ are continuous, there always exists τ such that

$$\|x_\tau\|_{\mathcal{L}_\infty} \le \rho\,, \quad \|u_\tau\|_{\mathcal{L}_\infty} \le \rho_u\,. \tag{2.182}$$

Then, it follows from the bounds in (2.182) and Lemma A.8.1 that the system in (2.163) can be rewritten over $[0,\tau]$ as

$$\begin{aligned} \dot{x}(t) &= A_m x(t) + b\left(\omega u(t) + \theta(t)\|x(t)\|_\infty + \sigma(t)\right), \quad x(0) = x_0\,, \\ y(t) &= c^\top x(t)\,, \end{aligned} \tag{2.183}$$

where $\theta(t)$ and $\sigma(t)$ are unknown signals satisfying

$$|\theta(t)| < \theta_b\,, \quad |\sigma(t)| < \Delta\,, \quad \forall\, t \in [0,\tau]\,. \tag{2.184}$$

$$|\dot{\theta}(t)| \le d_\theta(\rho,\rho_u)\,, \quad |\dot{\sigma}(t)| \le d_\sigma(\rho,\rho_u)\,, \quad \forall\, t \in [0,\tau]\,, \tag{2.185}$$

with $d_\theta(\rho,\rho_u) > 0$ and $d_\sigma(\rho,\rho_u) > 0$ being the bounds guaranteed by Lemma A.8.1.

Transient and Steady-State Performance

It follows from (2.172) and (2.183) that over $[0,\tau]$ the prediction error dynamics can be written as

$$\dot{\tilde{x}}(t) = A_m \tilde{x}(t) + b\big(\tilde{\omega}(t)u(t) + \tilde{\theta}(t)\|x(t)\|_\infty + \tilde{\sigma}(t)\big)\,, \quad \tilde{x}(0) = 0\,, \tag{2.186}$$

where

$$\tilde{\omega}(t) \triangleq \hat{\omega}(t) - \omega\,, \quad \tilde{\theta}(t) \triangleq \hat{\theta}(t) - \theta(t)\,, \quad \tilde{\sigma}(t) \triangleq \hat{\sigma}(t) - \sigma(t)\,. \tag{2.187}$$

Let $\tilde{\eta}(t) \triangleq \tilde{\omega}(t)u(t) + \tilde{\theta}^\top(t)\|x(t)\|_\infty + \tilde{\sigma}(t)$ and let $\tilde{\eta}(s)$ be the Laplace transform of it. Then, the error dynamics in (2.186) can be rewritten in the frequency domain as

$$\tilde{x}(s) = H(s)\tilde{\eta}(s)\,. \tag{2.188}$$

Before the main theorem, we prove the following lemma.

Lemma 2.4.2 For the system in (2.186), if $u(t)$ is continuous, and moreover the following bounds hold:

$$\|x_\tau\|_{\mathcal{L}_\infty} \le \rho\,, \quad \|u_\tau\|_{\mathcal{L}_\infty} \le \rho_u\,, \tag{2.189}$$

then

$$\|\tilde{x}_\tau\|_{\mathcal{L}_\infty} \le \sqrt{\frac{\theta_m(\rho,\rho_u)}{\lambda_{\min}(P)\Gamma}}\,,$$

where

$$\begin{aligned} \theta_m(\rho,\rho_u) \triangleq\; & 4\theta_b^2 + 4\Delta^2 + (\omega_u - \omega_l)^2 \\ & + 4\frac{\lambda_{\max}(P)}{\lambda_{\min}(Q)}\left(\theta_b d_\theta(\rho,\rho_u) + \Delta d_\sigma(\rho,\rho_u)\right). \end{aligned} \tag{2.190}$$

Proof. Consider the following Lyapunov function candidate:

$$V(\tilde{x}(t),\tilde{\omega}(t),\tilde{\theta}(t),\tilde{\sigma}(t)) = \tilde{x}^\top(t)P\tilde{x}(t) + \frac{1}{\Gamma}\left(\tilde{\omega}^2(t)+\tilde{\theta}^2(t)+\tilde{\sigma}^2(t)\right).$$

We can verify straightforwardly that

$$V(0) \leq \frac{(\omega_u-\omega_l)^2+4\theta_b^2+4\Delta^2}{\Gamma} \leq \frac{\theta_m(\rho,\rho_u)}{\Gamma}.$$

Further, we need to prove that

$$V(t) \leq \frac{\theta_m(\rho,\rho_u)}{\Gamma}, \quad \forall\, t\in[0,\tau].$$

It follows from (2.189) that the bounds in (2.185) hold for arbitrary $t\in[0,\tau]$, i.e.,

$$|\dot{\theta}(t)| \leq d_\theta(\rho,\rho_u), \quad |\dot{\sigma}(t)| \leq d_\sigma(\rho,\rho_u), \qquad \forall\, t\in[0,\tau]. \tag{2.191}$$

Let $t_1\in(0,\tau]$ be the first time instant of the discontinuity of either of the derivatives of $\dot{\theta}(t)$ and $\dot{\sigma}(t)$. Using the projection-based adaptation laws from (2.173), the following upper bound for $\dot{V}(t)$ can be obtained for $t\in[0,t_1)$:

$$\dot{V}(t) \leq -\tilde{x}^\top(t)Q\tilde{x}(t) + \frac{2}{\Gamma}\left|\tilde{\theta}(t)\dot{\theta}(t)+\tilde{\sigma}(t)\dot{\sigma}(t)\right|. \tag{2.192}$$

The projection algorithm ensures that for all $t\in[0,t_1)$,

$$\omega_l \leq \hat{\omega}(t) \leq \omega_u, \quad |\hat{\theta}(t)| \leq \theta_b, \quad |\hat{\sigma}(t)| \leq \Delta, \tag{2.193}$$

and therefore

$$\max_{t\in[0,t_1)}\left(\frac{1}{\Gamma}\left(\tilde{\omega}^2(t)+\tilde{\theta}^2(t)+\tilde{\sigma}^2(t)\right)\right) \leq \frac{(\omega_u-\omega_l)^2+4\theta_b^2+4\Delta^2}{\Gamma}. \tag{2.194}$$

If at arbitrary $t'\in[0,t_1)$

$$V(t') > \frac{\theta_m(\rho,\rho_u)}{\Gamma},$$

where $\theta_m(\rho,\rho_u)$ was defined in (2.190), then it follows from (2.194) that

$$\tilde{x}^\top(t')P\tilde{x}(t') > \frac{4\lambda_{\max}(P)}{\Gamma\lambda_{\min}(Q)}\left(\theta_b d_\theta(\rho,\rho_u)+\Delta d_\sigma(\rho,\rho_u)\right).$$

Hence

$$\tilde{x}^\top(t')Q\tilde{x}(t') \geq \frac{\lambda_{\min}(Q)}{\lambda_{\max}(P)}\tilde{x}^\top(t')P\tilde{x}(t') > 4\frac{\theta_b d_\theta(\rho,\rho_u)+\Delta d_\sigma(\rho,\rho_u)}{\Gamma}. \tag{2.195}$$

Moreover, it follows from (2.187) and (2.193) that for all $t\in[0,t_1)$

$$|\tilde{\theta}(t)| \leq 2\theta_b, \quad |\tilde{\sigma}(t)| \leq 2\Delta. \tag{2.196}$$

Since $\dot{\theta}(t)$ and $\dot{\sigma}(t)$ are continuous over $[0,t_1)$, the upper bounds in (2.191) and (2.196) lead to the following upper bound:

$$\frac{|\tilde{\theta}(t)\dot{\theta}(t)+\tilde{\sigma}(t)\dot{\sigma}(t)|}{\Gamma} \leq 2\frac{\theta_b d_\theta(\rho,\rho_u)+\Delta d_\sigma(\rho,\rho_u)}{\Gamma}. \tag{2.197}$$

If $V(t') > \frac{\theta_m(\rho,\rho_u)}{\Gamma}$, then from (2.192), (2.195), and (2.197) we have

$$\dot{V}(t') < 0 .$$

Thus, we have $V(t) \leq \frac{\theta_m(\rho,\rho_u)}{\Gamma}$ for all $t \in [0, t_1)$.

Since $\lambda_{\min}(P)\|\tilde{x}(t)\|^2 \leq \tilde{x}^\top(t)P\tilde{x}(t) \leq V(t)$, then from the continuity of $V(\cdot)$ we get the following upper bound for all $t \in [0, t_1]$:

$$\|\tilde{x}(t)\|_\infty \leq \|\tilde{x}(t)\| \leq \sqrt{\frac{\theta_m(\rho,\rho_u)}{\lambda_{\min}(P)\Gamma}} .$$

Let $t_2 \in (t_1, \tau]$ be the next time instant such that discontinuity of any of the derivatives $\dot{\theta}(t)$ and $\dot{\sigma}(t)$ occurs. Using similar derivations as above, we can prove that

$$\|\tilde{x}(t)\|_\infty \leq \sqrt{\frac{\theta_m(\rho,\rho_u)}{\lambda_{\min}(P)\Gamma}} , \quad t \in [t_1, t_2] .$$

Iterating the process until the time instant τ, we get

$$\|\tilde{x}_\tau\|_{\mathcal{L}_\infty} \leq \sqrt{\frac{\theta_m(\rho,\rho_u)}{\lambda_{\min}(P)\Gamma}} ,$$

which concludes the proof. □

Theorem 2.4.1 Consider the closed-loop reference system in (2.175) and the closed-loop system consisting of the system in (2.163) and the $\mathcal{L}_1$ adaptive controller in (2.172)–(2.174) subject to the $\mathcal{L}_1$-norm condition (2.165). If the adaptive gain is chosen to verify the design constraint

$$\Gamma \geq \frac{\theta_m(\rho,\rho_u)}{\lambda_{\min}(P)\gamma_0^2} , \tag{2.198}$$

then we have

$$\|\tilde{x}\|_{\mathcal{L}_\infty} \leq \gamma_0 , \tag{2.199}$$

$$\|x_{\text{ref}} - x\|_{\mathcal{L}_\infty} \leq \gamma_1 , \tag{2.200}$$

$$\|u_{\text{ref}} - u\|_{\mathcal{L}_\infty} \leq \gamma_2 , \tag{2.201}$$

where γ_1 and γ_2 are as defined in (2.167) and (2.170), respectively.

Proof. The proof is done by contradiction. Assume that the bounds (2.200) and (2.201) do not hold. Then, since $\|x_{\text{ref}}(0) - x(0)\|_\infty = 0 < \gamma_1$, $u_{\text{ref}}(0) - u(0) = 0$, and $x(t)$, $x_{\text{ref}}(t)$, $u(t)$, $u_{\text{ref}}(t)$ are continuous, there exists $\tau > 0$ such that

$$\|x_{\text{ref}}(t) - x(t)\|_\infty < \gamma_1 , \quad \|u_{\text{ref}}(t) - u(t)\|_\infty < \gamma_2, \quad \forall\, t \in [0, \tau)$$

and

$$\|x_{\text{ref}}(\tau) - x(\tau)\|_\infty = \gamma_1 , \quad \text{or} \quad \|u_{\text{ref}}(\tau) - u(\tau)\|_\infty = \gamma_2 ,$$

which implies that at least one of the following equalities holds:

$$\|(x_{\text{ref}} - x)_\tau\|_{\mathcal{L}_\infty} = \gamma_1, \quad \|(u_{\text{ref}} - u)_\tau\|_{\mathcal{L}_\infty} = \gamma_2. \tag{2.202}$$

Taking into consideration the definitions of ρ and ρ_u in (2.166) and (2.168), it follows from Lemma 2.4.1 and the equalities in (2.202) that

$$\|x_\tau\|_{\mathcal{L}_\infty} \leq \rho, \quad \|u_\tau\|_{\mathcal{L}_\infty} \leq \rho_u. \tag{2.203}$$

These bounds imply that the assumptions of Lemma 2.4.2 hold. Then, selecting the adaptive gain Γ according to the design constraint in (2.198), it follows that

$$\|\tilde{x}_\tau\|_{\mathcal{L}_\infty} \leq \gamma_0. \tag{2.204}$$

Let $\eta(t) \triangleq \theta(t)\|x(t)\|_\infty + \sigma(t)$. Then, it follows from (2.174) that

$$u(s) = -kD(s)(\omega u(s) + \eta(s) + \tilde{\eta}(s) - k_g r(s)),$$

which can be rewritten as

$$u(s) = -\frac{kD(s)}{1 + k\omega D(s)}(\eta(s) + \tilde{\eta}(s) - k_g r(s)) = -\frac{C(s)}{\omega}(\eta(s) + \tilde{\eta}(s) - k_g r(s)). \tag{2.205}$$

The response of the closed-loop system in the frequency domain consequently takes the form

$$x(s) = G(s)\eta(s) - H(s)C(s)\tilde{\eta}(s) + H(s)C(s)k_g r(s) + x_{\text{in}}(s).$$

This expression together with the response of the closed-loop reference system in (2.178) yields

$$x_{\text{ref}}(s) - x(s) = G(s)(\eta_{\text{ref}}(s) - \eta(s)) + H(s)C(s)\tilde{\eta}(s).$$

Moreover, the relationship in (2.188) leads to

$$x_{\text{ref}}(s) - x(s) = G(s)(\eta_{\text{ref}}(s) - \eta(s)) + C(s)\tilde{x}(s). \tag{2.206}$$

Since for all $t \in [0, \tau]$ the equalities

$$\eta_{\text{ref}}(t) - \eta(t) = f(t, x_{\text{ref}}(t)) - (\theta(t)\|x(t)\|_\infty + \sigma(t)) = f(t, x_{\text{ref}}(t)) - f(t, x(t))$$

hold, we have

$$\|(\eta_{\text{ref}} - \eta)_\tau\|_{\mathcal{L}_\infty} \leq \|(f(t, x_{\text{ref}}) - f(t, x))_\tau\|_{\mathcal{L}_\infty}.$$

Taking into account that $\|x(t)\|_\infty \leq \rho = \bar{\rho}_r(\rho_r)$, and also $\|x_{\text{ref}}(t)\|_\infty \leq \rho_r < \bar{\rho}_r(\rho_r)$ for all $t \in [0, \tau]$, Assumption 2.4.3 implies that for all $t \in [0, \tau]$,

$$|f(t, x_{\text{ref}}) - f(t, x)| \leq d_{f_x}(\bar{\rho}_r(\rho_r))\|x_{\text{ref}}(t) - x(t)\|_\infty.$$

From the redefinition in (2.164) it follows that $d_{f_x}(\bar{\rho}_r(\rho_r)) < L_{\rho_r}$, and hence

$$\|(\eta_{\text{ref}} - \eta)_\tau\|_{\mathcal{L}_\infty} \leq L_{\rho_r}\|(x_{\text{ref}} - x)_\tau\|_{\mathcal{L}_\infty}.$$

From this bound and the dynamics in (2.206), we get the following upper bound:

$$\|(x_{\text{ref}} - x)_\tau\|_{\mathcal{L}_\infty} \leq \|G(s)\|_{\mathcal{L}_1} L_{\rho_r} \|(x_{\text{ref}} - x)_\tau\|_{\mathcal{L}_\infty} + \|C(s)\|_{\mathcal{L}_1} \|\tilde{x}_\tau\|_{\mathcal{L}_\infty}.$$

Then, the upper bound in (2.204) and the $\mathcal{L}_1$-norm condition in (2.165) lead to the upper bound

$$\|(x_{\text{ref}} - x)_\tau\|_{\mathcal{L}_\infty} \leq \frac{\|C(s)\|_{\mathcal{L}_1}}{1 - \|G(s)\|_{\mathcal{L}_1} L_{\rho_r}} \gamma_0 = \gamma_1 - \beta < \gamma_1, \tag{2.207}$$

which contradicts the first equality in (2.202).

To show that the second equation in (2.202) also cannot hold, we notice that from (2.175) and (2.205) one can derive

$$u_{\text{ref}}(s) - u(s) = -\frac{C(s)}{\omega}(\eta_{\text{ref}}(s) - \eta(s)) + \frac{C(s)}{\omega}\tilde{\eta}(s).$$

Further, since Lemma A.12.1 implies

$$\frac{C(s)}{\omega}\tilde{\eta}(s) = \frac{1}{\omega} H_1(s)\tilde{x}(s),$$

it follows from Lemma 2.4.2 that

$$\|(u_{\text{ref}} - u)_\tau\|_{\mathcal{L}_\infty} \leq \left\|\frac{C(s)}{\omega}\right\|_{\mathcal{L}_1} L_{\rho_r} \|(x_{\text{ref}} - x)_\tau\|_{\mathcal{L}_\infty} + \left\|\frac{H_1(s)}{\omega}\right\|_{\mathcal{L}_1} \|\tilde{x}_\tau\|_{\mathcal{L}_\infty}.$$

The upper bounds in (2.204) and (2.207) and the definition of γ_2 in (2.170) lead to

$$\|(u_{\text{ref}} - u)_\tau\|_{\mathcal{L}_\infty} \leq \left\|\frac{C(s)}{\omega}\right\|_{\mathcal{L}_1} L_{\rho_r}(\gamma_1 - \beta) + \left\|\frac{H_1(s)}{\omega}\right\|_{\mathcal{L}_1} \gamma_0 < \gamma_2,$$

which contradicts the second equality in (2.202). This proves the bounds in (2.200)–(2.201). Thus, the bounds in (2.203) hold uniformly, which implies that the bound in (2.204) also holds uniformly. This proves the bound in (2.199). □

Remark 2.4.3 It follows from (2.198) that one can prescribe the arbitrary desired performance bound γ_0 by increasing the adaptive gain, which further implies from (2.167) and (2.170) that one can achieve arbitrarily small performance bounds γ_1 and γ_2 simultaneously.

Remark 2.4.4 Similar to the previous section, notice that letting $k \to \infty$ leads to $C(s) \to 1$, and thus the reference controller in the definition of the closed-loop reference system in (2.140) leads, in the limit, to perfect cancelation of uncertainties and recovers the performance of the ideal desired system. As before, one can check that setting $C(s) = 1$ takes away the uniform bound for the control signal, as the resulting $H_1(s)$ is an improper system.

Remark 2.4.5 Notice that the use of the parametrization $f(t, x(t)) = \theta(t)\|x(t)\|_\infty + \sigma(t)$ from Lemma A.8.1 led to the definition of a semiglobal positively invariant set, where the solutions lie. The obtained uniform performance bounds, in particular, question the need for neural-network-based approximation schemes for adaptive control, where the uncertainties are limited to state-dependent nonlinearities and the approximation properties are of existence nature, the convergence domains are local, and the obtained results are on ultimate boundedness.

2.4.4 Simulation Example: Wing Rock

In this section, we explore application of the $\mathcal{L}_1$ adaptive controller to wing rock, which can be described by a second-order nonlinear system. Wing rock is the limit cycle behavior in flight dynamics, caused by flow asymmetries in the result of nonlinear aerodynamic roll damping. It has been an active topic of research in the aerospace community over the past three decades (see [26] and the references therein). We test the performance of the $\mathcal{L}_1$ adaptive controller for various flight conditions. We also verify robustness of the closed-loop adaptive system to time delays. More details can be found in [26, 89].

Problem Formulation and Application of the $\mathcal{L}_1$ Adaptive Controller

An analytical model of wing rock for slender delta wings is given by [7, 65]

$$\begin{aligned}\ddot{\phi}(t) + \frac{a_0}{t_r^2}\phi(t) + \frac{a_1}{t_r}\dot{\phi}(t) + a_2|\dot{\phi}(t)|\dot{\phi}(t) + \frac{a_3}{t_r^2}\phi^3(t) + \frac{a_4}{t_r}\phi^2(t)\dot{\phi}(t) \\ + \frac{\omega}{t_r^2}u(t) + d(t,\phi(t),\dot{\phi}(t)) = 0,\end{aligned} \tag{2.208}$$

where $\phi(t) \in \mathbb{R}$ is the roll angle, $d(t,\phi(t),\dot{\phi}(t)) \in \mathbb{R}$ models disturbances and unknown nonlinearities, $u(t) \in \mathbb{R}$ is the antisymmetric aileron deflection, $\omega \in \mathbb{R}$ is the unknown control effectiveness, and t_r is the reference time conversion coefficient. Both the roll angle $\phi(t)$ and its derivative $\dot{\phi}(t)$ are assumed to be available for feedback. In this model, $\omega = 1$ corresponds to the nominal control efficiency. The aerodynamic coefficients a_0, a_1, a_2, a_3, a_4 and the control efficiency ω depend upon the angle of attack and are given in Table 2.3 for two fixed values of angle of attack α [65]. Since the control efficiency is not precisely known and depends upon the flight condition, it is treated as an unknown parameter in the design of the $\mathcal{L}_1$ adaptive controller. For the airspeed $V_f = 30$ m/s and the wingspan $b_w = 169$ mm, we have $t_r = b_w/(2V_f) = 0.0028$ s.

Table 2.3: Coefficients for wing rock motion.

α	a_0	a_1	a_2	a_3	a_4	ω
27.0 deg	0.0050	−0.0100	0.2000	−0.0025	0.0250	0.9000
35.0 deg	0.0060	−0.0120	0.2000	−0.0075	0.0400	1.2000

Letting $x \triangleq [\phi,\ \dot{\phi}]^\top$, the system in (2.208) can be rewritten in state-space form as

$$\begin{aligned}\dot{x}(t) &= Ax(t) + b(\omega u(t) + g_0(t,x(t))), \quad x(0) = x_0 \\ \phi(t) &= c^\top x(t),\end{aligned} \tag{2.209}$$

where

$$A \triangleq \begin{bmatrix} 0 & 1 \\ -a_0/t_r^2 & -a_1/t_r \end{bmatrix}, \quad b \triangleq \begin{bmatrix} 0 \\ 1/t_r^2 \end{bmatrix}, \quad c \triangleq \begin{bmatrix} 1 \\ 0 \end{bmatrix},$$
$$g_0(t,x) \triangleq -t_r^2 a_2|\dot{\phi}(t)|\dot{\phi}(t) - a_3\phi^3(t) - t_r a_4\phi^2\dot{\phi}(t) - t_r^2 d(t,\phi(t),\dot{\phi}(t)).$$

For this problem, we consider the following control law:

$$u(t) = u_m(t) + u_{\text{ad}}(t),$$
$$u_m(t) = -k_m^\top x(t),$$

where $k_m \in \mathbb{R}^2$ is the static feedback gain and $u_{\text{ad}}(t)$ is the adaptive control signal. Substituting this control law in (2.209), we obtain the following partially closed-loop dynamics:

$$\dot{x}(t) = \left(A - bk_m^\top\right)x(t) + b\left(\omega u_{\text{ad}}(t) + (1-\omega)k_m^\top x(t) + g_0(t, x(t))\right). \tag{2.210}$$

Further, denoting

$$A_m \triangleq \left(A - bk_m^\top\right), \quad g(t,x) \triangleq (1-\omega)k_m^\top x + g_0(t,x), \tag{2.211}$$

we can rewrite (2.210) as follows:

$$\begin{aligned} \dot{x}(t) &= A_m x(t) + b\left(\omega u_{\text{ad}}(t) + g(t, x(t))\right), \quad x(0) = x_0, \\ \phi(t) &= c^\top x(t). \end{aligned} \tag{2.212}$$

We select the static feedback gain k_m so that the state matrix A_m is Hurwitz and has its poles in desired locations. Then, letting

$$A_m \triangleq \begin{bmatrix} 0 & 1 \\ -a_{m1} & -a_{m2} \end{bmatrix},$$

it follows from (2.211) that

$$k_m = \begin{bmatrix} \frac{t_r^2 a_{m1} - a_0}{\omega} \\ \frac{t_r^2 a_{m2} - t_r a_1}{\omega} \end{bmatrix},$$

where a_{m1}, a_{m2} are the design parameters specifying the desired closed-loop dynamics. Notice that the dynamics in (2.212) can be cast into the form given in Section 2.4.1. Thus, we can define the adaptive control $u_{\text{ad}}(t)$ using (2.172), (2.173), and (2.174), subject to the $\mathcal{L}_1$-norm condition in (2.165).

Simulation Results

The phase portrait of the open-loop system for $d(t,\phi,\dot{\phi}) \equiv 0$ and for the angle of attack $\alpha = 27.0$ deg is given in Figure 2.36. One can see that the system's state trajectories contain a stable limit cycle. All trajectories starting inside the limit cycle converge to the limit cycle, while most of the trajectories outside the limit cycle are unstable.

Letting $a_{m1} = 50$, $a_{m2} = 14.14$ leads to $k_m = [-0.0046,\ 0.0001]$. The desired system dynamics are given by

$$\dot{x}_{\text{id}}(t) = A_m x_{\text{id}}(t) + bk_g r(t),$$
$$\phi_{\text{id}}(t) = c^\top x_{\text{id}}(t).$$

Further, let $k = 144$, and let the filter $D(s)$ be given by

$$D(s) = \frac{(s+500)(s+0.00400)^2}{s(s+368)(s+0.00439)^2}.$$

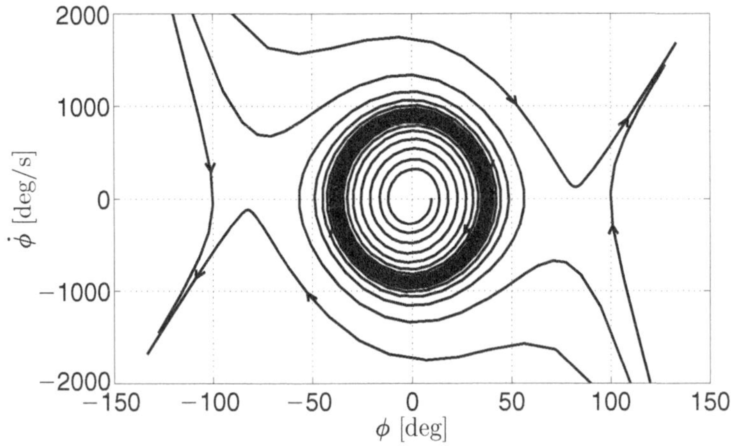

Figure 2.36: Phase portrait for free wing rock motion.

Also, let the projection bounds be $\Omega = [0.3,\ 2]$, $\Theta = [-10,\ 10]$, $\Delta = 10$, and let the adaptive gain be $\Gamma = 10^7$.

We consider two simulation scenarios for different values of angle of attack: $\alpha_1 = 27.0$ deg and $\alpha_2 = 35.0$ deg. In both scenarios we inject the disturbance

$$d(t,\phi,\dot{\phi}) = 10\sin(2\pi t) + 20\sin(\pi t)$$

into the plant dynamics and check the response of the closed-loop adaptive system to step reference signals of different amplitude.

The simulation results for the first scenario are given in Figures 2.37 and 2.38(a) and for the second scenario in Figures 2.39 and 2.38(b). In both cases we use the same controller without any retuning. Notice that the response of the closed-loop system is close to the desired system, which is scaled for different reference inputs. Moreover, the $\mathcal{L}_1$ adaptive controller is able to control the highly nonlinear system while working in different state-space regions with different stability properties (see Figure 2.38).

In Figures 2.37(b) and 2.39(b), we also plot the control history of the same system with $d(t,x,\dot{x}) \equiv 0$, along with the control signal of the closed-loop adaptive system. Comparison of the control signals shows that the control system with the adaptive controller generates adequate signals to compensate for disturbances, while in the absence of disturbances it produces clean control signal without oscillatory components.

In Section 2.2.5, we demonstrated that for LTI systems the $\mathcal{L}_1$ adaptive controller has a bounded-away-from-zero time-delay margin in the presence of fast adaptation. In the following simulations, we numerically verify a similar claim for the nonlinear wing rock example, using the insights from Section 2.2.5. Assume that the control input in (2.208) is replaced with the delayed signal $u_d(t)$, defined as

$$u_d(t) = \begin{cases} 0, & 0 \le t \le \tau, \\ u(t-\tau), & t > \tau, \end{cases}$$

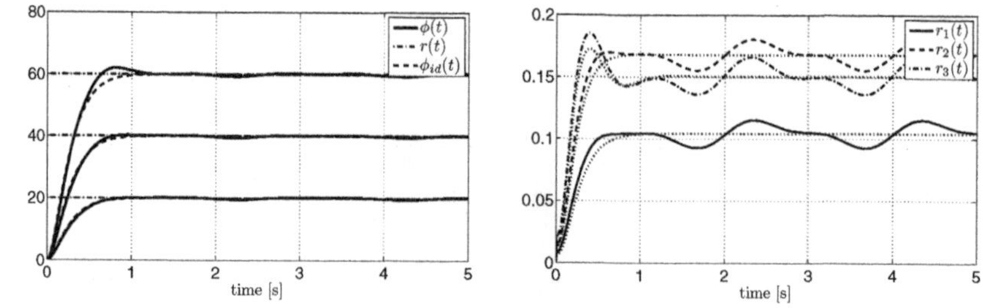

(a) Control system response $\phi(t)$, desired response $\phi_{\rm id}(t)$, and the reference command

(b) Control history $u(t)$ in the presence of uncertainties and without retuning (dotted)

Figure 2.37: Simulation results for $\alpha = 27.5$ deg.

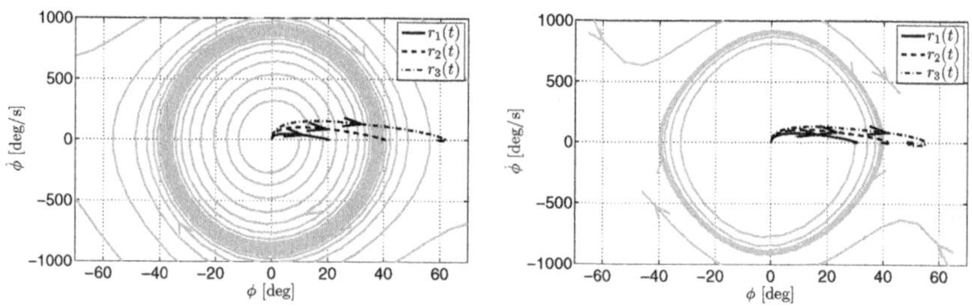

(a) $\alpha\ =\ 27.0$ deg

(b) $\alpha\ =\ 35.0$ deg

Figure 2.38: Phase portraits for both simulation scenarios.

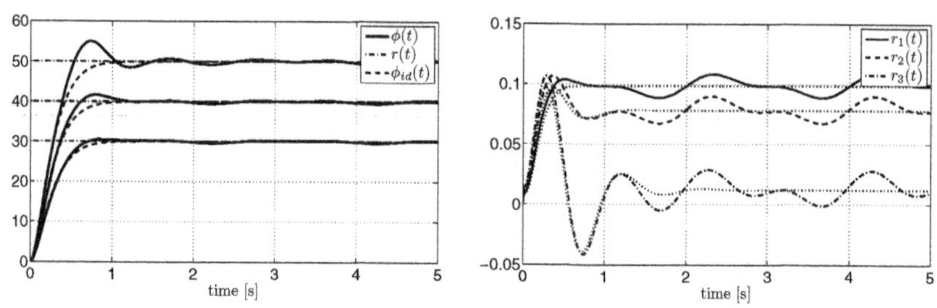

(a) Control system response $\phi(t)$, desired response $\phi_{\rm id}(t)$, and the reference command

(b) Control history $u(t)$ in the presence of uncertainties and without retuning (dotted)

Figure 2.39: Simulation results for $\alpha = 35.0$ deg.

where τ is the time delay. Figure 2.40 demonstrates the tracking performance for the first scenario without any retuning of the original $\mathcal{L}_1$ controller for $\tau = 5$ ms. Figure 2.41 demonstrates the simulation results for the second scenario with the same adaptive controller. We notice that the closed-loop system does not lose its stability in the presence of time delay. Also notice that the closed-loop performance in the presence of the time delay for both scenarios is almost the same as in the systems without time delay (Figures 2.37

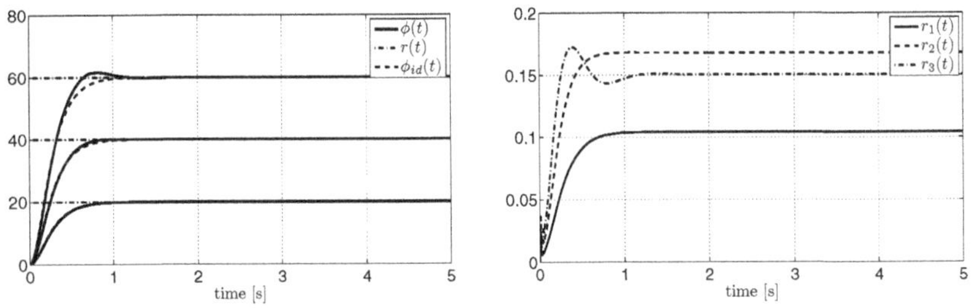

(a) Control system response $\phi(t)$, ideal response $\phi_{\text{id}}(t)$, and the reference command

(b) Control history $u(t)$

Figure 2.40: Time histories for $\alpha = 27.5$ deg in the presence of time delay of 5 ms.

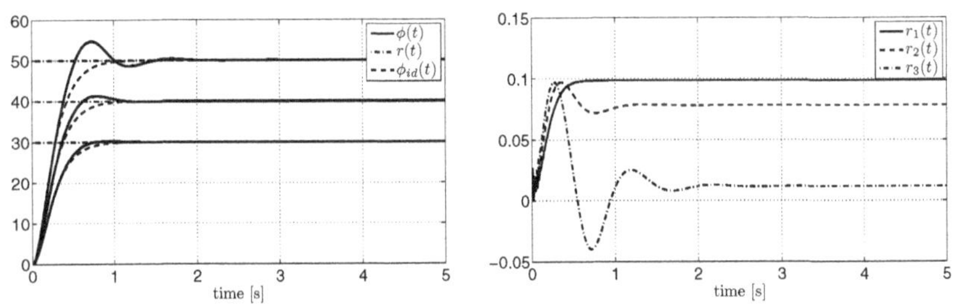

(a) Control system response $\phi(t)$, ideal response $\phi_{\text{id}}(t)$, and the reference command

(b) Control history $u(t)$

Figure 2.41: Time histories for $\alpha = 35.0$ deg in the presence of time delay of 5 ms.

and 2.39).

2.5 $\mathcal{L}_1$ Adaptive Controller in the Presence of Nonlinear Unmodeled Dynamics

In this section we integrate the tools from previous sections to expand the class of nonlinear systems, for which the $\mathcal{L}_1$ controller can be designed and analyzed with similar claims. We consider unmodeled actuator dynamics and also unmodeled internal dynamics in a nonlinear system with time- and state-dependent nonlinearities. We prove that with appropriate redefinition of the reference system the performance bounds from previous sections hold semiglobally [31].

2.5.1 Problem Formulation

In this section we present the $\mathcal{L}_1$ adaptive controller for nonlinear systems in the presence of unmodeled dynamics. Thus, consider the following class of systems:

$$\begin{aligned}
\dot{x}(t) &= A_m x(t) + b\big(\mu(t) + f(t, x(t), z(t))\big), \quad x(0) = x_0, \\
\dot{x}_z(t) &= g(t, x_z(t), x(t)), \quad x_z(0) = x_{z0}, \\
z(t) &= g_0(t, x_z(t)), \\
y(t) &= c^\top x(t),
\end{aligned} \tag{2.213}$$

where $x(t) \in \mathbb{R}^n$ is the system state (measured); $A_m \in \mathbb{R}^{n\times n}$ is a known Hurwitz matrix specifying the desired closed-loop dynamics; $b, c \in \mathbb{R}^n$ are known constant vectors; $y(t) \in \mathbb{R}$ is the system output; $f : \mathbb{R} \times \mathbb{R}^n \times \mathbb{R}^l \to \mathbb{R}$ is an unknown nonlinear map, which represents unknown system nonlinearities; $x_z(t) \in \mathbb{R}^m$ and $z(t) \in \mathbb{R}^l$ are the state and the output of unmodeled nonlinear dynamics; $g : \mathbb{R} \times \mathbb{R}^m \times \mathbb{R}^n \to \mathbb{R}^m$ and $g_0 : \mathbb{R} \times \mathbb{R}^m \to \mathbb{R}^l$ are unknown nonlinear maps continuous in their arguments; and $\mu(t) \in \mathbb{R}$ is the output of the following system:

$$\mu(s) = F(s)u(s),$$

where $u(t) \in \mathbb{R}$ is the control input and $F(s)$ is an unknown BIBO-stable and proper transfer function with known sign of its DC gain. The initial condition x_0 is assumed to be inside an arbitrarily large known set, i.e., $\|x_0\|_\infty \leq \rho_0 < \infty$ with known $\rho_0 > 0$. Let $X \triangleq [x^\top, z^\top]^\top$, and with a slight abuse of language let $f(t, X) \triangleq f(t, x, z)$.

Assumption 2.5.1 (Uniform boundedness of $f(t,0)$) There exists $B > 0$, such that $|f(t,0)| \leq B$ holds for all $t \geq 0$.

Assumption 2.5.2 (Semiglobal uniform boundedness of partial derivatives) For arbitrary $\delta > 0$ there exist positive constants $d_{f_x}(\delta) > 0$ and $d_{f_t}(\delta) > 0$ independent of time, such that for all $\|X\|_\infty \leq \delta$ the partial derivatives of $f(t, X)$ are piecewise-continuous and bounded,

$$\left\| \frac{\partial f(t,X)}{\partial X} \right\|_1 \leq d_{f_x}(\delta), \quad \left| \frac{\partial f(t,X)}{\partial t} \right| \leq d_{f_t}(\delta).$$

Assumption 2.5.3 (Stability of unmodeled dynamics) The x_z-dynamics are BIBO stable both with respect to initial conditions x_{z0} and input $x(t)$, i.e., there exist $L_1, L_2 > 0$ such that for all $t \geq 0$

$$\|z_t\|_{\mathcal{L}_\infty} \leq L_1 \|x_t\|_{\mathcal{L}_\infty} + L_2.$$

Assumption 2.5.4 (Partial knowledge of actuator dynamics) There exists $L_F > 0$, verifying $\|F(s)\|_{\mathcal{L}_1} \leq L_F$. Also, we assume that there exist known constants $\omega_l, \omega_u \in \mathbb{R}$, satisfying

$$0 < \omega_l \leq F(0) \leq \omega_u,$$

where, without loss of generality, we have assumed $F(0) > 0$. Finally, we assume (for design purposes) that we know a set $\mathbb{F}_\Delta$ of all admissible actuator dynamics.

The control objective is to design a full-state feedback adaptive controller to ensure that the output of the system $y(t)$ tracks a given bounded piecewise-continuous reference signal $r(t)$ with uniform and quantifiable performance bounds.

2.5.2 $\mathcal{L}_1$ Adaptive Control Architecture

Definitions and $\mathcal{L}_1$-Norm Sufficient Condition for Stability

Similar to Section 2.2.2, the design of the $\mathcal{L}_1$ adaptive controller proceeds by considering a feedback gain $k > 0$ and a strictly proper stable transfer function $D(s)$, which imply that

$$C(s) \triangleq \frac{kF(s)D(s)}{1+kF(s)D(s)} \tag{2.214}$$

is a strictly proper stable transfer function with DC gain $C(0) = 1$, for all $F(s) \in \mathbb{F}_\Delta$. Similar to the previous sections, let $x_{\text{in}}(s) \triangleq (s\mathbb{I} - A_m)^{-1}x_0$. Notice that from the fact that A_m is Hurwitz it follows that $\|x_{\text{in}}\|_{\mathcal{L}_\infty} \leq \rho_{\text{in}}$, where $\rho_{\text{in}} \triangleq \|s(s\mathbb{I} - A_m)^{-1}\|_{\mathcal{L}_1}\rho_0$. Further, let

$$L_\delta \triangleq \frac{\bar{\delta}(\delta)}{\delta} d_{f_x}(\bar{\delta}(\delta)), \quad \bar{\delta}(\delta) \triangleq \max\{\delta + \bar{\gamma}_1, L_1(\delta + \bar{\gamma}_1) + L_2\}, \tag{2.215}$$

where $d_{f_x}(\cdot)$ was introduced in Assumption 2.5.2, and $\bar{\gamma}_1 > 0$ is an arbitrary positive constant.

For the proofs of stability and performance bounds, the choice of k and $D(s)$ needs to ensure that for a given ρ_0, there exists $\rho_r > \rho_{\text{in}}$, such that the following $\mathcal{L}_1$-norm condition can be verified:

$$\|G(s)\|_{\mathcal{L}_1} < \frac{\rho_r - \|k_g C(s)H(s)\|_{\mathcal{L}_1}\|r\|_{\mathcal{L}_\infty} - \rho_{\text{in}}}{L_{\rho_r}\rho_r + B}, \tag{2.216}$$

where $G(s) \triangleq H(s)(1 - C(s))$, $H(s) \triangleq (s\mathbb{I} - A_m)^{-1}b$, and $k_g \triangleq -1/(c^\top A_m^{-1}b)$ is the feedforward gain required for tracking of step-reference command $r(t)$ with zero steady-state error.

To streamline the subsequent analysis of stability and performance bounds, we need to introduce some notation. Let $C_u(s) \triangleq C(s)/F(s)$. Notice that from the definition of $C(s)$ given in (2.214) and the fact that $D(s)$ is strictly proper and stable, while $F(s)$ is proper and stable, it follows that $C_u(s)$ is a strictly proper and stable transfer function. Further, similar to Section 2.3, we define

$$H_1(s) \triangleq C_u(s)\frac{1}{c_o^\top H(s)} c_o^\top, \tag{2.217}$$

where $c_o \in \mathbb{R}^n$ is a vector that renders $H_1(s)$ BIBO stable and proper. We refer to Lemma A.12.1 for the existence of such c_o.

We also define

$$\rho \triangleq \rho_r + \bar{\gamma}_1,$$

where

$$\gamma_1 \triangleq \frac{\|C(s)\|_{\mathcal{L}_1}}{1 - \|G(s)\|_{\mathcal{L}_1}L_{\rho_r}}\gamma_0 + \beta, \tag{2.218}$$

with β and γ_0 being arbitrary small positive constants, such that $\gamma_1 \leq \bar{\gamma}_1$. Moreover, let

$$\rho_u \triangleq \rho_{u_r} + \gamma_2,$$

where ρ_{ur} and γ_2 are defined as

$$\rho_{ur} \triangleq \|C_u(s)\|_{\mathcal{L}_1}(|k_g|\|r\|_{\mathcal{L}_\infty} + L_{\rho_r}\rho_r + B),$$
$$\gamma_2 \triangleq \|C_u(s)\|_{\mathcal{L}_1} L_{\rho_r}\gamma_1 + \|H_1(s)\|_{\mathcal{L}_1}\gamma_0. \tag{2.219}$$

Finally, using the conservative knowledge of $F(s)$, let

$$\Delta_1 \triangleq L_\rho L_2 + B + \epsilon, \tag{2.220}$$
$$\Delta_2 \triangleq \max_{F(s)\in\mathbb{F}_\Delta} \|F(s) - (\omega_l + \omega_u)/2\|_{\mathcal{L}_1}\rho_u, \tag{2.221}$$
$$\Delta \triangleq \Delta_1 + \Delta_2, \tag{2.222}$$
$$\rho_{\dot{u}} \triangleq \|ksD(s)\|_{\mathcal{L}_1}(\rho_u\omega_u + L_\rho\rho + \sigma_b + |k_g|\|r\|_{\mathcal{L}_\infty}),$$

where ϵ is an arbitrary positive constant.

Remark 2.5.1 In the following analysis we demonstrate that ρ_r and ρ characterize the positively invariant sets for the state of the closed-loop reference system (yet to be defined) and the state of the closed-loop adaptive system, respectively. We notice that, since $\bar{\gamma}_1$ can be set arbitrarily small, ρ can approximate ρ_r arbitrarily closely.

The elements of the $\mathcal{L}_1$ controller are introduced next.

State Predictor

We consider the following state predictor:

$$\dot{\hat{x}}(t) = A_m\hat{x}(t) + b(\hat{\omega}(t)u(t) + \hat{\theta}(t)\|x_t\|_{\mathcal{L}_\infty} + \hat{\sigma}(t)), \quad \hat{x}(0) = x_0,$$
$$\hat{y}(t) = c^\top\hat{x}(t), \tag{2.223}$$

where $\hat{x}(t) \in \mathbb{R}^n$ is the state of the predictor, while $\hat{\omega}(t), \hat{\theta}(t), \hat{\sigma}(t) \in \mathbb{R}$ are the adaptive estimates.

Adaptation Laws

The adaptive laws are defined via the projection operator as follows:

$$\dot{\hat{\omega}}(t) = \Gamma\text{Proj}(\hat{\omega}(t), -\tilde{x}^\top(t)Pbu(t)), \quad \hat{\omega}(0) = \hat{\omega}_0,$$
$$\dot{\hat{\theta}}(t) = \Gamma\text{Proj}(\hat{\theta}(t), -\tilde{x}^\top(t)Pb\|x_t\|_{\mathcal{L}_\infty}), \quad \hat{\theta}(0) = \hat{\theta}_0, \tag{2.224}$$
$$\dot{\hat{\sigma}}(t) = \Gamma\text{Proj}(\hat{\sigma}(t), -\tilde{x}^\top(t)Pb), \quad \hat{\sigma}(0) = \hat{\sigma}_0,$$

where $\tilde{x}(t) \triangleq \hat{x}(t) - x(t)$, $\Gamma \in \mathbb{R}^+$ is the adaptation gain, and $P = P^\top > 0$ is the solution of the algebraic Lyapunov equation $A_m^\top P + PA_m = -Q$ for arbitrary symmetric $Q = Q^\top > 0$. The projection operator ensures that $\hat{\omega}(t) \in \Omega \triangleq [\omega_l, \omega_u]$, $\hat{\theta}(t) \in \Theta = [-L_\rho, L_\rho]$, and also $|\hat{\sigma}(t)| \leq \Delta$.

Control Law

The control signal is generated as the output of the following (feedback) system:

$$u(s) = -kD(s)(\hat{\eta}(s) - k_g r(s)), \tag{2.225}$$

where $\hat{\eta}(t) \triangleq \hat{\omega}(t)u(t) + \hat{\theta}(t)\|x_t\|_{\mathcal{L}_\infty} + \hat{\sigma}(t)$, while k and $D(s)$ are as introduced in (2.214).

The $\mathcal{L}_1$ adaptive controller is defined via (2.223), (2.224), and (2.225), subject to the $\mathcal{L}_1$-norm condition in (2.216).

2.5.3 Analysis of the $\mathcal{L}_1$ Adaptive Controller

Closed-Loop Reference System

Consider the following closed-loop reference system:

$$\begin{aligned}
\dot{x}_{\text{ref}}(t) &= A_m x_{\text{ref}}(t) + b\big(\mu_{\text{ref}}(t) + f(t, x_{\text{ref}}(t), z(t))\big), \quad x_{\text{ref}}(0) = x_0, \\
\mu_{\text{ref}}(s) &= F(s) u_{\text{ref}}(s), \\
u_{\text{ref}}(s) &= C_u(s)(k_g r(s) - \eta_{\text{ref}}(s)), \\
y_{\text{ref}}(t) &= c^\top x_{\text{ref}}(t),
\end{aligned} \tag{2.226}$$

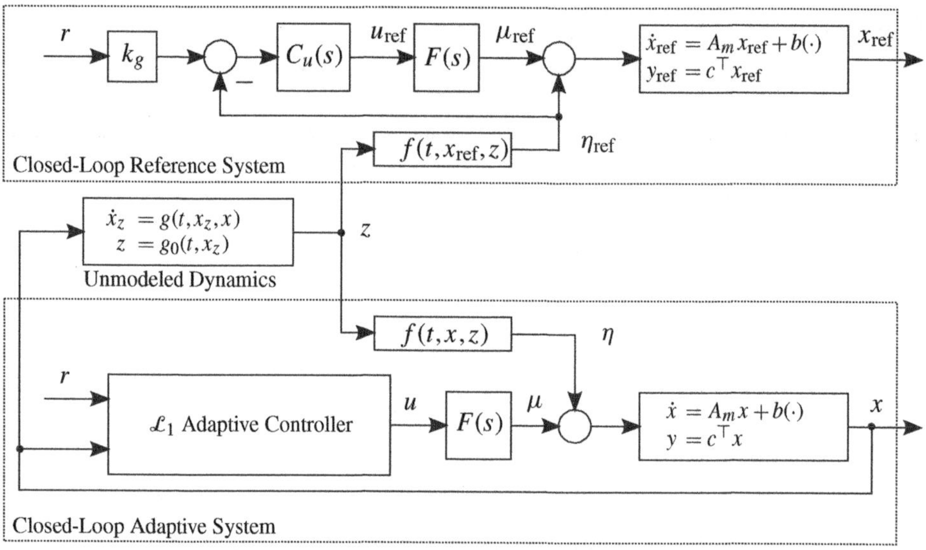

Figure 2.42: Closed-loop adaptive system and closed-loop reference system.

where $x_{\text{ref}}(t) \in \mathbb{R}^n$ is the state of the reference system, $u_{\text{ref}}(t) \in \mathbb{R}$ is the control input, and $\eta_{\text{ref}}(s)$ denotes the Laplace transform of $\eta_{\text{ref}}(t) \triangleq f(t, x_{\text{ref}}(t), z(t))$. Note that $z(t)$ is seen as an external disturbance to the closed-loop reference system. The block diagram of the closed-loop system with the $\mathcal{L}_1$ adaptive controller and the closed-loop reference system is illustrated in Figure 2.42. The next lemma proves the stability of this closed-loop reference system subject to an additional assumption on $z(t)$, which will later be verified in the proof of stability and performance bounds of the closed-loop adaptive system (Theorem 2.5.1).

Lemma 2.5.1 For the closed-loop reference system given in (2.226), subject to the $\mathcal{L}_1$-norm condition in (2.216), if for some $\tau \geq 0$

$$\|z_\tau\|_{\mathcal{L}_\infty} \leq L_1(\|x_{\text{ref}\tau}\|_{\mathcal{L}_\infty} + \gamma_1) + L_2, \tag{2.227}$$

then the following bounds hold:

$$\|x_{\text{ref}\tau}\|_{\mathcal{L}_\infty} < \rho_r, \quad \|u_{\text{ref}\tau}\|_{\mathcal{L}_\infty} < \rho_{ur}. \tag{2.228}$$

Proof. The closed-loop reference system in (2.226) can be rewritten as

$$x_{\text{ref}}(s) = G(s)\eta_{\text{ref}}(s) + H(s)C(s)k_g r(s) + x_{\text{in}}(s). \tag{2.229}$$

Lemmas A.6.2 and A.7.5 imply

$$\|x_{\text{ref}\tau}\|_{\mathcal{L}_\infty} \leq \|G(s)\|_{\mathcal{L}_1}\|\eta_{\text{ref}\tau}\|_{\mathcal{L}_\infty} + \|H(s)C(s)k_g\|_{\mathcal{L}_1}\|r\|_{\mathcal{L}_\infty} + \|x_{\text{in}}\|_{\mathcal{L}_\infty}. \tag{2.230}$$

Next, we use a contradictive argument to show stability of the closed-loop reference system. Assume that (2.228) does not hold. Then, since $\|x_{\text{ref}}(0)\|_\infty = \|x_0\|_\infty \leq \rho_0 < \rho_r$ and $x_{\text{ref}}(t)$ is continuous, there exists a time instant $\tau_1 \in (0, \tau]$, such that

$$\|x_{\text{ref}}(t)\|_\infty < \rho_r, \quad \forall\, t \in [0, \tau_1),$$

and

$$\|x_{\text{ref}}(\tau_1)\|_\infty = \rho_r\,,$$

which implies that

$$\|x_{\text{ref}\tau_1}\|_{\mathcal{L}_\infty} = \rho_r\,. \tag{2.231}$$

The bound in (2.227) implies that $\|z_{\tau_1}\|_{\mathcal{L}_\infty} \leq L_1(\rho_r+\gamma_1)+L_2$, which leads to

$$\|X_{\text{ref}\tau_1}\|_{\mathcal{L}_\infty} \leq \max\{\rho_r+\gamma_1,\, L_1(\rho_r+\gamma_1)+L_2\} \leq \bar{\rho}_r(\rho_r),$$

where we have used the definition of $\bar{\delta}(\delta)$ in (2.215). Assumption 2.5.2 implies that, for all $\|X_{\text{ref}}\|_\infty \leq \bar{\rho}_r(\rho_r)$, we have

$$|f(t,X_{\text{ref}})-f(t,0)| \leq d_{f_x}(\bar{\rho}_r(\rho_r))\|X_{\text{ref}}\|_\infty\,, \quad \forall\, t\in[0,\,\tau]\,.$$

Further, Assumption 2.5.1 and the definition of L_δ in (2.215) lead to

$$\|\eta_{\text{ref}\tau_1}\|_{\mathcal{L}_\infty} \leq d_{f_x}(\bar{\rho}_r(\rho_r))\bar{\rho}_r(\rho_r)+B = L_{\rho_r}\rho_r+B\,. \tag{2.232}$$

Thus, the bound in (2.230), together with the fact that $\|x_{\text{in}}\|_{\mathcal{L}_\infty} \leq \rho_{\text{in}}$, yields

$$\|x_{\text{ref}\tau_1}\|_{\mathcal{L}_\infty} \leq \|G(s)\|_{\mathcal{L}_1}(L_{\rho_r}\rho_r+B)+\|H(s)C(s)k_g\|_{\mathcal{L}_1}\|r\|_{\mathcal{L}_\infty}+\rho_{\text{in}}\,.$$

The condition in (2.216) can be solved for ρ_r to obtain the upper bound

$$\|G(s)\|_{\mathcal{L}_1}(L_{\rho_r}\rho_r+B)+\|k_gC(s)H(s)\|_{\mathcal{L}_1}\|r\|_{\mathcal{L}_\infty}+\rho_{\text{in}} < \rho_r\,,$$

which implies $\|x_{\text{ref}\tau_1}\|_{\mathcal{L}_\infty} < \rho_r$. This contradicts (2.231), which proves the first result in (2.228). As this bound is strict and holds uniformly for all $\tau_1 \in (0,\,\tau]$, one can rewrite the bound in (2.232) as a strict inequality,

$$\|\eta_{\text{ref}\tau}\|_{\mathcal{L}_\infty} < L_{\rho_r}\rho_r+B\,.$$

From the definition of the reference control signal in (2.226), it follows that

$$\begin{aligned}\|u_{\text{ref}\tau}\|_{\mathcal{L}_\infty} &\leq \|C_u(s)\|_{\mathcal{L}_1}(|k_g|\|r\|_{\mathcal{L}_\infty}+\|\eta_{\text{ref}\tau}\|_{\mathcal{L}_\infty})\\ &< \|C_u(s)\|_{\mathcal{L}_1}(|k_g|\|r\|_{\mathcal{L}_\infty}+L_{\rho_r}\rho_r+B) = \rho_{ur}\,,\end{aligned}$$

which completes the proof. □

Equivalent (Semi-)Linear Time-Varying System

In this section we transform the original nonlinear system with unmodeled dynamics in (2.213) into an equivalent (semi-)linear time-varying system with unknown time-varying parameters and disturbances. This transformation requires us to impose the following assumptions on the signals of the system: the control signal $u(t)$ is continuous, and moreover the following bounds hold

$$\|x_\tau\|_{\mathcal{L}_\infty} \leq \rho\,, \quad \|u_\tau\|_{\mathcal{L}_\infty} \leq \rho_u\,, \quad \|\dot{u}_\tau\|_{\mathcal{L}_\infty} \leq \rho_{\dot{u}}\,. \tag{2.233}$$

These assumptions will be verified later in the proof of Theorem 2.5.1. Next we construct the equivalent system in two steps.

First Equivalent System From the system dynamics in (2.213) and the first two bounds in (2.233), it follows that $\|\dot{x}_t\|_{\mathcal{L}_\infty}$ is bounded for all $t \in [0,\ \tau]$. Thus, Lemma A.9.1 implies that there exist continuous $\theta(t)$ and $\sigma_1(t)$ with (piecewise)-continuous derivatives, defined for $t \in [0,\ \tau]$, such that

$$\begin{aligned} |\theta(t)| &< L_\rho\,, & |\dot{\theta}(t)| &\leq d_\theta\,, \\ |\sigma_1(t)| &< \Delta_1\,, & |\dot{\sigma}_1(t)| &\leq d_{\sigma_1}\,, \end{aligned} \tag{2.234}$$

and

$$f(t, x(t), z(t)) = \theta(t)\|x_t\|_{\mathcal{L}_\infty} + \sigma_1(t)\,,$$

where L_ρ and Δ_1 are defined in (2.215) and (2.220), while the algorithm for computation of $d_\theta > 0$, $d_{\sigma_1} > 0$ is derived in the proof of Lemma A.9.1. Thus, the system in (2.213) can be rewritten over $t \in [0,\ \tau]$ as

$$\begin{aligned} \dot{x}(t) &= A_m x(t) + b\big(\mu(t) + \theta(t)\|x_t\|_{\mathcal{L}_\infty} + \sigma_1(t)\big)\,, \quad x(0) = x_0\,, \\ y(t) &= c^\top x(t)\,. \end{aligned} \tag{2.235}$$

Second Equivalent System Taking into account the assumption on $u(t)$ and its derivative in (2.233), and using Lemma A.10.1, we can rewrite the signal $\mu(t)$ as

$$\mu(t) = \omega u(t) + \sigma_2(t)\,,$$

where $\omega \in (\omega_l,\ \omega_u)$ is an unknown constant and $\sigma_2(t)$ is a continuous signal with (piecewise)-continuous derivative, defined over $t \in [0,\ \tau]$, such that

$$|\sigma_2(t)| \leq \Delta_2\,, \quad |\dot{\sigma}_2(t)| \leq d_{\sigma_2}\,,$$

with Δ_2 as introduced in (2.221) and $d_{\sigma_2} \triangleq \|F(s) - (\omega_l + \omega_u)/2\|_{\mathcal{L}_1} \rho_{\dot{u}}$. This implies that one can rewrite the system in (2.235) over $t \in [0,\ \tau]$ as follows:

$$\begin{aligned} \dot{x}(t) &= A_m x(t) + b\big(\omega u(t) + \theta(t)\|x_t\|_{\mathcal{L}_\infty} + \sigma(t)\big)\,, \quad x(0) = x_0\,, \\ y(t) &= c^\top x(t)\,, \end{aligned} \tag{2.236}$$

where $\theta(t)$ was introduced in (2.234), $\sigma(t) \triangleq \sigma_1(t) + \sigma_2(t)$ is an unknown continuous time-varying signal subject to $|\sigma(t)| < \Delta$, with Δ as introduced in (2.222), and $|\dot{\sigma}(t)| < d_\sigma$, with $d_\sigma \triangleq d_{\sigma_1} + d_{\sigma_2}$.

Transient and Steady-State Performance

Using the equivalent (semi-)linear system in (2.236), one can write the prediction error dynamics over $t \in [0,\ \tau]$ as

$$\dot{\tilde{x}}(t) = A_m \tilde{x}(t) + b\left(\tilde{\omega}(t)u(t) + \tilde{\theta}(t)\|x_t\|_{\mathcal{L}_\infty} + \tilde{\sigma}(t)\right), \quad \tilde{x}(0) = 0\,, \tag{2.237}$$

where $\tilde{\omega}(t) \triangleq \hat{\omega}(t) - \omega$, $\tilde{\theta}(t) \triangleq \hat{\theta}(t) - \theta(t)$, and $\tilde{\sigma}(t) \triangleq \hat{\sigma}(t) - \sigma(t)$.

Lemma 2.5.2 For the prediction-error dynamics in (2.237), if $u(t)$ is continuous, and moreover the following bounds hold:

$$\|x_\tau\|_{\mathcal{L}_\infty} \leq \rho\,, \quad \|u_\tau\|_{\mathcal{L}_\infty} \leq \rho_u\,, \quad \|\dot{u}_\tau\|_{\mathcal{L}_\infty} \leq \rho_{\dot{u}}\,,$$

then

$$\|\tilde{x}_\tau\|_{\mathcal{L}_\infty} \leq \sqrt{\frac{\theta_m(\rho, \rho_u, \rho_{\dot{u}})}{\lambda_{\min}(P)\Gamma}}, \tag{2.238}$$

where

$$\theta_m(\rho, \rho_u, \rho_{\dot{u}}) \triangleq (\omega_u - \omega_l)^2 + 4L_\rho^2 + 4\Delta^2 + 4\frac{\lambda_{\max}(P)}{\lambda_{\min}(Q)}(L_\rho d_\theta + \Delta d_\sigma).$$

Proof. Consider the following Lyapunov function candidate:

$$V(\tilde{x}(t), \tilde{\omega}(t), \tilde{\theta}(t), \tilde{\sigma}(t)) = \tilde{x}^\top(t) P \tilde{x}(t) + \frac{1}{\Gamma}(\tilde{\omega}^2(t) + \tilde{\theta}^2(t) + \tilde{\sigma}^2(t)).$$

Similar to the proof of Lemma 5.1.2, we can use the adaptation laws in (2.224) and Property B.2 of the projection operator to derive the following upper bound on the derivative of the Lyapunov function:

$$\dot{V}(t) \leq -\tilde{x}^\top(t) Q \tilde{x}(t) + \frac{2}{\Gamma}|\tilde{\theta}(t)\dot{\theta}(t) + \tilde{\sigma}(t)\dot{\sigma}(t)|.$$

Let $t_1 \in (0, \tau]$ be the time instant, when the first discontinuity of $\dot{\theta}(t)$ or $\dot{\sigma}(t)$ occurs, or $t_1 = \tau$ if there are no discontinuities. Notice that

$$\max_{t \in [0, t_1]} \frac{1}{\Gamma}(\tilde{\omega}^2(t) + \tilde{\theta}^2(t) + \tilde{\sigma}^2(t)) \leq \frac{(\omega_u - \omega_l)^2 + 4L_\rho^2 + 4\Delta^2}{\Gamma} < \frac{\theta_m(\rho, \rho_u, \rho_{\dot{u}})}{\Gamma}.$$

This, along with the fact that $\tilde{x}(0) = 0$, leads to

$$V(0) < \frac{\theta_m(\rho, \rho_u, \rho_{\dot{u}})}{\Gamma}. \tag{2.239}$$

If at arbitrary time $t_2 \in [0, t_1]$

$$V(t_2) > \frac{\theta_m(\rho, \rho_u, \rho_{\dot{u}})}{\Gamma},$$

then from the fact that

$$\begin{aligned} V(t_2) &\leq \tilde{x}^\top(t_2) P \tilde{x}(t_2) + \frac{1}{\Gamma}\left((\omega_u - \omega_l)^2 + 4L_\rho^2 + 4\sigma_b^2\right) \\ &= x^\top(t_2) P \tilde{x}(t_2) + \Gamma^{-1}\theta_m(\rho, \rho_u, \rho_{\dot{u}}) - 4\frac{\lambda_{\max}(P)}{\Gamma\lambda_{\min}(Q)}(L_\rho d_\theta + \Delta d_\sigma), \end{aligned}$$

it follows that

$$\tilde{x}^\top(t_2) P \tilde{x}(t_2) \geq 4\frac{\lambda_{\max}(P)}{\Gamma\lambda_{\min}(Q)}(L_\rho d_\theta + \Delta d_\sigma).$$

Further, one can write

$$\tilde{x}^\top(t_2) Q \tilde{x}(t_2) \geq \frac{\lambda_{\min}(Q)}{\lambda_{\max}(P)}\tilde{x}^\top(t_2) P(t_2) \tilde{x}(t_2) \geq \frac{4}{\Gamma}(L_\rho d_\theta + \Delta d_\sigma). \tag{2.240}$$

Notice also that

$$\frac{1}{\Gamma}|\tilde{\theta}(t)\dot{\theta}(t)+\tilde{\sigma}(t)\dot{\sigma}(t)| \leq \frac{2}{\Gamma}(L_{\rho}d_{\theta}+\Delta d_{\sigma}),$$

which, along with the bound in (2.240), leads to

$$\dot{V}(t_2) < 0.$$

Thus, the continuity of $V(t_2)$ along with the bound on the initial condition in (2.239) implies

$$V(t_2) \leq \frac{\theta_m(\rho,\rho_u,\rho_{\dot{u}})}{\Gamma}.$$

Continuity of $V(t)$ allows for repeating these derivations for all points of discontinuity of $\dot{\theta}(t)$ or $\dot{\sigma}(t)$, leading to the following bound for all $t \in [0,\tau]$:

$$V(t) \leq \frac{\theta_m(\rho,\rho_u,\rho_{\dot{u}})}{\Gamma}.$$

This further implies that

$$\|\tilde{x}(t)\| \leq \sqrt{\frac{\theta_m(\rho,\rho_u,\rho_{\dot{u}})}{\lambda_{\min}(P)\Gamma}}.$$

The result in (2.238) follows from the fact that this bound holds uniformly for all $t \in [0,\tau]$. □

The next theorem specifies the uniform performance bounds that the $\mathcal{L}_1$ adaptive controller, defined via (2.223)–(2.225) and subject to the $\mathcal{L}_1$-norm condition in (2.216), guarantees in both transient and steady-state.

Theorem 2.5.1 If the adaptive gain verifies the design constraint

$$\Gamma \geq \frac{\theta_m(\rho,\rho_u,\rho_{\dot{u}})}{\lambda_{\min}(P)\gamma_0^2}, \tag{2.241}$$

where γ_0 is as introduced in (2.219), then the following bounds hold:

$$\|u\|_{\mathcal{L}_\infty} \leq \rho_u, \tag{2.242}$$
$$\|x\|_{\mathcal{L}_\infty} \leq \rho, \tag{2.243}$$
$$\|\tilde{x}\|_{\mathcal{L}_\infty} \leq \gamma_0, \tag{2.244}$$
$$\|x_{\text{ref}}-x\|_{\mathcal{L}_\infty} \leq \gamma_1, \tag{2.245}$$
$$\|u_{\text{ref}}-u\|_{\mathcal{L}_\infty} \leq \gamma_2. \tag{2.246}$$

Proof. We prove (2.245) and (2.246) following a contradicting argument. Assume that (2.245) and (2.246) do not hold. Then, since

$$\|x_{\text{ref}}(0)-x(0)\|_\infty = 0, \quad \|u_{\text{ref}}(0)-u(0)\|_\infty = 0,$$

continuity of $x_{\text{ref}}(t)$, $x(t)$, $u_{\text{ref}}(t)$, $u(t)$ implies that there exists time instant $\tau > 0$, for which

$$\|x_{\text{ref}}(t)-x(t)\|_\infty < \gamma_1, \quad \|u_{\text{ref}}(t)-u(t)\|_\infty < \gamma_2, \qquad \forall t \in [0,\tau)$$

and

$$\|x_{\text{ref}}(\tau) - x(\tau)\|_\infty = \gamma_1\,, \quad \text{or} \quad \|u_{\text{ref}}(\tau) - u(\tau)\|_\infty = \gamma_2\,.$$

This consequently implies that at least one of the following equalities holds:

$$\|(x_{\text{ref}} - x)_\tau\|_{\mathcal{L}_\infty} = \gamma_1\,, \quad \|(u_{\text{ref}} - u)_\tau\|_{\mathcal{L}_\infty} = \gamma_2\,. \tag{2.247}$$

Assumption 2.5.3 and the first equality in (2.247) lead to

$$\|z_\tau\|_{\mathcal{L}_\infty} \leq L_1(\|x_{\text{ref}\tau}\|_{\mathcal{L}_\infty} + \gamma_1) + L_2\,. \tag{2.248}$$

Then, since all the conditions of Lemma 2.5.1 hold, the following bounds are valid:

$$\|x_{\text{ref}\tau}\|_{\mathcal{L}_\infty} < \rho_r\,, \quad \|u_{\text{ref}\tau}\|_{\mathcal{L}_\infty} < \rho_{ur}\,, \tag{2.249}$$

which in turn lead to

$$\|x_\tau\|_{\mathcal{L}_\infty} \leq \rho_r + \gamma_1 \leq \rho\,, \quad \|u_\tau\|_{\mathcal{L}_\infty} \leq \rho_{ur} + \gamma_2 = \rho_u\,. \tag{2.250}$$

Let $\tilde{\eta}(t)$ be defined as

$$\tilde{\eta}(t) \triangleq \tilde{\omega}(t)u(t) + \tilde{\theta}(t)\|x_t\|_{\mathcal{L}_\infty} + \tilde{\sigma}(t)\,.$$

Then for $t \in [0,\ \tau]$, we have

$$\hat{\eta}(t) = \mu(t) + \tilde{\eta}(t) + \eta(t)\,,$$

where $\eta(t) \triangleq f(t, x(t), z(t))$. Substituting the Laplace transform of $\hat{\eta}(t)$ into (2.225) yields

$$u(s) = -C_u(s)(\tilde{\eta}(s) + \eta(s) - k_g r(s))\,. \tag{2.251}$$

Further, the closed-loop response of the system in (2.213) with the $\mathcal{L}_1$ adaptive controller can be written in the frequency domain as

$$x(s) = G(s)\eta(s) - H(s)C(s)\tilde{\eta}(s) + H(s)C(s)k_g r(s) + x_{\text{in}}(s)\,.$$

Also, recall that in (2.229) the response of the closed-loop reference system is presented as

$$x_{\text{ref}}(s) = G(s)\eta_{\text{ref}}(s) + H(s)C(s)k_g r(s) + x_{\text{in}}(s)\,.$$

The two expressions above lead to

$$x_{\text{ref}}(s) - x(s) = G(s)(\eta_{\text{ref}}(s) - \eta(s)) + H(s)C(s)\tilde{\eta}(s)\,. \tag{2.252}$$

Further, consider the control law in (2.225). From the properties of the projection operator we have

$$\|\hat{\eta}_\tau\|_{\mathcal{L}_\infty} \leq \omega_u \rho_u + L_\rho \rho + \Delta\,,$$

and consequently

$$\|\dot{u}_\tau\| \leq \|ksD(s)\|_{\mathcal{L}_1}(\omega_u \rho_u + L_\rho \rho + \Delta + |k_g|\|r\|_{\mathcal{L}_\infty}) = \rho_{\dot{u}}\,.$$

These bounds imply that the conditions of Lemma 2.5.2 hold. Selecting the adaptive gain Γ according to the design constraint in (2.241), we get

$$\|\tilde{x}_\tau\|_{\mathcal{L}_\infty} \leq \gamma_0 . \tag{2.253}$$

It follows from (2.248) and the first bound in (2.249) that

$$\|z_\tau\|_{\mathcal{L}_\infty} \leq L_1(\rho_r+\gamma_1)+L_2 ,$$

which, along with (2.250), leads to

$$\|X_\tau\|_{\mathcal{L}_\infty} \leq \max\{\rho_r+\gamma_1, L_1(\rho_r+\gamma_1)+L_2\} \leq \bar{\rho}_r(\rho_r) .$$

Similarly, one can show that

$$\|X_{\mathrm{ref}\tau}\|_{\mathcal{L}_\infty} \leq \bar{\rho}_r(\rho_r) .$$

Thus, Assumption 2.5.2 yields the upper bound

$$\|(\eta_{\mathrm{ref}}-\eta)_\tau\|_{\mathcal{L}_\infty} \leq d_{f_x}(\bar{\rho}_r(\rho_r))\|(x_{\mathrm{ref}}-x)_\tau\|_{\mathcal{L}_\infty} ,$$

and since $d_{f_x}(\bar{\rho}_r(\rho_r)) < L_{\rho_r}$, it follows that

$$\|(\eta_{\mathrm{ref}}-\eta)_\tau\|_{\mathcal{L}_\infty} \leq L_{\rho_r}\|(x_{\mathrm{ref}}-x)_\tau\|_{\mathcal{L}_\infty} .$$

Moreover, the error dynamics in (2.237) can be rewritten in the frequency domain as $\tilde{x}(s) = H(s)\tilde{\eta}(s)$, and therefore it follows from (2.252) that

$$\|(x_{\mathrm{ref}}-x)_\tau\|_{\mathcal{L}_\infty} \leq \|G(s)\|_{\mathcal{L}_1} L_{\rho_r}\|(x_{\mathrm{ref}}-x)_\tau\|_{\mathcal{L}_\infty} + \|C(s)\|_{\mathcal{L}_1}\|\tilde{x}_\tau\|_{\mathcal{L}_\infty} ,$$

which leads to

$$\|(x_{\mathrm{ref}}-x)_\tau\|_{\mathcal{L}_\infty} \leq \frac{\|C(s)\|_{\mathcal{L}_1}}{1-\|G(s)\|_{\mathcal{L}_1} L_{\rho_r}}\|\tilde{x}_\tau\|_{\mathcal{L}_\infty} .$$

Recalling the bound in (2.253) yields

$$\|(x_{\mathrm{ref}}-x)_\tau\|_{\mathcal{L}_\infty} \leq \frac{\|C(s)\|_{\mathcal{L}_1}}{1-\|G(s)\|_{\mathcal{L}_1} L_{\rho_r}}\gamma_0 = \gamma_1-\beta < \gamma_1 . \tag{2.254}$$

Hence, we obtain a contradiction to the first equality in (2.247).

To show that the second equality in (2.247) also cannot hold, consider (2.226) and (2.251), which lead to

$$u_{\mathrm{ref}}(s)-u(s) = -C_u(s)(\eta_{\mathrm{ref}}(s)-\eta(s)) + C_u(s)\tilde{\eta}(s) .$$

Using the bound on $\|(\eta_{\mathrm{ref}}-\eta)_\tau\|_{\mathcal{L}_\infty}$, Lemma A.12.1, and the definition of $H_1(s)$ in (2.217), we can write

$$\|(u_{\mathrm{ref}}-u)_\tau\|_{\mathcal{L}_\infty} \leq \|C_u(s)\|_{\mathcal{L}_1} L_{\rho_r}\|(x_{\mathrm{ref}}-x)_\tau\|_{\mathcal{L}_\infty} + \|H_1(s)\|_{\mathcal{L}_1}\|\tilde{x}_\tau\|_{\mathcal{L}_\infty} .$$

Further, we can use the upper bound on $\|(x_{\mathrm{ref}}-x)_\tau\|_{\mathcal{L}_\infty}$ in (2.254) to obtain

$$\|(u_{\mathrm{ref}}-u)_\tau\|_{\mathcal{L}_\infty} \leq \|C_u(s)\|_{\mathcal{L}_1} L_{\rho_r}(\gamma_1-\beta) + \|H_1(s)\|_{\mathcal{L}_1}\gamma_0 < \gamma_2 , \tag{2.255}$$

which contradicts the second equality in (2.247).

This proves the bounds in (2.254)–(2.255). Then, the bounds in (2.250) hold uniformly, which implies that the bound in (2.253) also holds uniformly. This proves the bounds in (2.242)–(2.244). □

Remark 2.5.2 It follows from (2.241) that one can prescribe the arbitrary desired performance bound γ_0 by increasing the adaptive gain Γ, which further implies from (2.218) and (2.219) that one can achieve arbitrarily small performance bounds γ_1 and γ_2 simultaneously.

Remark 2.5.3 Similar to previous sections, notice that letting $k \to \infty$ leads to $C(s) \to 1$, and thus the reference controller in the definition of the closed-loop reference system in (2.226) leads, in the limit, to perfect cancelation of uncertainties and recovers the performance of the ideal desired system. As before, setting $C(s) = 1$ leads to an improper $H_1(s)$, and hence the uniform bound for the control signal fails to hold.

2.5.4 Simulation Example

To illustrate the results derived in this chapter, consider the following nonlinear system with unmodeled dynamics:

$$\begin{aligned} \dot{x}(t) &= \begin{bmatrix} 0 & 1 \\ -1 & -1.4 \end{bmatrix} x(t) + \begin{bmatrix} 0 \\ 1 \end{bmatrix} (\mu(t) + f(t, x(t), z(t))), \quad x(0) = x_0, \\ y(t) &= \begin{bmatrix} 1 & 0 \end{bmatrix} x(t), \end{aligned} \tag{2.256}$$

where

$$f(t, x(t), z(t)) = f_1(t, x(t), z(t)) = x_1(t) + 1.4x_2(t) + x_1^2(t) + x_2^2(t) + z^2(t),$$

and

$$\mu(s) = F(s)u(s), \qquad F(s) = F_1(s) = \frac{75}{s+75},$$

while the unmodeled dynamics are given by

$$z(s) = z_1(s) = \frac{s-1}{s^2+3s+2}\nu(s), \qquad \nu(t) = \nu_1(t) = \sin(0.2t)x_1(t) + x_2(t).$$

We implement the $\mathcal{L}_1$ adaptive controller according to (2.223), (2.224), and (2.225). In our design, we set the following parameters for the controller:

$$D(s) = \frac{1}{s}, \quad k = 50, \quad \Gamma = 100000.$$

For the adaptation law, we set the following projection bounds:

$$\Omega = [0.1,\ 3], \quad \Theta = [-5,\ 5], \quad \Delta = 100.$$

We consider bounded reference signals with $\|r\|_{\mathcal{L}_\infty} \le 0.2$. In order to track step-reference signals with zero steady-state error for the ideal system

$$y_{\text{id}}(s) \triangleq c^\top H(s) k_g r(s),$$

we set $k_g = 1$. Setting $Q = \mathbb{I}$ and solving the algebraic Lyapunov equation gives us

$$P = \begin{bmatrix} 1.4143 & 0.5000 \\ 0.5000 & 0.7143 \end{bmatrix}.$$

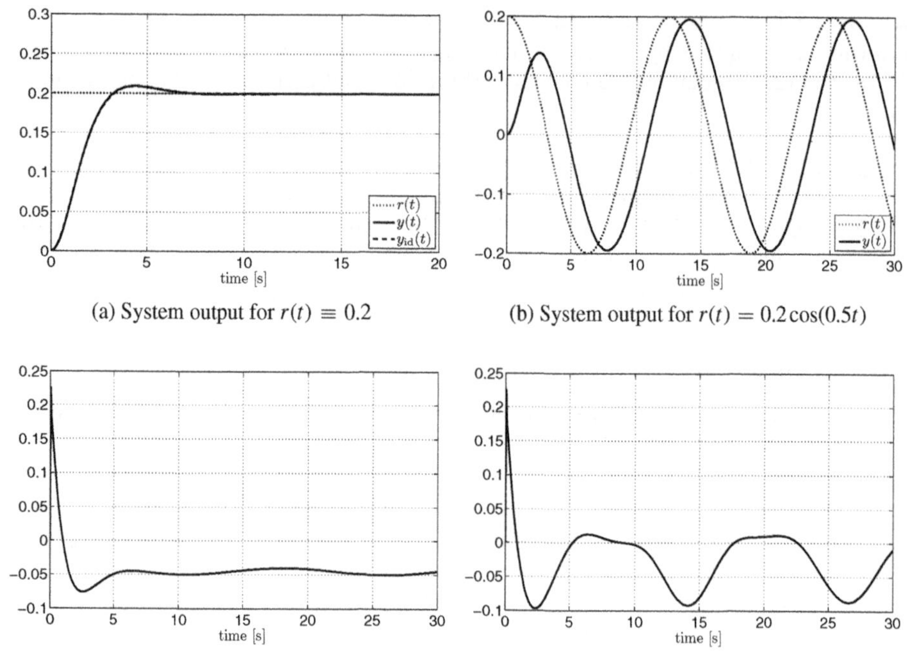

(a) System output for $r(t) \equiv 0.2$

(b) System output for $r(t) = 0.2\cos(0.5t)$

(c) Control history for $r(t) \equiv 0.2$

(d) Control history for $r(t) = 0.2\cos(0.5t)$

Figure 2.43: Performance of the $\mathcal{L}_1$ adaptive controller for $\mu(s) = F_1(s)u(s)$, $f(\cdot) = f_1(\cdot)$, $z(t) = z_1(t)$, and $\nu(t) = \nu_1(t)$.

Next we verify the sufficient condition in (2.216). Letting $\rho_r = 1.5$, we compute the conservative estimate for L_{ρ_r} according to (2.215) and obtain $L_{\rho_r} = 7.3$. From (2.256) we have

$$H(s) = \begin{bmatrix} \frac{1}{s^2+1.4s+1} & \frac{s}{s^2+1.4s+1} \end{bmatrix}^\top .$$

The stability condition in (2.216) takes the form

$$\|G(s)\|_{\mathcal{L}_1} L_{\rho_r} < 1 - \frac{\|k_g C(s)H(s)\|_{\mathcal{L}_1}\|r\|_{\mathcal{L}_\infty}}{\rho_r} - \frac{\|s(s\mathbb{I} - A_m)^{-1}\|_{\mathcal{L}_1}\rho_0}{\rho_r},$$

where we set $\rho_0 = 0.2$. Using the definition of $C(s)$ from (2.214) we compute numerically the $\mathcal{L}_1$-norms of $G(s)$ and $k_g C(s)H(s)$ and obtain the inequality

$$\|G(s)\|_{\mathcal{L}_1} L_{\rho_r} = 0.42 < 0.56,$$

which implies that the $\mathcal{L}_1$-norm condition is satisfied. Notice that this condition is satisfied for arbitrary $k > 48$. The choice of $k = 50$ is explained from the fact that a smaller bandwidth of $C(s)$ leads to better robustness (see Section 2.2.5).

Figure 2.43 shows the simulation results for the closed-loop system described above for the two reference signals $r(t) \equiv 0.2$ (step signal) and $r(t) = 0.2\cos(0.5t)$. One can

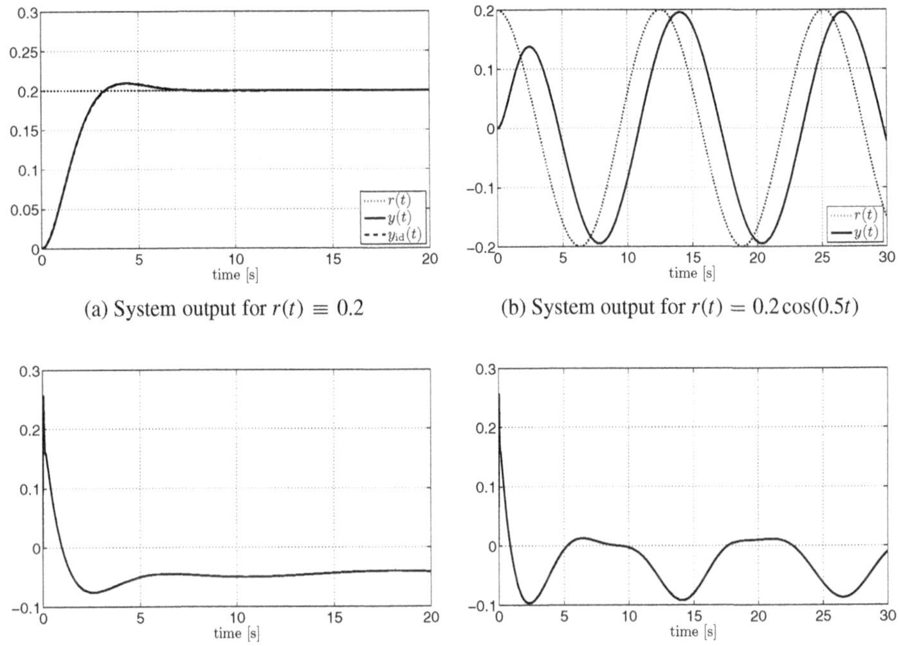

(a) System output for $r(t) \equiv 0.2$

(b) System output for $r(t) = 0.2\cos(0.5t)$

(c) Control history for $r(t) \equiv 0.2$

(d) Control history for $r(t) = 0.2\cos(0.5t)$

Figure 2.44: Performance of the $\mathcal{L}_1$ adaptive controller for $\mu(s) = F_2(s)u(s)$, $f(\cdot) = f_1(\cdot)$, $z(t) = z_1(t)$, and $\nu(t) = \nu_1(t)$.

see that the closed-loop adaptive system has good tracking performance in both transient and steady-state. The behavior of the system output $y(t)$ is close to the ideal $y_{\rm id}(t)$, which is given by a linear system. All performance specifications such as phase-lag for the sinusoidal signal, overshoot, and settling time for the step-reference signal are very close to the ideal system.

Next we change the unmodeled actuator dynamics to

$$F(s) = F_2(s) = \frac{50}{s+50}\,.$$

This change of the unmodeled dynamics does not violate the $\mathcal{L}_1$-norm condition in (2.216). The simulation results with these actuator dynamics and without any retuning of the adaptive controller are shown in Figure 2.44. One can see that the closed-loop system remains stable and there is no degradation in performance.

Next, we change the input to the unmodeled dynamics

$$\nu(t) = \nu_2(t) = \sin(5t)x_1(t) + x_2(t) + 1.4\sin(5t)\,.$$

Again, it can be verified that this change does not violate the $\mathcal{L}_1$-norm condition. The simulation results for this case without any retuning are given in Figure 2.45. While there are

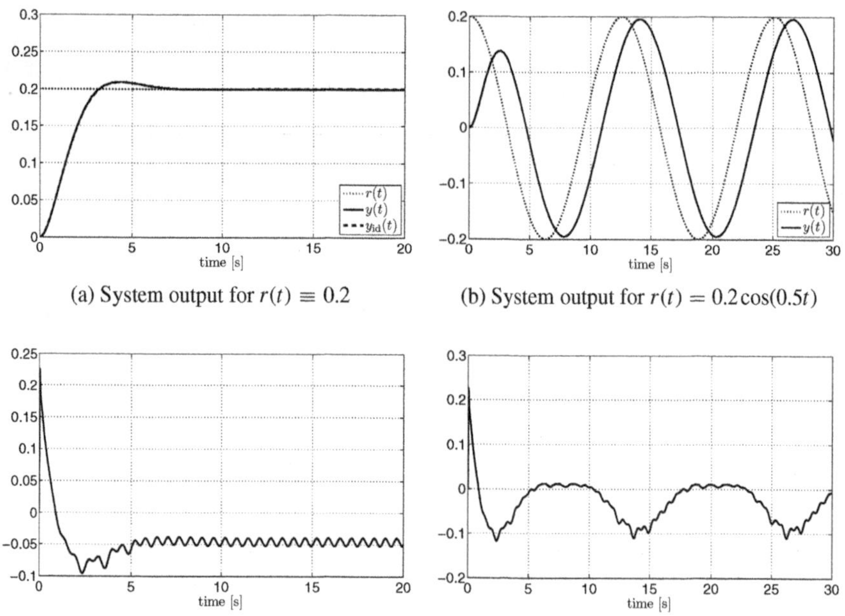

(a) System output for $r(t) \equiv 0.2$

(b) System output for $r(t) = 0.2\cos(0.5t)$

(c) Control history for $r(t) \equiv 0.2$

(d) Control history for $r(t) = 0.2\cos(0.5t)$

Figure 2.45: Performance of the $\mathcal{L}_1$ adaptive controller for $\mu(s) = F_1(s)u(s)$, $f(\cdot) = f_1(\cdot)$, $z(t) = z_1(t)$, and $\nu(t) = \nu_2(t)$.

almost no changes in the system output, we see that the control signal changes to compensate for the new type of the uncertainty.

Further, we change the unmodeled dynamics and the nonlinearities as follows:

$$z(s) = z_2(s) = \frac{1}{s^2 + 30s + 100}\nu_1(s),$$
$$f(t,x(t),z(t)) = f_2(t,x(t),z(t)) = x_1^2(t) + x_2^2(t) + x_1^2(t)x_2^2(t) + x_1(t)x_2(t) + z^2(t).$$

The simulation results for these cases are shown in Figures 2.46 and 2.47. We see that the performance of the system does not change significantly, and the conclusions made in the previous scenario also hold here.

Next, we test the $\mathcal{L}_1$ adaptive system performance for nonzero initialization error. We set the initial conditions of the state predictor different from the plant: $x_0 = [0.1,\ 0.1]$, $\hat{x}_0 = [1.5,\ -1]$. The simulation results are given in Figure 2.48. One can see that the estimation error is rapidly decaying.

In Section 2.2.5, we proved that the closed-loop system with the $\mathcal{L}_1$ adaptive controller has bounded-away-from-zero time-delay margin. Using the insights from that section, we test the robustness of the closed-loop system in this simulation example to time delays. For this purpose, we introduce a time delay of 20 ms at the system input and repeat the simulations for the system nonlinearities and unmodeled dynamics considered above. We do not retune the controller. The simulation results are shown in Figure 2.49. One can

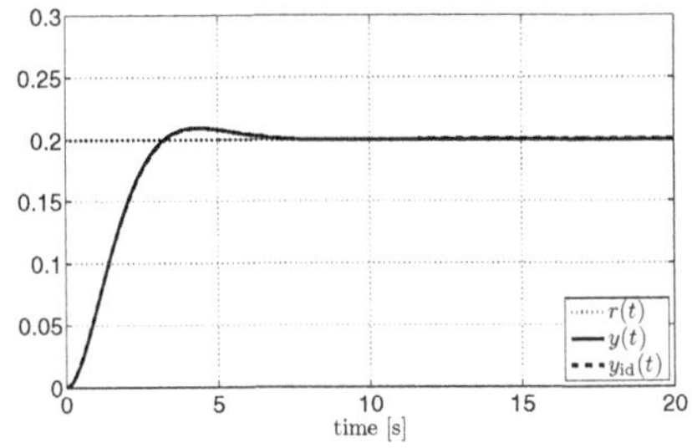

(a) System output for $r(t) \equiv 0.2$

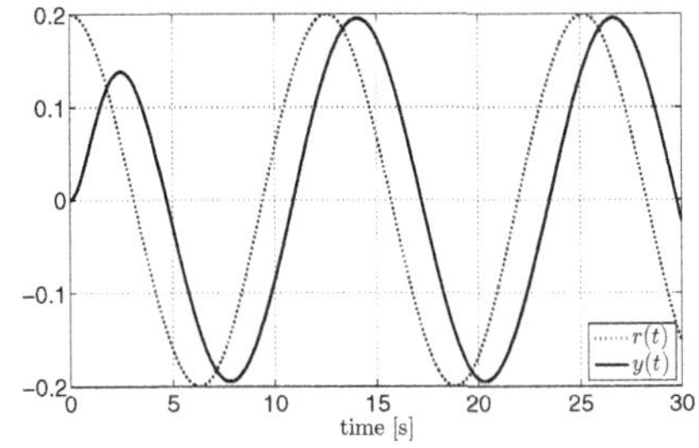

(b) System output for $r(t) = 0.2\cos(0.5t)$

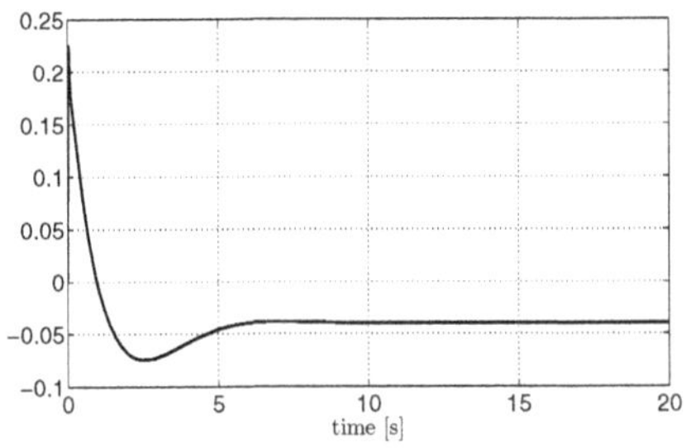

(c) Control history for $r(t) \equiv 0.2$

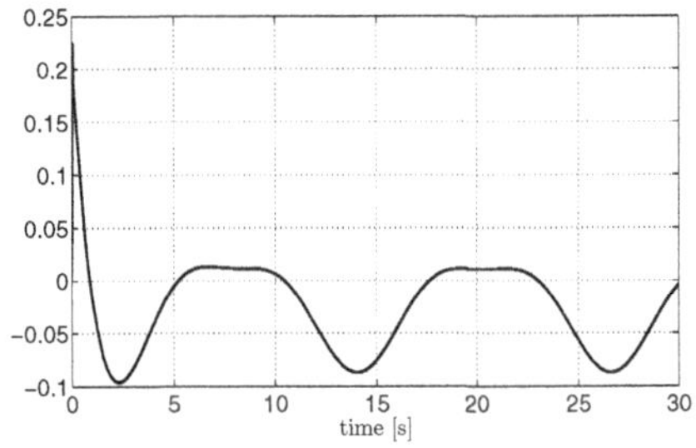

(d) Control history for $r(t) = 0.2\cos(0.5t)$

Figure 2.46: Performance of the $\mathcal{L}_1$ adaptive controller for $\mu(s) = F_1(s)u(s)$, $f(\cdot) = f_1(\cdot)$, $z(t) = z_2(t)$, and $\nu(t) = \nu_1(t)$.

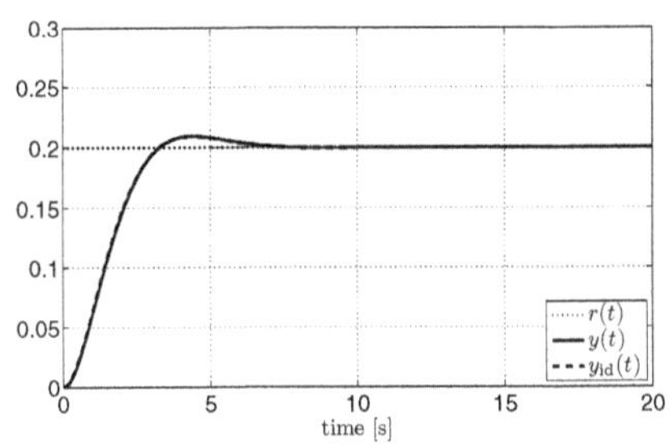

(a) System output for $r(t) \equiv 0.2$

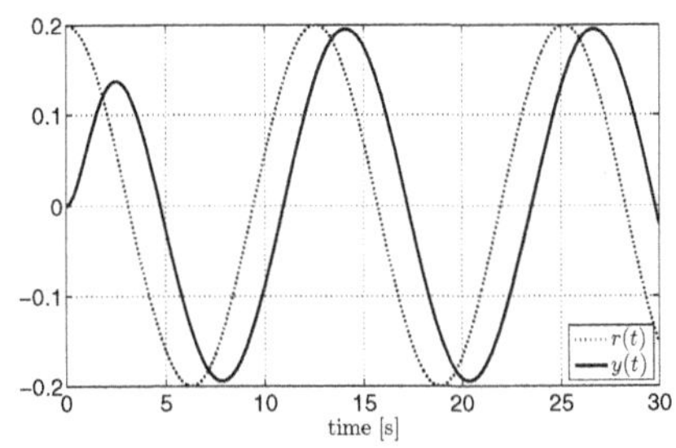

(b) System output for $r(t) = 0.2\cos(0.5t)$

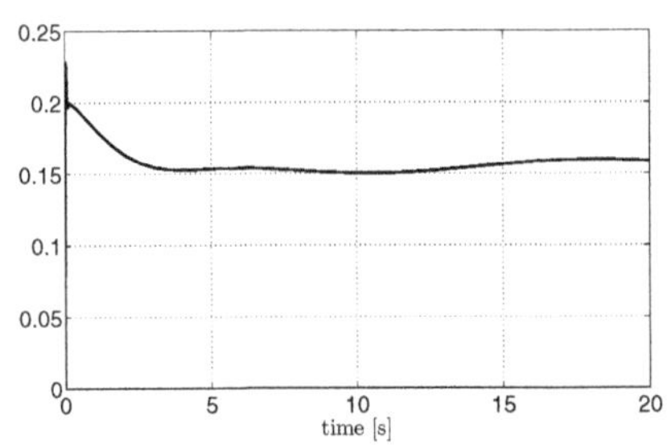

(c) Control history for $r(t) \equiv 0.2$

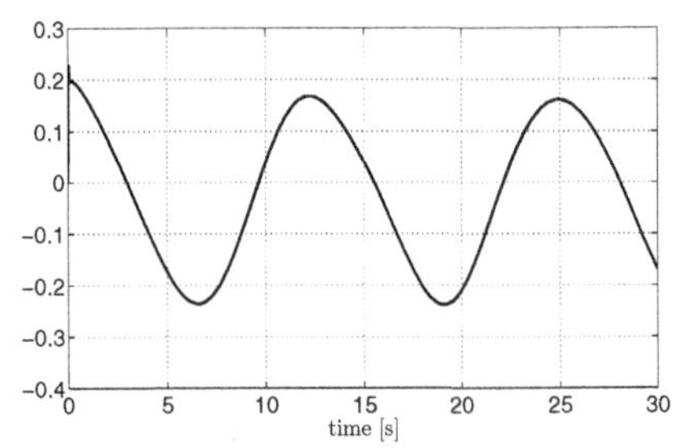

(d) Control history for $r(t) = 0.2\cos(0.5t)$

Figure 2.47: Performance of the $\mathcal{L}_1$ adaptive controller for $\mu(s) = F_1(s)u(s)$, $f(\cdot) = f_2(\cdot)$, $z(t) = z_2(t)$, and $\nu(t) = \nu_1(t)$.

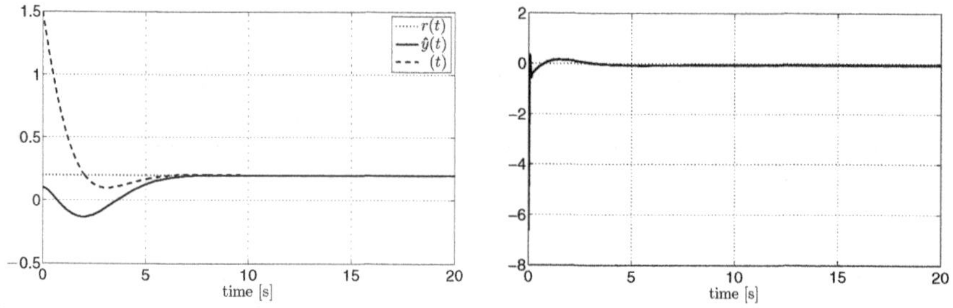

(a) System output

(b) Control history

Figure 2.48: Performance of the $\mathcal{L}_1$ adaptive controller for nonzero trajectory initialization error.

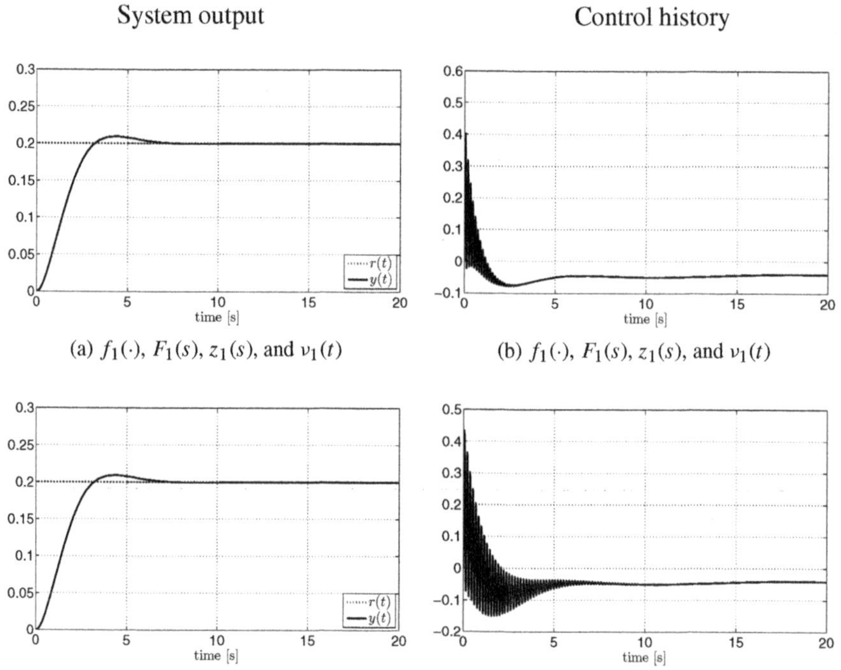

(a) $f_1(\cdot)$, $F_1(s)$, $z_1(s)$, and $\nu_1(t)$

(b) $f_1(\cdot)$, $F_1(s)$, $z_1(s)$, and $\nu_1(t)$

(c) $f_1(\cdot)$, $F_2(s)$, $z_1(s)$, and $\nu_1(t)$

(d) $f_1(\cdot)$, $F_2(s)$, $z_1(s)$, and $\nu_1(t)$

Figure 2.49: Performance of the $\mathcal{L}_1$ adaptive controller in the presence of a time delay of 20 ms.

see that the output of the closed-loop adaptive system for all considered cases of nonlinearities and unmodeled dynamics does not change significantly even in the presence of the time delay. However, one can observe some expected oscillations in the control channel, which indicate that the time delay of 20 ms is relatively close to the actual time-delay margin of the adaptive system.

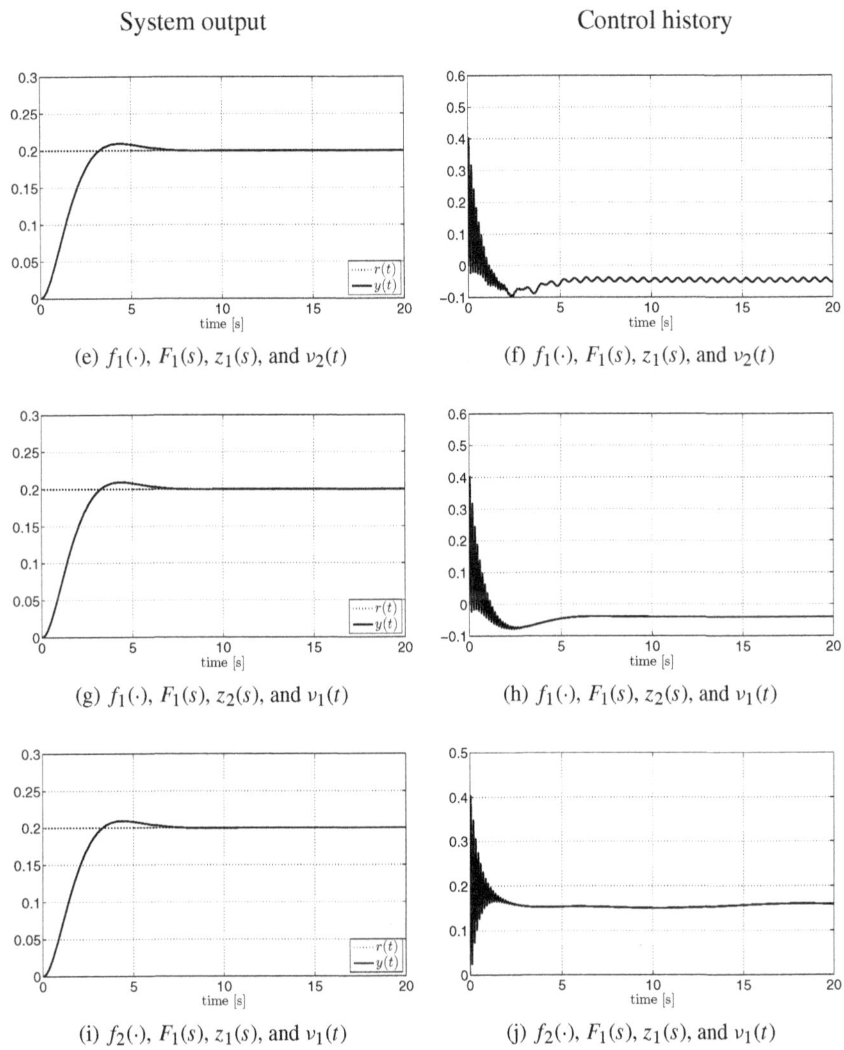

(e) $f_1(\cdot)$, $F_1(s)$, $z_1(s)$, and $v_2(t)$

(f) $f_1(\cdot)$, $F_1(s)$, $z_1(s)$, and $v_2(t)$

(g) $f_1(\cdot)$, $F_1(s)$, $z_2(s)$, and $v_1(t)$

(h) $f_1(\cdot)$, $F_1(s)$, $z_2(s)$, and $v_1(t)$

(i) $f_2(\cdot)$, $F_1(s)$, $z_1(s)$, and $v_1(t)$

(j) $f_2(\cdot)$, $F_1(s)$, $z_1(s)$, and $v_1(t)$

Figure 2.49: Performance of the $\mathcal{L}_1$ adaptive controller in the presence of a time delay of 20 ms.

2.6 Filter Design for Performance and Robustness Trade-Off

In this section we address the problem of designing the filter $C(s)$, which decides the trade-off between performance and robustness. We consider the system and the $\mathcal{L}_1$ adaptive controller given in Section 2.1. From the relationships in (2.7), (2.19) and (2.26), (2.27) it is straightforward to notice that, in addition to increasing the rate of adaptation Γ, one needs to select $C(s)$ to minimize $\|(1 - C(s))H(s)\|_{\mathcal{L}_1}$ for performance improvement. One simple choice was given by Lemma 2.1.5, which states that, by increasing the bandwidth of a first-order $C(s)$, the $\mathcal{L}_1$-norm of $(1-C(s))H(s)$ can be rendered arbitrarily small. It further follows from (2.80) that increasing the bandwidth of $C(s)$ will reduce the time-delay margin to zero. For the purpose of identifying the *optimal trade-off* between the two objectives, we consider the following two constrained optimization problems separately [108].

Problem 1 (Optimization of performance retaining a desired time–delay margin)

$$\begin{aligned} &\min_{C(s)} \|(1-C(s))H(s)\|_{\mathcal{L}_1} \\ \text{s.t.}\quad & C(0)=1 \text{ and } \mathcal{T}(L_o(s)) \geq \tau_{\text{gr}}, \end{aligned} \tag{2.257}$$

where $\tau_{\text{gr}} > 0$ is a desired lower bound on the time-delay margin. Recall that the loop transfer function $L_o(s)$ was defined in (2.80).

Problem 2 (Maximization of time-delay margin retaining a desired performance bound)

$$\begin{aligned} &\max_{C(s)} \mathcal{T}(L_o(s)) \\ \text{s.t.}\quad & C(0)=1 \text{ and } \|(1-C(s))H(s)\|_{\mathcal{L}_1} \leq \lambda_{\text{gp}}, \end{aligned} \tag{2.258}$$

where λ_{gp} is a given upper bound on the $\mathcal{L}_1$-norm of $(1-C(s))H(s)$, specifying the desired performance.

Let the state-space realization for $H(s)$ be given by

$$\dot{x}(t) = A_m x(t) + b u(t), \quad x(0) = 0, \tag{2.259}$$

where $x(t) \in \mathbb{R}^n$ is the system state vector; $u(t) \in \mathbb{R}$ is the control signal; $b, c \in \mathbb{R}^n$ are known constant vectors; and $A_m \in \mathbb{R}^{n\times n}$ is a known Hurwitz matrix specifying the desired closed-loop dynamics. Also, let the state-space realization for $C(s)$ be given by

$$\begin{aligned} \dot{x}_f(t) &= A_f x_f(t) + b_f u_f(t), \quad x_f(0) = 0, \\ y_f(t) &= c_f^\top x_f(t), \end{aligned} \tag{2.260}$$

where $x_f(t) \in \mathbb{R}^{n_f}$ is the state of the filter, $y_f(t) \in \mathbb{R}$ is the output of the filter, $u_f(t) \in \mathbb{R}$ is the input, $A_f \in \mathbb{R}^{n_f \times n_f}$ is a Hurwitz matrix, and $b_f, c_f \in \mathbb{R}^{n_f}$ are the input and the output vectors, respectively. The dimension n_f of the filter dynamics can be freely chosen by the user a priori and is one of the parameters in the foregoing optimization problems.

The objective of the algorithms presented below is to determine the "optimal" state-space realization matrices of the filter $C(s)$ in the context of the above-formulated problems. Since the optimization of the $\mathcal{L}_1$-norm of $(1-C(s))H(s)$ over the parameters of $C(s)$ is a nonconvex problem by definition, we provide an overview of methods that can be employed for investigation of this problem.

First we review some stochastic optimization methods, which have already proved to be useful in control-related nonconvex optimization problems [87, 123]. Then we formulate the above-presented optimization problems in terms of LMIs, which provide an opportunity to use MATLAB tools to obtain a possible (conservative, but guaranteed) solution for the filter realization.

2.6.1 Overview of Stochastic Optimization Algorithms

Random search algorithms, which are also called Monte Carlo methods or stochastic algorithms [160, 184], refer to computational methods that use repetitive random samplings to solve simulation problems in various applications. These algorithms are successfully used to tackle a wide class of optimization problems [148, 152]. We now illustrate the main idea of stochastic optimization algorithms by starting with the following standard constrained optimization problem:

$$\begin{aligned} &\min J(p) \\ \text{s.t.} \quad & p \in \mathcal{S}. \end{aligned} \tag{2.261}$$

Here p denotes the decision variable that captures the entries of the filter realization given by (A_f, b_f, c_f); $J(\cdot)$ represents the objective function, which is given by $\|(1-C(s))H(s)\|_{\mathcal{L}_1}$ for (2.257) or $\mathcal{T}(L_o(s))$ for (2.258); and $\mathcal{S}$ specifies the set of constraints, defined in (2.257) or (2.258) separately for each of the two cases.

The basic idea of the stochastic optimization algorithm for solving (2.261) is composed of the following two key steps:

1. Generate a sequence of random samples $p^{(i)}, \quad i = 1, 2, \dots$.
2. Obtain the optimal solution p^* from the samples.

Next we introduce several widely used variances of stochastic optimization methods with control applications.

Randomized Algorithms (RAs) In this approach, an optimal solution is selected from a set of samples $\{p^{(i)}, i = 1, \dots, n_s\}$, where n_s is the number of samples. The elements in the set are randomly independently identically distributed over the feasible solution set $\mathcal{S}$. The accuracy and the confidence levels of the solution depend upon the selected number of samples, which can be quantified as

$$n_s \geq \left\lceil \frac{\log(\frac{1}{\beta})}{\log(\frac{1}{1-\epsilon})} \right\rceil,$$

where $\epsilon \in (0,1)$ and $\beta \in (0,1)$ represent the desired accuracy and the confidence levels, respectively. Details of this algorithm are given in [87, 165, 166].

Adaptive Random Search Algorithms (ARSAs) Unlike RAs, in this method the optimization is done by iterations, where the sample point at the step $i+1$ is determined by

$$p^{(i+1)} = p^{(i)} + t^{(i)} d^{(i)},$$

where $t^{(i)}$ and $d^{(i)}$ represent the ith step size and a feasible direction. On each iteration of the algorithm, one performs a random search for $t^{(i)}$ and $d^{(i)}$, such that $p^{(i+1)} \in \mathcal{S}$ and $J\left(p^{(j+1)}\right) < J\left(p^{(j)}\right)$. One can refer to [11, 21, 184] for more details.

Particle Swarm Optimization Algorithms (PSOAs). PSOA is a population-based random search algorithm, which considers a set of potential solutions $\{p^{(i)}\} \subset \mathcal{S}$, called *particles*. The set of potential solutions is initialized with a population of random samples. During the optimization process the algorithm assigns a randomized velocity and/or acceleration to each particle based on comprehensive update law, which ensures that the population of particles remains in the feasible solution set. The optimal solution is determined as the overall concurrent best value in the population [38, 93, 154].

Meta-Control Methodologies (MCMs). A meta-control methodology combines the idea of well-known simulated annealing with population-based random search algorithms. This combination provides a way to escape possible local minima. The method supports dynamical tuning of the algorithm parameters based on the observed values at each iteration of the algorithm. Details can be found in [95].

Remark 2.6.1 Since it is impossible to compute the exact value of the time-delay margin $\mathcal{T}(L_o(s))$ in the optimization problems of interest (2.257) and (2.258) due to the presence of the unknown value θ in the transfer function $L_o(s)$, the optimization method needs to estimate the value of $\mathcal{T}(L_o(s))$. This can be done by employing a recently developed RA for sampling the uncertainty θ [166]. This approach considers the probabilistic worst-case or empirical mean performance, instead of relying on the deterministic worst-case values. Specifically, to check the feasibility of the constraint for guaranteed time-delay margin in (2.257), for all $\theta \in \Theta$, the method considers a set of randomly generated samples $\{\theta^{(j)}, j = 1,\dots,N_\theta\}$ and evaluates $\mathcal{T}(L_o(s)) \geq \tau_{\text{gr}}$ for all $\theta^{(j)}$, $j = 1,\dots,N_\theta$, where N_θ denotes the number of random samples for the uncertain parameter θ. Similarly, to solve the optimization problem in (2.258), the objective function $\mathcal{T}(L_o(s))$ must be replaced by $\min_{\{\theta^{(j)}\}} \mathcal{T}(L_o(s))$. A brief summary of these methods is given in [92].

2.6.2 LMI-Based Filter Design

$\mathcal{L}_1$-Norm Optimization

In this section, we investigate the constraint-free $\mathcal{L}_1$-norm minimization problem of the cascaded system $(1 - C(s))H(s)$. The $\mathcal{L}_1$-norm minimization problem is nonconvex by definition. We hence consider the $*$-norm, which serves as an upper bound of the $\mathcal{L}_1$-norm of a stable LTI system [2] (see Definition C.2.1 in Appendix). From now on we fix b_f to render our optimization problems convex and computationally tractable.

From (2.259) and (2.260), it follows that the state-space representation of the cascaded system $(1 - C(s))H(s)$ is given by

$$\dot{\xi}(t) = \begin{bmatrix} A_m & -bc_f^\top \\ \mathbf{0} & A_f \end{bmatrix} \xi(t) + \begin{bmatrix} b \\ b_f \end{bmatrix} u(t),$$
$$y(t) = \begin{bmatrix} \mathbb{I} & \mathbf{0} \end{bmatrix} \xi(t),$$

where $\xi(t) \in \mathbb{R}^{n+n_f}$ and $y(t) \in \mathbb{R}^n$ are the state and the output, respectively, and $\mathbb{I}$ and $\mathbf{0}$ are the identity and the zero matrices of appropriate dimensions.

Theorem 2.6.1 If there exist a scalar $\alpha \in \mathbb{R}^+$, a vector $q \in \mathbb{R}^{n_f}$, a matrix $M \in \mathbb{R}^{n_f \times n_f}$, and positive definite matrices $P_1 \in \mathbb{R}^{n \times n}$, $P_2 \in \mathbb{R}^{n_f \times n_f}$, solving the LMIs

$$\Phi(\alpha, P_1, P_2, M, q) \leq 0, \tag{2.262}$$
$$P_1 \leq \lambda_{gp}\mathbb{I}, \tag{2.263}$$

where

$$\Phi(\alpha, P_1, P_2, M, q) \triangleq \begin{bmatrix} \begin{bmatrix} Q_\alpha & -bq^\top \\ -qb^\top & M + M^\top + \alpha P_2 \end{bmatrix} & \begin{bmatrix} b \\ b_f \end{bmatrix} \\ \begin{bmatrix} b^\top & b_f^\top \end{bmatrix} & -\alpha \end{bmatrix},$$

for some $\lambda_{gp} > 0$ and $b_f \in \mathbb{R}^{n_f}$, where $Q_\alpha \triangleq A_m P_1 + P_1 A_m^\top + \alpha P_1$, then

$$\|(1 - C(s))H(s)\|_{\mathcal{L}_1} \leq \lambda_{gp},$$

and the parameters of the corresponding filter are given by

$$A_f = M P_2^{-1} \quad \text{and} \quad c_f = P_2^{-1} q.$$

Proof. First, consider a structured Lyapunov matrix

$$P_\alpha = \begin{bmatrix} P_1 & \mathbf{0} \\ \mathbf{0} & P_2 \end{bmatrix},$$

and its corresponding change of variables $q \triangleq P_2 c_f$ and $M \triangleq A_f P_2$. It is straightforward to see that the inequality in (2.262) is equivalent to

$$\begin{bmatrix} A_m & -bc_f^\top \\ \mathbf{0} & A_f \end{bmatrix} \begin{bmatrix} P_1 & \mathbf{0} \\ \mathbf{0} & P_2 \end{bmatrix} + \begin{bmatrix} P_1 & \mathbf{0} \\ \mathbf{0} & P_2 \end{bmatrix} \begin{bmatrix} A_m & -bc_f^\top \\ \mathbf{0} & A_f \end{bmatrix}^\top$$
$$+ \alpha \begin{bmatrix} P_1 & \mathbf{0} \\ \mathbf{0} & P_2 \end{bmatrix} + \frac{1}{\alpha} \begin{bmatrix} b \\ b_f \end{bmatrix} \begin{bmatrix} b \\ b_f \end{bmatrix}^\top \leq 0.$$

Note that the state-space representation of $(1 - C(s))H(s)$ bears the same form of the system in (C.2). Therefore, direct application of Theorem C.2.1 leads to the next result:

$$\|(1 - C(s))H(s)\|_{\mathcal{L}_1} \leq \left\| \begin{bmatrix} \mathbb{I} & \mathbf{0} \end{bmatrix} \begin{bmatrix} P_1 & \mathbf{0} \\ \mathbf{0} & P_2 \end{bmatrix} \begin{bmatrix} \mathbb{I} \\ \mathbf{0} \end{bmatrix} \right\|_2 = \|P_1\|_2.$$

Thus, $P_1 \leq \lambda_{gp}\mathbb{I}$ implies $\|(1 - C(s))H(s)\|_{\mathcal{L}_1} \leq \lambda_{gp}$. □

We further refer to this theorem to find the minimal upper bound of the $\mathcal{L}_1$-norm. Let $\lambda \geq \|(1 - C(s))H(s)\|_{\mathcal{L}_1}$ be the upper bound to be minimized. We replace the bound λ_{gp} by a decision variable λ in (2.263) and formulate the following optimization problem (*generalized eigenvalue problem* (GEVP); see Appendix C.1):

$$\begin{aligned} &\min_{\alpha, P_1 > 0, P_2 > 0, M, q} \quad \lambda, \\ &\text{s.t. (2.262) holds and } P_1 \leq \lambda \mathbb{I}. \end{aligned} \tag{2.264}$$

The *optimal* filter $C(s)$ is then realized via $(A_f = M P_2^{-1},\ b_f,\ c_f = P_2^{-1} q)$.

Remark 2.6.2 Since a necessary condition for the feasibility of (2.262) is $Q_\alpha \leq 0$, a feasible solution α for the GEVP (2.264) is bounded as $0 < \alpha \leq -2\Re(\lambda_{\max}(A_m))$.

Time-Delay Margin Optimization

Similarly, we develop LMI tools for constraint-free time-delay margin optimization. We consider the time-delay margin maximization problem for the $\mathcal{L}_1$ adaptive controller.

Let the state-space realization for $L_o(s)$ in the presence of time delay τ be given by

$$\begin{aligned}\dot{x}_l(t) &= \begin{bmatrix} A_f + b_f c_f^\top & b_f c_f^\top & \mathbf{0} \\ \mathbf{0} & A_f & b_f \theta^\top \\ \mathbf{0} & \mathbf{0} & A_m + b\theta^\top \end{bmatrix} x_l(t) \\ &+ \begin{bmatrix} \mathbf{0} & \mathbf{0} & \mathbf{0} \\ -b_f c_f^\top & -b_f c_f^\top & \mathbf{0} \\ -b c_f^\top & -b c_f^\top & \mathbf{0} \end{bmatrix} x_l(t-\tau),\end{aligned} \tag{2.265}$$

where $x_l(t) \in \mathbb{R}^{n+2n_f}$ is the state of $L_o(s)$.

Theorem 2.6.2 If for a fixed vector b_f and prespecified lower bound $\tau_{gr} > 0$ there exist a vector $q \in \mathbb{R}^{n_f}$, a matrix $M \in \mathbb{R}^{n_f \times n_f}$, and positive definite matrices $P_1 \in \mathbb{R}^{n_f \times n_f}$ and $P_2 \in \mathbb{R}^{n \times n}$ satisfying

$$\Psi(P_1, P_2, M, q, \tau_{qr}) \leq 0, \tag{2.266}$$

where

$$\Psi(P_1, P_2, M, q, \tau_{qr}) \triangleq \begin{bmatrix} Q_1 + Q_1^\top + Q_2 + Q_2^\top & Q_1^\top & Q_2^\top & Q_2 \\ Q_1 & -\frac{1}{\tau_{gr}} P & \mathbf{0} & \mathbf{0} \\ Q_2 & \mathbf{0} & -\frac{1}{\tau_{gr}} P & \mathbf{0} \\ Q_2^\top & \mathbf{0} & \mathbf{0} & -\frac{1}{2\tau_{gr}} P \end{bmatrix},$$

$$Q_1 \triangleq \begin{bmatrix} M + b_f q^\top & b_f q^\top & \mathbf{0} \\ \mathbf{0} & M & b_f \theta^\top P_2 \\ \mathbf{0} & \mathbf{0} & A_m P_2 + b\theta^\top P_2 \end{bmatrix},$$

$$Q_2 \triangleq \begin{bmatrix} \mathbf{0} & \mathbf{0} & \mathbf{0} \\ -b_f q^\top & -b_f q^\top & \mathbf{0} \\ -b q^\top & -b q^\top & \mathbf{0} \end{bmatrix},$$

$$P \triangleq \begin{bmatrix} P_1 & \mathbf{0} & \mathbf{0} \\ \mathbf{0} & P_1 & \mathbf{0} \\ \mathbf{0} & \mathbf{0} & P_2 \end{bmatrix},$$

then the system (2.265) is BIBS stable for arbitrary $0 \leq \tau \leq \tau_{gr}$.

Proof. The result of the theorem immediately follows from application of Lemma C.3.1 to the LTI system in (2.265). □

Notice that from Theorem 2.6.2 it follows that the time-delay margin of the system (2.265) is greater than τ_{gr}. This result is summarized in the following corollary.

Corollary 2.6.1 If there exist matrices $M \in \mathbb{R}^{n_f \times n_f}$, $q \in \mathbb{R}^{n_f}$ and positive definite matrices $P_1 \in \mathbb{R}^{n_f \times n_f}$ and $P_2 \in \mathbb{R}^{n \times n}$ satisfying the inequality in (2.266), then the time-delay margin $\mathcal{T}$ of the closed-loop system (2.1) with the $\mathcal{L}_1$ adaptive controller, defined via (2.4)–(2.6), is lower bounded by τ_{gr}. The realization of the corresponding filter $C(s)$ is then given by $(MP_1^{-1}, b_f, P_1^{-1}q)$.

Notice that τ_{gr} in (2.266) can be viewed as a generalized eigenvalue. We can formulate the following GEVP for the time-delay margin maximization, where we use τ_s to denote the optimization objective:

$$\begin{aligned} &\max_{q,\, M,\, P_1>0,\, P_2>0} \quad \tau_s \\ &\text{s.t.} \quad \Psi(P_1, P_2, M, q, \tau_s) \le 0 . \end{aligned} \tag{2.267}$$

Remark 2.6.3 The augmented matrix Q_1 in $\Psi(\cdot)$ contains the unknown parameter θ, which lies in the compact set Θ. This implies that the linear matrix inequalities (2.267) and (2.266) must be checked for all values of θ, thus rendering the numerical optimization intractable. However, the matrix Q_1 is affine in $\theta \in \Theta$, and because the uncertainty set Θ can be bounded by a polytope, one can use Corollary C.4.1 and check just a finite number of LMIs for the vertices of the polytope. If a solution for the filter realization is found for this finite number of LMIs, then it will also solve the original optimization problem.

Constrained Optimization for $\mathcal{L}_1$-Norm and Time-Delay Margin

In this section, we address the constrained optimization problems in (2.257) and (2.258) via LMI formulations. To address the $\mathcal{L}_1$-norm optimization problem in (2.264) in the presence of a constraint on the time-delay margin, we add the additional LMI condition from Theorem 2.6.2 to the problem statement. Similarly, for the constrained time-delay margin optimization, the GEVP optimization problem in (2.267) is used, together with the LMI condition for the $\mathcal{L}_1$-norm given in Theorem 2.6.1.

We start by proving the following theorem, which handles the constraint $C(0) = 1$ in the LMI optimization problem.

Theorem 2.6.3 Let the state-space realization of $C(s)$ be given as in (2.260). Let $P \in \mathbb{R}^{n_f \times n_f}$ be a nonsingular symmetric matrix, and also let $M \in \mathbb{R}^{n_f \times n_f}$ be a nonsingular matrix. Further, let $m \in \mathbb{R}^{n_f}$ be the last row of $M^\top$. If A_f, b_f, and c_f are chosen as

$$A_f \triangleq MP^{-1}, \qquad b_f \triangleq \begin{bmatrix} 0, & \cdots & 0, & -1 \end{bmatrix}^\top, \qquad c_f = P^{-1}m,$$

then we have

$$C(0) = 1 .$$

Proof. Let $p \in \mathbb{R}^{n_f}$ be the last column of M^{-1}. It is straightforward to verify that

$$C(0) = -c_f^\top A_f^{-1} b_f = -m^\top P^{-\top} P M^{-1} b_f = -m^\top M^{-1} b_f = m^\top q .$$

Then, from $MM^{-1} = \mathbb{I}$, it follows that $m^\top q = 1$, which leads to

$$C(0) = 1 .$$

Next we address the constrained problem in (2.257) for performance optimization retaining a desired time-delay margin.

Theorem 2.6.4 For a given desired lower bound for the time-delay margin τ_{gr} of the closed-loop $\mathcal{L}_1$ adaptive control system, if there exist scalars α, $\lambda \in \mathbb{R}$, matrices $P_0 =$

$P_0^\top > 0$, $P_1 = P_1^\top > 0$, $P_2 = P_2^\top > 0$ of appropriate dimensions, $M \in \mathbb{R}^{n_f \times n_f}$, and a vector $q \in \mathbb{R}^{n_f}$ solving the GEVP

$$\min_{\alpha,\, P_0,\, P_1,\, P_2,\, M,\, q} \quad \lambda$$

such that

$$\begin{aligned} P_0 &\leq \lambda \mathbb{I}, \\ \Phi(\alpha, P_0, P_1, M, q) &\leq 0, \end{aligned} \tag{2.268}$$

and

$$\Psi(P_0, P_2, M, q, \tau_{gr}) \leq 0, \tag{2.269}$$

where M, q, and b_f comply with the structure in Theorem 2.6.3 and $\Phi(\cdot)$ and $\Psi(\cdot)$ are as defined in (2.262) and (2.266), then the problem (2.257) is solved by via the following realization of the filter $C(s)$:

$$(A_f = M P_0^{-1},\; b_f,\; c_f = P_0^{-1} q).$$

Proof. Let λ be the upper bound of the $\mathcal{L}_1$-norm of $(1 - C(s))H(s)$ and let τ_{gr} be a given lower bound on the time-delay margin to be satisfied. Consider the $\mathcal{L}_1$-norm optimization problem (2.264) and the LMI condition in (2.266) along with the definitions for Φ in Theorem 2.6.1 and Ψ in Theorem 2.6.2, where M, q, and b_f comply with the structure in Theorem 2.6.3. By substituting P_0, M, and q into (2.264) and (2.266), we get (2.268) and (2.269), respectively. It follows from Theorem 2.6.3 that the choice of b_f, M, and q ensures that $C(0) = 1$. Then the optimization problem (2.264), together with the constraint (2.266), yields the filter $C(s)$ via the realization of

$$(A_f = M P_0^{-1},\; b_f,\; c_f = P_0^{-1} q),$$

which completes the proof. □

For the time-delay margin optimization problem in (2.258), we have the following result, where the proof is omitted since the idea is similar to Theorem 2.6.4.

Theorem 2.6.5 Let λ_{gp} be a desired upper bound of the $\mathcal{L}_1$-norm of $(1 - C(s))H(s)$. If there exist $\alpha \in \mathbb{R}$, matrices $P_0 = P_0^\top > 0$, $P_1 = P_1^\top > 0$, $P_2 = P_2^\top > 0$ of appropriate dimensions, $M \in \mathbb{R}^{n_f \times n_f}$ and a vector $q \in \mathbb{R}^{n_f}$ solving the GEVP

$$\max_{\alpha,\, P_0,\, P_1,\, P_2,\, M,\, q} \quad \tau_s,$$

subject to

$$\begin{aligned} P_0 &\leq \lambda_{gp} \mathbb{I}, \\ \Phi(\alpha, P_0, P_1, M, q) &\leq 0, \end{aligned}$$

and

$$\Psi(P_0, P_2, M, q, \tau_s) \leq 0,$$

where M, q, and b_f comply with the structure given in Theorem 2.6.3, and $\Phi(\cdot)$ and $\Psi(\cdot)$ are defined in (2.262) and (2.266), then the problem in (2.258) is solved by choosing the following realization of the filter $C(s)$:

$$(A_f = MP_0^{-1},\ b_f,\ c_f = P_0^{-1}q).$$

Remark 2.6.4 We note that the existence of a solution to the above LMIs and GEVPs is not guaranteed in the general case. Theorems 2.6.4 and 2.6.5, however, suggest systematic ways to solve computationally tractable convex optimization problems for (2.257) and (2.258).

Obviously, by resorting the tuning of $C(s)$ to the LMI algorithms presented above, we obtain overly conservative solution for the state-space realization of $C(s)$. The structure imposed on the block-diagonal Lyapunov matrices P_α and P toward rendering the optimization problems convex is the first step, which increases the degree of conservatism. The second step of introducing conservatism lies in the definition of the structure for the decision variables M and q, considered in Theorem 2.6.3.

Chapter 3

State Feedback in the Presence of Unmatched Uncertainties

This chapter presents the extension of the ideas from previous sections to systems in the presence of unmatched uncertainties. The class of uncertainties includes time- and state-dependent nonlinearities and unmodeled dynamics. We first present the $\mathcal{L}_1$ adaptive backstepping controller for strict-feedback systems and proceed by considering systems of a more general type, which cannot be addressed by recursive design methods. We present a new piecewise-constant adaptive law, which uses the sampling rate of the available CPU to update the parametric estimates. The sufficient conditions in this case can be interpreted to determine the performance limitations due to the cross-coupling dynamics. In particular, the solution presented in Section 3.3 has been flight tested on NASA's GTM (AirSTAR) and evaluated in mid- to high-fidelity simulations for the X-48B and X-29 aircraft.

3.1 $\mathcal{L}_1$ Adaptive Controller for Nonlinear Strict-Feedback Systems

This section presents the $\mathcal{L}_1$ adaptive control architecture for nonlinear strict-feedback systems in the presence of uncertain system input gain and unknown time-varying nonlinearities. Similar to earlier developments, we prove that the $\mathcal{L}_1$ adaptive control architecture ensures guaranteed transient response for system's input and output signals simultaneously.

3.1.1 Problem Formulation

Consider the following system:

$$\begin{aligned}
\dot{x}_1(t) &= f_1(t, x_1(t)) + x_2(t), \quad x_1(0) = x_{1_0}, \\
\dot{x}_2(t) &= f_2(t, x(t)) + \omega u(t), \quad x_2(0) = x_{2_0}, \\
y(t) &= x_1(t),
\end{aligned} \tag{3.1}$$

where $x_1(t),\ x_2(t) \in \mathbb{R}$ are the states of the system (measured), $x(t) \triangleq [x_1(t),\ x_2(t)]^\top$; $u(t) \in \mathbb{R}$ is the control signal; $\omega \in \mathbb{R}$ is an unknown parameter; and $f_1(t, x_1) : \mathbb{R} \times \mathbb{R} \to \mathbb{R}$ and $f_2(t, x) : \mathbb{R} \times \mathbb{R}^2 \to \mathbb{R}$ are unknown nonlinear maps continuous in their arguments. The initial condition $x_0 \triangleq [x_{1_0}, x_{2_0}]^\top$ is assumed to be inside an arbitrarily large known set, so that $\|x_0\|_\infty \le \rho_0 < \infty$ for some known $\rho_0 > 0$.

Assumption 3.1.1 (Partial knowledge of uncertain system input gain) Let

$$\omega \in \Omega \triangleq [\omega_l, \omega_u],$$

where $0 < \omega_l < \omega_u$ are known conservative bounds.

Assumption 3.1.2 (Uniform boundedness of $f_i(t,0)$) There exists $B > 0$, such that $|f_i(t,0)| \le B$, $i = 1,2$, holds for all $t \ge 0$.

Assumption 3.1.3 (Semiglobal uniform boundedness of partial derivatives) For $i = 1,2$, and arbitrary $\delta > 0$, there exist positive constants $d_{f_{xi}}(\delta) > 0$ and $d_{f_{ti}}(\delta) > 0$ independent of time such that for all $\|x(t)\|_\infty < \delta$, the partial derivatives of $f_i(\cdot)$ are piecewise-continuous and bounded:

$$\left|\frac{\partial f_1(t,x_1)}{\partial x_1}\right| \le d_{f_{x1}}(\delta), \quad \left|\frac{\partial f_1(t,x_1)}{\partial t}\right| \le d_{f_{t1}}(\delta),$$
$$\left\|\frac{\partial f_2(t,x)}{\partial x}\right\|_1 \le d_{f_{x2}}(\delta), \quad \left|\frac{\partial f_2(t,x)}{\partial t}\right| \le d_{f_{t2}}(\delta).$$

In this section we present the $\mathcal{L}_1$ adaptive (backstepping) controller, which ensures that the system output $y(t)$ tracks a given bounded twice continuously differentiable reference signal $r(t)$ with uniform and quantifiable performance bounds.

3.1.2 $\mathcal{L}_1$ Adaptive Control Architecture

Definitions and $\mathcal{L}_1$-Norm Sufficient Condition for Stability

The design of the $\mathcal{L}_1$ adaptive controller involves a stable low-pass filter $C_1(s)$ of relative degree of at least 2 with unity DC gain, $C_1(0) = 1$, and also a positive feedback gain $k > 0$ and a strictly proper transfer function $D(s)$, which lead, for all $\omega \in \Omega$, to a stable transfer function

$$C_2(s) \triangleq \frac{\omega k D(s)}{1 + \omega k D(s)},$$

with unity DC gain $C_2(0) = 1$. We assume zero initialization for the state-space realization of these filters. Using the above notation, let

$$C(s) \triangleq \begin{bmatrix} C_1(s) & 0 \\ 0 & C_2(s) \end{bmatrix}.$$

Also, let A_m and A_g be defined as

$$A_m \triangleq \begin{bmatrix} -a_1 & 0 \\ 0 & -a_2 \end{bmatrix}, \quad A_g \triangleq \begin{bmatrix} -a_1 & 1 \\ 0 & -a_2 \end{bmatrix},$$

where $a_1 > 0$ and $a_2 > 0$ are positive constants specifying the desired closed-loop dynamics.

Further, denote by $r_{\max}$ the (known) upper bound on the reference signals that we want to track. Similarly, let $\dot{r}_{\max}$ and $\ddot{r}_{\max}$ be the (known) upper bounds on their first and second derivatives. Let

$$\rho_{\text{in}} \triangleq \left\| s(s\mathbb{I} - A_g)^{-1} \right\|_{\mathcal{L}_1} ((1+a_1)(\rho_0 + r_{\max}) + \dot{r}_{\max}) . \tag{3.2}$$

Notice that from the fact that A_g is Hurwitz and $(s\mathbb{I} - A_g)^{-1}$ is a strictly proper transfer matrix, it follows that $\|s(s\mathbb{I} - A_g)^{-1}\|_{\mathcal{L}_1}$ is bounded.

Moreover, for every $\delta > 0$, let

$$L_{i\delta} \triangleq \frac{\bar{\delta}(\delta)}{\delta} d_{f_{xi}}(\bar{\delta}(\delta)), \quad \bar{\delta}(\delta) \triangleq \delta + \bar{\gamma}_1 , \tag{3.3}$$

where $d_{f_x}(\cdot)$ was introduced in Assumption 3.1.3 and $\bar{\gamma}_1$ is an arbitrarily small positive constant. Using the redefinitions in (3.3), let L_δ be defined as

$$L_\delta \triangleq \max\{L_{1\delta}, L_{2\delta}\} . \tag{3.4}$$

For the proofs of stability and performance bounds, the selection of $C_1(s)$, k, and $D(s)$ needs to ensure that, for given ρ_0, there exists $\rho_r > \rho_{\text{in}}$ such that the following $\mathcal{L}_1$-norm upper bound can be verified:

$$\|G(s)\|_{\mathcal{L}_1} < \frac{\rho_r - \beta_0(\rho_r)}{L^\star_{\rho_r} \rho_r + \beta_1(\rho_r)} , \tag{3.5}$$

where

$$G(s) \triangleq H(s)(\mathbb{I} - C(s)) , \quad H(s) \triangleq \left(s\mathbb{I} - A_g\right)^{-1} , \tag{3.6}$$

while $L^\star_{\rho_r}$, $\beta_0(\rho_r)$ and $\beta_1(\rho_r)$ are defined as

$$\begin{aligned} L^\star_{\rho_r} &\triangleq \left(1 + a_1 + \|C_1(s)\|_{\mathcal{L}_1} L_{\rho_r}\right) \kappa_1(\rho_r), \\ \beta_0(\rho_r) &\triangleq \|C_1(s)\|_{\mathcal{L}_1} B + \left(1 + \|C_1(s)\|_{\mathcal{L}_1} L_{\rho_r}\right) r_{\max} + \dot{r}_{\max} \\ &\quad + \left(1 + a_1 + \|C_1(s)\|_{\mathcal{L}_1} L_{\rho_r}\right) \rho_{\text{in}} , \\ \beta_1(\rho_r) &\triangleq \left(1 + a_1 + \|C_1(s)\|_{\mathcal{L}_1} L_{\rho_r}\right) \kappa_2 , \end{aligned} \tag{3.7}$$

with

$$\kappa_1(\rho_r) \triangleq \left(1 + a_1 + a_2 \|C_1(s)\|_{\mathcal{L}_1} + \|sC_1(s)\|_{\mathcal{L}_1}\right) L_{\rho_r} + a_2(1 + a_1) + a_1 , \tag{3.8}$$

$$\begin{aligned} \kappa_2 &\triangleq \left(1 + a_1 + a_2 \|C_1(s)\|_{\mathcal{L}_1} + \|sC_1(s)\|_{\mathcal{L}_1}\right) B + a_1 a_2 r_{\max} \\ &\quad + (a_1 + a_2)\dot{r}_{\max} + \ddot{r}_{\max} . \end{aligned} \tag{3.9}$$

To streamline the subsequent analysis of stability and performance bounds, we need to introduce some notations. Define

$$\rho \triangleq \rho_r + \bar{\gamma}_1 , \tag{3.10}$$

and let γ_1 be given by

$$\gamma_1 \triangleq \frac{\kappa_3(\rho_r)}{1-\|G(s)\|_{\mathcal{L}_1} L^\star_{\rho_r}}\gamma_0 + \beta , \tag{3.11}$$

where β and γ_0 are arbitrarily small positive constants such that $\gamma_1 \leq \bar{\gamma}_1$, and $\kappa_3(\rho_r)$ is defined as

$$\begin{aligned}\kappa_3(\rho_r) \triangleq &\left(1+a_1+\|C_1(s)\|_{\mathcal{L}_1} L_{\rho_r}\right)\left(\|G(s)\|_{\mathcal{L}_1}\kappa_4 + \|H(s)C(s)(s\mathbb{I}-A_m)\|_{\mathcal{L}_1}\right)\\ &+\|C_1(s)(s+a_1)\|_{\mathcal{L}_1}\end{aligned} \tag{3.12}$$

with

$$\kappa_4 \triangleq a_2\,\|C_1(s)(s+a_1)\|_{\mathcal{L}_1} + \|sC_1(s)(s+a_1)\|_{\mathcal{L}_1} . \tag{3.13}$$

Moreover, let

$$\rho_u \triangleq \rho_{u_r} + \gamma_2 , \tag{3.14}$$

where ρ_{ur} and γ_2 are given as

$$\rho_{ur} \triangleq \left\|\frac{C_2(s)}{\omega}\right\|_{\mathcal{L}_1}\left(\kappa_1(\rho_r)\rho_r + \kappa_2\right), \tag{3.15}$$

$$\gamma_2 \triangleq \left\|\frac{C_2(s)}{\omega}\right\|_{\mathcal{L}_1}\kappa_1(\rho_r)\gamma_1 + \kappa_5\gamma_0 , \tag{3.16}$$

with κ_5 being defined as

$$\kappa_5 \triangleq \left\|\frac{C_2(s)}{\omega}\right\|_{\mathcal{L}_1}\kappa_4 + \left\|\frac{C_2(s)}{\omega}(s+a_2)\right\|_{\mathcal{L}_1} .$$

Finally, let

$$\theta_b \triangleq \max\{d_{f_x 1}(\rho), d_{f_x 2}(\rho)\}, \qquad \Delta \triangleq B + \epsilon , \tag{3.17}$$

where $\epsilon > 0$ is an arbitrary constant.

The elements of $\mathcal{L}_1$ adaptive (backstepping) controller are introduced next.

State Predictor

We consider the following state predictor:

$$\begin{aligned}\dot{\hat{x}}_1(t) &= -a_1\tilde{x}_1(t) + \hat{\theta}_1(t)|x_1(t)| + \hat{\sigma}_1(t) + x_2(t), \quad \hat{x}_1(0) = x_{1_0} ,\\ \dot{\hat{x}}_2(t) &= -a_2\tilde{x}_2(t) + \hat{\theta}_2(t)\|x(t)\|_\infty + \hat{\sigma}_2(t) + \hat{\omega}(t)u(t), \quad \hat{x}_2(0) = x_{2_0} ,\end{aligned} \tag{3.18}$$

where $\hat{x}_1(t), \hat{x}_2(t) \in \mathbb{R}$ are the predictor states, $\hat{x}(t) \triangleq [\hat{x}_1(t), \hat{x}_2(t)]^\top$; $\tilde{x}(t) \triangleq [\tilde{x}_1(t), \tilde{x}_2(t)]^\top \triangleq [\hat{x}_1(t) - x_1(t), \hat{x}_2(t) - x_2(t)]^\top$ is the prediction error; and $\hat{\theta}_1(t)$, $\hat{\theta}_2(t)$, $\hat{\sigma}_1(t)$, $\hat{\sigma}_2(t)$, and $\hat{\omega}(t)$ are the adaptive estimates.

Adaptation Laws

The adaptive estimates are governed by

$$\begin{aligned}
\dot{\hat{\omega}}(t) &= \Gamma \mathrm{Proj}\left(\hat{\omega}(t), -\tilde{x}^\top(t)P[0,\ 1]^\top u(t)\right), \quad \hat{\omega}(0)=\hat{\omega}_0,\\
\dot{\hat{\theta}}_1(t) &= \Gamma \mathrm{Proj}\left(\hat{\theta}_1(t), -\tilde{x}^\top(t)P[1,\ 0]^\top |x_1(t)|\right), \quad \hat{\theta}_1(0)=\hat{\theta}_{1_0},\\
\dot{\hat{\theta}}_2(t) &= \Gamma \mathrm{Proj}\left(\hat{\theta}_2(t), -\tilde{x}^\top(t)P[0,\ 1]^\top \|x(t)\|_\infty\right), \quad \hat{\theta}_2(0)=\hat{\theta}_{2_0},\\
\dot{\hat{\sigma}}_1(t) &= \Gamma \mathrm{Proj}\left(\hat{\sigma}_1(t), -\tilde{x}^\top(t)P[1,\ 0]^\top\right), \quad \hat{\sigma}_1(0)=\hat{\sigma}_{1_0},\\
\dot{\hat{\sigma}}_2(t) &= \Gamma \mathrm{Proj}\left(\hat{\sigma}_2(t), -\tilde{x}^\top(t)P[0,\ 1]^\top\right), \quad \hat{\sigma}_2(0)=\hat{\sigma}_{2_0},
\end{aligned} \tag{3.19}$$

where $\Gamma \in \mathbb{R}^+$ is the adaptation gain, and the symmetric positive definite matrix $P = P^\top > 0$ solves the Lyapunov equation $A_m^\top P + PA_m = -Q$ for arbitrary $Q = Q^\top > 0$. The projection operator $\mathrm{Proj}(\cdot,\ \cdot)$ ensures that $|\hat{\theta}_i(t)| \le \theta_b$, $\hat{\sigma}_i(t) \le \Delta$, $i = 1,2$, and $\hat{\omega}(t) \in \Omega$, where θ_b and Δ were defined in (3.17).

Control Law

Let

$$\alpha(t) \triangleq -a_1(x_1(t)-r(t)) - \hat{\eta}_{1C}(t) + \dot{r}(t), \tag{3.20}$$

where $\hat{\eta}_{1C}(t)$ is the signal with Laplace transform $\hat{\eta}_{1C}(s) \triangleq C_1(s)\hat{\eta}_1(s)$ with $\hat{\eta}_1(t) \triangleq \hat{\theta}_1(t)|x_1(t)| + \hat{\sigma}_1(t)$. Then, the control signal is generated as the output of the (feedback) system

$$u(s) = -kD(s)\eta_u(s), \tag{3.21}$$

where $\eta_u(s)$ is the Laplace transform of the signal

$$\eta_u(t) \triangleq \hat{\eta}_2(t) + a_2(x_2(t) - \alpha(t)) - \dot{\alpha}(t),$$

with $\hat{\eta}_2(t) \triangleq \hat{\omega}(t)u(t) + \hat{\theta}_2(t)\|x(t)\|_\infty + \hat{\sigma}_2(t)$.

3.1.3 Analysis of the $\mathcal{L}_1$ Adaptive Controller

Equivalent (Semi-)Linear Time-Varying System

Next, we refer to Lemma A.8.1 to transform the nonlinear system in (3.1) into a (semi-)linear system with unknown time-varying parameters and disturbances. Since

$$\|x_0\|_\infty < \rho_r < \rho, \quad u(0) = 0,$$

and $x(t)$, $u(t)$ are continuous, there always exists τ such that

$$\|x_\tau\|_{\mathcal{L}_\infty} \le \rho, \quad \|u_\tau\|_{\mathcal{L}_\infty} \le \rho_u.$$

Then, Lemma A.8.1 implies that the nonlinear system in (3.1) can be rewritten over $t \in [0, \tau]$ as

$$\begin{aligned} \dot{x}_1(t) &= \theta_1(t)|x_1(t)| + \sigma_1(t) + x_2(t), \quad x_1(0) = x_{1_0}, \\ \dot{x}_2(t) &= \theta_2(t)\|x(t)\|_\infty + \sigma_2(t) + \omega u(t), \quad x_2(0) = x_{2_0}, \end{aligned} \tag{3.22}$$

where $\theta_i(t)$, $\sigma_i(t)$, $i = 1,2$, are unknown time-varying signals subject to

$$\begin{aligned} |\theta_i(t)| < \theta_b, \quad |\sigma_i(t)| < \Delta, \quad \forall\, t \in [0, \tau], \\ |\dot{\theta}_i(t)| < d_{\theta_i}(\rho, \rho_u), \quad |\dot{\sigma}_i(t)| \leq d_{\sigma_i}(\rho, \rho_u), \quad \forall\, t \in [0, \tau], \end{aligned}$$

with θ_b and Δ being defined in (3.17), and $d_{\theta_i}(\rho, \rho_u) > 0$ and $d_{\sigma_i}(\rho, \rho_u) > 0$ being the bounds guaranteed by Lemma A.8.1.

Closed-Loop Reference System

In this section, we characterize the closed-loop reference system that the $\mathcal{L}_1$ adaptive controller tracks in both transient and steady-state and prove its stability. Toward that end, consider the ideal nonadaptive version of the adaptive controller and define the *closed-loop reference system* as

$$\begin{aligned} \dot{x}_{1\text{ref}}(t) &= f_1(t, x_{1\text{ref}}(t)) + x_{2\text{ref}}(t), \quad x_{1\text{ref}}(0) = x_{10}, \\ \dot{x}_{2\text{ref}}(t) &= f_2(t, x_{\text{ref}}(t)) + \omega u_{\text{ref}}(t), \quad x_{2\text{ref}}(0) = x_{20}, \\ u_{\text{ref}}(s) &= -\frac{C_2(s)}{\omega}\eta_{u_{\text{ref}}}(s), \end{aligned} \tag{3.23}$$

where $x_{\text{ref}}(t) \triangleq [x_{1\text{ref}}(t),\ x_{2\text{ref}}(t)]^\top$, $x_{\text{ref}}(0) \triangleq x_0$, and

$$\eta_{u_{\text{ref}}}(t) \triangleq f_2(t, x_{\text{ref}}(t)) + a_2(x_{2\text{ref}}(t) - \alpha_{\text{ref}}(t)) - \dot{\alpha}_{\text{ref}}(t), \tag{3.24}$$

with $\alpha_{\text{ref}}(t)$ being defined as

$$\alpha_{\text{ref}}(t) \triangleq -a_1(x_{1\text{ref}}(t) - r(t)) - \eta_{1C\text{ref}}(t) + \dot{r}(t), \tag{3.25}$$

and with $\eta_{1C\text{ref}}(t)$ being the signal with the Laplace transform $\eta_{1C\text{ref}}(s) \triangleq C_1(s)\eta_{1\text{ref}}(s)$, where $\eta_{1\text{ref}}(t) \triangleq f_1(t, x_{1\text{ref}}(t))$. For convenience, we also define $\eta_{2\text{ref}}(t) \triangleq f_2(t, x_{\text{ref}}(t))$.

Lemma 3.1.1 For the closed-loop reference system in (3.23), subject to the $\mathcal{L}_1$-norm condition in (3.5), if $\|x_0\|_\infty < \rho_0$, then

$$\|x_{\text{ref}\tau}\|_{\mathcal{L}_\infty} < \rho_r, \quad \|u_{\text{ref}\tau}\|_{\mathcal{L}_\infty} < \rho_{ur}, \tag{3.26}$$

where ρ_r and ρ_{ur} were defined in (3.5) and (3.15), respectively.

Proof. Suppose that the bound on $\|x_{\text{ref}}\|_{\mathcal{L}_\infty}$ is not true. Then, because $\|x_0\|_\infty < \rho_r$, it follows from the continuity of the solution that there exists a time instant $\tau > 0$, such that

$$\|x_{\text{ref}}(t)\|_\infty < \rho_r, \quad \forall\, t \in [0, \tau),$$

and

$$\|x_{\rm ref}(\tau)\|_\infty = \rho_r\,,$$

which implies that

$$\|x_{\rm ref\tau}\|_{\mathcal{L}_\infty} = \rho_r\,. \tag{3.27}$$

Let

$$e_{\rm ref}(t) \triangleq \begin{bmatrix} e_{1\rm ref}(t) \\ e_{2\rm ref}(t) \end{bmatrix} \triangleq \begin{bmatrix} x_{1\rm ref}(t) - r(t) \\ x_{2\rm ref}(t) - \alpha_{\rm ref}(t) \end{bmatrix}. \tag{3.28}$$

It follows from (3.23) and (3.25) that

$$\begin{aligned} \dot e_{\rm ref}(t) &= \begin{bmatrix} \dot x_{1\rm ref}(t) - \dot r(t) \\ \dot x_{2\rm ref}(t) - \dot\alpha_{\rm ref}(t) \end{bmatrix} = \begin{bmatrix} \eta_{1\rm ref}(t) + x_{2\rm ref}(t) - \dot r(t) \\ \eta_{2\rm ref}(t) + \omega u_{\rm ref}(t) - \dot\alpha_{\rm ref}(t) \end{bmatrix} \\ &= A_g e_{\rm ref}(t) + \begin{bmatrix} \eta_{1\rm ref}(t) - \eta_{1C\rm ref}(t) \\ a_2 e_{2\rm ref}(t) + \eta_{2\rm ref}(t) - \dot\alpha_{\rm ref}(t) + \omega u_{\rm ref}(t) \end{bmatrix}. \end{aligned}$$

In frequency domain, this expression can be rewritten as

$$e_{\rm ref}(s) = H(s)\begin{bmatrix} \eta_{1\rm ref}(s) - \eta_{1C\rm ref}(s) \\ a_2 e_{2\rm ref}(s) + \eta_{2\rm ref}(s) - \eta_{\dot\alpha_{\rm ref}}(s) + \omega u_{\rm ref}(s) \end{bmatrix} + H(s) e_0\,, \tag{3.29}$$

where $\eta_{\dot\alpha_{\rm ref}}(s)$ is the Laplace transform of $\dot\alpha_{\rm ref}(t)$ and e_0 is the initial condition of $e_{\rm ref}(t)$, i.e., $e_0 \triangleq e_{\rm ref}(0)$. Note that e_0 can be upper bounded as follows:

$$\|e_0\|_\infty \le (1 + a_1)(\rho_0 + r_{\max}) + \dot r_{\max}\,.$$

Also, it follows from (3.23) and (3.24) that

$$u_{\rm ref}(s) = -\frac{C_2(s)}{\omega}\left(a_2 e_{2\rm ref}(s) + \eta_{2\rm ref}(s) - \eta_{\dot\alpha\rm ref}(s)\right). \tag{3.30}$$

Substituting (3.30) into (3.29), we have

$$e_{\rm ref}(s) = G(s)\zeta_{\rm ref}(s) + H(s)e_0\,, \tag{3.31}$$

where $\zeta_{\rm ref}(s)$ is the Laplace transform of the signal

$$\zeta_{\rm ref}(t) \triangleq \begin{bmatrix} \eta_{1\rm ref}(t) \\ a_2 e_{2\rm ref}(t) + \eta_{2\rm ref}(t) - \dot\alpha_{\rm ref}(t) \end{bmatrix}. \tag{3.32}$$

Next, we derive a bound for $\|e_{\rm ref\,\tau}\|_{\mathcal{L}_\infty}$. First, we note that

$$\|\zeta_{\rm ref\,\tau}\|_{\mathcal{L}_\infty} \le \max\left\{\|\eta_{1\rm ref\,\tau}\|_{\mathcal{L}_\infty}, \|\eta_{2\rm ref\,\tau}\|_{\mathcal{L}_\infty}\right\} + a_2\|e_{2\rm ref\,\tau}\|_{\mathcal{L}_\infty} + \|\dot\alpha_{\rm ref\,\tau}\|_{\mathcal{L}_\infty}\,,$$

and, since $e_{2\rm ref}(t) = x_{2\rm ref}(t) - \alpha_{\rm ref}(t)$, it follows that

$$\begin{aligned} \|\zeta_{\rm ref\,\tau}\|_{\mathcal{L}_\infty} \le{}& \max\left\{\|\eta_{1\rm ref\,\tau}\|_{\mathcal{L}_\infty}, \|\eta_{2\rm ref\,\tau}\|_{\mathcal{L}_\infty}\right\} + a_2\|x_{2\rm ref\,\tau}\|_{\mathcal{L}_\infty} \\ &+ a_2\|\alpha_{\rm ref\,\tau}\|_{\mathcal{L}_\infty} + \|\dot\alpha_{\rm ref\,\tau}\|_{\mathcal{L}_\infty}\,. \end{aligned} \tag{3.33}$$

Next, taking into consideration Assumptions 3.1.2 and 3.1.3, the equality in (3.27), together with the fact that $\rho_r < \bar{\rho}_r(\rho_r)$, yields the following upper bound:

$$\left\| \eta_{i\mathrm{ref}\,\tau} \right\|_{\mathcal{L}_\infty} \leq d_{f_{xi}}(\bar{\rho}_r(\rho_r))\bar{\rho}_r(\rho_r) + B\,, \qquad i = 1,2\,.$$

From the redefinition in (3.3), it follows that

$$\left\| \eta_{i\mathrm{ref}\,\tau} \right\|_{\mathcal{L}_\infty} \leq L_{i\rho_r}\rho_r + B\,, \qquad i = 1,2\,,$$

and the definition of L_δ in (3.4) leads to

$$\left\| \eta_{i\mathrm{ref}\,\tau} \right\|_{\mathcal{L}_\infty} \leq L_{\rho_r}\rho_r + B\,, \qquad i = 1,2\,. \tag{3.34}$$

Moreover, from the definition of $\alpha_{\mathrm{ref}}(t)$ in (3.25) and the bounds in (3.27) and (3.34), it follows that

$$\begin{aligned} \|\alpha_{\mathrm{ref}\,\tau}\|_{\mathcal{L}_\infty} &\leq \|C_1(s)\|_{\mathcal{L}_1} \left\| \eta_{1\mathrm{ref}\,\tau} \right\|_{\mathcal{L}_\infty} + a_1 \left(\|x_{1\mathrm{ref}\,\tau}\|_{\mathcal{L}_\infty} + \|r\|_{\mathcal{L}_\infty} \right) + \|\dot{r}\|_{\mathcal{L}_\infty} \\ &\leq \|C_1(s)\|_{\mathcal{L}_1} \left(L_{\rho_r}\rho_r + B \right) + a_1 \left(\rho_r + \|r\|_{\mathcal{L}_\infty} \right) + \|\dot{r}\|_{\mathcal{L}_\infty}\,. \end{aligned} \tag{3.35}$$

Further, from the definition of $\alpha_{\mathrm{ref}}(t)$ in (3.25) and the definition of the closed-loop reference system in (3.23), we have

$$\begin{aligned} \dot{\alpha}_{\mathrm{ref}}(t) &= -a_1 \left(\dot{x}_{1\mathrm{ref}}(t) - \dot{r}(t) \right) - \dot{\eta}_{1C\mathrm{ref}}(t) + \ddot{r}(t) \\ &= -a_1 \left(\eta_{1\mathrm{ref}}(t) + x_{2\mathrm{ref}}(t) - \dot{r}(t) \right) - \dot{\eta}_{1C\mathrm{ref}}(t) + \ddot{r}(t)\,, \end{aligned}$$

which, together with the bound in (3.34) and the equality in (3.27), leads to

$$\begin{aligned} \|\dot{\alpha}_{\mathrm{ref}\,\tau}\|_{\mathcal{L}_\infty} &\leq \left(a_1 + \|sC_1(s)\|_{\mathcal{L}_1} \right) \left\| \eta_{1\mathrm{ref}\,\tau} \right\|_{\mathcal{L}_\infty} + a_1 \left\| x_{2\mathrm{ref}\,\tau} \right\|_{\mathcal{L}_\infty} \\ &\quad + a_1 \|\dot{r}\|_{\mathcal{L}_\infty} + \|\ddot{r}\|_{\mathcal{L}_\infty} \\ &\leq \left(a_1 + \|sC_1(s)\|_{\mathcal{L}_1} \right) \left(L_{\rho_r}\rho_r + B \right) + a_1\rho_r + a_1 \|\dot{r}\|_{\mathcal{L}_\infty} + \|\ddot{r}\|_{\mathcal{L}_\infty}\,. \end{aligned} \tag{3.36}$$

Then, the bounds in (3.33), (3.34), (3.35), and (3.36) lead to

$$\begin{aligned} \left\| \zeta_{\mathrm{ref}\,\tau} \right\|_{\mathcal{L}_\infty} &\leq \Big(\left(1 + a_1 + a_2\|C_1(s)\|_{\mathcal{L}_1} + \|sC_1(s)\|_{\mathcal{L}_1} \right) L_{\rho_r} + a_2(1 + a_1) + a_1 \Big) \rho_r \\ &\quad + \Big(\left(1 + a_1 + a_2\|C_1(s)\|_{\mathcal{L}_1} + \|sC_1(s)\|_{\mathcal{L}_1} \right) B + a_1a_2\|r\|_{\mathcal{L}_\infty} \\ &\quad + (a_1 + a_2)\|\dot{r}\|_{\mathcal{L}_\infty} + \|\ddot{r}\|_{\mathcal{L}_\infty} \Big)\,, \end{aligned}$$

and the definitions of $\kappa_1(\rho_r)$ and κ_2 in (3.8) and (3.9) imply that

$$\left\| \zeta_{\mathrm{ref}\,\tau} \right\|_{\mathcal{L}_\infty} \leq \kappa_1(\rho_r)\rho_r + \kappa_2\,. \tag{3.37}$$

From this bound, the expression in (3.31), and the definition of ρ_{in} in (3.2), it follows that

$$\|e_{\mathrm{ref}\,\tau}\|_{\mathcal{L}_\infty} \leq \|G(s)\|_{\mathcal{L}_1} \left(\kappa_1(\rho_r)\rho_r + \kappa_2 \right) + \rho_{\mathrm{in}}\,. \tag{3.38}$$

Using the bound above, we next prove that the $\mathcal{L}_1$-norm condition in (3.5) leads to the first bound in (3.26). From the definition of $e_{\mathrm{ref}}(t)$ in (3.28) it follows that

$$\begin{bmatrix} x_{1\mathrm{ref}}(t) \\ x_{2\mathrm{ref}}(t) \end{bmatrix} = e_{\mathrm{ref}}(t) + \begin{bmatrix} r(t) \\ \alpha_{\mathrm{ref}}(t) \end{bmatrix},$$

which leads to the following upper bound:

$$\|x_{\text{ref}\,\tau}\|_{\mathcal{L}_\infty} \le \|e_{\text{ref}\,\tau}\|_{\mathcal{L}_\infty} + \|r\|_{\mathcal{L}_\infty} + \|\alpha_{\text{ref}\,\tau}\|_{\mathcal{L}_\infty} . \tag{3.39}$$

From the definition of $\alpha_{\text{ref}}(t)$ in (3.25) we have

$$\alpha_{\text{ref}}(t) = -a_1 e_{1\text{ref}}(t) - \eta_{1C\text{ref}}(t) + \dot{r}(t),$$

and consequently

$$\begin{aligned}\|\alpha_{\text{ref}\,\tau}\|_{\mathcal{L}_\infty} &\le a_1 \|e_{1\text{ref}\,\tau}\|_{\mathcal{L}_\infty} + \|C_1(s)\|_{\mathcal{L}_1}\left(L_{\rho_r}\|x_{1\text{ref}\,\tau}\|_{\mathcal{L}_\infty} + B\right) + \|\dot{r}\|_{\mathcal{L}_\infty}\\ &\le a_1 \|e_{1\text{ref}\,\tau}\|_{\mathcal{L}_\infty} + \|C_1(s)\|_{\mathcal{L}_1}\left(L_{\rho_r}\left(\|e_{1\text{ref}\,\tau}\|_{\mathcal{L}_\infty} + \|r\|_{\mathcal{L}_\infty}\right) + B\right) + \|\dot{r}\|_{\mathcal{L}_\infty}\\ &\le a_1 \|e_{\text{ref}\,\tau}\|_{\mathcal{L}_\infty} + \|C_1(s)\|_{\mathcal{L}_1}\left(L_{\rho_r}\left(\|e_{\text{ref}\,\tau}\|_{\mathcal{L}_\infty} + \|r\|_{\mathcal{L}_\infty}\right) + B\right) + \|\dot{r}\|_{\mathcal{L}_\infty} ,\end{aligned}$$

which, together with the bound in (3.39), leads to

$$\begin{aligned}\|x_{\text{ref}\,\tau}\|_{\mathcal{L}_\infty} \le{}& \left(1 + a_1 + \|C_1(s)\|_{\mathcal{L}_1} L_{\rho_r}\right)\|e_{\text{ref}\,\tau}\|_{\mathcal{L}_\infty}\\ &+ \left(\|C_1(s)\|_{\mathcal{L}_1}\left(L_{\rho_r}\|r\|_{\mathcal{L}_\infty} + B\right) + \|r\|_{\mathcal{L}_\infty} + \|\dot{r}\|_{\mathcal{L}_\infty}\right).\end{aligned}$$

Then, the upper bound on $\|e_{\text{ref}\,\tau}\|_{\mathcal{L}_\infty}$ in (3.38) yields

$$\begin{aligned}\|x_{\text{ref}\,\tau}\|_{\mathcal{L}_\infty} \le{}& \left(1 + a_1 + \|C_1(s)\|_{\mathcal{L}_1} L_{\rho_r}\right)\left(\|G(s)\|_{\mathcal{L}_1}(\kappa_1(\rho_r)\rho_r + \kappa_2) + \rho_{\text{in}}\right)\\ &+ \left(\|C_1(s)\|_{\mathcal{L}_1}\left(L_{\rho_r}\|r\|_{\mathcal{L}_\infty} + B\right) + \|r\|_{\mathcal{L}_\infty} + \|\dot{r}\|_{\mathcal{L}_\infty}\right),\end{aligned}$$

which implies that

$$\|x_{\text{ref}\,\tau}\|_{\mathcal{L}_\infty} \le \|G(s)\|_{\mathcal{L}_1}\left(L^\star_{\rho_r}\rho_r + \beta_1(\rho_r)\right) + \beta_0(\rho_r).$$

The condition in (3.5) can be solved for ρ_r to obtain the bound

$$\|G(s)\|_{\mathcal{L}_1}\left(L^\star_{\rho_r}\rho_r + \beta_1(\rho_r)\right) + \beta_0(\rho_r) < \rho_r,$$

which leads to

$$\|x_{\text{ref}\,\tau}\|_{\mathcal{L}_\infty} < \rho_r$$

and contradicts the equality in (3.27), thus proving the first bound in (3.26).

To prove the second bound in (3.26), we first note that, from the expressions in (3.30) and (3.32), it follows that

$$\|u_{\text{ref}\,\tau}\|_{\mathcal{L}_\infty} \le \left\|\frac{C_2(s)}{\omega}\right\|_{\mathcal{L}_1} \|\zeta_{\text{ref}\,\tau}\|_{\mathcal{L}_\infty} .$$

Since the bound on $\|x_{\text{ref}}\|_{\mathcal{L}_\infty}$ in (3.26) implies that the upper bound in (3.37) holds for all $t \in [0,\tau]$ with strict inequality, we have

$$\|u_{\text{ref}}\|_{\mathcal{L}_\infty} < \left\|\frac{C_2(s)}{\omega}\right\|_{\mathcal{L}_1} (\kappa_1(\rho_r)\rho_r + \kappa_2) = \rho_{ur} ,$$

which proves the bound on $\|u_{\text{ref}}\|_{\mathcal{L}_\infty}$ in (3.26). □

Transient and Steady-State Performance

Using (3.18) and (3.22) one can write the error dynamics over $t \in [t_0, \tau]$

$$\dot{\tilde{x}}(t) = A_m\tilde{x}(t) + \begin{bmatrix} \tilde{\theta}_1(t)|x_1(t)| + \tilde{\sigma}_1(t) \\ \tilde{\omega}(t)u(t) + \tilde{\theta}_2(t)\|x(t)\|_\infty + \tilde{\sigma}_2(t) \end{bmatrix}, \quad \tilde{x}(0) = 0, \tag{3.40}$$

where $\tilde{\omega}(t) \triangleq \hat{\omega}(t) - \omega$, $\tilde{\theta}_1(t) \triangleq \hat{\theta}_1(t) - \theta_1(t)$, $\tilde{\sigma}_1(t) \triangleq \hat{\sigma}_1(t) - \sigma_1(t)$, $\tilde{\theta}_2(t) \triangleq \hat{\theta}_2(t) - \theta_2(t)$, and $\tilde{\sigma}_2(t) \triangleq \hat{\sigma}_2(t) - \sigma_2(t)$.

Lemma 3.1.2 For the dynamics in (3.40), if

$$\|x_\tau\|_{\mathcal{L}_\infty} \leq \rho, \quad \|u_\tau\|_{\mathcal{L}_\infty} \leq \rho_u, \tag{3.41}$$

and the adaptive gain is chosen to satisfy the design constraint

$$\Gamma > \frac{\theta_m(\rho_r)}{\lambda_{\min}(P)\gamma_0^2}, \tag{3.42}$$

where γ_0 was introduced in (3.11) and $\theta_m(\rho_r)$ is defined as

$$\theta_m(\rho_r) = 8\theta_b^2 + 8\Delta^2 + (\omega_u - \omega_l)^2 + 4\frac{\lambda_{\max}(P)}{\lambda_{\min}(Q)}\left(\sum_{i=1}^{2}\left(\theta_b d_{\theta_i} + \Delta d_{\sigma_i}\right)\right), \tag{3.43}$$

the following bound holds:

$$\|\tilde{x}_\tau\|_{\mathcal{L}_\infty} < \gamma_0 .$$

Proof. First, we note that, if the bounds in (3.41) hold, then one can write the prediction error dynamics in (3.40). Next, consider the Lyapunov function candidate

$$V(\tilde{x}(t), \tilde{\theta}_i(t), \tilde{\sigma}_i(t), \tilde{\omega}(t)) = \tilde{x}^\top(t)P\tilde{x}(t) + \frac{1}{\Gamma}\left(\tilde{\theta}_1^2(t) + \tilde{\theta}_2^2(t) + \tilde{\sigma}_1^2(t) + \tilde{\sigma}_2^2(t) + \tilde{\omega}^2(t)\right), \tag{3.44}$$

and let $t_1 \in [0, \tau)$ be the first instant of the discontinuity of any of the derivatives $\dot{\theta}_i$ or $\dot{\sigma}_i$. Next we prove that

$$V(t) \leq \frac{\theta_m(\rho_r)}{\Gamma}, \quad \forall\, t \in [0,\ t_1].$$

The projection-based adaptive laws in (3.19) ensure that for arbitrary $t \in [0, t_1)$,

$$\dot{V}(t) \leq -\tilde{x}^\top(t)Q\tilde{x}(t) + \frac{2}{\Gamma}\left(|\tilde{\theta}_1(t)||\dot{\theta}_1(t)| + |\tilde{\theta}_2(t)||\dot{\theta}_2(t)| + |\tilde{\sigma}_1(t)||\dot{\sigma}_1(t)| + |\tilde{\sigma}_2(t)||\dot{\sigma}_2(t)|\right).$$

They also imply that

$$\max_{t \in [0,\ t_1)}\left(\tilde{\theta}_1^2(t) + \tilde{\theta}_2^2(t) + \tilde{\sigma}_1^2(t) + \tilde{\sigma}_2^2(t) + \tilde{\omega}^2(t)\right) \leq \left(4\left(2\theta_b^2 + 2\Delta^2\right) + (\omega_u - \omega_l)^2\right).$$

Since $\dot{\theta}_i(t)$ and $\dot{\sigma}_i(t)$ are continuous over $t \in [0,\ t_1)$, we conclude

$$\max_{t\in[0,\ t_1)} \Big(|\tilde{\theta}_1(t)||\dot{\theta}_1(t)| + |\tilde{\theta}_2(t)||\dot{\theta}_2(t)| + |\tilde{\sigma}_1(t)||\dot{\sigma}_1(t)| + |\tilde{\sigma}_2(t)||\dot{\sigma}_2(t)|\Big) \le 2\sum_{i=1}^{2}\left(\theta_b d_{\theta_i} + \Delta d_{\sigma_i}\right).$$

Next, notice that if at arbitrary time $t' \in [0, t_1)$, $V(t') > \frac{\theta_m(\rho_r)}{\Gamma}$, it follows from (3.43) and (3.44) that

$$\tilde{x}^\top(t')P\tilde{x}(t') > 4\frac{\lambda_{\max}(P)}{\lambda_{\min}(Q)}\left(\sum_{i=1}^{2}\left(\theta_b d_{\theta_i} + \Delta d_{\sigma_i}\right)\right),$$

and therefore

$$\tilde{x}^\top(t')Q\tilde{x}(t') > 4\sum_{i=1}^{2}\left(\theta_b d_{\theta_i} + \Delta d_{\sigma_i}\right).$$

This implies that $\dot{V}(t') < 0$. Since $V(0) < \frac{\theta_m(\rho_r)}{\Gamma}$, it follows that $V(t) \le \frac{\theta_m(\rho_r)}{\Gamma}$ for arbitrary $t \in [0,\ t_1)$, which yields

$$\|\tilde{x}(t)\|_\infty \le \sqrt{\frac{\theta_m(\rho_r)}{(\lambda_{\min}(P)\Gamma)}}, \quad \forall\, t \in [0, t_1).$$

The design constraint on Γ in (3.42) leads to

$$\|\tilde{x}(t)\|_\infty < \gamma_0, \quad \forall\, t \in [0, t_1).$$

If $t_2 \in (t_1,\ \tau]$ is the next instant of the discontinuity of the derivative, we can use similar derivations as above to prove that

$$\|\tilde{x}(\tau)\|_\infty < \gamma_0, \quad \forall\, t \in [t_1,\ t_2).$$

Iterating the process until the time instant t, we get

$$\|\tilde{x}_\tau\|_{\mathcal{L}_\infty} < \gamma_0,$$

which completes the proof. □

Theorem 3.1.1 For the system in (3.1), with the $\mathcal{L}_1$ adaptive controller defined via (3.18)–(3.21), subject to the $\mathcal{L}_1$-norm condition in (3.5) and the design constraint in (3.42), if $\|x_0\|_\infty < \rho_0$, then

$$\|x\|_{\mathcal{L}_\infty} \le \rho, \tag{3.45}$$

$$\|u\|_{\mathcal{L}_\infty} \le \rho_u, \tag{3.46}$$

$$\|\tilde{x}\|_{\mathcal{L}_\infty} \le \gamma_0, \tag{3.47}$$

$$\|x_{\text{ref}} - x\|_{\mathcal{L}_\infty} \le \gamma_1, \tag{3.48}$$

$$\|u_{\text{ref}} - u\|_{\mathcal{L}_\infty} \le \gamma_2, \tag{3.49}$$

where γ_1 and γ_2 were defined in (3.11) and (3.16), respectively.

Proof. The proof is done by contradiction. Assume that the bounds in (3.48) and (3.49) do not hold. Then, since $\|x_{\rm ref}(0)-x(0)\|_\infty = 0 < \gamma_1$, $\|u_{\rm ref}(0)-u(0)\|_\infty = 0 < \gamma_2$, and $x(t)$, $x_{\rm ref}(t)$, $u(t)$, and $u_{\rm ref}(t)$ are continuous, there exists a time $\tau > 0$ such that

$$\|x_{\rm ref}(\tau)-x(\tau)\|_\infty = \gamma_1\,, \qquad \text{or} \qquad \|u_{\rm ref}(\tau)-u(\tau)\|_\infty = \gamma_2\,,$$

while

$$\|x_{\rm ref}(t)-x(t)\|_\infty < \gamma_1\,, \quad \|u_{\rm ref}(t)-u(t)\|_\infty < \gamma_2\,, \quad \forall\, t \in [0,\tau)\,,$$

which implies that that at least one of the following equalities holds:

$$\|\,(x_{\rm ref}-x)_\tau\,\|_{\mathcal{L}_\infty} = \gamma_1\,, \qquad \|\,(u_{\rm ref}-u)_\tau\,\|_{\mathcal{L}_\infty} = \gamma_2\,. \tag{3.50}$$

Taking into consideration the definitions of ρ and ρ_u in (3.10) and (3.14), it follows from Lemma 3.1.1 and the equalities in (3.50) that

$$\|\,x_\tau\,\|_{\mathcal{L}_\infty} \le \rho\,, \quad \|\,u_\tau\,\|_{\mathcal{L}_\infty} \le \rho_u\,. \tag{3.51}$$

These bounds imply that the assumptions of Lemma 3.1.2 hold. Then, selecting the adaptive gain Γ according to the design constraint in (3.42), it follows that

$$\|\,\tilde{x}_\tau\,\|_{\mathcal{L}_\infty} \le \gamma_0\,. \tag{3.52}$$

Next, let $\eta_1(t)$ and $\eta_2(t)$ be defined as

$$\eta_1(t) \triangleq \theta_1(t)|x_1(t)| + \sigma_1(t)\,, \qquad \eta_2(t) \triangleq \theta_2(t)\|x(t)\|_\infty + \sigma_2(t)\,.$$

From the bounds in (3.51) it follows that for all $t \in [0,\tau]$ the following equalities hold:

$$\eta_1(t) \;=\; f_1(t,x_1(t))\,, \qquad \eta_2(t) \;=\; f_2(t,x(t))\,.$$

Also, let $e(t)$ be defined as

$$e(t) \triangleq \begin{bmatrix} x_1(t)-r(t) \\ x_2(t)-\alpha(t) \end{bmatrix},$$

and notice that, at time zero, since we assumed zero initialization of the filter $C_1(s)$ in the control law, we have $e(0) = e_{\rm ref}(0) = e_0$. Then, from the system dynamics in (3.1) and the definition of $\alpha(t)$ in (3.20), it follows that

$$\begin{aligned} \dot{e}(t) &= \begin{bmatrix} \dot{x}_1(t)-\dot{r}(t) \\ \dot{x}_2(t)-\dot{\alpha}(t) \end{bmatrix} = \begin{bmatrix} \eta_1(t)+x_2(t)-\dot{r}(t) \\ \eta_2(t)+\omega u(t)-\dot{\alpha}(t) \end{bmatrix} \\ &= A_g e(t) + \begin{bmatrix} \eta_1(t)-\hat{\eta}_{1C}(t) \\ a_2 e_2(t)+\eta_2(t)-\dot{\alpha}(t)+\omega u(t) \end{bmatrix}. \end{aligned}$$

The dynamics above can be written in the frequency domain as

$$e(s) = H(s)\begin{bmatrix} \eta_1(s)-\hat{\eta}_{1C}(s) \\ a_2 e_2(s)+\eta_2(s)-\eta_{\dot{\alpha}}(s)+\omega u(s) \end{bmatrix} + H(s)e_0\,, \tag{3.53}$$

where $\eta_{\dot{\alpha}}(s)$ is the Laplace transform of $\dot{\alpha}(t)$. Moreover, it follows from (3.21) that

$$u(s) = -\frac{C_2(s)}{\omega}\Big(a_2 e_2(s) + \eta_2(s) - \eta_{\dot{\alpha}}(s) + \tilde{\eta}_\omega(s) + \tilde{\eta}_2(s)\Big), \tag{3.54}$$

where $\tilde{\eta}_2(s)$ and $\tilde{\eta}_\omega(s)$ are the Laplace transforms of $\tilde{\eta}_2(t) \triangleq \tilde{\theta}_2(t)\|x(t)\|_\infty + \tilde{\sigma}_2(t)$ and $\tilde{\eta}_\omega(t) \triangleq \tilde{\omega}(t)u(t)$, respectively. Substituting (3.54) into (3.53) leads to

$$e(s) = G(s)\zeta(s) - H(s)C(s)\tilde{\zeta}(s) + H(s)e_0, \tag{3.55}$$

where $\zeta(s)$ and $\tilde{\zeta}(s)$ are the Laplace transforms of the signals $\zeta(t)$ and $\tilde{\zeta}(t)$ defined as

$$\zeta(t) \triangleq \begin{bmatrix} \eta_1(t) \\ a_2 e_2(t) + \eta_2(t) - \dot{\alpha}(t) \end{bmatrix},$$
$$\tilde{\zeta}(t) \triangleq \begin{bmatrix} \tilde{\eta}_1(t) \\ \tilde{\omega}(t)u(t) + \tilde{\eta}_2(t) \end{bmatrix},$$

with $\tilde{\eta}_1(t) \triangleq \tilde{\theta}_1(t)|x_1(t)| + \tilde{\sigma}_1(t)$. The expression in (3.55), together with the response of the closed-loop reference system in (3.31), yields

$$e_{\text{ref}}(s) - e(s) = G(s)(\zeta_{\text{ref}}(s) - \zeta(s)) + H(s)C(s)\tilde{\zeta}(s), \tag{3.56}$$

where $\zeta_{\text{ref}}(t)$ was introduced in (3.32). Also, the error dynamics in (3.40) lead to

$$\tilde{x}(s) = (s\mathbb{I} - A_m)^{-1}\tilde{\zeta}(s), \tag{3.57}$$

which implies that the expression in (3.56) can be rewritten as

$$e_{\text{ref}}(s) - e(s) = G(s)(\zeta_{\text{ref}}(s) - \zeta(s)) + H(s)C(s)(s\mathbb{I} - A_m)\tilde{x}(s). \tag{3.58}$$

Next, we derive an upper bound on $\|(e_{\text{ref}} - e)_\tau\|_{\mathcal{L}_\infty}$ in terms of $\|(x_{\text{ref}} - x)_\tau\|_{\mathcal{L}_\infty}$. From the definitions of $\zeta(t)$ and $\zeta_{\text{ref}}(t)$, it follows that

$$\zeta_{\text{ref}}(t) - \zeta(t) =$$
$$\begin{bmatrix} \eta_{1\text{ref}}(t) - \eta_1(t) \\ (\eta_{2\text{ref}}(t) - \eta_2(t)) + a_2(x_{2\text{ref}}(t) - x_2(t)) - a_2(\alpha_{\text{ref}}(t) - \alpha(t)) + (\dot{\alpha}_{\text{ref}}(t) - \dot{\alpha}(t)) \end{bmatrix}, \tag{3.59}$$

which implies that

$$\begin{aligned} \left\|(\zeta_{\text{ref}} - \zeta)_\tau\right\|_{\mathcal{L}_\infty} \le{}& \max\left\{\left\|(\eta_{1\text{ref}} - \eta_1)_\tau\right\|_{\mathcal{L}_\infty}, \left\|(\eta_{2\text{ref}} - \eta_2)_\tau\right\|_{\mathcal{L}_\infty}\right\} \\ &+ a_2\|(x_{2\text{ref}} - x_2)_\tau\|_{\mathcal{L}_\infty} + a_2\|(\alpha_{\text{ref}} - \alpha)_\tau\|_{\mathcal{L}_\infty} \\ &+ \|(\dot{\alpha}_{\text{ref}} - \dot{\alpha})_\tau\|_{\mathcal{L}_\infty}. \end{aligned} \tag{3.60}$$

Taking into account that $\|x(t)\|_\infty \le \rho = \bar{\rho}_r(\rho_r)$ and also $\|x_{\text{ref}}(t)\|_\infty \le \rho_r < \bar{\rho}_r(\rho_r)$ for all $t \in [0, \tau]$, Assumption 3.1.3 implies that

$$\left\|(\eta_{i\text{ref}} - \eta_i)_\tau\right\|_{\mathcal{L}_\infty} \le d_{f_{xi}}(\bar{\rho}_r(\rho_r))\,\|(x_{\text{ref}} - x)_\tau\|_{\mathcal{L}_\infty}, \qquad i = 1, 2.$$

From the definitions in (3.3) and (3.4), it follows that $d_{f_{xi}}(\bar{\rho}_r(\rho_r)) < L_{\rho_r}$, $i = 1,2$, and hence

$$\left\|(\eta_{i\text{ref}} - \eta_i)_\tau\right\|_{\mathcal{L}_\infty} \leq L_{\rho_r}\left\|(x_{\text{ref}} - x)_\tau\right\|_{\mathcal{L}_\infty}, \qquad i = 1,2. \tag{3.61}$$

Moreover, it follows from the definitions of $\alpha(t)$ and $\alpha_{\text{ref}}(t)$ in (3.20) and (3.25) that

$$\alpha_{\text{ref}}(s) - \alpha(s) = -a_1\left(x_{1\text{ref}}(s) - x_1(s)\right) - C_1(s)\left(\eta_{1\text{ref}}(s) - \eta_1(s)\right) + C_1(s)\tilde{\eta}_1(s),$$

and the error dynamics in (3.57) further imply that

$$\alpha_{\text{ref}}(s) - \alpha(s) = -a_1\left(x_{1\text{ref}}(s) - x_1(s)\right) - C_1(s)\left(\eta_{1\text{ref}}(s) - \eta_1(s)\right) + C_1(s)(s + a_1)\tilde{x}_1(s).$$

The above expression, together with the bounds in (3.61), yields

$$\begin{aligned}\left\|(\alpha_{\text{ref}} - \alpha)_\tau\right\|_{\mathcal{L}_\infty} \leq{}& a_1\left\|(x_{1\text{ref}} - x_1)_\tau\right\|_{\mathcal{L}_\infty} + \left\|C_1(s)\right\|_{\mathcal{L}_1} L_{\rho_r}\left\|(x_{\text{ref}} - x)_\tau\right\|_{\mathcal{L}_\infty}\\ &+ \left\|C_1(s)(s + a_1)\right\|_{\mathcal{L}_1}\left\|\tilde{x}_\tau\right\|_{\mathcal{L}_\infty}.\end{aligned}$$

Recalling that $\eta_{\dot{\alpha}}(s)$ and $\eta_{\dot{\alpha}_{\text{ref}}}(s)$ denote the Laplace transforms of the signals $\dot{\alpha}(t)$ and $\dot{\alpha}_{\text{ref}}(t)$, it follows from the definitions of $\alpha(t)$ and $\alpha_{\text{ref}}(t)$, together with the error dynamics in (3.57), that

$$\begin{aligned}\eta_{\dot{\alpha}_{\text{ref}}}(s) - \eta_{\dot{\alpha}}(s) ={}& -a_1\left(\eta_{1\text{ref}}(s) - \eta_1(s)\right) - a_1\left(x_{2\text{ref}}(s) - x_2(s)\right)\\ &+ sC_1(s)\left(\eta_{1\text{ref}}(s) - \eta_1(s)\right) + sC_1(s)(s + a_1)\tilde{x}_1(s),\end{aligned}$$

which, together with the bounds in (3.61), leads to the following upper bound:

$$\begin{aligned}\left\|(\dot{\alpha}_{\text{ref}} - \dot{\alpha})_\tau\right\|_{\mathcal{L}_\infty} \leq{}& a_1\left\|(x_{1\text{ref}} - x_1)_\tau\right\|_{\mathcal{L}_\infty} + a_1\left\|(x_{2\text{ref}} - x_2)_\tau\right\|_{\mathcal{L}_\infty}\\ &+ \left\|sC_1(s)\right\|_{\mathcal{L}_1}\left\|(x_{1\text{ref}} - x_1)_\tau\right\|_{\mathcal{L}_\infty}\\ &+ \left\|sC_1(s)(s + a_1)\right\|_{\mathcal{L}_1}\left\|\tilde{x}_\tau\right\|_{\mathcal{L}_\infty}.\end{aligned} \tag{3.62}$$

Then, the expression in (3.58), along with the bounds in (3.60)–(3.62) and the definitions of $\kappa_1(\rho_r)$ and κ_4 in (3.8) and (3.13), leads to

$$\begin{aligned}\left\|(e_{\text{ref}} - e)_\tau\right\|_{\mathcal{L}_\infty} \leq{}& \left\|G(s)\right\|_{\mathcal{L}_1}\left(\kappa_1(\rho_r)\left\|(x_{\text{ref}} - x)_\tau\right\|_{\mathcal{L}_\infty} + \kappa_4\left\|\tilde{x}_\tau\right\|_{\mathcal{L}_\infty}\right)\\ &+ \left\|H(s)C(s)(s\mathbb{I} - A_m)\right\|_{\mathcal{L}_1}\left\|\tilde{x}_\tau\right\|_{\mathcal{L}_\infty},\end{aligned}$$

or, equivalently,

$$\begin{aligned}\left\|(e_{\text{ref}} - e)_\tau\right\|_{\mathcal{L}_\infty} \leq{}& \left\|G(s)\right\|_{\mathcal{L}_1}\kappa_1(\rho_r)\left\|(x_{\text{ref}} - x)_\tau\right\|_{\mathcal{L}_\infty}\\ &+ \left(\left\|G(s)\right\|_{\mathcal{L}_1}\kappa_4 + \left\|H(s)C(s)(s\mathbb{I} - A_m)\right\|_{\mathcal{L}_1}\right)\left\|\tilde{x}_\tau\right\|_{\mathcal{L}_\infty}.\end{aligned} \tag{3.63}$$

Next, noting that

$$x_{\text{ref}}(t) - x(t) = \left(e_{\text{ref}}(t) - e(t)\right) + \begin{bmatrix} 0 \\ \alpha_{\text{ref}}(t) - \alpha(t) \end{bmatrix},$$

we can derive the following upper bound on $\|(x_{\rm ref}-x)_\tau\|_{\mathcal{L}_\infty}$:

$$\|(x_{\rm ref}-x)_\tau\|_{\mathcal{L}_\infty} \le \|(e_{\rm ref}-e)_\tau\|_{\mathcal{L}_\infty} + \|(\alpha_{\rm ref}-\alpha)_\tau\|_{\mathcal{L}_\infty}. \tag{3.64}$$

From the definitions of $\alpha(t)$ and $\alpha_{\rm ref}(t)$ in (3.20) and (3.25) some straightforward manipulations lead to

$$\begin{aligned}\|(\alpha_{\rm ref}-\alpha)_\tau\|_{\mathcal{L}_\infty} &\le a_1\|(e_{\rm ref}-e)_\tau\|_{\mathcal{L}_\infty} + \|C_1(s)\|_{\mathcal{L}_1} L_{\rho_r}\|(e_{\rm ref}-e)_\tau\|_{\mathcal{L}_\infty}\\ &\quad + \|C_1(s)(s+a_1)\|_{\mathcal{L}_1}\|\tilde{x}_\tau\|_{\mathcal{L}_\infty},\end{aligned}$$

which, together with the bound in (3.64), yields

$$\begin{aligned}\|(x_{\rm ref}-x)_\tau\|_{\mathcal{L}_\infty} &\le \left(1+a_1+\|C_1(s)\|_{\mathcal{L}_1} L_{\rho_r}\right)\|(e_{\rm ref}-e)_\tau\|_{\mathcal{L}_\infty}\\ &\quad + \|C_1(s)(s+a_1)\|_{\mathcal{L}_1}\|\tilde{x}_\tau\|_{\mathcal{L}_\infty}.\end{aligned}$$

We can now apply the bound obtained in (3.63) to the expression above to find

$$\|(x_{\rm ref}-x)_\tau\|_{\mathcal{L}_\infty} \le \|G(s)\|_{\mathcal{L}_1} L^\star_{\rho_r}\|(x_{\rm ref}-x)_\tau\|_{\mathcal{L}_\infty} + \kappa_3(\rho_r)\|\tilde{x}_\tau\|_{\mathcal{L}_\infty},$$

where we have used the definitions of $L^\star_{\rho_r}$ and $\kappa_3(\rho_r)$ in (3.7) and (3.12), respectively. Then, noting that the $\mathcal{L}_1$-norm condition in (3.5) ensures that $\|G(s)\|_{\mathcal{L}_1} L^\star_{\rho_r} < 1$, we can derive the bound

$$\|(x_{\rm ref}-x)_\tau\|_{\mathcal{L}_\infty} \le \frac{\kappa_3(\rho_r)}{1-\|G(s)\|_{\mathcal{L}_1} L^\star_{\rho_r}}\|\tilde{x}_\tau\|_{\mathcal{L}_\infty},$$

which, together with the bound in (3.52), leads to

$$\|(x_{\rm ref}-x)_\tau\|_{\mathcal{L}_\infty} \le \frac{\kappa_3(\rho_r)}{1-\|G(s)\|_{\mathcal{L}_1} L^\star_{\rho_r}}\gamma_0.$$

The definition of γ_1 in (3.11) implies that

$$\|(x_{\rm ref}-x)_\tau\|_{\mathcal{L}_\infty} \le \gamma_1 - \beta < \gamma_1, \tag{3.65}$$

which contradicts the first equality in (3.50).

On the other hand, it follows from (3.30) and (3.54) that

$$\begin{aligned}u_{\rm ref}(s)-u(s) = -\frac{C_2(s)}{\omega}\Big(&\left(\eta_{2_{\rm ref}}(s)-\eta_2(s)\right) + a_2\left(x_{2_{\rm ref}}(s)-x_2(s)\right)\\ &- a_2\left(\alpha_{\rm ref}(s)-\alpha(s)\right) - \left(\eta_{\dot{\alpha}_{\rm ref}}(s)-\eta_{\dot{\alpha}}(s)\right) - \left(\tilde{\eta}_\omega(s)+\tilde{\eta}_2(s)\right)\Big).\end{aligned}$$

Using the expression in (3.59) and the error dynamics in (3.57), one can show that

$$\|(u_{\rm ref}-u)_\tau\|_{\mathcal{L}_\infty} \le \left\|\frac{C_2(s)}{\omega}\right\|_{\mathcal{L}_1}\left\|(\zeta_{\rm ref}-\zeta)_\tau\right\|_{\mathcal{L}_\infty} + \left\|\frac{C_2(s)}{\omega}(s+a_2)\right\|_{\mathcal{L}_1}\|\tilde{x}_\tau\|_{\mathcal{L}_\infty},$$

which leads to

$$\begin{aligned}\|(u_{\rm ref}-u)_\tau\|_{\mathcal{L}_\infty} &\le \left\|\frac{C_2(s)}{\omega}\right\|_{\mathcal{L}_1}\left(\kappa_1(\rho_r)\|(x_{\rm ref}-x)_\tau\|_{\mathcal{L}_\infty} + \kappa_4\|\tilde{x}_\tau\|_{\mathcal{L}_\infty}\right)\\ &\quad + \left\|\frac{C_2(s)}{\omega}(s+a_2)\right\|_{\mathcal{L}_1}\|\tilde{x}_\tau\|_{\mathcal{L}_\infty}.\end{aligned}$$

Finally, the bounds in (3.52) and (3.65), along with the definition of γ_2 in (3.16), yield

$$\|(u_{\text{ref}} - u)_\tau\|_{\mathcal{L}_\infty} \leq \left\|\frac{C_2(s)}{\omega}\right\|_{\mathcal{L}_1} \kappa_1(\rho_r)(\gamma_1 - \beta) + \kappa_5\gamma_0 < \gamma_2,$$

which contradicts the second equality in (3.50). This proves the bounds in (3.48) and (3.49). Thus, the bounds in (3.51) hold uniformly, which implies that the bound in (3.52) also holds uniformly. This proves the bounds in (3.45)–(3.47). □

Remark 3.1.1 It follows from the definitions of γ_1 and γ_2 in (3.11) and (3.16) that one can achieve arbitrary desired performance bounds for the system's signals, both input and output, simultaneously, by increasing the adaptive gain.

Remark 3.1.2 To understand how the performance bounds can be used for ensuring uniform transient response with *desired* specifications, consider the *ideal* control law for the system in (3.1):

$$u_{\text{id}}(t) = \omega^{-1}(-a_2(x_{2\text{id}}(t) - \alpha_{\text{id}}(t)) - \eta_{2\text{id}}(t) + \dot{\alpha}_{\text{id}}(t)),$$
$$\alpha_{\text{id}}(t) = -a_1(x_{1\text{id}}(t) - r(t)) - \eta_{1\text{id}}(t) + \dot{r}(t),$$

with $\eta_{1\text{id}} = f_1(x_{1\text{id}}(t), t)$ and $\eta_{2\text{id}} = f_2(t, x_{\text{id}}(t))$. This *ideal* nonadaptive controller leads to the *desired* system response

$$\dot{e}_{\text{id}}(t) = A_g e_{\text{id}}(t), \tag{3.66}$$

where $e_{\text{id}}(t) \triangleq [(x_{1\text{id}}(t) - r(t)),\ (x_{2\text{id}}(t) - \alpha_{\text{id}}(t))]^\top$. In the closed-loop reference system (3.23), $u_{\text{id}}(t)$ is further low-pass filtered to have a guaranteed low-frequency range. Similar to Section 2.1.4, the response of the closed-loop reference system can be made as close as possible to (3.66) by reducing $\|G(s)\|_{\mathcal{L}_1}$. From the definition of $G(s)$ in (3.6), and noting that $(\mathbb{I} - C(s))$ has a diagonal structure with $(1 - C_1(s))$ and $(1 - C_2(s))$ as diagonal elements, we notice that $\|G(s)\|_{\mathcal{L}_1}$ can be made arbitrarily small by increasing the bandwidths of the low-pass filters $C_1(s)$ and $C_2(s)$.

3.1.4 Simulation Example

To verify numerically the results proved in this section, we consider the system given in (3.1). We perform simulations for the following scenarios:

- Scenario 1:

$$\omega = \omega_1 \triangleq 0.8,$$
$$f_1(t, x_1(t)) = f_1^1(t, x_1(t)) \triangleq 0.1x_1(t) + 0.5x_1^2(t) - 0.2\sin(0.1t),$$
$$f_2(t, x(t)) = f_2^1(t, x(t)) \triangleq x_1(t) + x_2(t) + x_1^2(t) + x_2^2(t) + \sin(t).$$

- Scenario 2:

$$\omega = \omega_2 \triangleq 1.1.$$
$$f_1(t, x_1(t)) = f_1^1(t, x_1(t)),$$
$$f_2(t, x(t)) = f_2^1(t, x(t)).$$

– Scenario 3:

$$\begin{aligned}
&\omega = \omega_1\,,\\
&f_1(t,x(t)) = f_1^2(t,x(t)) \triangleq -x_1(t) + 0.3x_1^2(t) + \sin(x_1(t)) - 0.1\cos(t)\,,\\
&f_2(t,x(t)) = f_2^1(t,x(t))\,.
\end{aligned}$$

– Scenario 4:

$$\begin{aligned}
&\omega = \omega_1\,,\\
&f_1(t,x(t)) = f_1^1(t,x(t))\,,\\
&f_2(t,x(t)) = f_2^2(t,x(t)) \triangleq -x_1(t) + x_2(t) + \sin(x_1(t))x_2^2(t) + x_1(t)x_2(t)\,,
\end{aligned}$$

– Scenario 5:

$$\begin{aligned}
&\omega = \omega_2\,,\\
&f_1(t,x(t)) = f_1^2(t,x(t))\,,\\
&f_2(t,x(t)) = f_2^2(t,x(t))\,.
\end{aligned}$$

In Scenario 1 we consider an underactuated system with significant nonlinearities and a sinusoidal disturbance. Scenarios 2–4 are derived from Scenario 1 by changing only one of the unknown quantities at a time; the uncertainty in the input gain is changed in Scenario 2, and the unknown matched and unmatched nonlinearities are changed in Scenarios 3 and 4. In Scenario 5 we change all unknown quantities simultaneously.

We implement the $\mathcal{L}_1$ adaptive controller according to (3.18), (3.19), and (3.21). For all five scenarios we use a single $\mathcal{L}_1$ controller design with the control parameters

$$k = 10\,, \quad D(s) = \frac{1}{s}\,, \quad C_1(s) = \frac{0.1}{(s+0.1)(s+1)}\,, \quad a_1 = 2\,, \quad a_2 = 2\,.$$

We set the projection bounds to be $\Omega = [0.1,\ 5]$, $L_\rho = 20$, and $\Delta = 50$ and the adaptation gain to $\Gamma = 100000$.

Figure 3.1 shows the simulation results for Scenarios 1 and 2; Figure 3.2 shows the results for Scenarios 1 and 3; Figure 3.3 shows the results for Scenarios 1 and 4. In these simulations, we set the system reference input to $r(t) = 0.5\cos(0.5t)$. From the results one can see that the fast adaptation ability of the $\mathcal{L}_1$ adaptive controller ensures uniform transient performance for different uncertainties. We notice that the system's output $y(t) = x_1(t)$ is not significantly affected by changing the dynamics. However, the control signal changes significantly to ensure adequate compensation for those.

Figure 3.4 shows the simulation results for Scenario 1 with reference signals of different amplitudes:

$$r_1(t) = 0.2\cos(0.5t)\,, \quad r_2(t) = 0.5\cos(0.5t)\,.$$

We observe that the system response scales with scaled reference inputs.

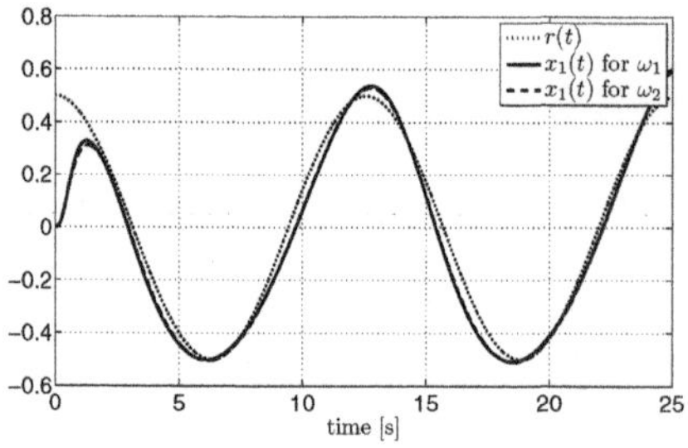

(a) $r(t)$ and $x_1(t)$

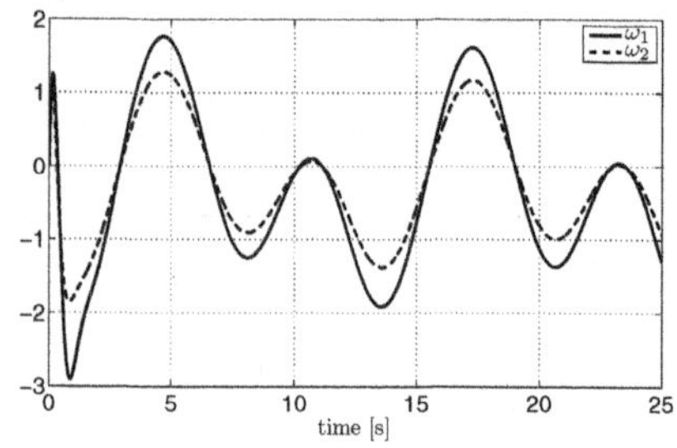

(b) Time history of $u(t)$

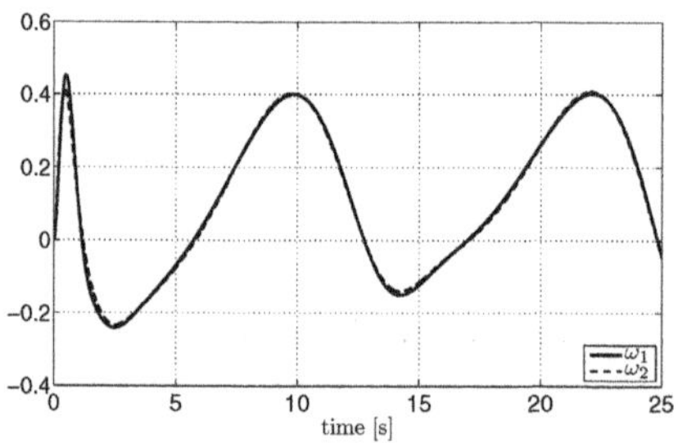

(c) $x_2(t)$

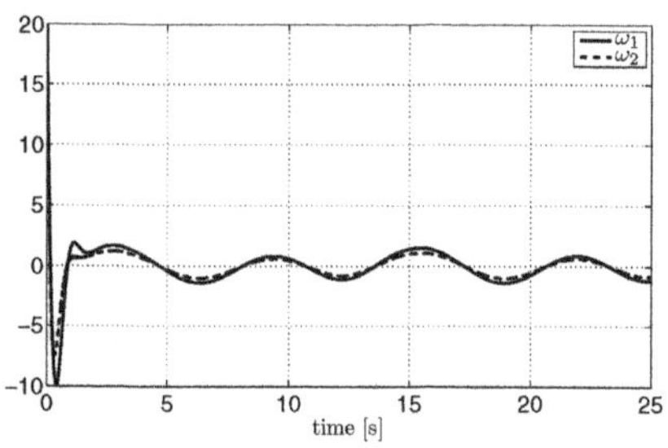

(d) Time history of $\dot{u}(t)$

Figure 3.1: Performance of the $\mathcal{L}_1$ controller for Scenarios 1 and 2.

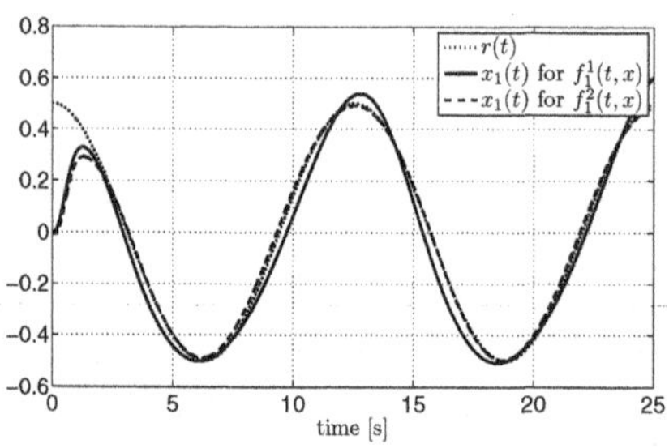

(a) $r(t)$ and $x_1(t)$

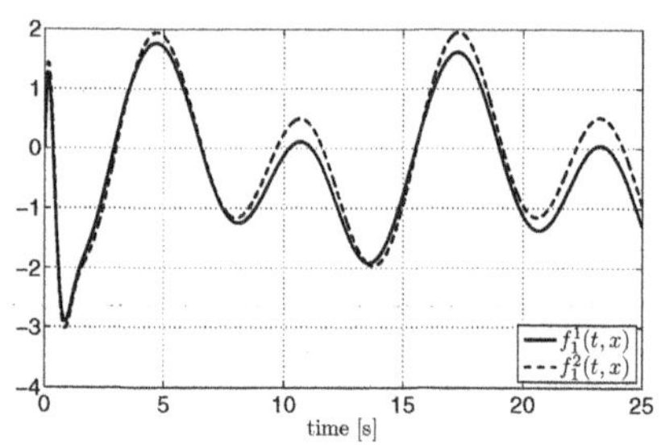

(b) Time history of $u(t)$

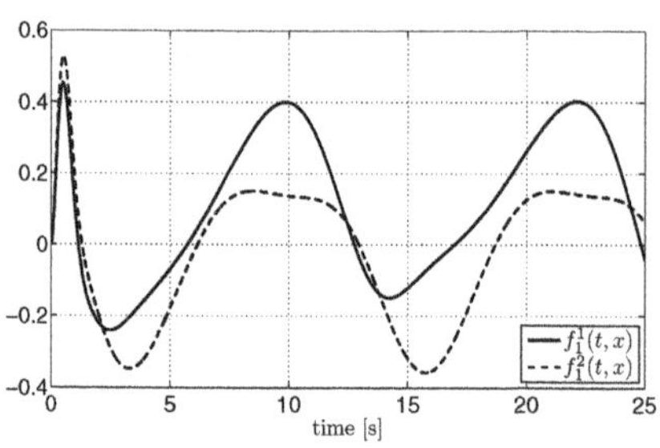

(c) $x_2(t)$

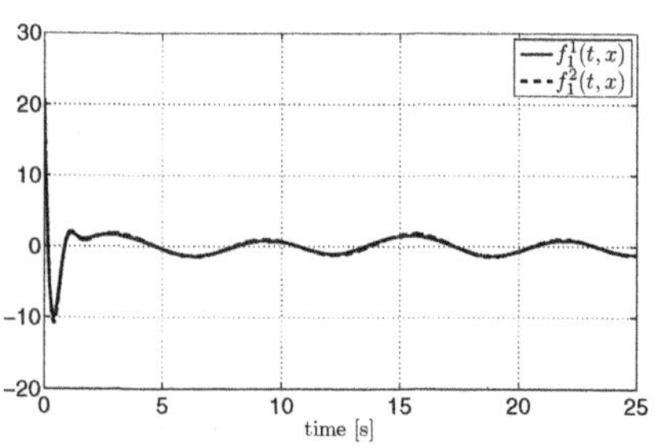

(d) Time history of $\dot{u}(t)$

Figure 3.2: Performance of the $\mathcal{L}_1$ controller for Scenarios 1 and 3.

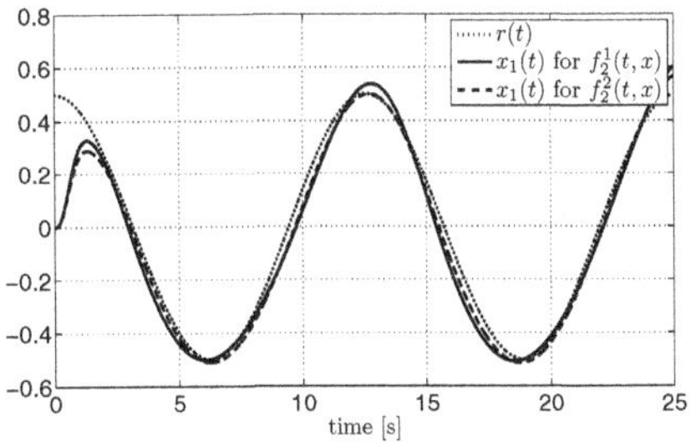

(a) $r(t)$ and $x_1(t)$

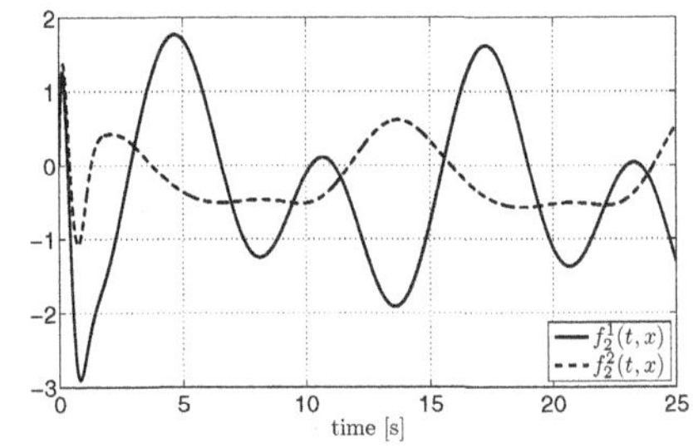

(b) Time history of $u(t)$

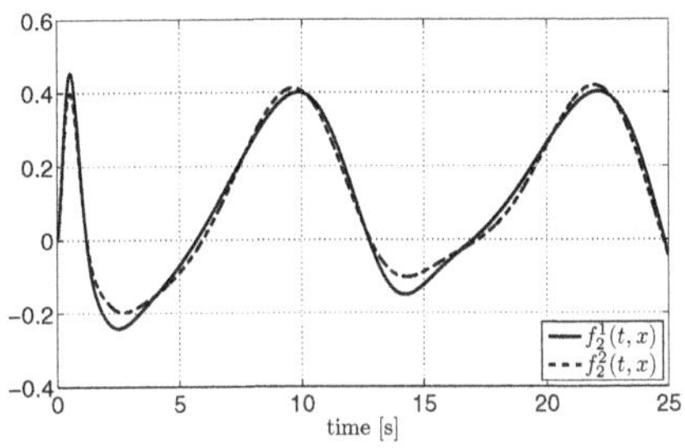

(c) $x_2(t)$

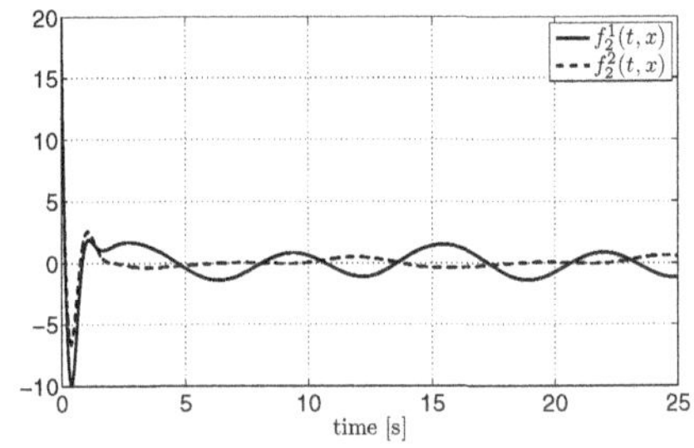

(d) Time history of $\dot{u}(t)$

Figure 3.3: Performance of the $\mathcal{L}_1$ controller for Scenarios 1 and 4.

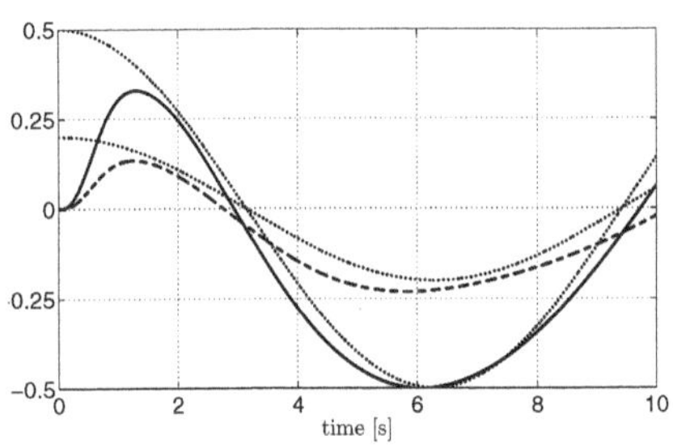

(a) $x_1(t)$ for $r_1(t)$ and $r_2(t)$

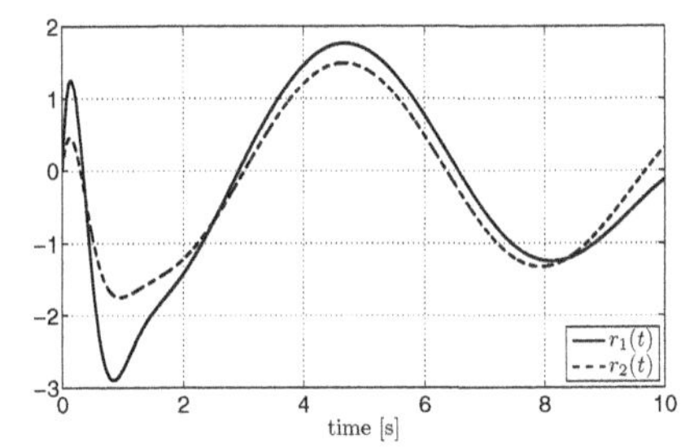

(b) Time history of $u(t)$

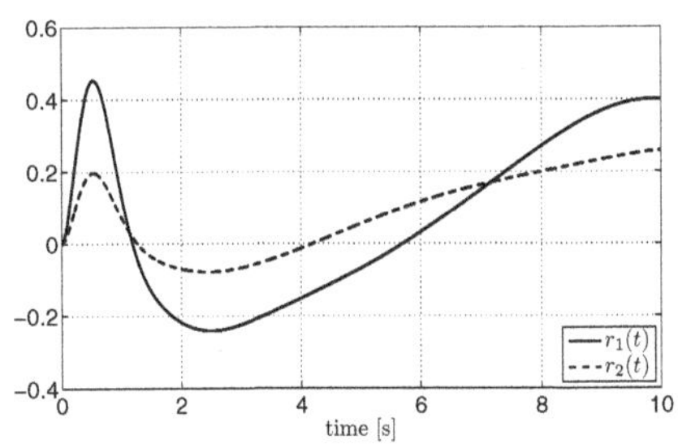

(c) $x_2(t)$

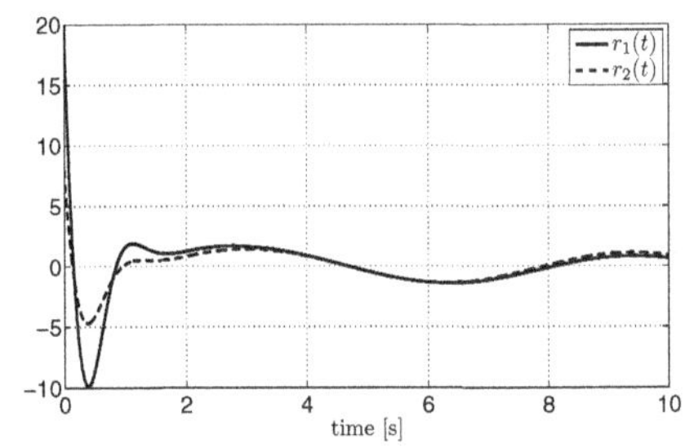

(d) Time history of $\dot{u}(t)$

Figure 3.4: Performance of the $\mathcal{L}_1$ controller for $r_1(t)$ and $r_2(t)$ reference signals (Scenario 1).

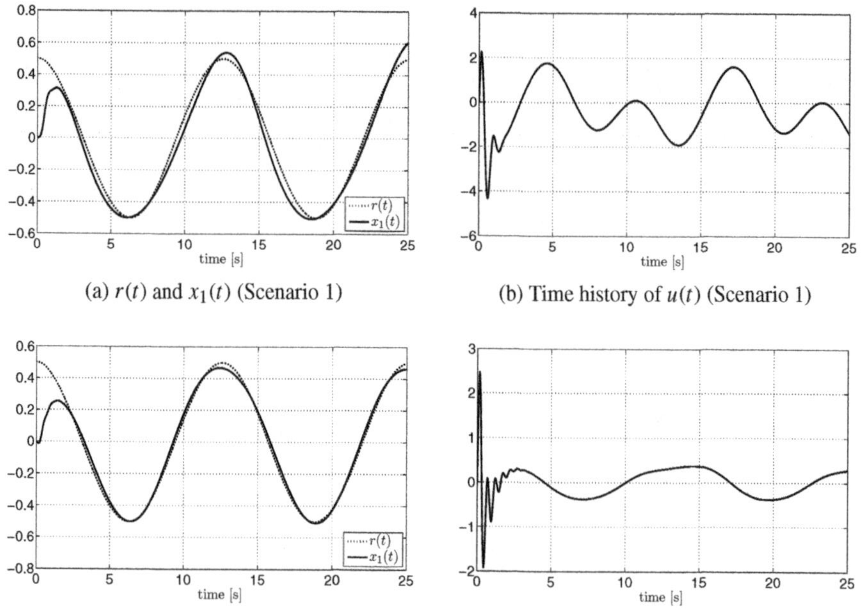

(a) $r(t)$ and $x_1(t)$ (Scenario 1)

(b) Time history of $u(t)$ (Scenario 1)

(c) $r(t)$ and $x_1(t)$ (Scenario 5)

(d) Time history of $u(t)$ (Scenario 5)

Figure 3.5: Performance of the $\mathcal{L}_1$ controller with time delay of 80 ms.

Finally, we numerically test the robustness of the $\mathcal{L}_1$ adaptive controller to time delays. Figure 3.5 shows the simulation results for Scenarios 1 and 5 in the presence of a time delay of 0.08 s. One can see that the system has some expected degradation in the performance but remains stable. Moreover, the system output in the presence of time delay remains close to the one in the absence of time delay for both cases of uncertainties.

3.2 $\mathcal{L}_1$ Adaptive Controller for Multi-Input Multi-Output Systems in the Presence of Unmatched Nonlinear Uncertainties

This section presents the $\mathcal{L}_1$ adaptive controller for a class of multi-input multi-output uncertain systems in the presence of uncertain system input gain and time- and state-dependent unknown nonlinearities, *without enforcing matching conditions*. The class of systems considered includes general unmatched uncertainties that cannot be addressed by recursive design methods developed for strict-feedback systems in Section 3.1, semi-strict-feedback systems, pure-feedback systems, and block-strict-feedback systems [100, 181]. We show that, subject to a set of mild assumptions, the system can be transformed into an equivalent (semi-)linear system with time-varying unknown parameters and disturbances. For the latter, we apply the $\mathcal{L}_1$ adaptive controller, which yields semiglobal performance results for the original nonlinear system. The adaptive algorithm ensures uniformly bounded transient response for the system's signals, both input and output, simultaneously, in addition to steady-state tracking.

3.2.1 Problem Formulation

Consider the following system dynamics:

$$\begin{aligned} \dot{x}(t) &= A_m x(t) + B_m \omega u(t) + f(t, x(t), z(t)), \quad x(0) = x_0, \\ \dot{x}_z(t) &= g(t, x_z(t), x(t)), \quad x_z(0) = x_{z0}, \\ z(t) &= g_o(t, x_z(t)), \\ y(t) &= C x(t), \end{aligned} \tag{3.67}$$

where $x(t) \in \mathbb{R}^n$ is the system state vector (measured); $u(t) \in \mathbb{R}^m$ is the control signal ($m \leq n$); $y(t) \in \mathbb{R}^m$ is the regulated output; A_m is a known Hurwitz $n \times n$ matrix that defines the desired dynamics for the closed-loop system; $B_m \in \mathbb{R}^{n \times m}$ is a known full-rank constant matrix, (A_m, B_m) is controllable; $C \in \mathbb{R}^{m \times n}$ is a known full-rank constant matrix, (A_m, C) is observable; $\omega \in \mathbb{R}^{m \times m}$ is the uncertain system input gain matrix; $z(t) \in \mathbb{R}^p$ and $x_z(t) \in \mathbb{R}^l$ are the output and the state vector of internal unmodeled dynamics; and $f : \mathbb{R} \times \mathbb{R}^n \times \mathbb{R}^p \to \mathbb{R}^n$, $g_o : \mathbb{R} \times \mathbb{R}^l \to \mathbb{R}^p$, and $g : \mathbb{R} \times \mathbb{R}^l \times \mathbb{R}^n \to \mathbb{R}^l$ are unknown nonlinear functions continuous in their arguments. The initial condition x_0 is assumed to be inside an arbitrarily large known set, i.e., $\|x_0\|_\infty \leq \rho_0 < \infty$ for some $\rho_0 > 0$.

The system in (3.67) can also be written in the form

$$\begin{aligned} \dot{x}(t) &= A_m x(t) + B_m (\omega u(t) + f_1(t, x(t), z(t))) + B_{um} f_2(t, x(t), z(t)), \quad x(0) = x_0, \\ \dot{x}_z(t) &= g(t, x_z(t), x(t)), \quad x_z(0) = x_{z0}, \\ z(t) &= g_o(t, x_z(t)), \\ y(t) &= C x(t), \end{aligned} \tag{3.68}$$

where $B_{um} \in \mathbb{R}^{n \times (n-m)}$ is a constant matrix such that $B_m^\top B_{um} = 0$ and also $\text{rank}([B_m, \ B_{um}]) = n$, while $f_1 : \mathbb{R} \times \mathbb{R}^n \times \mathbb{R}^p \to \mathbb{R}^m$ and $f_2 : \mathbb{R} \times \mathbb{R}^n \times \mathbb{R}^p \to \mathbb{R}^{(n-m)}$ are unknown nonlinear functions that verify

$$\begin{bmatrix} f_1(t, x(t), z(t)) \\ f_2(t, x(t), z(t)) \end{bmatrix} = \begin{bmatrix} B_m & B_{um} \end{bmatrix}^{-1} f(t, x(t), z(t)).$$

In this problem formulation, $f_1(\cdot)$ represents the matched component of the unknown nonlinearities, whereas $B_{um} f_2(\cdot)$ represents the unmatched uncertainties.

Let $X \triangleq [x^\top, \ z^\top]^\top$, and with a slight abuse of language let $f_i(t, X) \triangleq f_i(t, x, z)$, $i = 1, 2$. The system above verifies the following assumptions.

Assumption 3.2.1 (Boundedness of $f_i(t, 0)$) There exists $B_{i0} > 0$, such that $\|f_i(t, 0)\|_\infty \leq B_{i0}$ holds for all $t \geq 0$, and for $i = 1, 2$.

Assumption 3.2.2 (Semiglobal uniform boundedness of partial derivatives) For $i = 1, 2$ and arbitrary $\delta > 0$, there exist positive constants $d_{f_{xi}}(\delta) > 0$ and $d_{f_{ti}}(\delta) > 0$ independent of time, such that for all $\|X(t)\|_\infty < \delta$, the partial derivatives of $f_i(t, X)$ are piecewise-continuous and bounded:

$$\left\| \frac{\partial f_i(t, X)}{\partial X} \right\|_\infty \leq d_{f_{xi}}(\delta), \qquad \left\| \frac{\partial f_i(t, X)}{\partial t} \right\|_\infty \leq d_{f_{ti}}(\delta),$$

where the first is a matrix induced ∞-norm, and the second is a vector ∞-norm.

Assumption 3.2.3 (Stability of unmodeled dynamics) The x_z-dynamics are BIBO stable with respect to both initial conditions x_{z0} and input $x(t)$, i.e., there exist $L_z, B_z > 0$ such that for all $t \geq 0$

$$\|z_t\|_{\mathcal{L}_\infty} \leq L_z \|x_t\|_{\mathcal{L}_\infty} + B_z .$$

Assumption 3.2.4 (Partial knowledge of the system input gain) The system input gain matrix ω is assumed to be an unknown (nonsingular) strictly row-diagonally dominant matrix with $\mathrm{sgn}(\omega_{ii})$ known. Also, we assume that there exists a known compact convex set Ω such that $\omega \in \Omega \subset \mathbb{R}^{m\times m}$.

Assumption 3.2.5 (Stability of matched transmission zeros) The transmission zeros of the transfer matrix $H_m(s) = C(s\mathbb{I} - A_m)^{-1}B_m$ lie in the open left half plane.

The control objective is to design an adaptive state-feedback controller to ensure that $y(t)$ tracks the output response of a *desired system* $M(s)$ defined as

$$M(s) \triangleq C(s\mathbb{I} - A_m)^{-1} B_m K_g(s),$$

where $K_g(s)$ is a feedforward prefilter, to a given bounded piecewise-continuous reference signal $r(t)$ *in both transient* and *steady-state*, while all other signals remain bounded.

3.2.2 $\mathcal{L}_1$ Adaptive Control Architecture

Definitions and $\mathcal{L}_1$-Norm Sufficient Condition for Stability

To streamline the subsequent analysis, we need to introduce some notations. Let

$$\begin{aligned} H_{xm}(s) &\triangleq (s\mathbb{I}_n - A_m)^{-1} B_m , \\ H_{xum}(s) &\triangleq (s\mathbb{I}_n - A_m)^{-1} B_{um} , \\ H_m(s) &\triangleq C H_{xm}(s) = C(s\mathbb{I}_n - A_m)^{-1} B_m , \\ H_{um}(s) &\triangleq C H_{xum}(s) = C(s\mathbb{I}_n - A_m)^{-1} B_{um} , \end{aligned}$$

and let $x_{\text{in}}(t)$ be the signal with Laplace transform $x_{\text{in}}(s) \triangleq (s\mathbb{I}_n - A_m)^{-1}x_0$ and $\rho_{\text{in}} \triangleq \|s(s\mathbb{I} - A_m)^{-1}\|_{\mathcal{L}_1}\rho_0$. Since A_m is Hurwitz and x_0 is bounded, then $\|x_{\text{in}}\|_{\mathcal{L}_\infty} \leq \rho_{\text{in}}$.

Further, for every $\delta > 0$, let

$$L_{i\delta} \triangleq \frac{\bar{\delta}(\delta)}{\delta} d_{f_{xi}}(\bar{\delta}(\delta)), \quad \bar{\delta}(\delta) \triangleq \max\{\delta + \bar{\gamma}_1,\ L_z(\delta + \bar{\gamma}_1) + B_z\}, \tag{3.69}$$

where $d_{f_x}(\cdot)$ was introduced in Assumption 3.2.2 with $\bar{\gamma}_1$ being an arbitrary small positive constant.

The design of the $\mathcal{L}_1$ adaptive controller involves a feedback gain matrix $K \in \mathbb{R}^{m\times m}$ and an $m \times m$ strictly proper transfer matrix $D(s)$, which lead, for all $\omega \in \Omega$, to a strictly proper stable

$$C(s) \triangleq \omega K D(s)(\mathbb{I}_m + \omega K D(s))^{-1} , \tag{3.70}$$

with DC gain $C(0) = \mathbb{I}_m$. The choice of $D(s)$ needs to ensure also that $C(s)H_m^{-1}(s)$ is a proper stable transfer matrix. For a particular class of systems, a possible choice for $D(s)$ might be $D(s) = \frac{1}{s}\mathbb{I}_m$, which yields a strictly proper $C(s)$ of the form

$$C(s) = \omega K \left(s\mathbb{I}_m + \omega K\right)^{-1},$$

with the condition that the choice of K must ensure that $-\omega K$ is Hurwitz. It is easy to show that, if one lets $\lambda_{\max}(-\omega K)$ be the maximum (negative) eigenvalue of the Hurwitz matrix $-\omega K$, then one has

$$\lim_{\lambda_{\max}(-\omega K)\to -\infty} \|\mathbb{I}_m - C(s)\|_{\mathcal{L}_1} = 0.$$

For the proofs of stability and performance bounds, the choice of K and $D(s)$ also needs to ensure that, for a given ρ_0, there exists $\rho_r > \rho_{\text{in}}$ such that the following $\mathcal{L}_1$-norm condition holds:

$$\|G_m(s)\|_{\mathcal{L}_1} + \|G_{um}(s)\|_{\mathcal{L}_1}\ell_0 < \frac{\rho_r - \left\|H_{xm}(s)C(s)K_g(s)\right\|_{\mathcal{L}_1}\|r\|_{\mathcal{L}_\infty} - \rho_{\text{in}}}{L_{1\rho_r}\rho_r + B_0}, \tag{3.71}$$

where

$$G_m(s) \triangleq H_{xm}(s)(\mathbb{I}_m - C(s)), \tag{3.72}$$

$$G_{um}(s) \triangleq \left(\mathbb{I}_n - H_{xm}(s)C(s)H_m^{-1}(s)C\right)H_{xum}(s), \tag{3.73}$$

while

$$\ell_0 \triangleq \frac{L_{2\rho_r}}{L_{1\rho_r}}, \quad B_0 \triangleq \max\left\{B_{10}, \frac{B_{20}}{\ell_0}\right\},$$

and $K_g(s)$ is the (BIBO-stable) feedforward prefilter. Further, let ρ be defined as

$$\rho \triangleq \rho_r + \bar{\gamma}_1, \tag{3.74}$$

and let γ_1 be given by

$$\gamma_1 \triangleq \frac{\left\|H_{xm}(s)C(s)H_m^{-1}(s)C\right\|_{\mathcal{L}_1}}{1 - \|G_m(s)\|_{\mathcal{L}_1}L_{1\rho_r} - \|G_{um}(s)\|_{\mathcal{L}_1}L_{2\rho_r}}\gamma_0 + \beta, \tag{3.75}$$

where γ_0 and β are arbitrarily small positive constants such that $\gamma_1 \le \bar{\gamma}_1$. Next, let

$$\rho_u \triangleq \rho_{ur} + \gamma_2, \tag{3.76}$$

where ρ_{ur} and γ_2 are defined as

$$\begin{aligned}
\rho_{ur} &\triangleq \left\|\omega^{-1}C(s)\right\|_{\mathcal{L}_1}\left(L_{1\rho_r}\rho_r + B_{10}\right) + \left\|\omega^{-1}C(s)H_m^{-1}(s)H_{um}(s)\right\|_{\mathcal{L}_1}\left(L_{2\rho_r}\rho_r + B_{20}\right) \\
&\quad + \left\|\omega^{-1}C(s)K_g(s)\right\|_{\mathcal{L}_1}\|r\|_{\mathcal{L}_\infty}, \\
\gamma_2 &\triangleq \left(\left\|\omega^{-1}C(s)\right\|_{\mathcal{L}_1}L_{1\rho_r} + \left\|\omega^{-1}C(s)H_m^{-1}(s)H_{um}(s)\right\|_{\mathcal{L}_1}L_{2\rho_r}\right)\gamma_1 \\
&\quad + \left\|\omega^{-1}C(s)H_m^{-1}(s)C\right\|_{\mathcal{L}_1}\gamma_0.
\end{aligned} \tag{3.77}$$

Finally, let

$$\theta_{b_i} \triangleq L_{i\rho}, \quad \sigma_{b_i} \triangleq L_{i\rho}B_z + B_{i0} + \epsilon_i, \quad i = 1,2, \tag{3.78}$$

where $\epsilon_i > 0$, $i = 1, 2$, are arbitrary numbers.

The $\mathcal{L}_1$ adaptive control architecture is introduced below.

State Predictor

We consider the following state predictor:

$$\begin{aligned}\dot{\hat{x}}(t) &= A_m\hat{x}(t) + B_m(\hat{\omega}(t)u(t) + \hat{\theta}_1(t)\,\|x_t\|_{\mathcal{L}_\infty} + \hat{\sigma}_1(t)) \\ &\quad + B_{um}(\hat{\theta}_2(t)\,\|x_t\|_{\mathcal{L}_\infty} + \hat{\sigma}_2(t)), \qquad \hat{x}(0) = x_0, \\ \hat{y}(t) &= C\hat{x}(t),\end{aligned} \tag{3.79}$$

where $\hat{\omega}(t) \in \mathbb{R}^{m\times m}$, $\hat{\theta}_1(t) \in \mathbb{R}^m$, $\hat{\sigma}_1(t) \in \mathbb{R}^m$, $\hat{\theta}_2(t) \in \mathbb{R}^{n-m}$, and $\hat{\sigma}_2(t) \in \mathbb{R}^{n-m}$ are the adaptive estimates.

Adaptation Laws

The adaptation laws for $\hat{\omega}(t)$, $\hat{\theta}_1(t)$, $\hat{\sigma}_1(t)$, $\hat{\theta}_2(t)$, and $\hat{\sigma}_2(t)$ are defined as

$$\begin{aligned}\dot{\hat{\omega}}(t) &= \Gamma\text{Proj}(\hat{\omega}(t), -(\tilde{x}^\top(t)PB_m)^\top u^\top(t)), \quad \hat{\omega}(0) = \hat{\omega}_0, \\ \dot{\hat{\theta}}_1(t) &= \Gamma\text{Proj}(\hat{\theta}_1(t), -(\tilde{x}^\top(t)PB_m)^\top \|x_t\|_{\mathcal{L}_\infty}), \quad \hat{\theta}_1(0) = \hat{\theta}_{1_0}, \\ \dot{\hat{\sigma}}_1(t) &= \Gamma\text{Proj}(\hat{\sigma}_1(t), -(\tilde{x}^\top(t)PB_m)^\top), \quad \hat{\sigma}_1(0) = \hat{\sigma}_{1_0}, \\ \dot{\hat{\theta}}_2(t) &= \Gamma\text{Proj}(\hat{\theta}_2(t), -(\tilde{x}^\top(t)PB_{um})^\top \|x_t\|_{\mathcal{L}_\infty}), \quad \theta_2(0) = \hat{\theta}_{2_0}, \\ \dot{\hat{\sigma}}_2(t) &= \Gamma\text{Proj}(\hat{\sigma}_2(t), -(\tilde{x}^\top(t)PB_{um})^\top), \quad \hat{\sigma}_2(0) = \hat{\sigma}_{2_0},\end{aligned} \tag{3.80}$$

where $\tilde{x}(t) \triangleq \hat{x}(t) - x(t)$, $\Gamma \in \mathbb{R}^+$ is the adaptation gain, $P = P^\top > 0$ is the solution to the algebraic Lyapunov equation $A_m^\top P + PA_m = -Q$ for arbitrary $Q = Q^\top > 0$, and $\text{Proj}(\cdot,\cdot)$ denotes the projection operator defined in Definition B.3. The projection operator ensures that $\hat{\omega}(t) \in \Omega$, $\|\hat{\theta}_i(t)\|_\infty \le \theta_{b_i}$, $\|\hat{\sigma}_i(t)\|_\infty \le \sigma_{b_i}$, for $i = 1, 2$, where θ_{b_i} and σ_{b_i} are as defined in (3.78).

Control Law

The control law is generated as the output of the (feedback) system

$$u(s) = -KD(s)\hat{\eta}(s), \tag{3.81}$$

where $\hat{\eta}(s)$ is the Laplace transform of the signal

$$\hat{\eta}(t) \triangleq \hat{\omega}(t)u(t) + \hat{\eta}_1(t) + \hat{\eta}_{2m}(t) - r_g(t), \tag{3.82}$$

with $r_g(s) \triangleq K_g(s)r(s)$, $\hat{\eta}_{2m}(s) \triangleq H_m^{-1}(s)H_{um}(s)\hat{\eta}_2(s)$, and with $\hat{\eta}_1(t)$ and $\hat{\eta}_2(t)$ being defined as $\hat{\eta}_i(t) \triangleq \hat{\theta}_i(t)\,\|x_t\|_{\mathcal{L}_\infty} + \hat{\sigma}_i(t)$, $i = 1,2$.

Conventional design methods from multivariable control theory can be used to design the prefilter $K_g(s)$ to achieve desired decoupling properties (see, e.g., [54]). As an example, if one chooses $K_g(s)$ as the constant matrix $K_g = -(CA_m^{-1}B_m)^{-1}$, then the diagonal elements of the desired transfer matrix $M(s) = C(s\mathbb{I}_n - A_m)^{-1}B_m K_g$ have DC gain equal to one, while the off-diagonal elements have zero DC gain.

The $\mathcal{L}_1$ adaptive controller consists of (3.79)–(3.81), subject to the $\mathcal{L}_1$-norm condition in (3.71).

3.2.3 Analysis of the $\mathcal{L}_1$ Adaptive Controller

Closed-Loop Reference System

In this section, we characterize the closed-loop reference system that the $\mathcal{L}_1$ adaptive controller tracks both in transient and steady-state and prove its stability. Toward this end, we consider the ideal nonadaptive version of the adaptive controller and define the *closed-loop reference system* as

$$\begin{aligned} \dot{x}_{\text{ref}}(t) &= A_m x_{\text{ref}}(t) + B_m\left(\omega u_{\text{ref}}(t) + f_1(t, x_{\text{ref}}(t), z(t))\right) \\ &\quad + B_{um} f_2(t, x_{\text{ref}}(t), z(t)), \quad x_{\text{ref}}(0) = x_0, \\ u_{\text{ref}}(s) &= -\omega^{-1} C(s)\big(\eta_{1\text{ref}}(s) + H_m^{-1}(s) H_{um}(s)\eta_{2\text{ref}}(s) - K_g(s) r(s)\big), \\ y_{\text{ref}}(t) &= C x_{\text{ref}}(t), \end{aligned} \tag{3.83}$$

where $\eta_{1\text{ref}}(s)$ and $\eta_{2\text{ref}}(s)$ are the Laplace transforms of the signals $\eta_{i\text{ref}}(t) \triangleq f_i(t, x_{\text{ref}}(t), z(t))$, $i = 1, 2$.

Lemma 3.2.1 For the closed-loop reference system in (3.83), subject to the $\mathcal{L}_1$-norm condition (3.71), if $\|x_0\|_\infty \le \rho_0$ and

$$\| z_\tau \|_{\mathcal{L}_\infty} \le L_z(\| x_{\text{ref}\,\tau} \|_{\mathcal{L}_\infty} + \gamma_1) + B_z, \tag{3.84}$$

then

$$\| x_{\text{ref}\,\tau} \|_{\mathcal{L}_\infty} < \rho_r, \tag{3.85}$$

$$\| u_{\text{ref}\,\tau} \|_{\mathcal{L}_\infty} < \rho_{ur}. \tag{3.86}$$

Proof. It follows from (3.83) and the definitions of $G_m(s)$ and $G_{um}(s)$ in (3.72) and (3.73) that

$$x_{\text{ref}}(s) = G_m(s)\eta_{1\text{ref}}(s) + G_{um}(s)\eta_{2\text{ref}}(s) + H_{xm}(s)C(s)K_g(s)r(s) + x_{\text{in}}(s). \tag{3.87}$$

Then, for all $t \in [0, \tau]$ we have

$$\begin{aligned} \| x_{\text{ref}\,t} \|_{\mathcal{L}_\infty} &\le \|G_m(s)\|_{\mathcal{L}_1} \left\| \eta_{1\text{ref}\,t} \right\|_{\mathcal{L}_\infty} + \|G_{um}(s)\|_{\mathcal{L}_1} \left\| \eta_{2\text{ref}\,t} \right\|_{\mathcal{L}_\infty} \\ &\quad + \left\| H_{xm}(s)C(s)K_g(s) \right\|_{\mathcal{L}_1} \|r\|_{\mathcal{L}_\infty} + \rho_{\text{in}}. \end{aligned} \tag{3.88}$$

If the bound (3.85) is not true, since $\|x_{\text{ref}}(0)\|_\infty = \|x_0\|_\infty < \rho_r$ and $x_{\text{ref}}(t)$ is continuous, there exists a time $\tau_1 \in (0, \tau]$ such that

$$\begin{aligned} &\|x_{\text{ref}}(t)\|_\infty < \rho_r, \quad \forall t \in [0, \tau_1), \\ &\|x_{\text{ref}}(\tau_1)\|_\infty = \rho_r, \end{aligned}$$

which implies that

$$\left\| x_{\text{ref}\,\tau_1} \right\|_{\mathcal{L}_\infty} = \rho_r . \tag{3.89}$$

It follows from the assumption in (3.84) and the bound in (3.89) that

$$\left\| z_{\tau_1} \right\|_{\mathcal{L}_\infty} \le L_z(\rho_r + \gamma_1) + B_z ,$$

and hence, from the definition of $\bar{\delta}(\delta)$ in (3.69), we have

$$\left\| X_{\text{ref}\,\tau_1} \right\|_{\mathcal{L}_\infty} = \left\| \begin{bmatrix} x_{\text{ref}}^\top & z^\top \end{bmatrix}^\top_{\tau_1} \right\|_{\mathcal{L}_\infty} \le \bar{\rho}_r(\rho_r) = \max\{\rho_r + \bar{\gamma}_1, L_z(\rho_r + \bar{\gamma}_1) + B_z\} .$$

Then, it follows from Assumptions 3.2.1 and 3.2.2 that

$$\left\| \eta_{i\text{ref}\,\tau_1} \right\|_{\mathcal{L}_\infty} \le d_{f_{xi}}(\bar{\rho}_r(\rho_r)) \left\| X_{\text{ref}\,\tau_1} \right\|_{\mathcal{L}_\infty} + B_{i0} \le d_{f_{xi}}(\bar{\rho}_r(\rho_r))\bar{\rho}_r(\rho_r) + B_{i0} , \qquad i = 1,2,$$

and the redefinition in (3.69) leads to the following bounds:

$$\left\| \eta_{i\text{ref}\,\tau_1} \right\|_{\mathcal{L}_\infty} \le L_{i\rho_r}\rho_r + B_{i0} , \quad i = 1,2 . \tag{3.90}$$

These bounds, together with the upper bound in (3.88), lead to

$$\begin{aligned} \left\| x_{\text{ref}\,\tau_1} \right\|_{\mathcal{L}_\infty} \le{}& \|G_m(s)\|_{\mathcal{L}_1} \left(L_{1\rho_r}\rho_r + B_{10}\right) + \|G_{um}(s)\|_{\mathcal{L}_1} \left(L_{2\rho_r}\rho_r + B_{20}\right) \\ &+ \left\| H_{xm}(s)C(s)K_g(s) \right\|_{\mathcal{L}_1} \|r\|_{\mathcal{L}_\infty} + \rho_{\text{in}} . \end{aligned}$$

The condition in (3.71) can be solved for ρ_r to obtain the bound

$$\begin{aligned} &\left(\|G_m(s)\|_{\mathcal{L}_1} + \|G_{um}(s)\|_{\mathcal{L}_1} \ell_0\right) \left(L_{1\rho_r}\rho_r + B_0\right) \\ &\qquad + \left\| H_{xm}(s)C(s)K_g(s) \right\|_{\mathcal{L}_1} \|r\|_{\mathcal{L}_\infty} + \rho_{\text{in}} < \rho_r , \end{aligned}$$

which leads to

$$\left\| x_{\text{ref}\,\tau_1} \right\|_{\mathcal{L}_\infty} < \rho_r .$$

This contradicts the equality in (3.89), thus proving the bound in (3.85). This further implies that the upper bounds in (3.90) hold for all $t \in [0,\tau]$ with strict inequality, which in turn implies that

$$\left\| \eta_{1\text{ref}\,\tau} \right\|_{\mathcal{L}_\infty} < L_{1\rho_r}\rho_r + B_{10} , \qquad \left\| \eta_{2\text{ref}\,\tau} \right\|_{\mathcal{L}_\infty} < L_{2\rho_r}\rho_r + B_{20} .$$

The bound on $u_{\text{ref}}(t)$ follows from (3.83) and the two bounds above,

$$\begin{aligned} \|u_{\text{ref}\,\tau}\|_{\mathcal{L}_\infty} <{}& \left\| \omega^{-1}C(s) \right\|_{\mathcal{L}_1} \left(L_{1\rho_r}\rho_r + B_{10}\right) \\ &+ \left\| \omega^{-1}C(s)H_m^{-1}(s)H_{um}(s) \right\|_{\mathcal{L}_1} \left(L_{2\rho_r}\rho_r + B_{20}\right) \\ &+ \left\| \omega^{-1}C(s)K_g(s) \right\|_{\mathcal{L}_1} \|r\|_{\mathcal{L}_\infty} , \end{aligned}$$

which proves (3.86). □

Equivalent (Semi-)Linear Time-Varying System

In this section, we refer to Lemma A.9.1 to demonstrate that the nonlinear system with unmodeled dynamics in (3.68) can be transformed into a (semi-)linear system with unknown time-varying parameters and time-varying disturbances. According to Lemma A.9.1 for the system in (3.68), if $u(t)$ is continuous, and moreover the following bounds hold:

$$\|x_\tau\|_{\mathcal{L}_\infty} \le \rho\,, \quad \|u_\tau\|_{\mathcal{L}_\infty} \le \rho_u\,,$$

then, for all $t \in [0, \tau]$, there exist continuous $\theta_1(t) \in \mathbb{R}^m$, $\sigma_1(t) \in \mathbb{R}^m$, $\theta_2(t) \in \mathbb{R}^{n-m}$, and $\sigma_2(t) \in \mathbb{R}^{n-m}$ with (piecewise)-continuous derivative such that

$$\begin{aligned} &\|\theta_i(t)\|_\infty < \theta_{b_i} = \theta_{b_i}(\rho_r)\,, && \|\dot\theta_i(t)\|_\infty < d_{\theta_i} = d_{\theta_i}(\rho_r)\,,\\ &\|\sigma_i(t)\|_\infty < \sigma_{b_i} = \sigma_{b_i}(\rho_r)\,, && \|\dot\sigma_i(t)\|_\infty < d_{\sigma_i} = d_{\sigma_i}(\rho_r)\,, \end{aligned}$$

$$f_i(t, x(t), z(t)) = \theta_i(t)\,\|x_t\|_{\mathcal{L}_\infty} + \sigma_i(t)\,,$$

for $i = 1, 2$, where θ_{b_i} and σ_{b_i} were defined in (3.78). Thus the system in (3.68) can be rewritten over $t \in [0, \tau]$ as

$$\begin{aligned} \dot x(t) &= A_m x(t) + B_m\left(\omega u(t) + \theta_1(t)\,\|x_t\|_{\mathcal{L}_\infty} + \sigma_1(t)\right)\\ &\quad + B_{um}\left(\theta_2(t)\,\|x_t\|_{\mathcal{L}_\infty} + \sigma_2(t)\right)\,, \quad x(0) = x_0\,,\\ y(t) &= Cx(t)\,. \end{aligned} \tag{3.91}$$

Transient and Steady-State Performance

Let $\theta_m(\rho_r)$ be defined as

$$\begin{aligned} \theta_m(\rho_r) \triangleq\; & 4\left(\max_{\omega\in\Omega} \mathrm{tr}(\omega^\top\omega) + \left(\theta_{b_1}^2 + \sigma_{b_1}^2\right)m + \left(\theta_{b_2}^2 + \sigma_{b_2}^2\right)(n-m)\right)\\ & + 4\frac{\lambda_{\max}(P)}{\lambda_{\min}(Q)}\left(\left(\theta_{b_1}d_{\theta_1} + \sigma_{b_1}d_{\sigma_1}\right)m + \left(\theta_{b_2}d_{\theta_2} + \sigma_{b_2}d_{\sigma_2}\right)(n-m)\right)\,. \end{aligned} \tag{3.92}$$

Also, let

$$\tilde\omega(t) = \hat\omega(t) - \omega\,, \quad \tilde\theta_i(t) = \hat\theta_i(t) - \theta_i(t)\,, \quad \tilde\sigma_i(t) = \hat\sigma_i(t) - \sigma_i(t)\,, \quad i = 1, 2\,.$$

Using the above notations, the following error dynamics can be derived from (3.79) and (3.91):

$$\dot{\tilde x}(t) = A_m\tilde x(t) + B_m(\tilde\omega(t)u(t) + \tilde\eta_1(t)) + B_{um}\tilde\eta_2(t)\,, \qquad \tilde x(0) = 0\,, \tag{3.93}$$

where $\tilde\eta_i(t) \triangleq \hat\eta_i(t) - \eta_i(t)$, with $\eta_i(t) \triangleq \theta_i(t)\,\|x_t\|_{\mathcal{L}_\infty} + \sigma_i(t)$, $i = 1, 2$.

Next, we show that if the adaptive gain Γ is chosen to verify the lower bound

$$\Gamma > \frac{\theta_m(\rho_r)}{\lambda_{\min}(P)\gamma_0^2}\,, \tag{3.94}$$

and the projection is confined to the bounds

$$\hat\omega(t) \in \Omega\,, \quad \|\hat\theta_i(t)\|_\infty \le \theta_{b_i}\,, \quad \|\hat\sigma_i(t)\|_\infty \le \sigma_{b_i}\,, \quad i = 1, 2\,, \tag{3.95}$$

then the prediction error $\tilde{x}(t)$ between the state of the system and the state predictor can be systematically reduced in both transient and steady-state by increasing the adaptation gain. The following lemma summarizes this result.

Lemma 3.2.2 Let the adaptive gain be lower bounded as in (3.94) and the projection be confined to the bounds in (3.95). Given the system in (3.68) and the $\mathcal{L}_1$ adaptive controller defined via (3.79)–(3.81) subject to the $\mathcal{L}_1$-norm condition in (3.71), if

$$\|x_\tau\|_{\mathcal{L}_\infty} \leq \rho, \quad \|u_\tau\|_{\mathcal{L}_\infty} \leq \rho_u, \tag{3.96}$$

then we have

$$\|\tilde{x}_\tau\|_{\mathcal{L}_\infty} < \gamma_0,$$

where γ_0 was introduced in (3.75).

Proof. It follows from the assumption in (3.96) and Lemma A.9.1 that the system in (3.68) can be rewritten as in (3.91) for all $t \in [0,\tau]$, with

$$\|\theta_i(t)\|_\infty < \theta_{b_i}(\rho_r), \quad \|\dot{\theta}_i(t)\|_\infty < d_{\theta_i}(\rho_r), \tag{3.97}$$

$$\|\sigma_i(t)\|_\infty < \sigma_{b_i}(\rho_r), \quad \|\dot{\sigma}_i(t)\|_\infty < d_{\sigma_i}(\rho_r), \quad \forall\, t \in [0,\tau]. \tag{3.98}$$

Consider the following Lyapunov function candidate:

$$\begin{aligned} V(\tilde{x}(t),\tilde{\omega}(t),\tilde{\theta}_i(t),\tilde{\sigma}_i(t)) &= \tilde{x}^\top(t)P\tilde{x}(t) \\ &+ \frac{1}{\Gamma}\left(\mathrm{tr}(\tilde{\omega}^\top(t)\tilde{\omega}(t)) + \sum_{i=1}^{2}\left(\tilde{\theta}_i^\top(t)\tilde{\theta}_i(t) + \tilde{\sigma}_i^\top(t)\tilde{\sigma}_i(t)\right)\right). \end{aligned} \tag{3.99}$$

Next we prove that

$$V(t) \leq \frac{\theta_m(\rho_r)}{\Gamma}, \quad \forall\, t \in [0,\tau].$$

Toward that end, first notice that

$$V(0) \leq \frac{4}{\Gamma}\left(\max_{\omega\in\Omega}\mathrm{tr}(\omega^\top\omega) + \theta_{b_1}^2 m + \sigma_{b_1}^2 m + \theta_{b_2}^2(n-m) + \sigma_{b_2}^2(n-m)\right) \leq \frac{\theta_m(\rho_r)}{\Gamma}.$$

Let $\tau_1 \in (0,\tau]$ be the first time instant of discontinuity of either of the derivatives of $\theta_i(t)$ and $\sigma_i(t)$. Using the projection-based adaptive laws in (3.80), one has for arbitrary $t \in [0,\tau_1)$ the following upper bound:

$$\dot{V}(t) \leq -\tilde{x}^\top(t)Q\tilde{x}(t) + \frac{2}{\Gamma}\left|\sum_{i=1}^{2}\left(\tilde{\theta}_i^\top(t)\dot{\theta}_i(t) + \tilde{\sigma}_i^\top(t)\dot{\sigma}_i(t)\right)\right|.$$

Since $\dot{\theta}_i(t)$ and $\dot{\sigma}_i(t)$ are continuous for all $t \in [0,\tau_1)$, the upper bounds in (3.97) and (3.98) lead to

$$\dot{V}(t) \leq -\tilde{x}^\top(t)Q\tilde{x}(t) + \frac{4}{\Gamma}\left(\theta_{b_1}d_{\theta_1}m + \sigma_{b_1}d_{\sigma_1}m + \theta_{b_2}d_{\theta_2}(n-m) + \sigma_{b_2}d_{\sigma_2}(n-m)\right). \tag{3.100}$$

The projection operator ensures that for all $t \in [0, \tau_1)$

$$\hat{\omega}(t) \in \Omega, \quad \|\hat{\theta}_i(t)\|_\infty \le \theta_{b_i}, \quad \|\hat{\sigma}_i(t)\|_\infty \le \sigma_{b_i}, \quad i = 1, 2,$$

and therefore

$$\begin{aligned} &\max_{t\in[0,\tau_1)} \left(\operatorname{tr}(\tilde{\omega}^\top(t)\tilde{\omega}(t)) + \sum_{i=1}^{2} \left(\tilde{\theta}_i^\top(t)\tilde{\theta}_i(t) + \tilde{\sigma}_i^\top(t)\tilde{\sigma}_i(t) \right) \right) \\ &\qquad \le 4\Big(\max_{\omega\in\Omega} \operatorname{tr}(\omega^\top \omega) + \left(\theta_{b_1}^2 + \sigma_{b_1}^2\right) m + \left(\theta_{b_2}^2 + \sigma_{b_2}^2\right)(n-m) \Big). \end{aligned}$$

If at arbitrary $\tau' \in (0, \tau_1)$ we have $V(\tau') > \frac{\theta_m(\rho_r)}{\Gamma}$, then it follows from the Lyapunov function in (3.99), the definition of $\theta_m(\rho_r)$ in (3.92), and the bound in (3.101) that

$$\tilde{x}^\top(\tau')P\tilde{x}(\tau') > \frac{4}{\Gamma}\frac{\lambda_{\max}(P)}{\lambda_{\min}(Q)} \left(\theta_{b_1} d_{\theta_1} m + \sigma_{b_1} d_{\sigma_1} m + \theta_{b_2} d_{\theta_2}(n-m) + \sigma_{b_2} d_{\sigma_2}(n-m) \right),$$

and hence

$$\begin{aligned} \tilde{x}^\top(\tau')Q\tilde{x}(\tau') &\ge \frac{\lambda_{\min}(Q)}{\lambda_{\max}(P)} \tilde{x}^\top(\tau')P\tilde{x}(\tau') \\ &> \frac{4}{\Gamma}\left(\left(\theta_{b_1} d_{\theta_1} + \sigma_{b_1} d_{\sigma_1}\right) m + \left(\theta_{b_2} d_{\theta_2} + \sigma_{b_2} d_{\sigma_2}\right)(n-m) \right). \end{aligned} \tag{3.101}$$

Thus, if $V(\tau') > \frac{\theta_m(\rho_r)}{\Gamma}$, then from (3.100) and (3.101) we have

$$\dot{V}(\tau') < 0. \tag{3.102}$$

It follows from (3.102) that $V(t) \le \frac{\theta_m(\rho_r)}{\Gamma}$ for arbitrary $t \in [0, \tau_1)$. Since

$$\lambda_{\min}(P)\|\tilde{x}(t)\|_2^2 \le \tilde{x}^\top(t)P\tilde{x}(t) \le V(t),$$

then for arbitrary $t \in [0, \tau_1)$

$$\|\tilde{x}(t)\|_\infty^2 \le \|\tilde{x}(t)\|_2^2 \le \frac{\theta_m(\rho_r)}{\lambda_{\min}(P)\Gamma}.$$

Since $V(t)$ is continuous, we further have

$$\|\tilde{x}(t)\|_\infty \le \sqrt{\frac{\theta_m(\rho_r)}{\lambda_{\min}(P)\Gamma}}.$$

Continuity of $\theta_i(t)$, $\sigma_i(t)$, $\hat{\omega}(t)$, $\hat{\theta}_i(t)$, and $\hat{\sigma}_i(t)$ implies that

$$V(\tau_1) \le \frac{\theta_m(\rho_r)}{\Gamma}.$$

Next, let $\tau_2 \in (\tau_1, \tau]$ be the next time instant such that the discontinuity of any of the derivatives of $\theta_i(t)$ and $\sigma_i(t)$ occurs. Using similar derivations as above, we can prove that

$$\|\tilde{x}(t)\|_\infty \le \sqrt{\frac{\theta_m(\rho_r)}{\lambda_{\min}(P)\Gamma}}, \quad \forall\, t \in (\tau_1, \tau_2].$$

Iterating this process until the time instant τ, we get

$$\|\tilde{x}_\tau\|_{\mathcal{L}_\infty} \leq \sqrt{\frac{\theta_m(\rho_r)}{\lambda_{\min}(P)\Gamma}},$$

and the choice for the adaptive gain Γ in (3.94) leads to

$$\|\tilde{x}_\tau\|_{\mathcal{L}_\infty} < \gamma_0,$$

which concludes the proof. □

We note that the closed-loop reference system is not implementable since it uses the unknown system input gain matrix ω, the unknown signal $z(t)$, and the unknown functions f_1 and f_2. This auxiliary closed-loop system is used only for analysis purposes and is not involved in the implementation of the $\mathcal{L}_1$ adaptive controller. In the following theorem, we prove the stability and derive the performance bounds of the actual closed-loop adaptive system with the $\mathcal{L}_1$ adaptive controller with respect to this reference system.

Theorem 3.2.1 Let the adaptive gain be lower bounded as in (3.94) and the projection be confined to the bounds in (3.95). Given the closed-loop system with the $\mathcal{L}_1$ adaptive controller defined via (3.79)–(3.81), subject to the $\mathcal{L}_1$-norm condition in (3.71), and the closed-loop reference system in (3.83), if

$$\|x_0\|_\infty \leq \rho_0,$$

then we have

$$\|x\|_{\mathcal{L}_\infty} \leq \rho, \tag{3.103}$$

$$\|u\|_{\mathcal{L}_\infty} \leq \rho_u, \tag{3.104}$$

$$\|\tilde{x}\|_{\mathcal{L}_\infty} \leq \gamma_0, \tag{3.105}$$

$$\|x_{\text{ref}} - x\|_{\mathcal{L}_\infty} \leq \gamma_1, \tag{3.106}$$

$$\|u_{\text{ref}} - u\|_{\mathcal{L}_\infty} \leq \gamma_2, \tag{3.107}$$

$$\|y_{\text{ref}} - y\|_{\mathcal{L}_\infty} \leq \|C\|_\infty \gamma_1, \tag{3.108}$$

where γ_1 and γ_2 were defined in (3.75) and (3.77), respectively.

Proof. Assume that the bounds in (3.106) and (3.107) do not hold. Then, since $\|x_{\text{ref}}(0) - x(0)\|_\infty = 0 < \gamma_1$, $\|u_{\text{ref}}(0) - u(0)\|_\infty = 0 < \gamma_2$, and $x(t)$, $x_{\text{ref}}(t)$, $u(t)$, and $u_{\text{ref}}(t)$ are continuous, there exists τ such that

$$\|x_{\text{ref}}(\tau) - x(\tau)\|_\infty = \gamma_1 \text{ or}$$
$$\|u_{\text{ref}}(\tau) - u(\tau)\|_\infty = \gamma_2,$$

while

$$\|x_{\text{ref}}(t) - x(t)\|_\infty < \gamma_1, \quad \|u_{\text{ref}}(t) - u(t)\|_\infty < \gamma_2, \quad \forall\, t \in [0, \tau).$$

This implies that that at least one of the following equalities holds:

$$\|(x_{\text{ref}} - x)_\tau\|_{\mathcal{L}_\infty} = \gamma_1, \quad \|(u_{\text{ref}} - u)_\tau\|_{\mathcal{L}_\infty} = \gamma_2. \tag{3.109}$$

It follows from Assumption 3.2.3 that

$$\|z_\tau\|_{\mathcal{L}_\infty} \leq L_z\left(\|x_{\text{ref}\,\tau}\|_{\mathcal{L}_\infty} + \gamma_1\right) + B_z\,. \tag{3.110}$$

Then, Lemma 3.2.1 implies that

$$\|x_{\text{ref}\,\tau}\|_{\mathcal{L}_\infty} \leq \rho_r\,, \quad \|u_{\text{ref}\,\tau}\|_{\mathcal{L}_\infty} \leq \rho_{ur}\,. \tag{3.111}$$

Using the definitions of ρ and ρ_u in (3.74) and (3.76), it follows from the bounds in (3.109) and (3.111) that

$$\begin{aligned}\|x_\tau\|_{\mathcal{L}_\infty} &\leq \rho_r + \gamma_1 \leq \rho\,,\\ \|u_\tau\|_{\mathcal{L}_\infty} &\leq \rho_{ur} + \gamma_2 \leq \rho_u\,.\end{aligned}$$

Hence, if one chooses the adaptive gain according to (3.94) and the projection is confined to the bounds in (3.95), Lemma 3.2.2 implies that

$$\|\tilde{x}_\tau\|_{\mathcal{L}_\infty} < \gamma_0\,. \tag{3.112}$$

Next, let $\tilde{\eta}(t) \triangleq \tilde{\omega}(t)u(t) + \tilde{\eta}_1(t) + \tilde{\eta}_{2m}(t)$, where $\tilde{\eta}_{2m}(t)$ is the signal with the Laplace transform $\tilde{\eta}_{2m}(s) \triangleq H_m^{-1}(s)H_{um}(s)\tilde{\eta}_2(s)$. It follows from (3.81) that

$$u(s) = -KD(s)\Big(\omega u(s) + \eta_1(s) + H_m^{-1}(s)H_{um}(s)\eta_2(s) - K_g(s)r(s) + \tilde{\eta}(s)\Big)\,,$$

where $\eta_1(s)$, $\eta_2(s)$, and $\tilde{\eta}(s)$ are the Laplace transforms of the signals $\eta_1(t)$, $\eta_2(t)$, and $\tilde{\eta}(t)$, respectively. Consequently

$$u(s) = -KD(s)(\mathbb{I}_m + \omega KD(s))^{-1}\Big(\eta_1(s) + H_m^{-1}(s)H_{um}(s)\eta_2(s) - K_g(s)r(s) + \tilde{\eta}(s)\Big)\,,$$

which leads to

$$\omega u(s) = -\omega KD(s)(\mathbb{I}_m + \omega KD(s))^{-1}\Big(\eta_1(s) + H_m^{-1}(s)H_{um}(s)\eta_2(s) - K_g(s)r(s) + \tilde{\eta}(s)\Big)\,. \tag{3.113}$$

Using the definition of $C(s)$ in (3.70), one can write

$$\omega u(s) = -C(s)\Big(\eta_1(s) + H_m^{-1}(s)H_{um}(s)\eta_2(s) - K_g(s)r(s) + \tilde{\eta}(s)\Big)\,,$$

and the system in (3.68) consequently takes the form

$$\begin{aligned}x(s) = {}& G_m(s)\eta_1(s) + G_{um}(s)\eta_2(s) - H_{xm}(s)C(s)\tilde{\eta}(s)\\ &+ H_{xm}(s)C(s)K_g(s)r(s) + x_{\text{in}}(s)\,.\end{aligned} \tag{3.114}$$

Next, from (3.87) and (3.114) we have

$$\begin{aligned}x_{\text{ref}}(s) - x(s) = {}& G_m(s)(\eta_{1\text{ref}}(s) - \eta_1(s)) + G_{um}(s)(\eta_{2\text{ref}}(s) - \eta_2(s))\\ &+ H_{xm}(s)C(s)\tilde{\eta}(s)\,.\end{aligned}$$

Moreover, it follows from the error dynamics in (3.93) that

$$H_m^{-1}(s)C\tilde{x}(s) = \tilde{\eta}_\omega(s) + \tilde{\eta}_1(s) + \tilde{\eta}_{2m}(s) = \tilde{\eta}(s)\,,$$

with $\tilde{\eta}_{\omega}(s)$ being the Laplace transform of the signal $\tilde{\eta}_{\omega}(t) \triangleq \tilde{\omega}(t)u(t)$, which leads to

$$x_{\text{ref}}(s) - x(s) = G_m(s)(\eta_{1\text{ref}}(s) - \eta_1(s)) + G_{um}(s)(\eta_{2\text{ref}}(s) - \eta_2(s)) + H_{xm}(s)C(s)H_m^{-1}(s)C\tilde{x}(s).$$

Therefore, we have

$$\|(x_{\text{ref}} - x)_\tau\|_{\mathcal{L}_\infty} \le \|G_m(s)\|_{\mathcal{L}_1} \left\|(\eta_{1\text{ref}} - \eta_1)_\tau\right\|_{\mathcal{L}_\infty} + \|G_{um}(s)\|_{\mathcal{L}_1} \left\|(\eta_{2\text{ref}} - \eta_2)_\tau\right\|_{\mathcal{L}_\infty} + \left\|H_{xm}(s)C(s)H_m^{-1}(s)C\right\|_{\mathcal{L}_1} \|\tilde{x}_\tau\|_{\mathcal{L}_\infty}. \tag{3.115}$$

Substituting (3.111) into (3.110) one obtains

$$\|z_\tau\|_{\mathcal{L}_\infty} \le L_z(\rho_r + \gamma_1) + B_z,$$

and hence, from the definition of $\bar{\delta}(\delta)$ in (3.69), we have

$$\|X_\tau\|_{\mathcal{L}_\infty} \le \max\{\rho_r + \gamma_1,\ L_z(\rho_r + \gamma_1) + B_z\} \le \bar{\rho}_r(\rho_r),$$
$$\|X_{\text{ref}\,\tau}\|_{\mathcal{L}_\infty} \le \max\{\rho_r,\ L_z(\rho_r + \gamma_1) + B_z\} \le \bar{\rho}_r(\rho_r).$$

Since for all $t \in [0, \tau]$, the following equalities hold:

$$\eta_{i\text{ref}}(t) - \eta_i(t) = f_i(t, X_{\text{ref}}(t)) - (\theta_i(t)\|x_t\|_{\mathcal{L}_\infty} + \sigma_i(t)) = f_i(t, X_{\text{ref}}(t)) - f_i(t, X(t)), \qquad i = 1, 2,$$

Assumption 3.2.2 implies that, for $i = 1, 2$, we have

$$\left\|(\eta_{i\text{ref}} - \eta_i)_\tau\right\|_{\mathcal{L}_\infty} \le d_{fxi}(\bar{\rho}_r(\rho_r))\,\|(X_{\text{ref}} - X)_\tau\|_{\mathcal{L}_\infty} = d_{fxi}(\bar{\rho}_r(\rho_r))\,\|(x_{\text{ref}} - x)_\tau\|_{\mathcal{L}_\infty}. \tag{3.116}$$

Then, from (3.115) we have

$$\|(x_{\text{ref}} - x)_\tau\|_{\mathcal{L}_\infty} \le \|G_m(s)\|_{\mathcal{L}_1} d_{fx1}(\bar{\rho}_r(\rho_r))\,\|(x_{\text{ref}} - x)_\tau\|_{\mathcal{L}_\infty} + \|G_{um}(s)\|_{\mathcal{L}_1} d_{fx2}(\bar{\rho}_r(\rho_r))\,\|(x_{\text{ref}} - x)_\tau\|_{\mathcal{L}_\infty} + \left\|H_{xm}(s)C(s)H_m^{-1}(s)C\right\|_{\mathcal{L}_1} \|\tilde{x}_\tau\|_{\mathcal{L}_\infty}.$$

From the redefinition in (3.69) it follows that $d_{fx1}(\bar{\rho}_r(\rho_r)) < L_{1\rho_r}$ and $d_{fx2}(\bar{\rho}_r(\rho_r)) < L_{2\rho_r}$, and, therefore, we obtain

$$\|(x_{\text{ref}} - x)_\tau\|_{\mathcal{L}_\infty} \le \|G_m(s)\|_{\mathcal{L}_1} L_{1\rho_r}\,\|(x_{\text{ref}} - x)_\tau\|_{\mathcal{L}_\infty} + \|G_{um}(s)\|_{\mathcal{L}_1} L_{2\rho_r}\,\|(x_{\text{ref}} - x)_\tau\|_{\mathcal{L}_\infty} + \left\|H_{xm}(s)C(s)H_m^{-1}(s)C\right\|_{\mathcal{L}_1} \|\tilde{x}_\tau\|_{\mathcal{L}_\infty}.$$

The upper bound in (3.112) and the $\mathcal{L}_1$-norm condition in (3.71) lead to the upper bound

$$\|(x_{\text{ref}} - x)_\tau\|_{\mathcal{L}_\infty} \le \frac{\left\|H_{xm}(s)C(s)H_m^{-1}(s)C\right\|_{\mathcal{L}_1}}{1 - \|G_m(s)\|_{\mathcal{L}_1} L_{1\rho_r} - \|G_{um}(s)\|_{\mathcal{L}_1} L_{2\rho_r}} \gamma_0,$$

which along with the definition of γ_1 in (3.75) leads to

$$\|(x_{\rm ref}-x)_\tau\|_{\mathcal{L}_\infty} \leq \gamma_1 - \beta < \gamma_1 . \tag{3.117}$$

On the other hand, it follows from (3.83) and (3.113) that

$$\begin{aligned} u_{\rm ref}(s) - u(s) = &-\omega^{-1}C(s)(\eta_{1\rm ref}(s) - \eta_1(s)) \\ &-\omega^{-1}C(s)H_m^{-1}(s)H_{um}(s)(\eta_{2\rm ref}(s) - \eta_2) + \omega^{-1}C(s)H_m^{-1}(s)C\tilde{x}(s). \end{aligned}$$

One can write

$$\begin{aligned} \|(u_{\rm ref}-u)_\tau\|_{\mathcal{L}_\infty} \leq &\left\|\omega^{-1}C(s)\right\|_{\mathcal{L}_1} \|(\eta_{1\rm ref}-\eta_1)_\tau\|_{\mathcal{L}_\infty} \\ &+\left\|\omega^{-1}C(s)H_m^{-1}(s)H_{um}(s)\right\|_{\mathcal{L}_1} \|(\eta_{2\rm ref}-\eta_2)_\tau\|_{\mathcal{L}_\infty} \\ &+\left\|\omega^{-1}C(s)H_m^{-1}(s)C\right\|_{\mathcal{L}_1} \|\tilde{x}_\tau\|_{\mathcal{L}_\infty} , \end{aligned}$$

and the bound in (3.116) leads to

$$\begin{aligned} \|(u_{\rm ref}-u)_\tau\|_{\mathcal{L}_\infty} \leq &\left\|\omega^{-1}C(s)\right\|_{\mathcal{L}_1} L_{1\rho_r} \|(x_{\rm ref}-x)_\tau\|_{\mathcal{L}_\infty} \\ &+\left\|\omega^{-1}C(s)H_m^{-1}(s)H_{um}(s)\right\|_{\mathcal{L}_1} L_{2\rho_r} \|(x_{\rm ref}-x)_\tau\|_{\mathcal{L}_\infty} \\ &+\left\|\omega^{-1}C(s)H_m^{-1}(s)C\right\|_{\mathcal{L}_1} \|\tilde{x}_\tau\|_{\mathcal{L}_\infty} . \end{aligned}$$

The bounds (3.112) and (3.117) and the definition of γ_2 in (3.77) lead to

$$\begin{aligned} \|(u_{\rm ref}-u)_\tau\|_{\mathcal{L}_\infty} \leq &\Big(\left\|\omega^{-1}C(s)\right\|_{\mathcal{L}_1} L_{1\rho_r} \\ &+\left\|\omega^{-1}C(s)H_m^{-1}(s)H_{um}(s)\right\|_{\mathcal{L}_1} L_{2\rho_r}\Big)(\gamma_1-\beta) \\ &+\left\|\omega^{-1}C(s)H_m^{-1}(s)C\right\|_{\mathcal{L}_1} \gamma_0 < \gamma_2 . \end{aligned} \tag{3.118}$$

Finally, we note that the upper bounds in (3.117) and (3.118) contradict the equalities in (3.109), which proves the bounds in (3.106) and (3.107). The results in (3.103)–(3.105) and (3.108) follow directly from the bounds in (3.111) and (3.112) and from the fact that $y_{\rm ref}(t) - y(t) = C(x_{\rm ref}(t) - x(t))$. □

Remark 3.2.1 Thus, the tracking error between $y(t)$ and $y_{\rm ref}(t)$, as well as $u(t)$ and $u_{\rm ref}(t)$, is uniformly bounded by a constant inverse proportional to the square root of the adaptive gain. This implies that in both transient and steady-state one can achieve arbitrary close tracking performance for both signals simultaneously by increasing Γ. To understand how these bounds can be used to ensure transient response with *desired* specifications, we consider the *ideal* control signal for the system in (3.68),

$$u_{\rm id}(s) = -\omega^{-1}\left(\eta_1(s) + H_m^{-1}(s)H_{um}(s)\eta_2(s) - K_g(s)r(s)\right), \tag{3.119}$$

which leads to the desired system output response

$$y_{id}(s) = H_m(s)K_g(s)r(s) \tag{3.120}$$

by canceling the uncertainties exactly. In the closed-loop reference system in (3.83), $u_{id}(t)$ is further low-pass filtered by $C(s)$ to have guaranteed low-frequency range. Thus the closed-loop reference system has a different response as compared to (3.120) achieved with (3.119). Similar to Section 2.1.4, the response of $y_{ref}(t)$ can be made as close as possible to (3.120) by reducing $(\|G_m(s)\|_{\mathcal{L}_1} + \|G_{um}(s)\|_{\mathcal{L}_1}\ell_0)$ arbitrarily. In the absence of unmatched uncertainties, we can make $\|G_m(s)\|_{\mathcal{L}_1}$ arbitrarily small by increasing the bandwidth of the low-pass filter $C(s)$. However, for the general case with unmatched uncertainties, the design of K and $D(s)$ which satisfy (3.71) is an open problem. We note also that the presence of unmatched uncertainties may limit the choice of the desired state matrix A_m.

3.2.4 Simulation Example

Consider the system

$$\begin{aligned}
\dot{x}(t) &= (A_m + A_\Delta)x(t) + B_m\omega u(t) + f_\Delta(t, x(t), z(t)),\\
y(t) &= Cx(t), \quad x(0) = x_0,
\end{aligned}$$

where

$$A_m = \begin{bmatrix} -1 & 0 & 0 \\ 0 & 0 & 1 \\ 0 & -1 & -1.8 \end{bmatrix}, \quad B_m = \begin{bmatrix} 1 & 0 \\ 0 & 0 \\ 1 & 1 \end{bmatrix},$$

$$C = \begin{bmatrix} 1 & 0 & 0 \\ 0 & 1 & 0 \end{bmatrix},$$

while $A_\Delta \in \mathbb{R}^{3\times 3}$ and $\omega \in \mathbb{R}^{2\times 2}$ are unknown constant matrices satisfying

$$\|A_\Delta\|_\infty \le 1, \quad \omega \in \begin{bmatrix} [0.6, 1.2] & [-0.2, 0.2] \\ [-0.2, 0.2] & [0.6, 1.2] \end{bmatrix} = \Omega,$$

and f_Δ is the (unknown) nonlinear function

$$f_\Delta(t, x, z) = \begin{bmatrix} \frac{k_1}{3}x^\top x + \tanh(\frac{k_2}{2}x_1)x_1 + k_3 z \\ \frac{k_4}{2}\operatorname{sech}(x_2)x_2 + \frac{k_5}{5}x_3^2 + \frac{k_6}{2}(1 - e^{-\lambda t}) + \frac{k_7}{2}z \\ k_8 x_3 \cos(\omega_u t) + k_9 z^2 \end{bmatrix},$$

with $k_i \in [-1, 1]$, $i = 1, \ldots, 9$, and $\lambda, \omega_u \in \mathbb{R}^+$. The internal unmodeled dynamics are given by

$$\begin{aligned}
\dot{x}_{z1}(t) &= x_{z2}(t),\\
\dot{x}_{z2}(t) &= -x_{z1}(t) + 0.8\left(1 - x_{z1}^2(t)\right)x_{z2}(t),\\
z(t) &= 0.1\,(x_{z1}(t) - x_{z2}(t)) + z_u(t),\\
z_u(s) &= \frac{-s+1}{\frac{s^2}{0.1^2} + \frac{0.8s}{0.1} + 1}\begin{bmatrix} 1 & -2 & 1 \end{bmatrix} x(s),
\end{aligned}$$

with $[x_{z1}(0), x_{z2}(0)] = [x_{z1_0}, x_{z2_0}]$. The control objective is to design a control $u(t)$ so that the output $y(t)$ of the system tracks the output of the desired model $M(s)$ in response to bounded reference inputs $r(t)$ ($\|r\|_{\mathcal{L}_\infty} \le 1$).

In the implementation of the $\mathcal{L}_1$ controller, we set

$$Q = \mathbb{I}_3, \quad \Gamma = 80000, \quad K = \begin{bmatrix} 8 & 0 \\ 0 & 8 \end{bmatrix},$$
$$D(s) = \frac{1}{s(\frac{s}{25}+1)(\frac{s}{70}+1)(\frac{s^2}{40^2}+\frac{1.8s}{40}+1)}\mathbb{I}_2,$$
$$K_g(s) \equiv K_g = -(CA_m^{-1}B_m)^{-1} = \begin{bmatrix} 1 & 0 \\ -1 & 1 \end{bmatrix}.$$

We choose conservatively $L_{1\rho} = L_{2\rho} = 40$ and $B_{10} = B_{20} = 5$, and thus the projection bounds can be chosen as

$$\hat{\theta}_1(t) \in [-40, 40]\mathbf{1}_m, \quad \hat{\sigma}_1(t) \in [-5, 5]\mathbf{1}_m,$$
$$\hat{\theta}_2(t) \in [-40, 40]\mathbf{1}_{(n-m)}, \quad \hat{\sigma}_2(t) \in [-5, 5]\mathbf{1}_{(n-m)},$$
$$\hat{\omega}_{11}(t),\ \hat{\omega}_{22}(t) \in [0.25, 3], \quad \hat{\omega}_{12}(t),\ \hat{\omega}_{21}(t) \in [-0.2, 0.2],$$

where $\mathbf{1}_r \in \mathbb{R}^r$ represents the vector with all elements 1.

To illustrate the performance of the $\mathcal{L}_1$ adaptive controller we consider five different scenarios:

- Scenario 1:

$$A_\Delta = \begin{bmatrix} 0.2 & -0.2 & -0.3 \\ -0.2 & -0.2 & 0.6 \\ -0.1 & 0 & -0.9 \end{bmatrix}, \quad \omega = \begin{bmatrix} 0.6 & -0.2 \\ 0.2 & 1.2 \end{bmatrix},$$
$$k_1 = -1,\ k_2 = 1,\ k_3 = 0,\ k_4 = 1,\ k_5 = 0,$$
$$k_6 = 0.2,\ \lambda = 0.3,\ k_7 = 1,\ k_8 = 0.6,\ \omega_u = 5,\ k_9 = -0.7.$$

- Scenario 2:

$$A_\Delta = \begin{bmatrix} 0.2 & -0.3 & 0.5 \\ 0 & 0 & 0 \\ -0.1 & 0.4 & 0.5 \end{bmatrix}, \quad \omega = \begin{bmatrix} 1 & 0 \\ 0 & 1 \end{bmatrix},$$
$$k_1 = 0,\ k_2 = 0,\ k_3 = 0,\ k_4 = 0,\ k_5 = 0,$$
$$k_6 = 0,\ k_7 = 0,\ k_8 = 0,\ k_9 = 0.$$

- Scenario 3:

$$A_\Delta = \begin{bmatrix} 0.1 & -0.4 & 0.5 \\ -0.5 & 0.5 & 0 \\ -0.2 & 0.3 & 0.5 \end{bmatrix}, \quad \omega = \begin{bmatrix} 0.8 & -0.1 \\ 0.1 & 0.8 \end{bmatrix},$$
$$k_1 = 0,\ k_2 = 0,\ k_3 = 0,\ k_4 = 0,\ k_5 = 0,$$
$$k_6 = 0,\ k_7 = 0,\ k_8 = 0,\ k_9 = 0.$$

– Scenario 4:

$$A_\Delta = \begin{bmatrix} 0.1 & 0.4 & 0.2 \\ -0.4 & -0.2 & 0.3 \\ -0.2 & 0.6 & -0.1 \end{bmatrix}, \quad \omega = \begin{bmatrix} 0.6 & 0 \\ 0.1 & 1.2 \end{bmatrix},$$
$$k_1 = 1,\ k_2 = 1,\ k_3 = 1,\ k_4 = 1,\ k_5 = 1,$$
$$k_6 = 1,\ \lambda = 0.1,\ k_7 = 1,\ k_8 = 1,\ \omega_u = 1,\ k_9 = 1.$$

– Scenario 5:

$$A_\Delta = \begin{bmatrix} 0.2 & -0.2 & -0.3 \\ 0.1 & -0.4 & 0.3 \\ -0.1 & 0 & -0.9 \end{bmatrix}, \quad \omega = \begin{bmatrix} 0.9 & -0.2 \\ 0.1 & 1.1 \end{bmatrix},$$
$$k_1 = 1,\ k_2 = 1,\ k_3 = 1,\ k_4 = 1,\ k_5 = -1,$$
$$k_6 = -1,\ \lambda = 0.3,\ k_7 = 1,\ k_8 = -1,\ \omega_u = 1,\ k_9 = 1.$$

All the scenarios above consider a significant amount of unmatched uncertainties in the dynamics of the system, except for Scenario 2, in which the uncertainty (affecting only the state matrix of the system) remains matched. Also, in Scenarios 2 and 3 the uncertainties affect only the state matrix and appear therefore in a linear fashion; instead, Scenarios 1, 4, and 5 consider nonlinear uncertain dynamics, both matched and unmatched. Moreover, Scenarios 1, 3, 4, and 5 include uncertainty in the system input gain matrix; in particular, Scenarios 1 and 5 consider significant coupling between the control channels, while Scenario 3 introduces a 20% reduction in the control efficiency in both control channels.

Figures 3.6 and 3.7 show, respectively, the response of the closed-loop system for Scenario 1 with the $\mathcal{L}_1$ adaptive controller (i) to a series of doublets of different amplitudes in the different channels, and (ii) to the sinusoidal reference signals $r(t) = [\sin(\frac{\pi}{3}t),\ 0.2 + 0.8\cos(\frac{\pi}{6}t)]$ and $r(t) = [0.5\sin(\frac{\pi}{3}t),\ 0.1 + 0.4\cos(\frac{\pi}{6}t)]$. One can observe that the $\mathcal{L}_1$ adaptive controller leads to scaled system output for scaled reference signals, similar to linear systems.

Figure 3.8 presents the closed-loop response to the same doublets as in Figure 3.6 but now for Scenarios 2, 3, 4, and 5. The $\mathcal{L}_1$ controller guarantees smooth and uniform transient performance in the presence of different nonlinearities affecting both the matched and the unmatched channel. Note that the control signals required to track the reference signals and compensate for the uncertainties are significantly different for each scenario. Also, note that, despite the large adaptation rate, the control signals are well in the low-frequency range. In order to show the benefits of the compensation for unmatched uncertainties, we repeat the same four scenarios (Scenarios 2–5) *without* the unmatched component in the control law (term $\hat{\eta}_{2m}(t)$ in equation (3.82)). We notice that, in this case, the $\mathcal{L}_1$ adaptive controller reduces to the MIMO version of the $\mathcal{L}_1$ controller introduced in Section 2.5. The results are shown in Figure 3.9. Since Scenario 2 only considered matched uncertainties, the performance in this case remains the same. For the other three scenarios, however, the closed-loop performance degrades significantly, especially for the second output $y_2(t)$.

Finally, it is important to emphasize that, for all the simulations provided above (Figures 3.6–3.9), the $\mathcal{L}_1$ adaptive controller has *not* been redesigned or retuned and that a *single set* of control parameters has been used for all the scenarios.

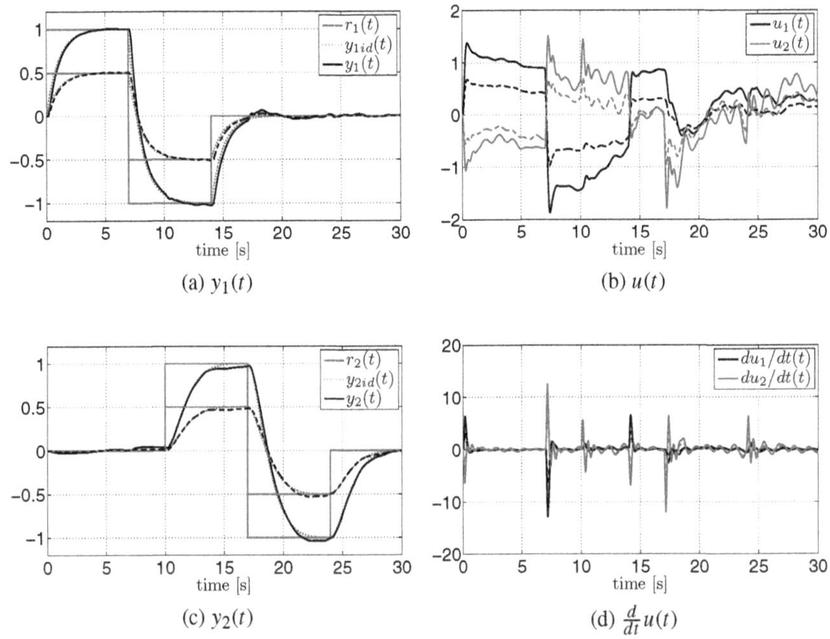

Figure 3.6: Scenario 1. Performance of the $\mathcal{L}_1$ adaptive controller for doublet commands.

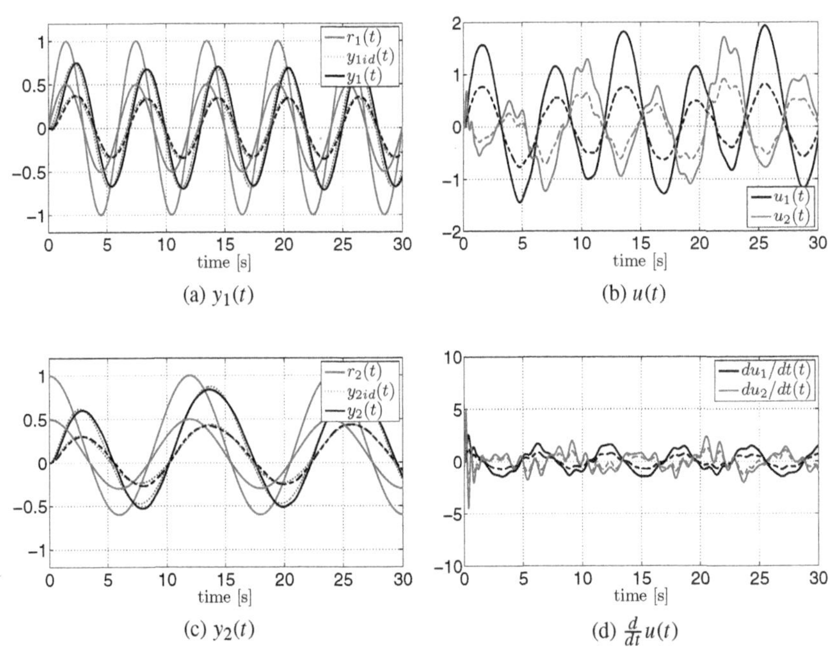

Figure 3.7: Scenario 1. Performance of the $\mathcal{L}_1$ adaptive controller for sinusoidal reference signals.

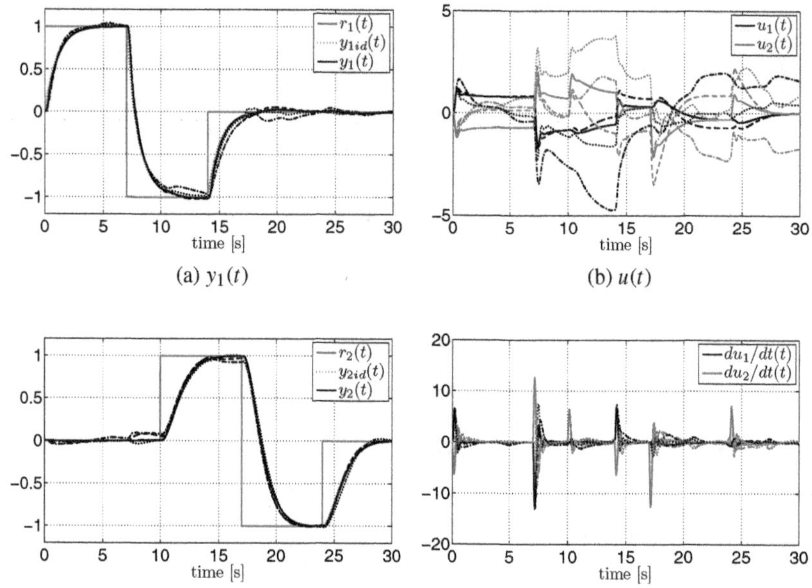

(a) $y_1(t)$ (b) $u(t)$

(c) $y_2(t)$ (d) $\frac{d}{dt}u(t)$

Figure 3.8: Scenarios 2 (solid), 3 (dashed), 4 (dash-dot), and 5 (dotted). Performance of the $\mathcal{L}_1$ adaptive controller for doublet commands.

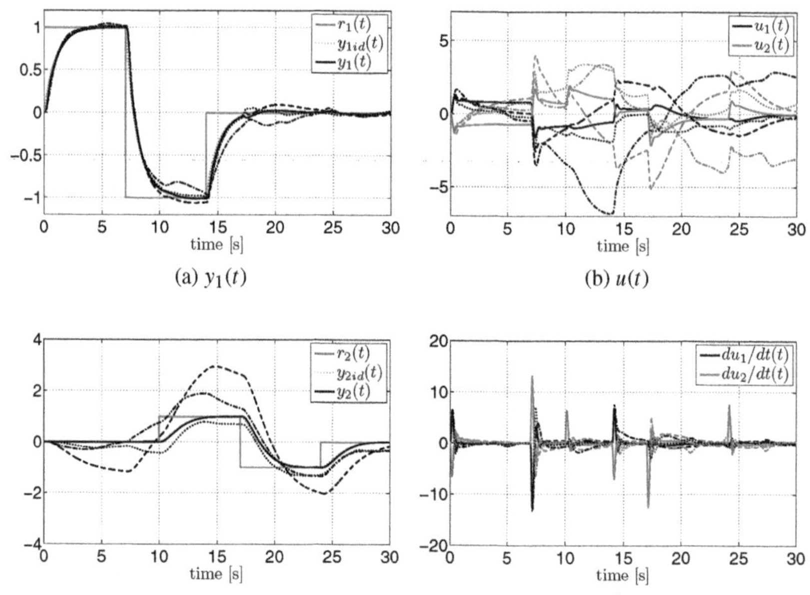

(a) $y_1(t)$ (b) $u(t)$

(c) $y_2(t)$ (d) $\frac{d}{dt}u(t)$

Figure 3.9: Scenarios 2 (solid), 3 (dashed), 4 (dash-dot), and 5 (dotted). Performance of the $\mathcal{L}_1$ adaptive controller for doublet commands *without* compensation for unmatched uncertainties.

3.3 Piecewise-Constant Adaptive Laws for $\mathcal{L}_1$ Adaptive Control in the Presence of Unmatched Nonlinear Uncertainties

This section presents a different estimation scheme for the $\mathcal{L}_1$ adaptive control architecture developed in the previous section. We consider again multi-input multi-output uncertain systems in the presence of uncertain system input gain and time- and state-dependent unknown nonlinearities, *without enforcing matching conditions*. In particular, the $\mathcal{L}_1$ adaptive controller developed in this section uses a fast estimation scheme based on a piecewise-constant adaptive law, first introduced in [33], and whose adaptation rate can be directly associated with the sampling rate of the available CPU. Similar to the previous section, the class of considered systems includes general unmatched uncertainties. The adaptive algorithm guarantees semiglobal uniform performance bounds for the system's signals, both input and output, simultaneously, and thus ensures uniform transient response in addition to steady-state tracking. This extension of the $\mathcal{L}_1$ adaptive controller, summarized in [180], has been successfully applied to the design of Flight Control Systems for NASA's GTM (AirSTAR) [59] and for the Boeing X-48B [106]. Results from piloted simulation evaluations on NASA's GTM are included in Section 6.1.2.

3.3.1 Problem Formulation

Consider the following system dynamics:

$$\begin{aligned} \dot{x}(t) &= A_m x(t) + B_m \omega u(t) + f(t, x(t), z(t)), \quad x(0) = x_0, \\ \dot{x}_z(t) &= g(t, x_z(t), x(t)), \quad x_z(0) = x_{z0}, \\ z(t) &= g_o(t, x_z(t)), \\ y(t) &= Cx(t), \end{aligned} \tag{3.121}$$

where $x(t) \in \mathbb{R}^n$ is the system state vector (measured); $u(t) \in \mathbb{R}^m$ is the control signal $(m \le n)$; $y(t) \in \mathbb{R}^m$ is the regulated output; A_m is a known Hurwitz $n \times n$ matrix that defines the desired dynamics for the closed-loop system; $B_m \in \mathbb{R}^{n \times m}$ is a known full-rank constant matrix, (A_m, B_m) is controllable; $C \in \mathbb{R}^{m \times n}$ is a known full-rank constant matrix, (A_m, C) is observable; $\omega \in \mathbb{R}^{m \times m}$ is the uncertain system input gain matrix; $z(t) \in \mathbb{R}^p$ and $x_z(t) \in \mathbb{R}^l$ are the output and the state vector of internal unmodeled dynamics; and $f : \mathbb{R} \times \mathbb{R}^n \times \mathbb{R}^p \to \mathbb{R}^n$, $g_o : \mathbb{R} \times \mathbb{R}^l \to \mathbb{R}^p$, and $g : \mathbb{R} \times \mathbb{R}^l \times \mathbb{R}^n \to \mathbb{R}^l$ are unknown nonlinear functions satisfying the standard assumptions on existence and uniqueness of solutions. The initial condition x_0 is assumed to be inside an arbitrarily large known set, i.e., $\|x_0\|_\infty \le \rho_0 < \infty$ for some $\rho_0 > 0$.

Again, we note that the system in (3.121) can also be written in the form

$$\begin{aligned} \dot{x}(t) &= A_m x(t) + B_m(\omega u(t) + f_1(t, x(t), z(t))) + B_{um} f_2(t, x(t), z(t)), \quad x(0) = x_0, \\ \dot{x}_z(t) &= g(t, x_z(t), x(t)), \quad x_z(0) = x_{z0}, \\ z(t) &= g_o(t, x_z(t)), \\ y(t) &= Cx(t), \end{aligned} \tag{3.122}$$

where $B_{um} \in \mathbb{R}^{n \times (n-m)}$ is a constant matrix such that $B_m^\top B_{um} = 0$ and also $\text{rank}([B_m,\ B_{um}]) = n$, while $f_1 : \mathbb{R} \times \mathbb{R}^n \times \mathbb{R}^p \to \mathbb{R}^m$ and $f_2 : \mathbb{R} \times \mathbb{R}^n \times \mathbb{R}^p \to \mathbb{R}^{(n-m)}$

are unknown nonlinear functions that verify

$$\begin{bmatrix} f_1(t,x(t),z(t)) \\ f_2(t,x(t),z(t)) \end{bmatrix} = B^{-1} f(t,x(t),z(t)), \quad B \triangleq \begin{bmatrix} B_m, & B_{um} \end{bmatrix}. \tag{3.123}$$

In this problem formulation, $f_1(\cdot)$ represents the matched component of the uncertainties, whereas $B_{um} f_2(\cdot)$ represents the unmatched component.

Let $X \triangleq [x^\top,\ z^\top]^\top$, and with a slight abuse of language let $f_i(t,X) \triangleq f_i(t,x,z)$, $i = 1,2$. The system above verifies the following assumptions.

Assumption 3.3.1 (Boundedness of $f_i(t,0)$) There exists $B_{i0} > 0$, such that $\|f_i(t,0)\|_\infty \le B_{i0}$ holds for all $t \ge 0$ and for $i = 1,2$.

Assumption 3.3.2 (Semiglobal Lipschitz condition) For arbitrary $\delta > 0$, there exist positive K_{1_δ}, K_{2_δ}, such that

$$\|f_i(t,X_1) - f_i(t,X_2)\|_\infty \le K_{i_\delta} \|X_1 - X_2\|_\infty, \quad i = 1,2,$$

for all $\|X_j\|_\infty \le \delta$, $j = 1,2$, uniformly in t.

Assumption 3.3.3 (Stability of unmodeled dynamics) The x_z-dynamics are BIBO stable with respect to both initial conditions x_{z0} and input $x(t)$, i.e., there exist L_z, $B_z > 0$ such that for all $t \ge 0$

$$\|z_t\|_{\mathcal{L}_\infty} \le L_z \|x_t\|_{\mathcal{L}_\infty} + B_z .$$

Assumption 3.3.4 (Partial knowledge of the system input gain) The system input gain matrix ω is assumed to be an unknown (nonsingular) strictly row-diagonally dominant matrix with $\mathrm{sgn}(\omega_{ii})$ known. Also, we assume that there exists a known compact convex set Ω, such that $\omega \in \Omega \subset \mathbb{R}^{m\times m}$, and that a nominal system input gain $\omega_0 \in \Omega$ is known.

Assumption 3.3.5 (Stability of matched transmission zeros) The transmission zeros of the transfer matrix $H_m(s) = C(s\mathbb{I} - A_m)^{-1} B_m$ lie in the open left half plane.

As in the previous section, the control objective is to design an adaptive state feedback controller to ensure that $y(t)$ tracks the output response of a *desired system* $M(s)$ defined as

$$M(s) \triangleq C\,(s\mathbb{I} - A_m)^{-1}\, B_m K_g(s),$$

where $K_g(s)$ is a feedforward prefilter, to a given bounded piecewise-continuous reference signal $r(t)$ *in both transient* and *steady-state*, while all other signals remain bounded.

3.3.2 $\mathcal{L}_1$ Adaptive Control Architecture

Definitions and $\mathcal{L}_1$-Norm Sufficient Condition for Stability

Let

$$\begin{aligned} H_{xm}(s) &\triangleq (s\mathbb{I}_n - A_m)^{-1} B_m, \\ H_{xum}(s) &\triangleq (s\mathbb{I}_n - A_m)^{-1} B_{um}, \\ H_m(s) &\triangleq C H_{xm}(s) = C\,(s\mathbb{I}_n - A_m)^{-1} B_m, \\ H_{um}(s) &\triangleq C H_{xum}(s) = C\,(s\mathbb{I}_n - A_m)^{-1} B_{um}, \end{aligned}$$

and also let $x_{\text{in}}(t)$ be the signal with Laplace transform $x_{\text{in}}(s) \triangleq (s\mathbb{I}_n - A_m)^{-1}x_0$, and $\rho_{\text{in}} \triangleq \|s(s\mathbb{I} - A_m)^{-1}\|_{\mathcal{L}_1}\rho_0$. Since A_m is Hurwitz and x_0 is finite, then $\|x_{\text{in}}\|_{\mathcal{L}_\infty} \leq \rho_{\text{in}}$. Further, for every $\delta > 0$, let

$$L_{i\delta} \triangleq \frac{\bar{\delta}(\delta)}{\delta} K_{i\bar{\delta}(\delta)}, \quad \bar{\delta}(\delta) \triangleq \max\{\delta + \bar{\gamma}_1, \ L_z(\delta + \bar{\gamma}_1) + B_z\}, \tag{3.124}$$

where K_{i_δ} was introduced in Assumption 3.3.2, and $\bar{\gamma}_1$ is an arbitrarily small positive constant.

The design of the $\mathcal{L}_1$ adaptive controller involves a feedback gain matrix $K \in \mathbb{R}^{m\times m}$ and an $m \times m$ strictly proper transfer matrix $D(s)$, which lead, for all $\omega \in \Omega$, to a strictly proper stable

$$C(s) \triangleq \omega K D(s)(\mathbb{I}_m + \omega K D(s))^{-1} \tag{3.125}$$

with DC gain $C(0) = \mathbb{I}_m$. The choice of $D(s)$ needs to ensure also that $C(s)H_m^{-1}(s)$ is a proper stable transfer matrix.

For the proofs of stability and performance bounds, the choice of K and $D(s)$ also needs to ensure that, for a given ρ_0, there exists $\rho_r > \rho_{\text{in}}$ such that the following $\mathcal{L}_1$-norm condition holds:

$$\|G_m(s)\|_{\mathcal{L}_1} + \|G_{um}(s)\|_{\mathcal{L}_1}\ell_0 < \frac{\rho_r - \left\|H_{xm}(s)C(s)K_g(s)\right\|_{\mathcal{L}_1}\|r\|_{\mathcal{L}_\infty} - \rho_{\text{in}}}{L_{1\rho_r}\rho_r + B_0}, \tag{3.126}$$

where

$$G_m(s) \triangleq H_{xm}(s)(\mathbb{I}_m - C(s)),$$
$$G_{um}(s) \triangleq \left(\mathbb{I}_n - H_{xm}(s)C(s)H_m^{-1}(s)C\right)H_{xum}(s),$$

while

$$\ell_0 \triangleq \frac{L_{2\rho_r}}{L_{1\rho_r}}, \quad B_0 \triangleq \max\left\{B_{10}, \ \frac{B_{20}}{\ell_0}\right\},$$

and $K_g(s)$ is the (BIBO-stable) feedforward prefilter. Further, let ρ be defined as

$$\rho \triangleq \rho_r + \bar{\gamma}_1, \tag{3.127}$$

and let γ_1 be given by

$$\gamma_1 \triangleq \frac{\left\|H_{xm}(s)C(s)H_m^{-1}(s)C\right\|_{\mathcal{L}_1}}{1 - \|G_m(s)\|_{\mathcal{L}_1}L_{1\rho_r} - \|G_{um}(s)\|_{\mathcal{L}_1}L_{2\rho_r}}\bar{\gamma}_0 + \beta, \tag{3.128}$$

where $\bar{\gamma}_0$ and β are arbitrarily small positive constants such that $\gamma_1 \leq \bar{\gamma}_1$. Let

$$\rho_u \triangleq \rho_{ur} + \gamma_2, \tag{3.129}$$

where ρ_{ur} and γ_2 are defined as

$$\begin{aligned}
\rho_{ur} &\triangleq \left\|\omega^{-1}C(s)\right\|_{\mathcal{L}_1}(L_{1\rho_r}\rho_r + B_{10}) + \left\|\omega^{-1}C(s)H_m^{-1}(s)H_{um}(s)\right\|_{\mathcal{L}_1}(L_{2\rho_r}\rho_r + B_{20}) \\
&\quad + \left\|\omega^{-1}C(s)K_g(s)\right\|_{\mathcal{L}_1}\|r\|_{\mathcal{L}_\infty}, \\
\gamma_2 &\triangleq \left(\left\|\omega^{-1}C(s)\right\|_{\mathcal{L}_1}L_{1\rho_r} + \left\|\omega^{-1}C(s)H_m^{-1}(s)H_{um}(s)\right\|_{\mathcal{L}_1}L_{2\rho_r}\right)\gamma_1 \\
&\quad + \left\|\omega^{-1}C(s)H_m^{-1}(s)C\right\|_{\mathcal{L}_1}\bar{\gamma}_0.
\end{aligned} \tag{3.130}$$

Further, let $T_s > 0$ be the *adaptation sampling time*, which can be associated with the sampling rate of the available CPU, and let $\varsigma(T_s)$ be

$$\varsigma(T_s) \triangleq \kappa_1(T_s)\Delta_1 + \kappa_2(T_s)\Delta_2,$$

where $\kappa_1(T_s)$ and $\kappa_2(T_s)$ are defined as

$$\kappa_1(T_s) \triangleq \int_0^{T_s} \left\| e^{A_m(T_s-\tau)} B_m \right\|_2 d\tau, \tag{3.131}$$

$$\kappa_2(T_s) \triangleq \int_0^{T_s} \left\| e^{A_m(T_s-\tau)} B_{um} \right\|_2 d\tau, \tag{3.132}$$

while Δ_1 and Δ_2 are given by

$$\Delta_1 \triangleq (\max_{\omega\in\Omega}\{\|\omega-\omega_0\|_2\}\rho_u + L_{1\rho}\rho + B_{10})\sqrt{m}, \tag{3.133}$$

$$\Delta_2 \triangleq \left(L_{2\rho}\rho + B_{20}\right)\sqrt{n-m},$$

with ω_0 being the best available guess of ω introduced in Assumption 3.3.4. Also, let $\alpha_1(t)$, $\alpha_2(t)$, $\alpha_3(t)$, and $\alpha_4(t)$ be defined as

$$\alpha_1(t) \triangleq \left\| e^{A_m t} \right\|_2,$$

$$\alpha_2(t) \triangleq \int_0^t \left\| e^{A_m(t-\tau)} \Phi^{-1}(T_s) e^{A_m T_s} \right\|_2 d\tau,$$

$$\alpha_3(t) \triangleq \int_0^t \left\| e^{A_m(t-\tau)} B_m \right\|_2 d\tau,$$

$$\alpha_4(t) \triangleq \int_0^t \left\| e^{A_m(t-\tau)} B_{um} \right\|_2 d\tau, \tag{3.134}$$

where $\Phi(T_s)$ is an $n \times n$ matrix defined as

$$\Phi(T_s) \triangleq A_m^{-1}\left(e^{A_m T_s} - \mathbb{I}_n\right). \tag{3.135}$$

Let

$$\bar{\alpha}_1(T_s) \triangleq \max_{t\in[0,\,T_s]} \alpha_1(t), \quad \bar{\alpha}_2(T_s) \triangleq \max_{t\in[0,T_s]} \alpha_2(t), \tag{3.136}$$

$$\bar{\alpha}_3(T_s) \triangleq \max_{t\in[0,\,T_s]} \alpha_3(t), \quad \bar{\alpha}_4(T_s) \triangleq \max_{t\in[0,\,T_s]} \alpha_4(t). \tag{3.137}$$

Finally, let

$$\gamma_0(T_s) \triangleq (\bar{\alpha}_1(T_s) + \bar{\alpha}_2(T_s))\varsigma(T_s) + \bar{\alpha}_3(T_s)\Delta_1 + \bar{\alpha}_4(T_s)\Delta_2. \tag{3.138}$$

Lemma 3.3.1 The following limiting relationship is true:

$$\lim_{T_s\to 0} \gamma_0(T_s) = 0.$$

Proof. We note that since $\bar{\alpha}_1(T_s)$, $\bar{\alpha}_2(T_s)$, Δ_1, and Δ_2 are bounded, it is enough to prove that

$$\lim_{T_s \to 0} \varsigma(T_s) = 0, \tag{3.139}$$

$$\lim_{T_s \to 0} \bar{\alpha}_3(T_s) = 0, \quad \lim_{T_s \to 0} \bar{\alpha}_4(T_s) = 0. \tag{3.140}$$

From the definition of $\kappa_1(T_s)$ and $\kappa_2(T_s)$ in (3.131)–(3.132) we have

$$\lim_{T_s \to 0} \kappa_1(T_s) = 0, \quad \lim_{T_s \to 0} \kappa_2(T_s) = 0,$$

and, then, since Δ_1 and Δ_2 are bounded, we have

$$\lim_{T_s \to 0} \varsigma(T_s) = 0,$$

which proves (3.139). Since $\alpha_3(t)$ and $\alpha_4(t)$ are continuous, it follows from (3.137) that

$$\lim_{T_s \to 0} \bar{\alpha}_3(T_s) = \lim_{t \to 0} \alpha_3(t) = 0, \quad \lim_{T_s \to 0} \bar{\alpha}_4(T_s) = \lim_{t \to 0} \alpha_4(t) = 0,$$

which prove (3.140). Boundedness of $\bar{\alpha}_1(T_s)$, $\bar{\alpha}_2(T_s)$, Δ_1, and Δ_2 implies that

$$\lim_{T_s \to 0} \left((\bar{\alpha}_1(T_s) + \bar{\alpha}_2(T_s)) \varsigma(T_s) + \bar{\alpha}_3(T_s)\Delta_1 + \bar{\alpha}_4(T_s)\Delta_2 \right) = 0,$$

which completes the proof. □

The $\mathcal{L}_1$ adaptive control architecture is introduced next.

State Predictor

We consider the following state predictor:

$$\begin{aligned} \dot{\hat{x}}(t) &= A_m \hat{x}(t) + B_m \left(\omega_0 u(t) + \hat{\sigma}_1(t) \right) + B_{um} \hat{\sigma}_2(t), \quad \hat{x}(0) = x_0, \\ \hat{y}(t) &= C \hat{x}(t), \end{aligned} \tag{3.141}$$

where $\hat{\sigma}_1(t) \in \mathbb{R}^m$ and $\hat{\sigma}_2(t) \in \mathbb{R}^{n-m}$ are the adaptive estimates.

Adaptation Laws

The adaptation laws for $\hat{\sigma}_1(t)$ and $\hat{\sigma}_2(t)$ are defined as

$$\begin{aligned} \begin{bmatrix} \hat{\sigma}_1(t) \\ \hat{\sigma}_2(t) \end{bmatrix} &= \begin{bmatrix} \hat{\sigma}_1(iT_s) \\ \hat{\sigma}_2(iT_s) \end{bmatrix}, \quad t \in [iT_s, (i+1)T_s), \\ \begin{bmatrix} \hat{\sigma}_1(iT_s) \\ \hat{\sigma}_2(iT_s) \end{bmatrix} &= - \begin{bmatrix} \mathbb{I}_m & 0 \\ 0 & \mathbb{I}_{n-m} \end{bmatrix} B^{-1} \Phi^{-1}(T_s) \mu(iT_s), \end{aligned} \tag{3.142}$$

for $i = 0, 1, 2, \ldots$, where B and $\Phi(T_s)$ were introduced in (3.123) and (3.135), respectively, while

$$\mu(iT_s) = e^{A_m T_s} \tilde{x}(iT_s), \quad \tilde{x}(t) = \hat{x}(t) - x(t).$$

Control Law

The control signal is generated as the output of the (feedback) system

$$u(s) = -KD(s)\hat{\eta}(s), \tag{3.143}$$

where $\hat{\eta}(s)$ is the Laplace transform of the signal

$$\hat{\eta}(t) \triangleq \omega_0 u(t) + \hat{\eta}_1(t) + \hat{\eta}_{2m}(t) - r_g(t), \tag{3.144}$$

with $r_g(s) \triangleq K_g(s)r(s)$, $\hat{\eta}_{2m}(s) \triangleq H_m^{-1}(s)H_{um}(s)\hat{\eta}_2(s)$, and with $\hat{\eta}_1(t) \triangleq \hat{\sigma}_1(t)$ and $\hat{\eta}_2(t) \triangleq \hat{\sigma}_2(t)$.

As before, we repeat that conventional design methods from multivariable control theory can be used to design the prefilter $K_g(s)$ to achieve desired decoupling properties. As an example, if one chooses $K_g(s)$ as the constant matrix $K_g = -(CA_m^{-1}B_m)^{-1}$, then the diagonal elements of the desired transfer matrix $M(s) = C(s\mathbb{I}_n - A_m)^{-1}B_m K_g$ have DC gain equal to one, while the off-diagonal elements have zero DC gain.

The $\mathcal{L}_1$ adaptive controller consists of (3.141)–(3.143), subject to the $\mathcal{L}_1$-norm condition in (3.126).

3.3.3 Analysis of the $\mathcal{L}_1$ Adaptive Controller

Closed-Loop Reference System

We consider the same *closed-loop reference system* as in Section 3.2:

$$\begin{aligned}
\dot{x}_{\text{ref}}(t) &= A_m x_{\text{ref}}(t) + B_m\left(\omega u_{\text{ref}}(t) + f_1(t, x_{\text{ref}}(t), z(t))\right) \\
&\quad + B_{um} f_2(t, x_{\text{ref}}(t), z(t)), \quad x_{\text{ref}}(0) = x_0, \\
u_{\text{ref}}(s) &= -\omega^{-1}C(s)\big(\eta_{1\text{ref}}(s) + H_m^{-1}(s)H_{um}(s)\eta_{2\text{ref}}(s) - K_g(s)r(s)\big), \\
y_{\text{ref}}(t) &= Cx_{\text{ref}}(t),
\end{aligned} \tag{3.145}$$

where $\eta_{i\text{ref}}(s)$ is the Laplace transform of $\eta_{i\text{ref}}(t) \triangleq f_i(t, x_{\text{ref}}(t), z(t))$, $i = 1, 2$.

Lemma 3.3.2 For the closed-loop reference system in (3.145), subject to the $\mathcal{L}_1$-norm condition (3.126), if $\|x_0\|_\infty \le \rho_0$ and

$$\| z_t \|_{\mathcal{L}_\infty} \le L_z(\| x_{\text{ref}\,t} \|_{\mathcal{L}_\infty} + \gamma_1) + B_z,$$

then

$$\begin{aligned}
\| x_{\text{ref}\,t} \|_{\mathcal{L}_\infty} &< \rho_r, \\
\| u_{\text{ref}\,t} \|_{\mathcal{L}_\infty} &< \rho_{ur}.
\end{aligned}$$

Proof. The proof of this lemma is similar to the proof of Lemma 3.2.1 in Section 3.2 and is therefore omitted. □

Transient and Steady-State Performance

The error dynamics can be derived from (3.122) and (3.141),

$$\dot{\tilde{x}}(t) = A_m\tilde{x}(t) + B_m\tilde{\eta}_1(t) + B_{um}\tilde{\eta}_2(t), \quad \tilde{x}(0) = 0, \tag{3.146}$$

where

$$\tilde{\eta}_1(t) \triangleq \hat{\sigma}_1(t) - ((\omega - \omega_0)u(t) + \eta_1(t)), \tag{3.147}$$

$$\tilde{\eta}_2(t) \triangleq \hat{\sigma}_2(t) - \eta_2(t), \tag{3.148}$$

with $\eta_i(t) = f_i(t, x(t), z(t))$, $i = 1, 2$.

Next we show that if T_s is chosen to ensure that

$$\gamma_0(T_s) < \bar{\gamma}_0, \tag{3.149}$$

then the tracking error between the state of the system and the state predictor can be systematically reduced in both transient and steady-state by reducing T_s.

Lemma 3.3.3 Let the adaptation rate be chosen to satisfy the design constraint in (3.149). Given the system in (3.122) and the $\mathcal{L}_1$ adaptive controller defined via (3.141)–(3.143), subject to the $\mathcal{L}_1$-norm condition in (3.126), if

$$\|x_\tau\|_{\mathcal{L}_\infty} \leq \rho, \quad \|u_\tau\|_{\mathcal{L}_\infty} \leq \rho_u, \tag{3.150}$$

we have

$$\|\tilde{x}_\tau\|_{\mathcal{L}_\infty} < \bar{\gamma}_0,$$

where $\bar{\gamma}_0$ is as introduced in (3.128).

Proof. If the bounds in (3.150) hold, then it follows from Assumption 3.3.3 that $\| z_\tau \|_{\mathcal{L}_\infty} \leq L_z\rho + B_z$, which leads to

$$\| X_\tau \|_{\mathcal{L}_\infty} \leq \bar{\rho}(\rho).$$

From Assumptions 3.3.2 and 3.3.1, and the redefinition in (3.124), one finds

$$\left\| \eta_{1\tau} \right\|_{\mathcal{L}_\infty} \leq L_{1\rho}\rho + B_{10}.$$

This implies that

$$\|\eta_1(t)\|_2 \leq \left(L_{1\rho}\rho + B_{10}\right)\sqrt{m}, \quad \forall t \in [0, \tau]. \tag{3.151}$$

Similarly, one can show that

$$\|\eta_2(t)\|_2 \leq \left(L_{2\rho}\rho + B_{20}\right)\sqrt{n-m}, \quad \forall t \in [0, \tau]. \tag{3.152}$$

It follows from the error dynamics in (3.146) that

$$\begin{aligned}\tilde{x}(iT_s+t) = \mathrm{e}^{A_m t}\tilde{x}(iT_s) + &\int_{iT_s}^{iT_s+t} \mathrm{e}^{A_m(iT_s+t-\xi)} B_m \hat{\sigma}_1(iT_s)d\xi \\ &+ \int_{iT_s}^{iT_s+t} \mathrm{e}^{A_m(iT_s+t-\xi)} B_{um}\hat{\sigma}_2(iT_s)d\xi \\ &- \int_{iT_s}^{iT_s+t} \mathrm{e}^{A_m(iT_s+t-\xi)} B_m\left((\omega-\omega_0)u(\xi)+\eta_1(\xi)\right)d\xi \\ &- \int_{iT_s}^{iT_s+t} \mathrm{e}^{A_m(iT_s+t-\xi)} B_{um}\eta_2(\xi)d\xi\,,\end{aligned}$$

which can be rewritten as

$$\begin{aligned}\tilde{x}(iT_s+t) = \mathrm{e}^{A_m t}\tilde{x}(iT_s) + &\int_0^t \mathrm{e}^{A_m(t-\xi)} B \begin{bmatrix} \hat{\sigma}_1(iT_s) \\ \hat{\sigma}_2(iT_s) \end{bmatrix} d\xi \\ &- \int_0^t \mathrm{e}^{A_m(t-\xi)} B_m\left((\omega-\omega_0)u(iT_s+\xi)+\eta_1(iT_s+\xi)\right)d\xi \\ &- \int_0^t \mathrm{e}^{A_m(t-\xi)} B_{um}\eta_2(iT_s+\xi)d\xi\,.\end{aligned}$$

Define the signals $\zeta_1(iT_s+t)$ and $\zeta_2(iT_s+t)$ as

$$\begin{aligned}\zeta_1(iT_s+t) &\triangleq \mathrm{e}^{A_m t}\tilde{x}(iT_s) + \int_0^t \mathrm{e}^{A_m(t-\xi)} B \begin{bmatrix} \hat{\sigma}_1(iT_s) \\ \hat{\sigma}_2(iT_s) \end{bmatrix} d\xi\,, \\ \zeta_2(iT_s+t) &\triangleq -\int_0^t \mathrm{e}^{A_m(t-\xi)} B_{um}\eta_2(iT_s+\xi)d\xi \\ &\quad - \int_0^t \mathrm{e}^{A_m(t-\xi)} B_m\left((\omega-\omega_0)u(iT_s+\xi)+\eta_1(iT_s+\xi)\right)d\xi\,.\end{aligned}$$

Next, we prove that

$$\|\tilde{x}(iT_s)\|_2 \le \varsigma(T_s)\,, \quad \forall\, iT_s \le \tau\,. \tag{3.153}$$

Because $\tilde{x}(0)=0$, it follows that $\|\tilde{x}(0)\|_2 < \varsigma(T_s)$. Consider now the time interval $[jT_s,(j+1)T_s)$, with $(j+1)T_s < \tau$. The prediction error at the sampling instant $(j+1)T_s$ is given by

$$\tilde{x}((j+1)T_s) = \zeta_1((j+1)T_s) + \zeta_2((j+1)T_s)\,,$$

with

$$\zeta_1((j+1)T_s) = \mathrm{e}^{A_m T_s}\tilde{x}(jT_s) + \int_0^{T_s} \mathrm{e}^{A_m(T_s-\xi)} B \begin{bmatrix} \hat{\sigma}_1(jT_s) \\ \hat{\sigma}_2(jT_s) \end{bmatrix} d\xi\,, \tag{3.154}$$

$$\begin{aligned}\zeta_2((j+1)T_s) = &-\int_0^{T_s} \mathrm{e}^{A_m(T_s-\xi)} B_{um}\eta_2(jT_s+\xi)d\xi \\ &- \int_0^{T_s} \mathrm{e}^{A_m(T_s-\xi)} B_m\left((\omega-\omega_0)u(jT_s+\xi)+\eta_1(jT_s+\xi)\right)d\xi\,.\end{aligned}$$

Substituting the adaptive law (3.142) into (3.154) leads to

$$\zeta_1((j+1)T_s) = 0\,.$$

Then, it follows that

$$\tilde{x}((j+1)T_s) = \zeta_2((j+1)T_s)\,,$$

and the bounds in (3.151) and (3.152) together with the definitions of $\kappa_1(T_s)$, $\kappa_2(T_s)$, Δ_1, and Δ_2, imply that

$$\|\tilde{x}((j+1)T_s)\|_2 \le \kappa_1(T_s)\Delta_1 + \kappa_2(T_s)\Delta_2 = \varsigma(T_s)\,.$$

This confirms the upper bound for arbitrary $(j+1)T_s \le \tau$, and hence, the upper bound in (3.153) holds for all $iT \le \tau$.

For all $iT + t \le \tau$, with $t \in (0,\ T_s]$, we can write

$$\begin{aligned}\tilde{x}(iT_s+t) = {} & \mathrm{e}^{A_m t}\tilde{x}(iT_s) + \int_0^t \mathrm{e}^{A_m(t-\xi)} B \begin{bmatrix} \hat{\sigma}_1(iT_s) \\ \hat{\sigma}_2(iT_s) \end{bmatrix} d\xi \\ & - \int_0^t \mathrm{e}^{A_m(t-\xi)} B_m \left((\omega-\omega_0)u(iT_s+\xi) + \eta_1(iT_s+\xi)\right) d\xi \\ & - \int_0^t \mathrm{e}^{A_m(t-\xi)} B_{um}\eta_2(iT_s+\xi) d\xi\,.\end{aligned}$$

The bounds in (3.151) and (3.152) and the definitions of Δ_1, Δ_2, $\alpha_1(\cdot)$, $\alpha_2(\cdot)$, $\alpha_3(\cdot)$, and $\alpha_4(\cdot)$ in (3.133)–(3.134) imply that

$$\|\tilde{x}(iT_s+t)\|_2 \le \alpha_1(t)\,\|\tilde{x}(iT_s)\|_2 + \alpha_2(t)\,\|\tilde{x}(iT_s)\|_2 + \alpha_3(t)\Delta_1 + \alpha_4(t)\Delta_2\,.$$

Next, for all $iT_s + \bar{t} \le \tau$, the upper bound in (3.153) and the definitions of $\bar{\alpha}_1(T_s)$, $\bar{\alpha}_2(T_s)$, $\bar{\alpha}_3(T_s)$, $\bar{\alpha}_4(T_s)$ in (3.136)–(3.137) lead to

$$\|\tilde{x}(iT_s+t)\|_2 \le (\bar{\alpha}_1(T_s) + \bar{\alpha}_2(T_s))\,\varsigma(T_s) + \bar{\alpha}_3(T_s)\Delta_1 + \bar{\alpha}_4(T_s)\Delta_2\,.$$

Further, because the right-hand side coincides with the definition of $\gamma_0(T_s)$ in (3.138), we have $\|\tilde{x}(t)\|_2 \le \gamma_0(T_s)$, which holds for all $t \in [0,\tau]$, which, along with the design constraint in (3.149), yields

$$\|\tilde{x}(t)\|_2 < \bar{\gamma}_0\,, \quad \forall\, t \in [0,\tau]\,,$$

and consequently implies that

$$\|\,\tilde{x}_\tau\|_{\mathcal{L}_\infty} < \bar{\gamma}_0\,.$$

The proof is complete. □

Similar to the previous section, in the following theorem we prove the stability and derive the performance bounds of the adaptive closed-loop system with the $\mathcal{L}_1$ adaptive controller. Although the proof of this result is similar to the proof of Theorem 3.2.1 in Section 3.2, we present it for the sake of completeness.

Theorem 3.3.1 Let the adaptation rate be chosen to satisfy (3.149). Given the closed-loop system with the $\mathcal{L}_1$ controller defined via (3.141)–(3.143), subject to the $\mathcal{L}_1$-norm condition in (3.126), and the closed-loop reference system in (3.145), if

$$\|x_0\|_\infty \leq \rho_0 ,$$

then we have

$$\|x\|_{\mathcal{L}_\infty} \leq \rho , \tag{3.155}$$
$$\|u\|_{\mathcal{L}_\infty} \leq \rho_u , \tag{3.156}$$
$$\|\tilde{x}\|_{\mathcal{L}_\infty} \leq \bar{\gamma}_0 , \tag{3.157}$$
$$\|x_{\rm ref} - x\|_{\mathcal{L}_\infty} \leq \gamma_1 , \tag{3.158}$$
$$\|u_{\rm ref} - u\|_{\mathcal{L}_\infty} \leq \gamma_2 , \tag{3.159}$$
$$\|y_{\rm ref} - y\|_{\mathcal{L}_\infty} \leq \|C\|_\infty \gamma_1 . \tag{3.160}$$

Proof. Assume that the bounds in (3.158) and (3.159) do not hold. Then, since $\|x_{\rm ref}(0) - x(0)\|_\infty = 0 < \gamma_1$, $\|u_{\rm ref}(0) - u(0)\|_\infty = 0 < \gamma_2$, and $x(t)$, $x_{\rm ref}(t)$, $u(t)$, and $u_{\rm ref}(t)$ are continuous, there exists τ such that

$$\|x_{\rm ref}(\tau) - x(\tau)\|_\infty = \gamma_1 \text{ or}$$
$$\|u_{\rm ref}(\tau) - u(\tau)\|_\infty = \gamma_2 ,$$

while

$$\|x_{\rm ref}(t) - x(t)\|_\infty < \gamma_1 , \quad \|u_{\rm ref}(t) - u(t)\|_\infty < \gamma_2 , \quad \forall\, t \in [0, \tau) .$$

This implies that at least one of the following equalities holds:

$$\|(x_{\rm ref} - x)_\tau\|_{\mathcal{L}_\infty} = \gamma_1 , \quad \|(u_{\rm ref} - u)_\tau\|_{\mathcal{L}_\infty} = \gamma_2 . \tag{3.161}$$

It follows from Assumption 3.3.3 that

$$\|z_\tau\|_{\mathcal{L}_\infty} \leq L_z \left(\|x_{\rm ref\,\tau}\|_{\mathcal{L}_\infty} + \gamma_1\right) + B_z . \tag{3.162}$$

Then, Lemma 3.3.2 implies that

$$\|x_{\rm ref\,\tau}\|_{\mathcal{L}_\infty} \leq \rho_r , \quad \|u_{\rm ref\,\tau}\|_{\mathcal{L}_\infty} \leq \rho_{ur} . \tag{3.163}$$

Using the definitions of ρ and ρ_u in (3.127) and (3.129), it follows from the bounds in (3.161) and (3.163) that

$$\|x_\tau\|_{\mathcal{L}_\infty} \leq \rho_r + \gamma_1 \leq \rho , \tag{3.164}$$
$$\|u_\tau\|_{\mathcal{L}_\infty} \leq \rho_{ur} + \gamma_2 = \rho_u . \tag{3.165}$$

Hence, if one chooses the adaptation sampling time T_s according to (3.149), Lemma 3.3.3 implies that

$$\|\tilde{x}_\tau\|_{\mathcal{L}_\infty} < \bar{\gamma}_0 . \tag{3.166}$$

Next, let $\tilde{\eta}(t) \triangleq \tilde{\eta}_1(t) + \tilde{\eta}_{2m}(t)$, with $\tilde{\eta}_{2m}(t)$ being the signal with its Laplace transform $\tilde{\eta}_{2m}(s) \triangleq H_m^{-1}(s)H_{um}(s)\tilde{\eta}_2(s)$, where $\tilde{\eta}_1(t)$ and $\tilde{\eta}_2(t)$ are as introduced in (3.147) and (3.148). It follows from (3.143) that

$$u(s) = -K\,D(s)\Big(\omega u(s) + \eta_1(s) + H_m^{-1}(s)H_{um}(s)\eta_2(s) - K_g(s)r(s) + \tilde{\eta}(s)\Big),$$

where $\eta_1(s)$, $\eta_2(s)$, and $\tilde{\eta}(s)$ are the Laplace transforms of the signals $\eta_1(t)$, $\eta_2(t)$, and $\tilde{\eta}(t)$, respectively. Consequently

$$u(s) = -K\,D(s)(\mathbb{I}_m + \omega K\,D(s))^{-1}\left(\eta_1(s) + H_m^{-1}(s)H_{um}(s)\eta_2(s) - K_g(s)r(s) + \tilde{\eta}(s)\right),$$

which leads to

$$\omega u(s) = -\omega K\,D(s)(\mathbb{I}_m + \omega K\,D(s))^{-1}\left(\eta_1(s) + H_m^{-1}(s)H_{um}(s)\eta_2(s) - K_g(s)r(s) + \tilde{\eta}(s)\right). \tag{3.167}$$

Using the definition of $C(s)$ in (3.125), one can write

$$\omega u(s) = -C(s)\Big(\eta_1(s) + H_m^{-1}(s)H_{um}(s)\eta_2(s) - K_g(s)r(s) + \tilde{\eta}(s)\Big),$$

and the system in (3.122) consequently takes the form

$$\begin{aligned} x(s) = {} & G_m(s)\eta_1(s) + G_{um}(s)\eta_2(s) - H_{xm}(s)C(s)\tilde{\eta}(s) \\ & + H_{xm}(s)C(s)K_g(s)r(s) + x_{\text{in}}(s). \end{aligned} \tag{3.168}$$

Next, from the definition of the closed-loop reference system in (3.145) and (3.168) we have

$$\begin{aligned} x_{\text{ref}}(s) - x(s) = {} & G_m(s)(\eta_{1\text{ref}}(s) - \eta_1(s)) + G_{um}(s)(\eta_{2\text{ref}}(s) - \eta_2(s)) \\ & + H_{xm}(s)C(s)\tilde{\eta}(s). \end{aligned}$$

Moreover, it follows from the error dynamics in (3.146) that

$$H_m^{-1}(s)C\tilde{x}(s) = \tilde{\eta}_1(s) + \tilde{\eta}_{2m}(s) = \tilde{\eta}(s),$$

which leads to

$$\begin{aligned} x_{\text{ref}}(s) - x(s) = {} & G_m(s)(\eta_{1\text{ref}}(s) - \eta_1(s)) \\ & + G_{um}(s)(\eta_{2\text{ref}}(s) - \eta_2(s)) + H_{xm}(s)C(s)H_m^{-1}(s)C\tilde{x}(s). \end{aligned}$$

Therefore, we have

$$\begin{aligned} \|(x_{\text{ref}} - x)_\tau\|_{\mathcal{L}_\infty} \le {} & \|G_m(s)\|_{\mathcal{L}_1} \left\|(\eta_{1\text{ref}} - \eta_1)_\tau\right\|_{\mathcal{L}_\infty} + \|G_{um}(s)\|_{\mathcal{L}_1} \left\|(\eta_{2\text{ref}} - \eta_2)_\tau\right\|_{\mathcal{L}_\infty} \\ & + \left\|H_{xm}(s)C(s)H_m^{-1}(s)C\right\|_{\mathcal{L}_1} \|\tilde{x}_\tau\|_{\mathcal{L}_\infty}. \end{aligned} \tag{3.169}$$

Substituting (3.163) in (3.162) one obtains

$$\|z_\tau\|_{\mathcal{L}_\infty} \le L_z(\rho_r + \gamma_1) + B_z,$$

and hence, from the definition of $\bar{\delta}(\delta)$ in (3.124), we have

$$\|X_\tau\|_{\mathcal{L}_\infty} \le \max\{\rho_r+\gamma_1,\ L_z(\rho_r+\gamma_1)+B_z\} \le \bar{\rho}_r(\rho_r),$$
$$\|X_{\text{ref}\,\tau}\|_{\mathcal{L}_\infty} \le \max\{\rho_r,\ L_z(\rho_r+\gamma_1)+B_z\} \le \bar{\rho}_r(\rho_r).$$

Assumption 3.3.2 implies that, for $i=1,2$, we have

$$\left\|(\eta_{i\text{ref}}-\eta_i)_\tau\right\|_{\mathcal{L}_\infty} \le K_{i\bar{\rho}_r(\rho_r)}\|(X_{\text{ref}}-X)_\tau\|_{\mathcal{L}_\infty} = K_{i\bar{\rho}_r(\rho_r)}\|(x_{\text{ref}}-x)_\tau\|_{\mathcal{L}_\infty}. \tag{3.170}$$

Then, from (3.169) we have

$$\begin{aligned}\|(x_{\text{ref}}-x)_\tau\|_{\mathcal{L}_\infty} \le{}& \|G_m(s)\|_{\mathcal{L}_1} K_{1\bar{\rho}_r(\rho_r)}\|(x_{\text{ref}}-x)_\tau\|_{\mathcal{L}_\infty}\\ &+\|G_{um}(s)\|_{\mathcal{L}_1} K_{2\bar{\rho}_r(\rho_r)}\|(x_{\text{ref}}-x)_\tau\|_{\mathcal{L}_\infty}\\ &+\left\|H_{xm}(s)C(s)H_m^{-1}(s)C\right\|_{\mathcal{L}_1}\|\tilde{x}_\tau\|_{\mathcal{L}_\infty}.\end{aligned}$$

From the redefinition in (3.124), it follows that $K_{1\bar{\rho}_r(\rho_r)} < L_{1\rho_r}$ and $K_{2\bar{\rho}_r(\rho_r)} < L_{2\rho_r}$, and therefore, we obtain

$$\begin{aligned}\|(x_{\text{ref}}-x)_\tau\|_{\mathcal{L}_\infty} \le{}& \|G_m(s)\|_{\mathcal{L}_1} L_{1\rho_r}\|(x_{\text{ref}}-x)_\tau\|_{\mathcal{L}_\infty}\\ &+\|G_{um}(s)\|_{\mathcal{L}_1} L_{2\rho_r}\|(x_{\text{ref}}-x)_\tau\|_{\mathcal{L}_\infty}\\ &+\left\|H_{xm}(s)C(s)H_m^{-1}(s)C\right\|_{\mathcal{L}_1}\|\tilde{x}_\tau\|_{\mathcal{L}_\infty}.\end{aligned}$$

The upper bound in (3.166) and the $\mathcal{L}_1$-norm condition in (3.126) lead to the upper bound

$$\|(x_{\text{ref}}-x)_\tau\|_{\mathcal{L}_\infty} \le \frac{\left\|H_{xm}(s)C(s)H_m^{-1}(s)C\right\|_{\mathcal{L}_1}}{1-\|G_m(s)\|_{\mathcal{L}_1}L_{1\rho_r}-\|G_{um}(s)\|_{\mathcal{L}_1}L_{2\rho_r}}\gamma_0,$$

which along with the definition of γ_1 in (3.128) leads to

$$\|(x_{\text{ref}}-x)_\tau\|_{\mathcal{L}_\infty} \le \gamma_1-\beta < \gamma_1. \tag{3.171}$$

On the other hand, it follows from (3.145) and (3.167) that

$$\begin{aligned}u_{\text{ref}}(s)-u(s) ={}& -\omega^{-1}C(s)(\eta_{1\text{ref}}(s)-\eta_1(s))\\ &-\omega^{-1}C(s)H_m^{-1}(s)H_{um}(s)(\eta_{2\text{ref}}(s)-\eta_2)+\omega^{-1}C(s)H_m^{-1}(s)C\tilde{x}(s).\end{aligned}$$

One can write

$$\begin{aligned}\|(u_{\text{ref}}-u)_\tau\|_{\mathcal{L}_\infty} \le{}& \left\|\omega^{-1}C(s)\right\|_{\mathcal{L}_1}\left\|(\eta_{1\text{ref}}-\eta_1)_\tau\right\|_{\mathcal{L}_\infty}\\ &+\left\|\omega^{-1}C(s)H_m^{-1}(s)H_{um}(s)\right\|_{\mathcal{L}_1}\left\|(\eta_{2\text{ref}}-\eta_2)_\tau\right\|_{\mathcal{L}_\infty}\\ &+\left\|\omega^{-1}C(s)H_m^{-1}(s)C\right\|_{\mathcal{L}_1}\|\tilde{x}_\tau\|_{\mathcal{L}_\infty},\end{aligned}$$

and the bound in (3.170) leads to

$$\begin{aligned}\|(u_{\text{ref}}-u)_\tau\|_{\mathcal{L}_\infty} \le{}& \left\|\omega^{-1}C(s)\right\|_{\mathcal{L}_1} L_{1\rho_r}\|(x_{\text{ref}}-x)_\tau\|_{\mathcal{L}_\infty}\\ &+\left\|\omega^{-1}C(s)H_m^{-1}(s)H_{um}(s)\right\|_{\mathcal{L}_1} L_{2\rho_r}\|(x_{\text{ref}}-x)_\tau\|_{\mathcal{L}_\infty}\\ &+\left\|\omega^{-1}C(s)H_m^{-1}(s)C\right\|_{\mathcal{L}_1}\|\tilde{x}_\tau\|_{\mathcal{L}_\infty}.\end{aligned}$$

The bounds (3.166) and (3.171) and the definition of γ_2 in (3.130) lead to

$$\begin{aligned}\|(u_{\text{ref}}-u)_\tau\|_{\mathcal{L}_\infty} \le \Big(&\left\|\omega^{-1}C(s)\right\|_{\mathcal{L}_1} L_{1\rho_r} \\ &+\left\|\omega^{-1}C(s)H_m^{-1}(s)H_{um}(s)\right\|_{\mathcal{L}_1} L_{2\rho_r}\Big)(\gamma_1-\beta) \\ &+\left\|\omega^{-1}C(s)H_m^{-1}(s)C\right\|_{\mathcal{L}_1}\bar{\gamma}_0 < \gamma_2 .\end{aligned} \tag{3.172}$$

Finally, we note that the upper bounds in (3.171) and (3.172) contradict the equalities in (3.161), which proves the bounds in (3.158) and (3.159). The results in (3.155)–(3.157) and (3.160) follow directly from the bounds in (3.164)–(3.166) and from the fact that $y(t)-y_{\text{ref}}(t)=C(x(t)-x_{\text{ref}}(t))$. □

Remark 3.3.1 Thus, the tracking error between $y(t)$ and $y_{\text{ref}}(t)$, as well as $u(t)$ and $u_{\text{ref}}(t)$, is uniformly bounded by an arbitrary small constant, which implies that in both transient and steady-state one can achieve arbitrarily close tracking performance for both signals simultaneously by reducing T_s. To understand how these bounds can be used for ensuring transient response with *desired* specifications, we consider the *ideal* control signal for the system in (3.122),

$$u_{\text{id}}(s)=-\omega^{-1}\left(\eta_1(s)+H_m^{-1}(s)H_{um}(s)\eta_2(s)-K_g(s)r(s)\right), \tag{3.173}$$

which leads to the desired system output response

$$y_{\text{id}}(s)=H_m(s)K_g(s)r(s), \tag{3.174}$$

by canceling the uncertainties exactly. In the closed-loop reference system in (3.145), $u_{\text{id}}(t)$ is further low-pass filtered by $C(s)$ to have guaranteed low-frequency range. Thus the closed-loop reference system has a different response as compared to (3.174) achieved with (3.173). Similar to Section 3.2, the response of $y_{\text{ref}}(t)$ can be made as close as possible to (3.174) by reducing $(\|G_m(s)\|_{\mathcal{L}_1}+\|G_{um}(s)\|_{\mathcal{L}_1}\ell_0)$ arbitrarily. In the absence of unmatched uncertainties, we can make $\|G_m(s)\|_{\mathcal{L}_1}$ arbitrarily small by increasing the bandwidth of the low-pass filter $C(s)$. However, for the general case with unmatched uncertainties, the design of K and $D(s)$ that satisfy (3.126) is an open problem. We note also that the presence of unmatched uncertainties may limit the choice of the desired state matrix A_m.

Remark 3.3.2 It is important to notice that the performance bounds given by (3.158)–(3.160) are exactly the same as the ones in (3.106)–(3.108), analyzed in Section 3.2 using projection-based adaptive laws. These results make clear that the key elements for the derivation of the performance bounds of the $\mathcal{L}_1$ adaptive controller are (i) an appropriate filtering structure in the control channel and (ii) the use of a predictor-based fast estimation scheme. Therefore, one would like to conjecture that other estimation algorithms from the literature can also be successfully employed in the design of $\mathcal{L}_1$ adaptive control architectures with similar performance bounds.

Remark 3.3.3 Similar to Section 2.1.6, the state predictor in (3.141) can be modified as follows:

$$\begin{aligned}\dot{\hat{x}}(t)&=A_m\hat{x}(t)+B_m\left(\omega_0 u(t)+\hat{\sigma}_1(t)\right)+B_{um}\hat{\sigma}_2(t)-K_{\text{sp}}\tilde{x}(t), \qquad \hat{x}(0)=x_0,\\ \hat{y}(t)&=C\hat{x}(t),\end{aligned}$$

where the constant matrix $K_{sp} \in \mathbb{R}^{n\times n}$ is used to assign *faster poles for the prediction error dynamics*, defined via $A_s \triangleq A_m - K_{sp}$. This modification of the state predictor adds damping to the adaptation loop inside the $\mathcal{L}_1$ adaptive controller and can be used to tune the frequency response and robustness margins of the closed-loop adaptive system. In general, this modification may require an increase of the adaptation sampling rate. The proofs of stability and performance bounds for the $\mathcal{L}_1$ adaptive controller presented in this section with the modified state predictor can be found in [180].

3.3.4 Simulation Example

In this section, we use the example in Section 3.2.4 in order to demonstrate that a similar level of performance can be achieved with this new adaptive law (a different fast estimation scheme). For the sake of completeness we repeat the system dynamics and the simulation parameters. Thus, consider again the system

$$\begin{aligned} \dot{x}(t) &= (A_m + A_\Delta)x(t) + B_m\omega u(t) + f_\Delta(t, x(t), z(t)), \quad x(0) = x_0, \\ y(t) &= Cx(t), \end{aligned}$$

where

$$A_m = \begin{bmatrix} -1 & 0 & 0 \\ 0 & 0 & 1 \\ 0 & -1 & -1.8 \end{bmatrix}, B_m = \begin{bmatrix} 1 & 0 \\ 0 & 0 \\ 1 & 1 \end{bmatrix},$$

$$C = \begin{bmatrix} 1 & 0 & 0 \\ 0 & 1 & 0 \end{bmatrix},$$

while $A_\Delta \in \mathbb{R}^{3\times 3}$ and $\omega \in \mathbb{R}^{2\times 2}$ are unknown constant matrices satisfying

$$\|A_\Delta\|_\infty \le 1, \quad \omega \in \begin{bmatrix} [0.6,\ 1.2] & [-0.2,\ 0.2] \\ [-0.2,\ 0.2] & [0.6,\ 1.2] \end{bmatrix} = \Omega,$$

and f_Δ is the (unknown) nonlinear function

$$f_\Delta(t, x, z) = \begin{bmatrix} \frac{k_1}{3}x^\top x + \tanh(\frac{k_2}{2}x_1)x_1 + k_3 z \\ \frac{k_4}{2}\operatorname{sech}(x_2)x_2 + \frac{k_5}{5}x_3^2 + \frac{k_6}{2}(1 - e^{-\lambda t}) + \frac{k_7}{2}z \\ k_8 x_3 \cos(\omega_u t) + k_9 z^2 \end{bmatrix},$$

for $k_i \in [-1,\ 1]$, $i = 1, \ldots, 9$, and $\lambda, \omega_u \in \mathbb{R}^+$. The internal unmodeled dynamics are given by

$$\begin{aligned} \dot{x}_{z1}(t) &= x_{z2}(t), \\ \dot{x}_{z2}(t) &= -x_{z1}(t) + 0.8\left(1 - x_{z1}^2(t)\right)x_{z2}(t), \\ z(t) &= 0.1\,(x_{z1}(t) - x_{z2}(t)) + z_u(t) \\ z_u(s) &= \frac{-s+1}{\frac{s^2}{0.1^2} + \frac{0.8s}{0.1} + 1}\begin{bmatrix} 1 & -2 & 1 \end{bmatrix} x(s), \end{aligned}$$

with $[x_{z1}(0),\ x_{z2}(0)] = [x_{z1_0},\ x_{z2_0}]$. The control objective is to design a control $u(t)$ so that the output $y(t)$ of the system tracks the output of the desired model $M(s)$ to bounded reference inputs $r(t)$ ($\|r\|_{\mathcal{L}_\infty} \le 1$).

In the implementation of the $\mathcal{L}_1$ controller, we set

$$T_s = \frac{1}{100}\,\mathrm{s}, \quad \omega_0 = \mathbb{I}_2, \quad K = \begin{bmatrix} 8 & 0 \\ 0 & 8 \end{bmatrix},$$
$$D(s) = \frac{1}{s(\frac{s}{25}+1)(\frac{s}{70}+1)(\frac{s^2}{40^2}+\frac{1.8s}{40}+1)}\mathbb{I}_2,$$
$$K_g(s) \equiv K_g = -(CA_m^{-1}B_m)^{-1} = \begin{bmatrix} 1 & 0 \\ -1 & 1 \end{bmatrix}.$$

To illustrate the performance of the $\mathcal{L}_1$ adaptive controller we consider the same five scenarios that we used in Section 3.2.4:

- Scenario 1:

$$A_\Delta = \begin{bmatrix} 0.2 & -0.2 & -0.3 \\ -0.2 & -0.2 & 0.6 \\ -0.1 & 0 & -0.9 \end{bmatrix}, \quad \omega = \begin{bmatrix} 0.6 & -0.2 \\ 0.2 & 1.2 \end{bmatrix},$$
$$k_1 = -1,\ k_2 = 1,\ k_3 = 0,\ k_4 = 1,\ k_5 = 0,$$
$$k_6 = 0.2,\ \lambda = 0.3,\ k_7 = 1,\ k_8 = 0.6,\ \omega_u = 5,\ k_9 = -0.7.$$

- Scenario 2:

$$A_\Delta = \begin{bmatrix} 0.2 & -0.3 & 0.5 \\ 0 & 0 & 0 \\ -0.1 & 0.4 & 0.5 \end{bmatrix}, \quad \omega = \begin{bmatrix} 1 & 0 \\ 0 & 1 \end{bmatrix},$$
$$k_1 = 0,\ k_2 = 0,\ k_3 = 0,\ k_4 = 0,\ k_5 = 0,$$
$$k_6 = 0,\ k_7 = 0,\ k_8 = 0,\ k_9 = 0.$$

- Scenario 3:

$$A_\Delta = \begin{bmatrix} 0.1 & -0.4 & 0.5 \\ -0.5 & 0.5 & 0 \\ -0.2 & 0.3 & 0.5 \end{bmatrix}, \quad \omega = \begin{bmatrix} 0.8 & -0.1 \\ 0.1 & 0.8 \end{bmatrix},$$
$$k_1 = 0,\ k_2 = 0,\ k_3 = 0,\ k_4 = 0,\ k_5 = 0,$$
$$k_6 = 0,\ k_7 = 0,\ k_8 = 0,\ k_9 = 0.$$

- Scenario 4:

$$A_\Delta = \begin{bmatrix} 0.1 & 0.4 & 0.2 \\ -0.4 & -0.2 & 0.3 \\ -0.2 & 0.6 & -0.1 \end{bmatrix}, \quad \omega = \begin{bmatrix} 0.6 & 0 \\ 0.1 & 1.2 \end{bmatrix},$$
$$k_1 = 1,\ k_2 = 1,\ k_3 = 1,\ k_4 = 1,\ k_5 = 1,$$
$$k_6 = 1,\ \lambda = 0.1,\ k_7 = 1,\ k_8 = 1,\ \omega_u = 1,\ k_9 = 1.$$

– Scenario 5:

$$A_\Delta = \begin{bmatrix} 0.2 & -0.2 & -0.3 \\ 0.1 & -0.4 & 0.3 \\ -0.1 & 0 & -0.9 \end{bmatrix}, \quad \omega = \begin{bmatrix} 0.9 & -0.2 \\ 0.1 & 1.1 \end{bmatrix},$$
$$k_1 = 1,\ k_2 = 1,\ k_3 = 1,\ k_4 = 1,\ k_5 = -1,$$
$$k_6 = -1,\ \lambda = 0.3,\ k_7 = 1,\ k_8 = -1,\ \omega_u = 1,\ k_9 = 1.$$

A detailed explanation of these scenarios was presented in Section 3.2.4. For the sake of completeness, we repeat the same explanation here. All the scenarios above consider a significant amount of unmatched uncertainties in the dynamics of the system, except for Scenario 2, in which the uncertainty (affecting only the state matrix of the system) remains matched. Also, in Scenarios 2 and 3 the uncertainties affect only the state matrix and appear therefore in a linear fashion; instead, Scenarios 1, 4, and 5 consider nonlinear uncertain dynamics, both matched and unmatched. Moreover, Scenarios 1, 3, 4, and 5 include uncertainty in the system input gain matrix; in particular, Scenarios 1 and 5 consider significant coupling between the control channels, while Scenario 3 introduces a 20% reduction in the control efficiency in both control channels.

Figures 3.10 and 3.11 show, respectively, the response of the closed-loop system for Scenario 1 with the $\mathcal{L}_1$ adaptive controller (i) to a series of doublets of different amplitudes in the different channels and (ii) to the sinusoidal reference signals $r(t) = [\sin(\frac{\pi}{6}t),\ 0.2 + 0.8\cos(\frac{\pi}{3}t)]$ and $r(t) = [0.5\sin(\frac{\pi}{6}t),\ 0.1 + 0.4\cos(\frac{\pi}{3}t)]$. One can observe that the $\mathcal{L}_1$ adaptive controller leads to scaled system output for scaled reference signals, similar to linear systems. Moreover, although the reference signals are different from the ones in Section 3.2.4, we can see that the $\mathcal{L}_1$ architecture with the new fast estimation scheme achieves a similar level of performance as the $\mathcal{L}_1$ controller with projection-based adaptation laws.

Next, we demonstrate that the $\mathcal{L}_1$ adaptive controller is able to guarantee a similar transient response of the closed-loop system for different (admissible) uncertainties in the plant. Figure 3.12 presents the closed-loop response to the same doublets as in Figure 3.10, but now for Scenarios 2, 3, 4, and 5. The $\mathcal{L}_1$ controller guarantees smooth and uniform transient performance in the presence of different nonlinearities affecting both the matched and the unmatched channel. Note that the control signals required to track the reference signals and compensate for the uncertainties are significantly different for each scenario. Also note that despite the high adaptation sampling rate, the control signals are well within the low-frequency range. In order to show the benefits of the compensation for unmatched uncertainties, we repeat the same four scenarios (Scenarios 2–5) *without* the unmatched component in the control law (term $\hat{\eta}_{2m}(t)$ in equation (3.144)). The results are shown in Figure 3.13. Since Scenario 2 considered only matched uncertainties, the performance in this case remains the same. For the other three scenarios, however, the closed-loop performance degrades significantly, especially for the second output $y_2(t)$. In particular, one can observe that for Scenario 3, the system becomes unstable.

Finally, it is important to emphasize that, for all the simulations provided above (Figures 3.10–3.13), the $\mathcal{L}_1$ adaptive controller has *not* been redesigned or retuned, and that a *single set* of control parameters has been used for all the scenarios. Moreover, the results in this section and in Section 3.2.4 illustrate that the design of the control law in $\mathcal{L}_1$ adaptive architectures (that is, the design of k, $D(s)$, and $K_g(s)$) is independent of the estimation scheme used, as long as the latter is estimating *fast*.

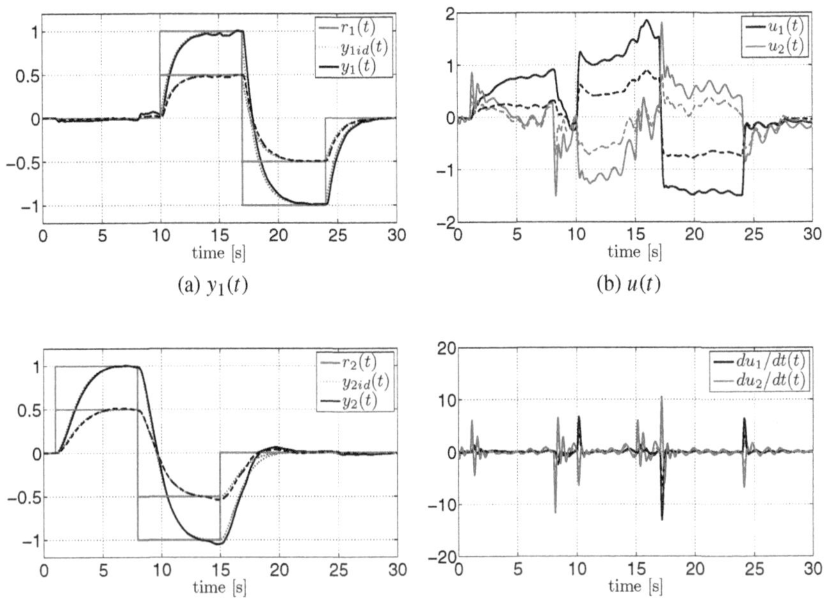

(a) $y_1(t)$ (b) $u(t)$ (c) $y_2(t)$ (d) $\frac{d}{dt}u(t)$

Figure 3.10: Scenario 1. Performance of the $\mathcal{L}_1$ adaptive controller for doublet commands.

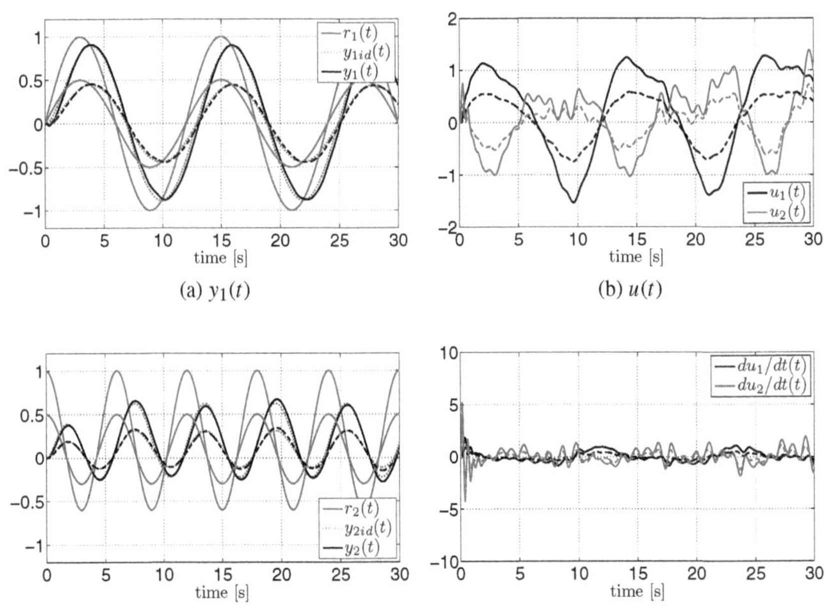

(a) $y_1(t)$ (b) $u(t)$ (c) $y_2(t)$ (d) $\frac{d}{dt}u(t)$

Figure 3.11: Scenario 1. Performance of the $\mathcal{L}_1$ adaptive controller for sinusoidal reference signals.

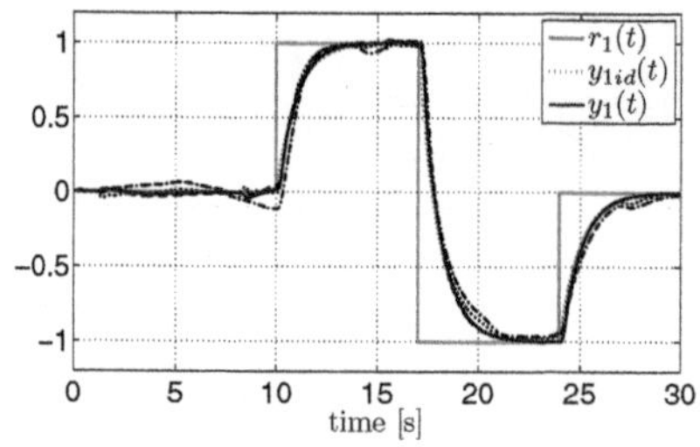

(a) $y_1(t)$

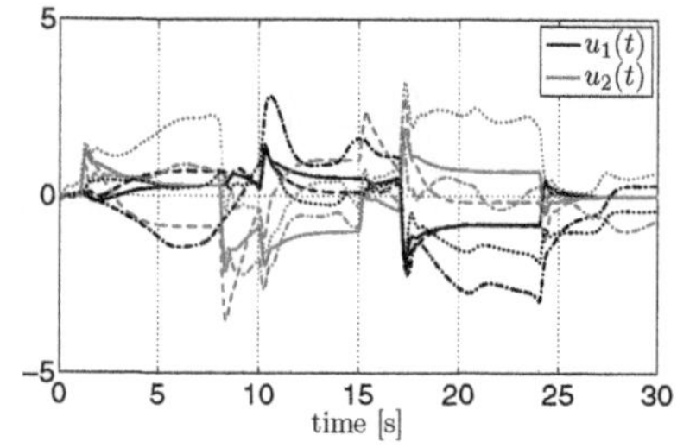

(b) $u(t)$

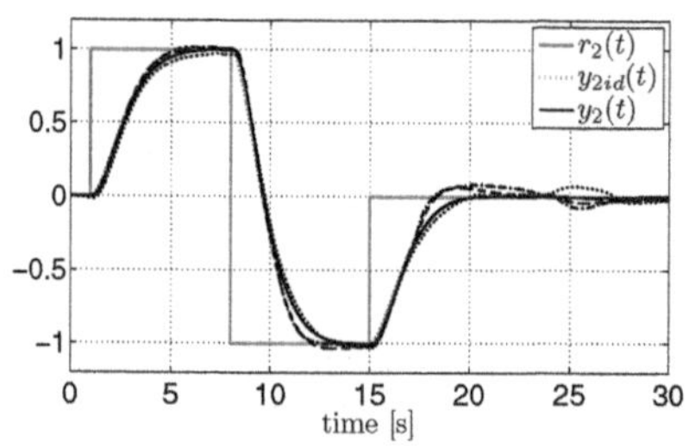

(c) $y_2(t)$

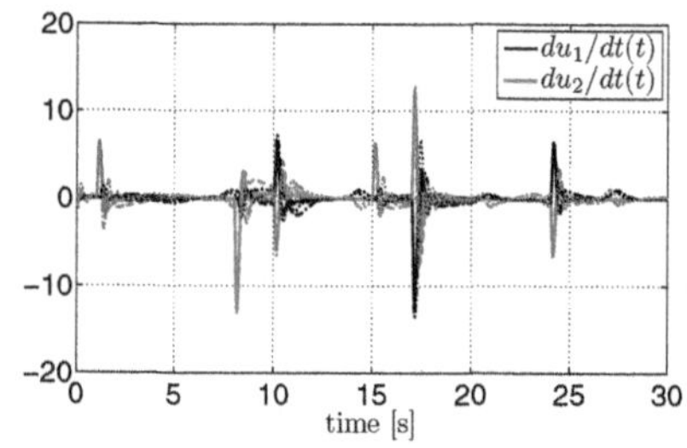

(d) $\frac{d}{dt}u(t)$

Figure 3.12: Scenarios 2 (solid), 3 (dashed), 4 (dash-dot), and 5 (dotted). Performance of the $\mathcal{L}_1$ adaptive controller for doublet commands.

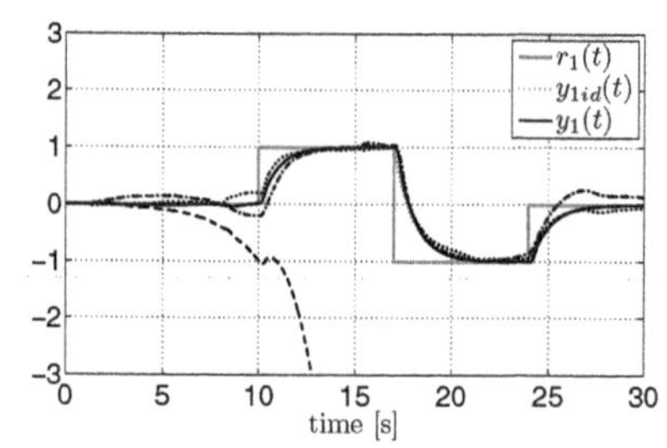

(a) $y_1(t)$

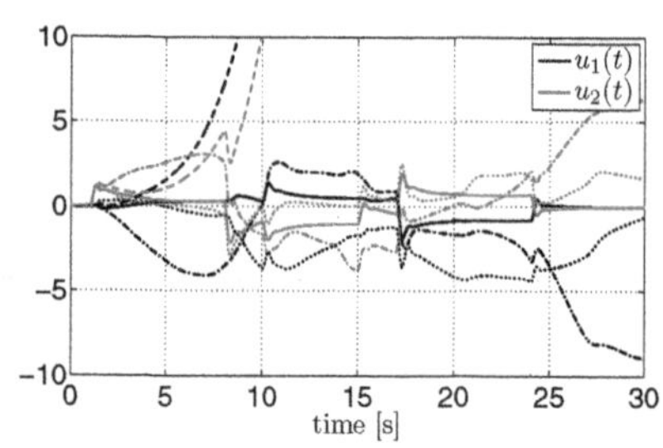

(b) $u(t)$

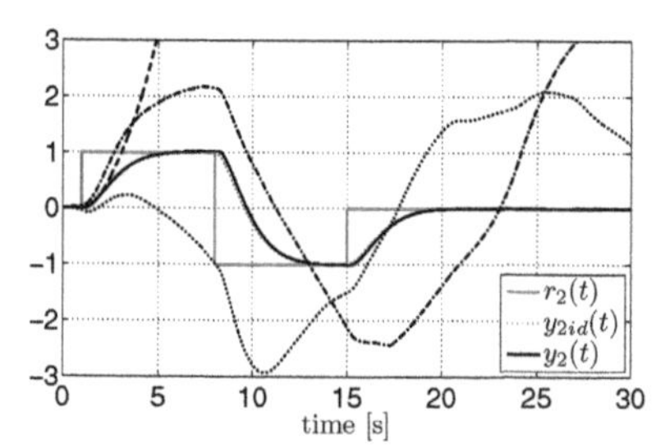

(c) $y_2(t)$

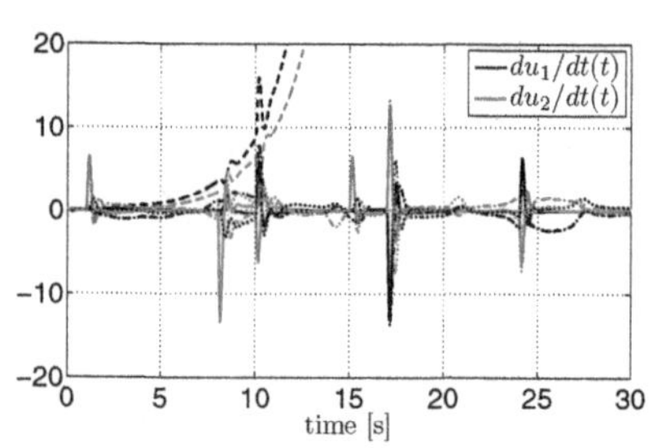

(d) $\frac{d}{dt}u(t)$

Figure 3.13: Scenarios 2 (solid), 3 (dashed), 4 (dash-dot), and 5 (dotted). Performance of the $\mathcal{L}_1$ adaptive controller for doublet commands *without* compensation for unmatched uncertainties.

Chapter 4

Output Feedback

This chapter extends the results to output feedback. We limit the discussion to single-input single-output (SISO) systems in the presence of unknown time-varying nonlinearities. We state the problem formulation in the frequency domain and consider two cases of performance specifications. The first case corresponds to strictly positive real (SPR) reference systems, and for simplicity we reduce the design and analysis here to a first-order reference system. In the second section, we present the solution for an arbitrary strictly proper, minimum phase and stable reference system. In this case, we use the piecewise-constant adaptive law to establish the desired performance bounds. The major difference of the analysis in output feedback as compared to that in state feedback is that the uncertainty is not decoupled in the $\mathcal{L}_1$-norm condition and enters directly into the underlying transfer function, for which the $\mathcal{L}_1$-norm must be computed. This adds an additional constraint on the choice of the filter and the desired reference system to ensure that this transfer function is stable and has bounded $\mathcal{L}_1$-norm. The two-cart benchmark example is used for numerical illustration. Chapter 6 summarizes flight test results obtained using the architectures presented in this chapter.

4.1 $\mathcal{L}_1$ Adaptive Output Feedback Controller for First-Order Reference Systems

This section presents the $\mathcal{L}_1$ adaptive output feedback controller for a SISO system of unknown dimension in the presence of unmodeled dynamics and time-varying disturbances. The methodology ensures uniformly bounded transient response for system's both signals, input and output, simultaneously, as compared to the same signals of a first-order stable reference system. The $\mathcal{L}_\infty$-norm bounds for the error signals between the closed-loop adaptive system and the closed-loop reference LTI system can be systematically reduced by increasing the adaptation gain [32]. The flight test results using this solution are summarized in Chapter 6.

4.1.1 Problem Formulation

Consider the SISO system

$$y(s) = A(s)(u(s) + d(s)), \tag{4.1}$$

where $u(s)$ is the Laplace transform of the system's input signal $u(t)$; $y(s)$ is the Laplace transform of the system's output signal $y(t)$; $A(s)$ is a strictly proper unknown transfer function; $d(s)$ is the Laplace transform of the time-varying nonlinear uncertainties and disturbances, denoted by $d(t) \triangleq f(t, y(t))$; and $f : \mathbb{R} \times \mathbb{R} \to \mathbb{R}$ is an unknown map, subject to the following assumptions.

Assumption 4.1.1 (Lipschitz continuity) There exist constants $L > 0$ and $L_0 > 0$, possibly arbitrarily large, such that the following inequalities hold uniformly in t:

$$|f(t, y_1) - f(t, y_2)| \leq L|y_1 - y_2|, \quad |f(t, y)| \leq L|y| + L_0.$$

Assumption 4.1.2 (Uniform boundedness of the rate of variation of uncertainties) There exist constants $L_1 > 0$, $L_2 > 0$, and $L_3 > 0$, possibly arbitrarily large, such that for all $t \geq 0$,

$$|\dot{d}(t)| \leq L_1|\dot{y}(t)| + L_2|y(t)| + L_3.$$

The control objective is to design an adaptive output feedback controller $u(t)$ such that the system output $y(t)$ tracks the given bounded piecewise-continuous reference input $r(t)$ following a desired reference model $M(s)$. In this section, we consider a first-order reference system, i.e.,

$$M(s) = \frac{m}{s + m}, \qquad m > 0. \tag{4.2}$$

4.1.2 $\mathcal{L}_1$ Adaptive Control Architecture

Definitions and $\mathcal{L}_1$-Norm Sufficient Condition for Stability

We note that the system in (4.1) can be rewritten in terms of the reference system, defined by $M(s)$, as

$$y(s) = M(s)(u(s) + \sigma(s)), \tag{4.3}$$

where the uncertainties due to $A(s)$ and $d(s)$ are lumped into the signal $\sigma(s)$, which is given by

$$\sigma(s) = \frac{(A(s) - M(s))u(s) + A(s)d(s)}{M(s)}. \tag{4.4}$$

The design of $\mathcal{L}_1$ adaptive controller proceeds by considering a strictly proper filter $C(s)$ with $C(0) = 1$, such that

$$H(s) \triangleq \frac{A(s)M(s)}{C(s)A(s) + (1 - C(s))M(s)} \quad \text{is stable} \tag{4.5}$$

and the following $\mathcal{L}_1$-norm condition holds:

$$\|G(s)\|_{\mathcal{L}_1} L < 1, \tag{4.6}$$

where $G(s) \triangleq H(s)(1 - C(s))$.

Letting

$$A(s) = \frac{A_n(s)}{A_d(s)}, \quad C(s) = \frac{C_n(s)}{C_d(s)}, \quad M(s) = \frac{M_n(s)}{M_d(s)}, \tag{4.7}$$

it follows from (4.5) that

$$H(s) = \frac{C_d(s)M_n(s)A_n(s)}{M_d(s)C_n(s)A_n(s) + (C_d(s) - C_n(s))M_n(s)A_d(s)}. \tag{4.8}$$

We note that a strictly proper $C(s)$ implies that the orders of $C_d(s) - C_n(s)$ and $C_d(s)$ are equal. Since the order of $A_d(s)$ is higher than the order of $A_n(s)$, we note that the transfer function $H(s)$ is strictly proper.

Next, we introduce some notations that will be useful for the proofs of stability and performance bounds. First, define

$$H_0(s) \triangleq \frac{A(s)}{C(s)A(s) + (1 - C(s))M(s)}, \tag{4.9}$$

$$H_1(s) \triangleq \frac{(A(s) - M(s))C(s)}{C(s)A(s) + (1 - C(s))M(s)}, \tag{4.10}$$

$$H_2(s) \triangleq \frac{C(s)H(s)}{M(s)}, \tag{4.11}$$

$$H_3(s) \triangleq -\frac{M(s)C(s)}{C(s)A(s) + (1 - C(s))M(s)}. \tag{4.12}$$

Using (4.7) in (4.9)–(4.12), we have

$$H_0(s) = \frac{C_d(s)A_n(s)M_d(s)}{H_d(s)},$$

$$H_1(s) = \frac{C_n(s)A_n(s)M_d(s) - C_n(s)A_d(s)M_n(s)}{H_d(s)}, \tag{4.13}$$

$$H_3(s) = -\frac{C_n(s)A_d(s)M_n(s)}{H_d(s)}, \tag{4.14}$$

where

$$H_d(s) \triangleq C_n(s)A_n(s)M_d(s) + M_n(s)A_d(s)(C_d(s) - C_n(s)).$$

Since $\deg(C_d(s) - C_n(s))$ is larger than $\deg(C_n(s))$, then $\deg\big(M_n(s)A_d(s)(C_d(s) - C_n(s))\big)$ is larger than $\deg(C_n(s)A_d(s)M_n(s))$. Since $\deg(A_d(s))$ is larger than $\deg(A_n(s))$, while $\deg(M_d(s)) - \deg(M_n(s)) = 1$, we note that $\deg(M_n(s)A_d(s)(C_d(s) - C_n(s)))$ is higher than $\deg(C_n(s)A_n(s)M_d(s))$. Therefore, $H_1(s)$ is strictly proper. We note from (4.8) and (4.13) that $H_1(s)$ has the same denominator as $H(s)$, and it follows from (4.5) that $H_1(s)$ is stable. Using similar arguments, it can be verified that $H_0(s)$ and H_3 are proper and stable transfer functions.

Let ρ_r be defined as

$$\rho_r \triangleq \frac{\|H(s)C(s)\|_{\mathcal{L}_1}\|r\|_{\mathcal{L}_\infty} + \|G(s)\|_{\mathcal{L}_1}L_0}{1 - \|G(s)\|_{\mathcal{L}_1}L}. \tag{4.15}$$

Also, let

$$\Delta \triangleq \|H_1(s)\|_{\mathcal{L}_1}\|r\|_{\mathcal{L}_\infty} + \|H_0(s)\|_{\mathcal{L}_1}(L\rho_r + L_0) + \left(\|H_1(s)/M(s)\|_{\mathcal{L}_1} + \|H_0(s)\|_{\mathcal{L}_1}\frac{\|H_2(s)\|_{\mathcal{L}_1}}{1-\|G(s)\|_{\mathcal{L}_1}L}L\right)\bar{\gamma}_0,$$

where $\bar{\gamma}_0 > 0$ is an arbitrary constant. It can be verified easily that $H_2(s)$ is strictly proper and stable and hence $\|H_2(s)\|_{\mathcal{L}_1}$ is bounded. Since $H_1(s)$ is stable and strictly proper, we note that $\|H_1(s)/M(s)\|_{\mathcal{L}_1}$ exists and, hence, Δ is bounded.

Further, let

$$\begin{aligned}\beta_1 &\triangleq 4\Delta\|H_0(s)\|_{\mathcal{L}_1}\left(\beta_{01}L_1 + \frac{\|H_2(s)\|_{\mathcal{L}_1}}{1-\|G(s)\|_{\mathcal{L}_1}L}L_2\right),\\ \beta_2 &\triangleq 4\Delta\Big(\|sH_1(s)\|_{\mathcal{L}_1}\left(\|r\|_{\mathcal{L}_\infty}+2\Delta\right) + \|H_0(s)\|_{\mathcal{L}_1}(\beta_{02}L_1 + L_2\rho_r + L_3)\Big),\end{aligned} \tag{4.16}$$

where β_{01} and β_{02} are defined as

$$\begin{aligned}\beta_{01} &\triangleq \|sH(s)(1-C(s))\|_{\mathcal{L}_1}\frac{\|H_2(s)\|_{\mathcal{L}_1}}{1-\|G(s)\|_{\mathcal{L}_1}L}L,\\ \beta_{02} &\triangleq \|sH(s)C(s)\|_{\mathcal{L}_1}\left(\|r\|_{\mathcal{L}_\infty}+2\Delta\right) + \|sH(s)(1-C(s))\|_{\mathcal{L}_1}(L\rho_r + L_0).\end{aligned} \tag{4.17}$$

Since $H(s)$ and $H_1(s)$ are strictly proper and stable, we note that $\|sH_1(s)\|_{\mathcal{L}_1}$, $\|sH(s)C(s)\|_{\mathcal{L}_1}$, and $\|sH(s)(1-C(s))\|_{\mathcal{L}_1}$ are bounded.

Finally, choose arbitrary $P \in \mathbb{R}^+$, and let $Q \triangleq 2mP$, while

$$\begin{aligned}\beta_3 &\triangleq \frac{P}{Q}\beta_1 = \frac{\beta_1}{2m},\\ \beta_4 &\triangleq 4\Delta^2 + \frac{P}{Q}\beta_2 = 4\Delta^2 + \frac{\beta_2}{2m}.\end{aligned} \tag{4.18}$$

The elements of $\mathcal{L}_1$ adaptive controller are introduced below.

Output Predictor

We consider the following output predictor:

$$\dot{\hat{y}}(t) = -m\hat{y}(t) + m\left(u(t) + \hat{\sigma}(t)\right), \quad \hat{y}(0) = 0, \tag{4.19}$$

where $\hat{\sigma}(t)$ is the adaptive estimate.

Adaptation Law

The adaptation of $\hat{\sigma}(t)$ is defined as

$$\dot{\hat{\sigma}}(t) = \Gamma\text{Proj}(\hat{\sigma}(t),\ -mP\tilde{y}(t)), \quad \hat{\sigma}(0) = 0, \tag{4.20}$$

where $\tilde{y}(t) \triangleq \hat{y}(t) - y(t)$ is the error signal between the output of the system in (4.3) and the predictor in (4.19), $\Gamma \in \mathbb{R}^+$ is the adaptation rate subject to the lower bound

$$\Gamma > \max\left\{\frac{\alpha\beta_3^2}{(\alpha-1)^2\beta_4 P}, \frac{\alpha\beta_4}{P\bar{\gamma}_0^2}\right\}, \tag{4.21}$$

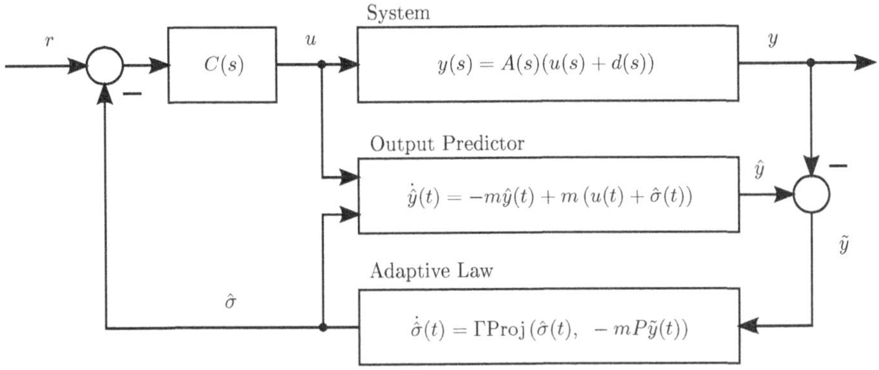

Figure 4.1: Closed-loop system with $\mathcal{L}_1$ adaptive controller.

with $\alpha > 1$ being an arbitrary constant, and the projection is performed with the following bound:

$$|\hat{\sigma}(t)| \leq \Delta, \quad \forall\, t \geq 0. \tag{4.22}$$

Control Law

The control signal is generated according to the following law, assuming zero initialization for $C(s)$:

$$u(s) = C(s)(r(s) - \hat{\sigma}(s)). \tag{4.23}$$

The complete $\mathcal{L}_1$ adaptive controller consists of (4.19), (4.20), and (4.23) subject to the $\mathcal{L}_1$-norm condition in (4.6). The closed-loop system is illustrated in Figure 4.1.

Remark 4.1.1 The sufficient condition for stability in (4.6) restricts the class of systems $A(s)$ in (4.1) that can be stabilized by the controller architecture in this section. However, as discussed in Section 4.1.4, the class of such systems is not empty.

4.1.3 Analysis of the $\mathcal{L}_1$ Adaptive Output Feedback Controller

Closed-Loop Reference System

Consider the following closed-loop reference system:

$$y_{\mathrm{ref}}(s) = M(s)(u_{\mathrm{ref}}(s) + \sigma_{\mathrm{ref}}(s)), \tag{4.24}$$

$$u_{\mathrm{ref}}(s) = C(s)(r(s) - \sigma_{\mathrm{ref}}(s)), \tag{4.25}$$

where

$$\sigma_{\mathrm{ref}}(s) = \frac{(A(s) - M(s))u_{\mathrm{ref}}(s) + A(s)d_{\mathrm{ref}}(s)}{M(s)}, \tag{4.26}$$

and $d_{\text{ref}}(s)$ is the Laplace transform of $d_{\text{ref}}(t) \triangleq f(t, y_{\text{ref}}(t))$. We note that there is no algebraic loop involved in the definition of $\sigma(s)$, $u(s)$ and $\sigma_{\text{ref}}(s)$, $u_{\text{ref}}(s)$.

The next lemma establishes the stability of the closed-loop reference system in (4.24)–(4.25).

Lemma 4.1.1 Let $C(s)$ and $M(s)$ verify the $\mathcal{L}_1$-norm condition in (4.6). Then the closed-loop reference system in (4.24)–(4.25) is BIBO stable.

Proof. It follows from (4.25) and (4.26) that

$$u_{\text{ref}}(s) = C(s)r(s) - \frac{C(s)((A(s) - M(s))u_{\text{ref}}(s) + A(s)d_{\text{ref}}(s))}{M(s)},$$

and hence

$$u_{\text{ref}}(s) = \frac{C(s)M(s)r(s) - C(s)A(s)d_{\text{ref}}(s)}{C(s)A(s) + (1 - C(s))M(s)}. \tag{4.27}$$

From (4.24)–(4.26) we have

$$y_{\text{ref}}(s) = A(s)(u_{\text{ref}}(s) + d_{\text{ref}}(s)). \tag{4.28}$$

Substituting (4.27) into (4.28), it follows from (4.5) that

$$\begin{aligned} y_{\text{ref}}(s) &= A(s)\left(\frac{C(s)M(s)r(s) - C(s)A(s)d_{\text{ref}}(s)}{C(s)A(s) + (1 - C(s))M(s)} + d_{\text{ref}}(s)\right) \\ &= A(s)M(s)\frac{C(s)r(s) + (1 - C(s))d_{\text{ref}}(s)}{C(s)A(s) + (1 - C(s))M(s)} \\ &= H(s)(C(s)r(s) + (1 - C(s))d_{\text{ref}}(s)). \end{aligned} \tag{4.29}$$

Since $H(s)$ and $G(s)$ are strictly proper and stable, the following upper bound holds:

$$\left\| y_{\text{ref}\,t} \right\|_{\mathcal{L}_\infty} \le \|H(s)C(s)\|_{\mathcal{L}_1} \|r\|_{\mathcal{L}_\infty} + \|G(s)\|_{\mathcal{L}_1} \left(L \left\| y_{\text{ref}\,t} \right\|_{\mathcal{L}_\infty} + L_0\right).$$

Then, the $\mathcal{L}_1$-norm condition in (4.6), together with the definition of ρ_r in (4.15), implies that

$$\left\| y_{\text{ref}\,t} \right\|_{\mathcal{L}_\infty} \le \rho_r, \tag{4.30}$$

which holds uniformly, and hence $\|y_{\text{ref}}\|_{\mathcal{L}_\infty}$ is bounded. Therefore, the closed-loop reference system in (4.24)–(4.25) is BIBO stable. □

Remark 4.1.2 We notice that the *ideal* control signal

$$u_{\text{id}}(t) = r(t) - \sigma_{\text{ref}}(t)$$

is the one that leads to the desired system response

$$y_{\text{id}}(s) = M(s)r(s)$$

by canceling the uncertainties exactly. Thus, the reference system in (4.24)–(4.25) has a different response as compared to the *ideal* one. It cancels only the uncertainties within the bandwidth of $C(s)$, which can be selected to be compatible with the control channel specifications. This is exactly what one can hope to achieve with any feedback in the presence of uncertainties.

Transient and Steady-State Performance

In this section, we analyze the stability of the closed-loop adaptive system with the $\mathcal{L}_1$ adaptive controller and derive the uniform performance bounds. Toward this end, let γ_0 be given by

$$\gamma_0 \triangleq \sqrt{\frac{\alpha\beta_4}{\Gamma P}}\,. \tag{4.31}$$

Then, it follows from (4.21) that $\gamma_0 < \bar{\gamma}_0$, and hence

$$\Delta \geq \|H_1(s)\|_{\mathcal{L}_1}\|r\|_{\mathcal{L}_\infty} + \|H_0(s)\|_{\mathcal{L}_1}(L\rho_r + L_0) + \left(\left\|\frac{H_1(s)}{M(s)}\right\|_{\mathcal{L}_1} + L\|H_0(s)\|_{\mathcal{L}_1}\frac{\|H_2(s)\|_{\mathcal{L}_1}}{1-\|G(s)\|_{\mathcal{L}_1}L}\right)\gamma_0\,.$$

Theorem 4.1.1 Consider the system in (4.1) and the $\mathcal{L}_1$ adaptive controller in (4.19), (4.20), and (4.23), subject to the $\mathcal{L}_1$-norm condition in (4.6). If the adaptive gain is chosen to satisfy the design constraint in (4.21), then the following bounds hold:

$$\|\tilde{y}\|_{\mathcal{L}_\infty} < \gamma_0\,, \tag{4.32}$$
$$\|y_{\text{ref}} - y\|_{\mathcal{L}_\infty} \leq \gamma_1\,, \tag{4.33}$$
$$\|u_{\text{ref}} - u\|_{\mathcal{L}_\infty} \leq \gamma_2\,, \tag{4.34}$$

where $\tilde{y}(t) \triangleq \hat{y}(t) - y(t)$, γ_0 is defined in (4.31), and

$$\gamma_1 \triangleq \frac{\|H_2(s)\|_{\mathcal{L}_1}}{1-\|G(s)\|_{\mathcal{L}_1}L}\gamma_0\,,$$
$$\gamma_2 \triangleq \|H_2(s)\|_{\mathcal{L}_1}L\gamma_1 + \|H_3(s)/M(s)\|_{\mathcal{L}_1}\gamma_0\,.$$

Proof. Let $\tilde{\sigma}(t) \triangleq \hat{\sigma}(t) - \sigma(t)$, where $\sigma(t)$ is defined in (4.4). It follows from (4.23) that

$$u(s) = C(s)r(s) - C(s)(\sigma(s) + \tilde{\sigma}(s))\,, \tag{4.35}$$

and the system in (4.3) consequently takes the form

$$y(s) = M(s)\Big(C(s)r(s) + (1-C(s))\sigma(s) - C(s)\tilde{\sigma}(s)\Big)\,. \tag{4.36}$$

Substituting (4.35) into (4.4), it follows from the definitions of $H(s)$, $H_0(s)$, and $H_1(s)$ in (4.5), (4.9), and (4.10) that

$$\sigma(s) = H_1(s)(r(s) - \tilde{\sigma}(s)) + H_0(s)d(s)\,. \tag{4.37}$$

Substituting (4.37) into (4.36), we have

$$\begin{aligned} y(s) &= M(s)(C(s)+H_1(s)(1-C(s)))(r(s)-\tilde{\sigma}(s)) \\ &\quad + H_0(s)M(s)(1-C(s))d(s). \end{aligned} \tag{4.38}$$

It can be verified from (4.5) and (4.10) that

$$M(s)(C(s)+H_1(s)(1-C(s))) = H(s)C(s) \quad \text{and} \quad H(s) = H_0(s)M(s),$$

and, hence, the expression in (4.38) can be rewritten as

$$y(s) = H(s)(C(s)r(s) - C(s)\tilde{\sigma}(s)) + H(s)(1-C(s))d(s). \tag{4.39}$$

Let $e(t) \triangleq y_{\text{ref}}(t) - y(t)$. From (4.29) and (4.39), one has

$$e(s) = H(s)(1-C(s))d_e(s) + H(s)C(s)\tilde{\sigma}(s),$$

where $d_e(s)$ is introduced to denote the Laplace transform of $d_e(t) \triangleq f(t, y_{\text{ref}}(t)) - f(t, y(t))$. Moreover, it follows from (4.3) and (4.19) that

$$\tilde{y}(s) = M(s)\tilde{\sigma}(s), \tag{4.40}$$

which implies that

$$C(s)H(s)\tilde{\sigma}(s) = \frac{C(s)H(s)}{M(s)} M(s)\tilde{\sigma}(s) = \frac{C(s)H(s)}{M(s)} \tilde{y}(s).$$

Lemma A.7.1 and Assumption 4.1.1 give the following upper bound:

$$\|e_t\|_{\mathcal{L}_\infty} \leq \|H(s)(1-C(s))\|_{\mathcal{L}_1} L \|e_t\|_{\mathcal{L}_\infty} + \|H_2(s)\|_{\mathcal{L}_1} \|\tilde{y}_t\|_{\mathcal{L}_\infty},$$

which leads to

$$\|e_t\|_{\mathcal{L}_\infty} \leq \frac{\|H_2(s)\|_{\mathcal{L}_1}}{1 - \|G(s)\|_{\mathcal{L}_1} L} \|\tilde{y}_t\|_{\mathcal{L}_\infty}. \tag{4.41}$$

First, we prove the bound in (4.32) by contradiction. Since $\tilde{y}(0) = 0$ and $\tilde{y}(t)$ is continuous, then assuming the opposite implies that there exists t' such that

$$\begin{aligned} &\|\tilde{y}(t)\| < \gamma_0, \quad \forall\, 0 \leq t < t', \\ &\|\tilde{y}(t')\| = \gamma_0, \end{aligned}$$

which leads to

$$\|\tilde{y}_{t'}\|_{\mathcal{L}_\infty} = \gamma_0. \tag{4.42}$$

Since $y(t) = y_{\text{ref}}(t) - e(t)$, it follows from (4.30) and (4.42) that

$$\|y_{t'}\|_{\mathcal{L}_\infty} \leq \|y_{\text{ref}t'}\|_{\mathcal{L}_\infty} + \|e_{t'}\|_{\mathcal{L}_\infty} \leq \rho_r + \frac{\|H_2(s)\|_{\mathcal{L}_1}}{1 - \|G(s)\|_{\mathcal{L}_1} L} \gamma_0. \tag{4.43}$$

It follows from (4.37) and (4.40) that

$$\sigma(s) = H_1(s)r(s) - \frac{H_1(s)}{M(s)}\tilde{y}(s) + H_0(s)d(s),$$

and hence, the equality in (4.42) implies that

$$\|\sigma_{t'}\|_{\mathcal{L}_\infty} \leq \|H_1(s)\|_{\mathcal{L}_1}\|r\|_{\mathcal{L}_\infty} + \left\|\frac{H_1(s)}{M(s)}\right\|_{\mathcal{L}_1}\gamma_0 + \|H_0(s)\|_{\mathcal{L}_1}(L\|y_{t'}\|_{\mathcal{L}_\infty} + L_0).$$

Along with (4.43) this leads to

$$\|\sigma_{t'}\|_{\mathcal{L}_\infty} \leq \Delta. \tag{4.44}$$

Consider the following candidate Lyapunov function:

$$V(\tilde{y}(t), \tilde{\sigma}(t)) = P\tilde{y}^2(t) + \Gamma^{-1}\tilde{\sigma}^2(t). \tag{4.45}$$

The adaptive law in (4.20) implies that for all $0 \leq t \leq t'$,

$$\dot{V}(t) \leq -Q\tilde{y}^2(t) + 2\Gamma^{-1}|\tilde{\sigma}(t)\dot{\sigma}(t)|. \tag{4.46}$$

It follows from (4.37) that

$$\sigma_d(s) = sH_1(s)(r(s) - \tilde{\sigma}(s)) + H_0(s)d_d(s), \tag{4.47}$$

where $\sigma_d(s)$ and $d_d(s)$ are the Laplace transformations of $\dot{\sigma}(t)$ and $\dot{d}(t)$, respectively. From (4.22) and (4.44), we have

$$\|\tilde{\sigma}_{t'}\|_{\mathcal{L}_\infty} \leq 2\Delta. \tag{4.48}$$

It follows from (4.43) that

$$\|d_{t'}\|_{\mathcal{L}_\infty} \leq L\rho_r + \frac{\|H_2(s)\|_{\mathcal{L}_1}}{1 - \|G(s)\|_{\mathcal{L}_1}L}L\gamma_0 + L_0. \tag{4.49}$$

From the definitions of β_{01} and β_{02} in (4.17), (4.39), and (4.49), we have $\|\dot{y}_{t'}\|_{\mathcal{L}_\infty} \leq \beta_{01}\gamma_0 + \beta_{02}$. It follows from Assumption 4.1.2 that

$$\|\dot{d}_{t'}\|_{\mathcal{L}_\infty} \leq L_2\|y_{t'}\|_{\mathcal{L}_\infty} + L_1(\beta_{01}\gamma_0 + \beta_{02}) + L_3. \tag{4.50}$$

From (4.43), (4.47), (4.50), and the definitions of β_1 and β_2 in (4.16), it follows that

$$\|\dot{\sigma}_{t'}\|_{\mathcal{L}_\infty} \leq \frac{\beta_1\gamma_0 + \beta_2}{4\Delta}. \tag{4.51}$$

Therefore, from (4.46), (4.48), and (4.51) we have

$$\dot{V}(t) \leq -Q\tilde{y}^2(t) + \Gamma^{-1}\Big(\beta_1\gamma_0 + \beta_2\Big), \quad \forall\, 0 \leq t \leq t'. \tag{4.52}$$

The projection algorithm ensures that $|\hat{\sigma}(t)| \leq \Delta$ for all $t \geq 0$, and therefore

$$\max_{t' \geq t \geq 0} \Gamma^{-1}\tilde{\sigma}^2(t) \leq 4\Delta^2/\Gamma. \tag{4.53}$$

Let $\theta_{\max} \triangleq \beta_3\gamma_0 + \beta_4$, where β_3 and β_4 are defined in (4.18). If at arbitrary $t \in [0,\ t']$, $V(t) > \theta_{\max}/\Gamma$, then it follows from (4.45) and (4.53) that

$$P\tilde{y}^2(t) > \frac{P(\beta_1\gamma_0 + \beta_2)}{\Gamma Q},$$

and hence

$$Q\tilde{y}^2(t) = (Q/P)P\tilde{y}^2(t) > (\beta_1\gamma_0 + \beta_2)/\Gamma. \tag{4.54}$$

From (4.52) and (4.54) it follows that if $V(t) > \theta_{\max}/\Gamma$ for some $t \in [0,\ t']$, then

$$\dot{V}(t) < 0. \tag{4.55}$$

Since $\tilde{y}(0) = 0$, we can verify that $V(0) \leq \left(\beta_3\gamma_0 + \beta_4\right)/\Gamma$. It follows from (4.55) that

$$V(t) \leq \theta_{\max}/\Gamma, \quad \forall\, 0 \leq t \leq t'. \tag{4.56}$$

Since $P\|\tilde{y}(t)\|^2 \leq V(t)$, then it follows from (4.56) that

$$|\tilde{y}(t)|^2 \leq \frac{\beta_3\gamma_0 + \beta_4}{\Gamma P}, \quad \forall\, 0 \leq t \leq t',$$

and from the definition of γ_0 in (4.31), we get

$$|\tilde{y}(t)|^2 \leq \left(\frac{\beta_3}{\alpha\beta_4}\sqrt{\frac{\alpha\beta_4}{\Gamma P}} + \frac{1}{\alpha}\right)\gamma_0^2, \quad \forall\, 0 \leq t \leq t'.$$

Then, the design constraint in (4.21) leads to

$$|\tilde{y}(t)|^2 < \gamma_0^2, \quad \forall\, 0 \leq t \leq t',$$

which contradicts the assumption in (4.42), and thus (4.32) holds. It follows from (4.6), (4.32), and (4.41) that

$$\|e_t\|_{\mathcal{L}_\infty} \leq \frac{\|H_2(s)\|_{\mathcal{L}_1}}{1 - \|G(s)\|_{\mathcal{L}_1} L}\gamma_0,$$

which holds uniformly for all $t \geq 0$ and therefore leads to the bound in (4.33).

On the other hand, it follows from (4.4) and (4.35) that

$$u(s) = \frac{M(s)(C(s)r(s) - C(s)\tilde{\sigma}(s)) - C(s)A(s)d(s)}{C(s)A(s) + (1 - C(s))M(s)}.$$

To prove the bound in (4.34), we notice that from (4.27) one can derive

$$\begin{aligned} u_{\mathrm{ref}}(s) - u(s) &= -H_2(s)d_e(s) - H_3(s)\tilde{\sigma}(s) \\ &= -H_2(s)d_e(s) - (H_3(s)/M(s))M(s)\tilde{\sigma}(s). \end{aligned} \tag{4.57}$$

It follows from (4.40) and (4.57) that

$$\|u_{\mathrm{ref}} - u\|_{\mathcal{L}_\infty} \leq \|H_2(s)\|_{\mathcal{L}_1} L \|y_{\mathrm{ref}} - y\|_{\mathcal{L}_\infty} + \|H_3(s)/M(s)\|_{\mathcal{L}_1} \|\tilde{y}\|_{\mathcal{L}_\infty},$$

which leads to (4.34). □

Thus, the tracking error between $y(t)$ and $y_{\text{ref}}(t)$, as well as between $u(t)$ and $u_{\text{ref}}(t)$, is uniformly bounded by a constant inverse proportional to $\sqrt{\Gamma}$. This implies that, during the transient phase, one can achieve arbitrarily close tracking performance for both signals simultaneously by increasing Γ.

Remark 4.1.3 We notice that if we set $C(s)=1$, then the bound in (4.34) is not well defined, since the second term degenerates into the $\mathcal{L}_1$-norm of an improper transfer function.

4.1.4 Design for the $\mathcal{L}_1$-Norm Condition

In this section, we discuss the classes of systems that can satisfy (4.5) via the choice of $M(s)$ and $C(s)$. For simplicity, we consider first-order $C(s)$, given by

$$C(s)=\frac{\omega}{s+\omega}, \tag{4.58}$$

and first-order $M(s)$, such as that in (4.2). It follows from (4.2) and (4.58) that

$$H(s)=\frac{m(s+\omega)A_n(s)}{\omega(s+m)A_n(s)+msA_d(s)}.$$

Stability of $H(s)$ is equivalent to stabilization of $A(s)$ by a PI controller $K_{PI}(s)$ of the structure

$$K_{PI}(s)=\frac{\omega(s+m)}{ms}, \tag{4.59}$$

where m and ω are the same as in (4.2) and (4.58). The loop-transfer function of the cascaded system $A(s)$ with the PI controller will be

$$L_{PI}(s)=\frac{\omega(s+m)}{ms}A(s),$$

leading to the following closed-loop system:

$$H_{PI}(s)=\frac{\omega(s+m)A_n(s)}{\omega(s+m)A_n(s)+msA_d(s)}. \tag{4.60}$$

Hence, the stability of $H(s)$ is equivalent to that of (4.60), and the problem can be reduced to identifying the class of systems $A(s)$ that can be stabilized by a PI controller. It also permits the use of root locus methods for checking the stability of $H(s)$ via the loop-transfer function $L_{PI}(s)$. We note that the PI controller $K_{PI}(s)$ in (4.59) adds a pole at the origin and a zero at $-m$ to the loop-transfer function, while ω/m is the proportional gain of the controller. In the absence of nonlinearity, one has $L=0$ in (4.6), and hence the stability of the closed-loop system follows from the stability of $H(s)$, which can be verified using methods from $\mathcal{H}_\infty$ robust control.

Minimum Phase Systems with Relative Degree 1 or 2

Consider a minimum phase system $H(s)$ with relative degree 1. Notice that the zeros of $L_{PI}(s)$ are located in the open left half plane. As the gain ω/m increases, it follows from the classical control theory that all the closed-loop poles approach the open-loop zeros except 1, which tends to ∞ along the negative real axis. This implies that all the closed-loop poles are located in the open left half plane. Hence, the transfer function in (4.60) is stable, and so is $H(s)$.

For a minimum phase system $H(s)$ with relative degree 2, with the increase of the gain ω/m there are two closed-loop poles approaching ∞ along the direction of $-\pi/2$ and $\pi/2$ in the complex plane. Let the abscissa of the intersection of the asymptotes and the real axis be δ. We note that the two infinite poles approach $\delta \pm j\infty$. If the choice of $M(s)$ ensures negative δ, then the closed-loop system can be stabilized by increasing the loop gain. Therefore, by choosing appropriate $M(s)$, we can ensure stability of minimum phase systems with relative degree 1 or 2.

Other Systems

We note that nonminimum phase systems can also be stabilized by a PI controller. However, the choice of m and ω is not straightforward. Reference [67] has addressed the problem of the filter design using standard root locus analysis from classical control. It has been shown that, if the system is nonminimum phase, then the choice of the low-pass filter and the reference system that would verify stability of $H(s)$ might be very limited.

Remark 4.1.4 We notice that, in light of the above discussion, a PI controller stabilizing $A(s)$ might also stabilize the system in the presence of the nonlinear uncertainty $f(t, y(t))$. However, the transient performance cannot be quantified in the presence of unknown $A(s)$. The $\mathcal{L}_1$ adaptive controller instead will generate different control signals $u(t)$ (always in the low-frequency range) for different unknown systems to ensure uniform transient performance for $y(t)$.

In the simulation example below, we demonstrate the application of the $\mathcal{L}_1$ adaptive controller for an unknown nonminimum phase system in the presence of unknown nonlinear uncertainties.

4.1.5 Simulation Example

As an illustrative example, consider the system in (4.1) with

$$A(s) = \frac{s^2 - 0.5s + 0.5}{s^3 - s^2 - 2s + 8}.$$

We note that $A(s)$ has both poles and zeros in the right half plane and hence it is an unstable nonminimum phase system. We consider the $\mathcal{L}_1$ adaptive controller, defined via (4.19), (4.20), and (4.23), with $m = 3$, $\omega = 10$, $\Gamma = 500$. We set $\Delta = 100$. First, we consider the response of the closed-loop system to a step-reference signal for $d(t) \equiv 0$. The simulation results are shown in Figure 4.2. Next, we consider

$$f(t, y(t)) = \sin(0.1t)y(t) + 2\sin(0.1t)$$

and apply the same controller without retuning. The system response and the control signal are plotted in Figures 4.3(a) and 4.3(b). Further, we consider a time-varying reference input $r(t) = 0.5\sin(0.3t)$ and notice that, without any retuning of the controller, the system response and the control signal behave as expected (Figures 4.3(c) and 4.3(d)). Finally,

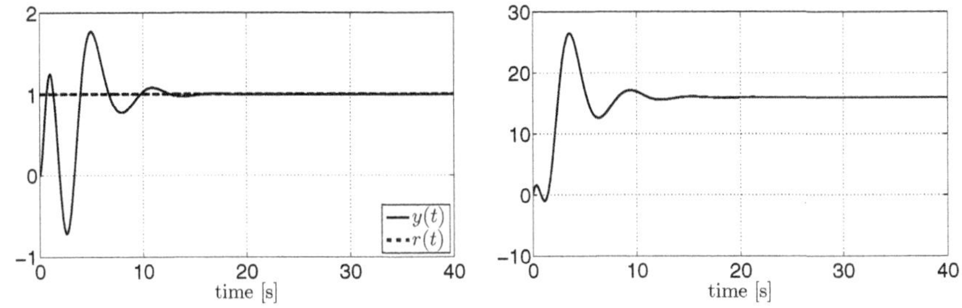

(a) $y(t)$ (solid) and $r(t)$ (dashed) (b) Time history of $u(t)$

Figure 4.2: Performance of the $\mathcal{L}_1$ controller for $f(t, y(t)) \equiv 0$.

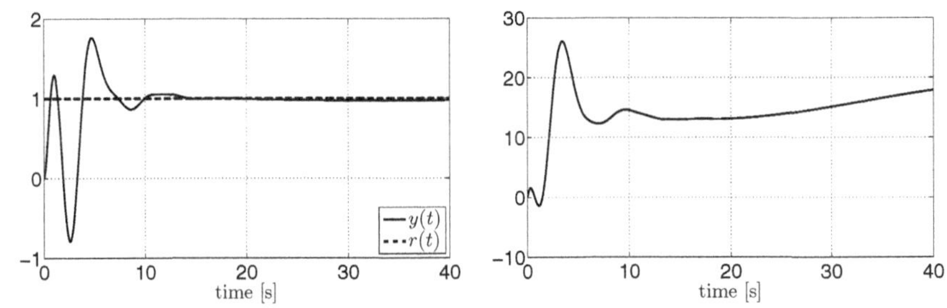

(a) $y(t)$ (solid) and $r(t)$ (dashed) (b) Time history of $u(t)$

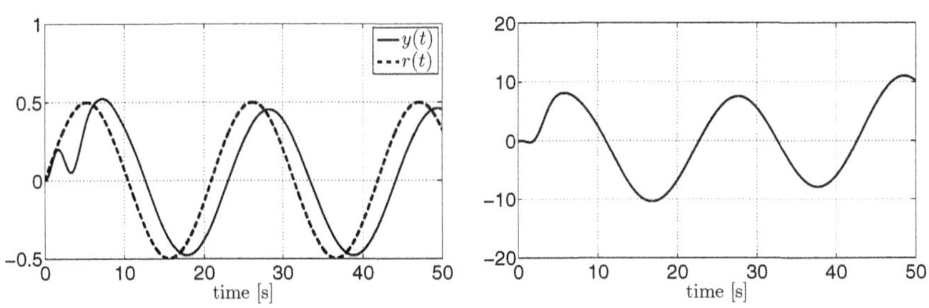

(c) $y(t)$ (solid) and $r(t)$ (dashed) (d) Time history of $u(t)$

Figure 4.3: Performance of the $\mathcal{L}_1$ controller for $f(t, y) = \sin(0.1t)y + 2\sin(0.1t)$.

Figure 4.4 presents the closed-loop system response and the control signal for a different uncertainty,

$$f(t, y(t)) = \sin(0.1t)y(t) + 2\sin(0.4t),$$

without any retuning of the controller.

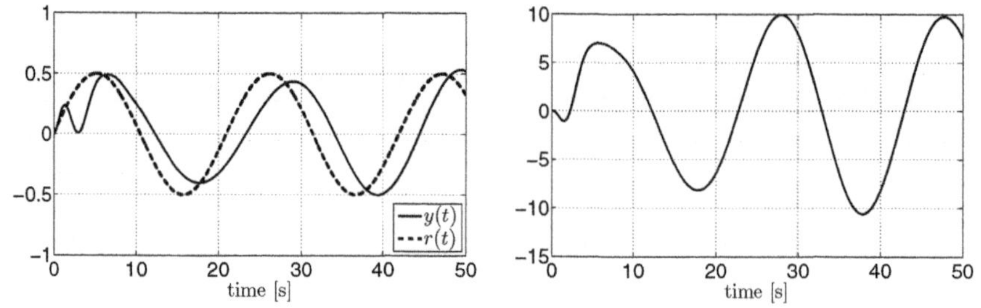

(a) $y(t)$ (solid) and $r(t)$ (dashed)

(b) Time history of $u(t)$

Figure 4.4: Performance of the $\mathcal{L}_1$ controller for $f(t,y) = \sin(0.1t)y + 2\sin(0.4t)$.

4.2 $\mathcal{L}_1$ Adaptive Output Feedback Controller for Non-SPR Reference Systems

This section presents an extension of the $\mathcal{L}_1$ adaptive output feedback controller, which achieves performance specifications defined by a non-SPR reference system. This extension is possible by invoking the piecewise-constant adaptive law, discussed in Section 3.3. The performance bounds between the closed-loop reference system and the closed-loop $\mathcal{L}_1$ adaptive system can be rendered arbitrarily small by reducing the step size of integration. The sampling time of the adaptive law can be set according to the available sampling rate of the CPU [33]. This solution has been tested in a midfidelity model of a generic flexible Crew Launch Vehicle provided by NASA [88].

4.2.1 Problem Formulation

Consider the following SISO system:

$$y(s) = A(s)(u(s) + d(s)), \quad y(0) = 0, \tag{4.61}$$

where $u(t) \in \mathbb{R}$ is the input; $y(t) \in \mathbb{R}$ is the system output; $A(s)$ is a strictly proper unknown transfer function of unknown relative degree n_r, for which only a known lower bound $1 < d_r \leq n_r$ is available; $d(s)$ is the Laplace transform of the time-varying uncertainties and disturbances $d(t) = f(t, y(t))$; and $f : \mathbb{R} \times \mathbb{R} \to \mathbb{R}$ is an unknown map, subject to the following assumption.

Assumption 4.2.1 (Lipschitz continuity) There exist constants $L > 0$ and $L_0 > 0$, such that

$$|f(t,y_1) - f(t,y_2)| \leq L|y_1 - y_2|, \quad |f(t,y)| \leq L|y| + L_0$$

hold uniformly in $t \geq 0$, where the numbers L and L_0 can be arbitrarily large.

Let $r(t)$ be a given bounded continuous reference input signal. The control objective is to design an adaptive output feedback controller $u(t)$ such that the system output $y(t)$ tracks the reference input $r(t)$ following a desired reference model $M(s)$, where $M(s)$ is a minimum-phase stable transfer function of relative degree $d_r > 1$.

4.2.2 $\mathcal{L}_1$ Adaptive Control Architecture

Definitions and $\mathcal{L}_1$-Norm Sufficient Condition for Stability

Similar to Section 4.1.2, we can rewrite the system in (4.61) as

$$y(s) = M(s)(u(s) + \sigma(s)), \quad y(0) = 0, \tag{4.62}$$

$$\sigma(s) = \frac{(A(s) - M(s))u(s) + A(s)d(s)}{M(s)}. \tag{4.63}$$

Let $(A_m, b_m, c_m^\top)$ be a minimal realization of $M(s)$, i.e., $(A_m, b_m, c_m^\top)$ is controllable and observable, and A_m is Hurwitz. The system in (4.62) can be rewritten as

$$\begin{aligned} \dot{x}(t) &= A_m x(t) + b_m(u(t) + \sigma(t)), \quad x(0) = 0, \\ y(t) &= c_m^\top x(t). \end{aligned} \tag{4.64}$$

The design of the $\mathcal{L}_1$ adaptive controller proceeds by considering a strictly proper system $C(s)$ of relative degree d_r, with $C(0) = 1$. Further, similar to Section 4.1.2, the selection of $C(s)$ and $M(s)$ must ensure that

$$H(s) \triangleq \frac{A(s)M(s)}{C(s)A(s) + (1 - C(s))M(s)} \quad \text{is stable} \tag{4.65}$$

and that the following $\mathcal{L}_1$-norm condition holds:

$$\|G(s)\|_{\mathcal{L}_1} L < 1, \tag{4.66}$$

where $G(s) \triangleq H(s)(1 - C(s))$.

Letting

$$A(s) = \frac{A_n(s)}{A_d(s)}, \quad C(s) = \frac{C_n(s)}{C_d(s)}, \quad M(s) = \frac{M_n(s)}{M_d(s)}, \tag{4.67}$$

where the numerators and the denominators are all polynomials of s, it follows from (4.65) that

$$H(s) = \frac{C_d(s)M_n(s)A_n(s)}{H_d(s)}, \tag{4.68}$$

where

$$H_d(s) \triangleq C_n(s)A_n(s)M_d(s) + M_n(s)A_d(s)(C_d(s) - C_n(s)). \tag{4.69}$$

A strictly proper $C(s)$ implies that the orders of $C_d(s) - C_n(s)$ and $C_d(s)$ are the same. Since the order of $A_d(s)$ is higher than the order of $A_n(s)$, the transfer function $H(s)$ is strictly proper.

Next, let

$$H_0(s) \triangleq \frac{A(s)}{C(s)A(s)+(1-C(s))M(s)}, \tag{4.70}$$

$$H_1(s) \triangleq \frac{(A(s)-M(s))C(s)}{C(s)A(s)+(1-C(s))M(s)}, \tag{4.71}$$

$$H_2(s) \triangleq \frac{H(s)C(s)}{M(s)}, \tag{4.72}$$

$$H_3(s) \triangleq -\frac{M(s)C(s)}{(C(s)A(s)+(1-C(s))M(s))}. \tag{4.73}$$

Using the expressions from (4.67) and (4.69), we can rewrite the equations for $H_0(s)$ and $H_1(s)$ as

$$H_0(s) = \frac{C_d(s)A_n(s)M_d(s)}{H_d(s)},$$
$$H_1(s) = \frac{C_n(s)A_n(s)M_d(s) - C_n(s)A_d(s)M_n(s)}{H_d(s)}. \tag{4.74}$$

Since $\deg(C_d(s) - C_n(s))$ is larger than $\deg(C_n(s))$ by d_r, then $\deg(M_n(s)A_d(s)(C_d(s)-C_n(s)))$ is larger than $\deg(C_n(s)A_d(s)M_n(s))$ by d_r. Since the $\deg(A_d(s))$ is larger than $\deg(A_n(s))$ by $n_r \geq d_r$, while $\deg(M_n(s))$ is larger than $\deg(M_d(s))$ by d_r, then $\deg(M_n(s)A_d(s)(C_d(s) - C_n(s)))$ is larger than $\deg(C_n(s)A_n(s)M_d(s))$. Therefore, $H_1(s)$ is strictly proper with relative degree d_r. We notice from (4.68) and (4.74) that $H_1(s)$ has the same denominator as $H(s)$, and therefore it follows from (4.66) that $H_1(s)$ is stable. Using similar arguments, it can be verified that $H_0(s)$ is proper and stable. Similarly, $H_2(s)$ is strictly proper and stable.

Also, let

$$\Delta \triangleq \|H_1(s)\|_{\mathcal{L}_1}\|r\|_{\mathcal{L}_\infty} + \|H_0(s)\|_{\mathcal{L}_1}(L\rho_r + L_0)$$
$$+ \left(\left\|\frac{H_1(s)}{M(s)}\right\|_{\mathcal{L}_1} + \|H_0(s)\|_{\mathcal{L}_1}\frac{\|H_2(s)\|_{\mathcal{L}_1}}{1-\|G(s)\|_{\mathcal{L}_1}L}L\right)\bar{\gamma}_0, \tag{4.75}$$

where $\bar{\gamma}_0 > 0$ is an arbitrary constant. Since both $H_1(s)$ and $M(s)$ are stable and strictly proper with relative degree d_r, and $M(s)$ is minimum phase, $H_1(s)/M(s)$ is stable and proper. Hence, $\|H_1(s)/M(s)\|_{\mathcal{L}_1}$ is bounded. Therefore Δ is also bounded. Further, since A_m is Hurwitz, there exists $P = P^\top > 0$ that satisfies the algebraic Lyapunov equation

$$A_m^\top P + PA_m = -Q, \quad \text{for arbitrary} \quad Q = Q^\top > 0.$$

From the properties of P, it follows that there exits nonsingular $\sqrt{P}$ such that

$$P = (\sqrt{P})^\top \sqrt{P}.$$

Given the vector $c_m^\top(\sqrt{P})^{-1}$, let D be a $(n-1)\times n$ matrix that contains the null space of $c_m^\top(\sqrt{P})^{-1}$, i.e.,

$$D(c_m^\top(\sqrt{P})^{-1})^\top = 0, \tag{4.76}$$

and further let

$$\Lambda \triangleq \begin{bmatrix} c_m^\top \\ D\sqrt{P} \end{bmatrix}.$$

From the definition of the null space, it follows that

$$\Lambda(\sqrt{P})^{-1} = \begin{bmatrix} c_m^\top(\sqrt{P})^{-1} \\ D \end{bmatrix}$$

is full rank, and hence Λ^{-1} exists.

Lemma 4.2.1 For arbitrary $\xi \triangleq [\begin{smallmatrix} y \\ z \end{smallmatrix}] \in \mathbb{R}^n$, where $y \in \mathbb{R}$ and $z \in \mathbb{R}^{n-1}$, there exist $p_1 > 0$ and positive definite $P_2 \in \mathbb{R}^{(n-1)\times(n-1)}$ such that

$$\xi^\top(\Lambda^{-1})^\top P \Lambda^{-1}\xi = p_1 y^2 + z^\top P_2 z.$$

Proof. Using $P = (\sqrt{P})^\top\sqrt{P}$, one can write

$$\xi^\top(\Lambda^{-1})^\top P \Lambda^{-1}\xi = \xi^\top(\sqrt{P}\Lambda^{-1})^\top(\sqrt{P}\Lambda^{-1})\xi.$$

We notice that

$$\Lambda(\sqrt{P})^{-1} = \begin{bmatrix} c_m^\top(\sqrt{P})^{-1} \\ D \end{bmatrix}.$$

Next, let

$$q_1 = (c_m^\top(\sqrt{P})^{-1})(c_m^\top(\sqrt{P})^{-1})^\top, \quad Q_2 = DD^\top.$$

From the expression in (4.76) we have

$$(\Lambda(\sqrt{P})^{-1})(\Lambda(\sqrt{P})^{-1})^\top = \begin{bmatrix} q_1 & 0 \\ 0 & Q_2 \end{bmatrix}.$$

Nonsingularity of Λ and $\sqrt{P}$ implies that $(\Lambda(\sqrt{P})^{-1})(\Lambda(\sqrt{P})^{-1})^\top$ is nonsingular as well, and therefore Q_2 is also nonsingular. Hence,

$$\begin{aligned}(\sqrt{P}\Lambda^{-1})^\top(\sqrt{P}\Lambda^{-1}) &= \left((\Lambda(\sqrt{P})^{-1})(\Lambda(\sqrt{P})^{-1})^\top\right)^{-1} \\ &= \begin{bmatrix} q_1^{-1} & 0 \\ 0 & Q_2^{-1} \end{bmatrix}.\end{aligned}$$

Denoting $p_1 \triangleq q_1^{-1}$ and $P_2 \triangleq Q_2^{-1}$ completes the proof. □

Let T_s be an arbitrary positive constant, which can be associated with the sampling rate of the available CPU, and $\mathbf{1}_1 = [1, 0, \ldots, 0]^\top \in \mathbb{R}^n$. Let

$$\mathbf{1}_1^\top e^{\Lambda A_m \Lambda^{-1} t} = [\eta_1(t), \eta_2^\top(t)], \tag{4.77}$$

where $\eta_1(t) \in \mathbb{R}$ and $\eta_2(t) \in \mathbb{R}^{n-1}$ contain the first and the 2-to-n elements of the row vector $\mathbf{1}_1^\top e^{\Lambda A_m \Lambda^{-1} t}$, and let

$$\kappa(T_s) \triangleq \int_0^{T_s} |\mathbf{1}_1^\top e^{\Lambda A_m \Lambda^{-1}(T_s - \tau)} \Lambda b_m| d\tau. \tag{4.78}$$

Also, let $\varsigma(T_s)$ be defined as

$$\varsigma(T_s) \triangleq \|\eta_2(T_s)\|\sqrt{\frac{\alpha}{\lambda_{\max}(P_2)}} + \kappa(T_s)\Delta\,,$$
$$\alpha \triangleq \lambda_{\max}(\Lambda^{-\top}P\Lambda^{-1})\left(\frac{2\Delta\|\Lambda^{-\top}Pb_m\|}{\lambda_{\min}(\Lambda^{-\top}Q\Lambda^{-1})}\right)^2. \tag{4.79}$$

Further, let $\Phi(T_s)$ be the $n \times n$ matrix

$$\Phi(T_s) \triangleq \int_0^{T_s} e^{\Lambda A_m \Lambda^{-1}(T_s-\tau)}\Lambda d\tau\,. \tag{4.80}$$

Next, we introduce the functions

$$\beta_1(T_s) \triangleq \max_{t\in[0,\,T_s]} |\eta_1(t)|\,, \quad \beta_2(T_s) \triangleq \max_{t\in[0,\,T_s]} \|\eta_2(t)\|\,, \tag{4.81}$$

and also

$$\beta_3(T_s) \triangleq \max_{t\in[0,\,T_s]} \eta_3(t)\,, \quad \beta_4(T_s) = \max_{t\in[0,\,T_s]} \eta_4(t)\,, \tag{4.82}$$

where

$$\eta_3(t) \triangleq \int_0^t |\mathbf{1}_1^\top e^{\Lambda A_m \Lambda^{-1}(t-\tau)}\Lambda\Phi^{-1}(T_s)e^{\Lambda A_m \Lambda^{-1}T_s}\mathbf{1}_1|d\tau\,,$$
$$\eta_4(t) \triangleq \int_0^t |\mathbf{1}_1^\top e^{\Lambda A_m \Lambda^{-1}(t-\tau)}\Lambda b_m|d\tau\,.$$

Finally, let

$$\gamma_0(T_s) \triangleq \beta_1(T_s)\varsigma(T_s) + \beta_2(T_s)\sqrt{\frac{\alpha}{\lambda_{\max}(P_2)}} + \beta_3(T_s)\varsigma(T_s) + \beta_4(T_s)\Delta\,. \tag{4.83}$$

Lemma 4.2.2 The following limiting relationship is true:

$$\lim_{T_s\to 0} \gamma_0(T_s) = 0\,.$$

Proof. Notice that since $\beta_1(T_s), \beta_3(T_s)$, α, and Δ are bounded, it is sufficient to prove that

$$\lim_{T_s\to 0} \varsigma(T_s) = 0\,, \tag{4.84}$$
$$\lim_{T_s\to 0} \beta_2(T_s) = 0\,, \tag{4.85}$$
$$\lim_{T_s\to 0} \beta_4(T_s) = 0\,. \tag{4.86}$$

Since

$$\lim_{T_s\to 0} \mathbf{1}_1^\top e^{\Lambda A_m \Lambda^{-1}T_s} = \mathbf{1}_1^\top\,,$$

then

$$\lim_{T_s\to 0} \eta_2(T_s) = \mathbf{0}_{n-1},$$

which implies

$$\lim_{T_s\to 0} \|\eta_2(T_s)\| = 0.$$

Further, it follows from the definition of $\kappa(T_s)$ in (4.78) that

$$\lim_{T_s\to 0} \kappa(T_s) = 0.$$

Since α and Δ are bounded, we have

$$\lim_{T_s\to 0} \varsigma(T_s) = 0,$$

which proves (4.84). Since $\eta_2(t)$ is continuous, it follows from (4.81) that

$$\lim_{T_s\to 0} \beta_2(T_s) = \lim_{t\to 0} \|\eta_2(t)\|.$$

Since

$$\lim_{t\to 0} \mathbf{1}_1^\top e^{\Lambda A_m \Lambda^{-1} t} = \mathbf{1}_1^\top,$$

we have

$$\lim_{t\to 0} \|\eta_2(t)\| = 0,$$

which proves (4.85). Similarly

$$\lim_{T_s\to 0} \beta_4(T_s) = \lim_{t\to 0} \|\eta_4(t)\| = 0,$$

which proves (4.86). Boundedness of α, $\beta_3(T_s)$, and Δ implies

$$\lim_{T_s\to 0}\left(\beta_1(T_s)\varsigma(T_s) + \beta_2(T_s)\sqrt{\frac{\alpha}{\lambda_{\max}(P_2)}} + \beta_3(T_s)\varsigma(T_s) + \beta_4(T_s)\Delta\right) = 0,$$

which completes the proof. □

The elements of $\mathcal{L}_1$ adaptive controller are introduced next.

Output Predictor

We consider the following output predictor:

$$\begin{aligned} \dot{\hat{x}}(t) &= A_m \hat{x}(t) + b_m u(t) + \hat{\sigma}(t), \quad \hat{x}(0) = 0, \\ \hat{y}(t) &= c_m^\top \hat{x}(t), \end{aligned} \tag{4.87}$$

where $\hat{\sigma}(t) \in \mathbb{R}^n$ is the vector of adaptive parameters. Notice that while $\sigma(t) \in \mathbb{R}$ in (4.64) is matched, the uncertainty estimation $\hat{\sigma}(t) \in \mathbb{R}^n$ in (4.87) is unmatched.

Adaptation Laws

Letting $\tilde{y}(t) \triangleq \hat{y}(t) - y(t)$, the update law for $\hat{\sigma}(t)$ is given by

$$\begin{aligned} \hat{\sigma}(t) &= \hat{\sigma}(iT_s), \quad t \in [iT_s, (i+1)T_s), \\ \hat{\sigma}(iT_s) &= -\Phi^{-1}(T_s)\mu(iT_s), \quad i = 0,1,2,\ldots, \end{aligned} \tag{4.88}$$

where $\Phi(T_s)$ was defined in (4.80) and

$$\mu(iT_s) = e^{\Lambda A_m \Lambda^{-1} T_s} \mathbf{1}_1 \tilde{y}(iT_s), \quad i = 0,1,2,\ldots.$$

Control Law

The control signal is defined as

$$u(s) = C(s)r(s) - \frac{C(s)}{c_m^\top (s\mathbb{I} - A_m)^{-1} b_m} c_m^\top (s\mathbb{I} - A_m)^{-1} \hat{\sigma}(s), \tag{4.89}$$

where $C(s)$ was first introduced in (4.65). The $\mathcal{L}_1$ adaptive controller consists of (4.87), (4.88), and (4.89), subject to the $\mathcal{L}_1$-norm condition in (4.66).

4.2.3 Analysis of the $\mathcal{L}_1$ Adaptive Controller

Closed-Loop Reference System

Consider the following closed-loop reference system:

$$y_{\text{ref}}(s) = M(s)(u_{\text{ref}}(s) + \sigma_{\text{ref}}(s)), \tag{4.90}$$

$$u_{\text{ref}}(s) = C(s)(r(s) - \sigma_{\text{ref}}(s)), \tag{4.91}$$

where

$$\sigma_{\text{ref}}(s) = \frac{(A(s) - M(s))u_{\text{ref}}(s) + A(s)d_{\text{ref}}(s)}{M(s)}, \tag{4.92}$$

and $d_{\text{ref}}(t) \triangleq f(t, y_{\text{ref}}(t))$.

Lemma 4.2.3 Let $C(s)$ and $M(s)$ verify the $\mathcal{L}_1$-norm condition in (4.66). Then, the closed-loop reference system in (4.90)–(4.91) is BIBO stable.

Proof. It follows from (4.92) and (4.91) that

$$u_{\text{ref}}(s) = \frac{C(s)M(s)r(s) - C(s)A(s)d_{\text{ref}}(s)}{C(s)A(s) + (1 - C(s))M(s)}, \tag{4.93}$$

while from (4.90)–(4.92) one can derive

$$y_{\text{ref}}(s) = H(s)\Big(C(s)r(s) + (1 - C(s))d_{\text{ref}}(s)\Big). \tag{4.94}$$

Since $H(s)$ is strictly proper and stable, $G(s) = H(s)(1 - C(s))$ is also strictly proper and stable, and therefore

$$\|y_{\text{ref}}\|_{\mathcal{L}_\infty} \leq \|H(s)C(s)\|_{\mathcal{L}_1} \|r\|_{\mathcal{L}_\infty} + \|G(s)\|_{\mathcal{L}_1} (L\|y_{\text{ref}}\|_{\mathcal{L}_\infty} + L_0).$$

Using the $\mathcal{L}_1$-norm condition in (4.66), it follows that

$$\|y_{\text{ref}t}\|_{\mathcal{L}_\infty} \le \rho_r \triangleq \frac{\|H(s)C(s)\|_{\mathcal{L}_1}\|r\|_{\mathcal{L}_\infty} + \|G(s)\|_{\mathcal{L}_1}L_0}{1 - \|G(s)\|_{\mathcal{L}_1}L} < \infty, \tag{4.95}$$

which holds uniformly, and hence, $\|y_{\text{ref}}\|_{\mathcal{L}_\infty}$ is bounded. This completes the proof. □

Transient and Steady-State Performance

We will now proceed with the derivation of the performance bounds. Toward this end, let $\tilde{x}(t) \triangleq \hat{x}(t) - x(t)$. Then, the error dynamics between (4.64) and (4.87) are given by

$$\begin{aligned} \dot{\tilde{x}}(t) &= A_m\tilde{x}(t) + \hat{\sigma}(t) - b_m\sigma(t), \quad \tilde{x}(0) = 0, \\ \tilde{y}(t) &= c_m^\top \tilde{x}(t). \end{aligned} \tag{4.96}$$

Lemma 4.2.4 Consider the system in (4.61) with the $\mathcal{L}_1$ adaptive controller and the closed-loop reference system in (4.90)–(4.91). The following upper bound holds:

$$\|(y_{\text{ref}} - y)_t\|_{\mathcal{L}_\infty} \le \frac{\|H_2(s)\|_{\mathcal{L}_1}}{1 - \|G(s)\|_{\mathcal{L}_1}L}\|\tilde{y}_t\|_{\mathcal{L}_\infty}.$$

Proof. Let

$$\tilde{\sigma}(s) \triangleq \frac{C(s)}{c_m^\top(s\mathbb{I} - A_m)^{-1}b_m} c_m^\top(s\mathbb{I} - A_m)^{-1}\hat{\sigma}(s) - C(s)\sigma(s). \tag{4.97}$$

It follows from (4.89) that

$$u(s) = C(s)r(s) - C(s)\sigma(s) - \tilde{\sigma}(s), \tag{4.98}$$

and the system in (4.62) consequently takes the form

$$y(s) = M(s)\Big(C(s)r(s) + (1 - C(s))\sigma(s) - \tilde{\sigma}(s)\Big). \tag{4.99}$$

Substituting $u(s)$ from (4.98) into (4.63) gives

$$\sigma(s) = \frac{\big(A(s) - M(s)\big)\big(C(s)r(s) - C(s)\sigma(s) - \tilde{\sigma}(s)\big) + A(s)d(s)}{M(s)},$$

and hence

$$\sigma(s) = \frac{\Big(A(s) - M(s)\Big)\Big(C(s)r(s) - \tilde{\sigma}(s)\Big) + A(s)d(s)}{M(s) + C(s)(A(s) - M(s))}. \tag{4.100}$$

Using the definitions of $H_0(s)$ and $H_1(s)$ in (4.70) and (4.71), we can write

$$\sigma(s) = H_1(s)r(s) - \frac{H_1(s)}{C(s)}\tilde{\sigma}(s) + H_0(s)d(s). \tag{4.101}$$

Substitution into (4.99) leads to

$$y(s) = M(s)\Big(C(s) + H_1(s)(1 - C(s))\Big)\left(r(s) - \frac{\tilde{\sigma}(s)}{C(s)}\right) + H_0(s)M(s)\Big(1 - C(s)\Big)d(s).$$

Recalling the definition of $H(s)$ from (4.65), one can verify that

$$M(s)(C(s) + H_1(s)(1 - C(s))) = H(s)C(s) \quad \text{and} \quad H(s) = H_0(s)M(s),$$

which implies that

$$y(s) = H(s)\Big(C(s)r(s) - \tilde{\sigma}(s)\Big) + H(s)(1 - C(s))d(s).$$

Letting $e(t) \triangleq y(t)_{\text{ref}} - y(t)$ and denoting by $d_e(s)$ the Laplace transform of $d_e(t) \triangleq f(t, y_{\text{ref}}(t)) - f(t, y(t))$, we can use the expression for $y_{\text{ref}}(s)$ in (4.94) to derive

$$e(s) = H(s)((1 - C(s))d_e(s) + \tilde{\sigma}(s)).$$

Lemma A.7.1 and Assumption 4.2.1 give the following upper bound:

$$\|e_t\|_{\mathcal{L}_\infty} \le \|H(s)(1 - C(s))\|_{\mathcal{L}_1} L \|e_t\|_{\mathcal{L}_\infty} + \|\eta_{\sigma t}\|_{\mathcal{L}_\infty}, \tag{4.102}$$

where $\eta_\sigma(t)$ is the signal with Laplace transform $\eta_\sigma(s) \triangleq H(s)\tilde{\sigma}(s)$. Using the expression for $\tilde{\sigma}(s)$ from (4.97), along with the expression for $y(s)$ from (4.62), and taking into consideration that

$$\hat{y}(s) = M(s)u(s) + c_m^\top(s\mathbb{I} - A_m)^{-1}\hat{\sigma}(s),$$

it follows that

$$\begin{aligned} \tilde{y}(s) &= c_m^\top(s\mathbb{I} - A_m)^{-1}\hat{\sigma}(s) - M(s)\sigma(s) \\ &= \frac{M(s)}{C(s)}\frac{C(s)}{M(s)}c_m^\top(s\mathbb{I} - A_m)^{-1}\hat{\sigma}(s) - \frac{M(s)}{C(s)}C(s)\sigma(s) = \frac{M(s)}{C(s)}\tilde{\sigma}(s). \end{aligned} \tag{4.103}$$

This implies that $\eta_\sigma(s)$ can be rewritten as

$$\eta_\sigma(s) = \frac{C(s)H(s)}{M(s)}\frac{M(s)}{C(s)}\tilde{\sigma}(s) = H_2(s)\tilde{y}(s),$$

and hence

$$\|\eta_{\sigma_t}\|_{\mathcal{L}_\infty} \le \|H_2(s)\|_{\mathcal{L}_1}\|\tilde{y}_t\|_{\mathcal{L}_\infty}.$$

Substituting this upper bound back into (4.102) completes the proof. □

Next, notice that using the definitions from (4.67), the transfer function $H_3(s)$ in (4.73) can be rewritten as

$$H_3(s) = \frac{-C_n(s)A_d(s)M_n(s)}{H_d(s)}, \tag{4.104}$$

where $H_d(s)$ was introduced in (4.69). Since $\deg(C_d(s) - C_n(s)) - \deg(C_n(s)) = d_r$, it can be checked straightforwardly that $H_3(s)$ is strictly proper. We notice from (4.68) and (4.104) that $H_3(s)$ has the same denominator as $H(s)$, and therefore it follows from (4.66) that $H_3(s)$ is stable. Since $H_3(s)$ is strictly proper and stable with relative degree d_r, $H_3(s)/M(s)$ is stable and proper and therefore its $\mathcal{L}_1$-norm is bounded.

Finally, consider the state transformation

$$\tilde{\xi} = \Lambda \tilde{x}\,.$$

It follows from (4.96) that

$$\dot{\tilde{\xi}}(t) = \Lambda A_m \Lambda^{-1} \tilde{\xi}(t) + \Lambda \hat{\sigma}(t) - \Lambda b_m \sigma(t), \quad \tilde{\xi}(0) = 0, \tag{4.105}$$

$$\tilde{y}(t) = \tilde{\xi}_1(t), \tag{4.106}$$

where $\tilde{\xi}_1(t)$ is the first element of $\tilde{\xi}(t)$.

Theorem 4.2.1 Consider the system in (4.61) and the $\mathcal{L}_1$ adaptive controller in (4.87), (4.88), and (4.89) subject to the $\mathcal{L}_1$-norm condition in (4.66). If we choose T_s to ensure

$$\gamma_0(T_s) < \bar{\gamma}_0\,, \tag{4.107}$$

where $\bar{\gamma}_0$ is an arbitrary positive constant introduced in (4.75), then

$$\|\tilde{y}\|_{\mathcal{L}_\infty} < \bar{\gamma}_0, \tag{4.108}$$

$$\|y_{\text{ref}} - y\|_{\mathcal{L}_\infty} \le \gamma_1, \quad \|u_{\text{ref}} - u\|_{\mathcal{L}_\infty} \le \gamma_2 \tag{4.109}$$

with

$$\gamma_1 \triangleq \frac{\|H_2(s)\|_{\mathcal{L}_1}}{1 - \|G(s)\|_{\mathcal{L}_1} L}\bar{\gamma}_0, \quad \gamma_2 \triangleq \|H_2(s)\|_{\mathcal{L}_1} L \gamma_1 + \left\| \frac{H_3(s)}{M(s)} \right\|_{\mathcal{L}_1} \bar{\gamma}_0\,.$$

Proof. First, we prove the bound in (4.108) by a contradiction argument. Since $\tilde{y}(0) = 0$ and $\tilde{y}(t)$ is continuous, then assuming the opposite implies that there exists t' such that

$$|\tilde{y}(t)| < \bar{\gamma}_0, \quad \forall\, 0 \le t < t',$$

$$|\tilde{y}(t')| = \bar{\gamma}_0,$$

which leads to

$$\|\tilde{y}_{t'}\|_{\mathcal{L}_\infty} = \bar{\gamma}_0\,. \tag{4.110}$$

Since $y(t) = y_{\text{ref}}(t) - e(t)$, the upper bound in (4.95) can be used to derive the following bound:

$$\begin{aligned} \|y_{t'}\|_{\mathcal{L}_\infty} &\le \|y_{\text{ref}t'}\|_{\mathcal{L}_\infty} + \|e_{t'}\|_{\mathcal{L}_\infty} \\ &\le \rho_r + \frac{\|C(s)H(s)/M(s)\|_{\mathcal{L}_1}}{1 - \|G(s)\|_{\mathcal{L}_1} L}\bar{\gamma}_0\,. \end{aligned} \tag{4.111}$$

Also, it follows from (4.101) and (4.103) that

$$\sigma(s) = H_1(s) r(s) - H_1(s)\tilde{y}(s)/M(s) + H_0(s) d(s),$$

and hence, the equality in (4.110) implies that

$$\|\sigma_{t'}\|_{\mathcal{L}_\infty} \leq \|H_1(s)\|_{\mathcal{L}_1}\|r\|_{\mathcal{L}_\infty} + \|H_1(s)/M(s)\|_{\mathcal{L}_1}\bar{\gamma}_0 + \|H_0(s)\|_{\mathcal{L}_1}(L\|y_{t'}\|_{\mathcal{L}_\infty} + L_0).$$

This, along with (4.111), leads to

$$\|\sigma_{t'}\|_{\mathcal{L}_\infty} \leq \Delta. \tag{4.112}$$

It follows from (4.105) that

$$\begin{aligned}\tilde{\xi}(iT_s+t) &= e^{\Lambda A_m \Lambda^{-1} t}\tilde{\xi}(iT_s) + \int_{iT_s}^{iT_s+t} e^{\Lambda A_m \Lambda^{-1}(iT_s+t-\tau)}\Lambda\hat{\sigma}(iT_s)d\tau \\ &\quad - \int_{iT_s}^{iT_s+t} e^{\Lambda A_m \Lambda^{-1}(iT_s+t-\tau)}\Lambda b_m \sigma(\tau)d\tau \\ &= e^{\Lambda A_m \Lambda^{-1} t}\tilde{\xi}(iT_s) + \int_0^t e^{\Lambda A_m \Lambda^{-1}(t-\tau)}\Lambda\hat{\sigma}(iT_s)d\tau \\ &\quad - \int_0^t e^{\Lambda A_m \Lambda^{-1}(t-\tau)}\Lambda b_m \sigma(iT_s+\tau)d\tau.\end{aligned} \tag{4.113}$$

Since

$$\tilde{\xi}(iT_s+t) = \begin{bmatrix} \tilde{y}(iT_s+t) \\ 0 \end{bmatrix} + \begin{bmatrix} 0 \\ \tilde{z}(iT_s+t) \end{bmatrix},$$

where $\tilde{z}(t) \triangleq [\tilde{\xi}_2(t),\ \tilde{\xi}_3(t), \ldots,\ \tilde{\xi}_n(t)]$, it follows from (4.113) that $\tilde{\xi}(\cdot)$ can be decomposed as

$$\tilde{\xi}(iT_s+t) = \chi(iT_s+t) + \zeta(iT_s+t), \tag{4.114}$$

where

$$\begin{aligned}\chi(iT_s+t) &\triangleq e^{\Lambda A_m \Lambda^{-1} t}\begin{bmatrix} \tilde{y}(iT_s) \\ 0 \end{bmatrix} + \int_0^t e^{\Lambda A_m \Lambda^{-1}(t-\tau)}\Lambda\hat{\sigma}(iT_s)d\tau, \\ \zeta(iT_s+t) &\triangleq e^{\Lambda A_m \Lambda^{-1} t}\begin{bmatrix} 0 \\ \tilde{z}(iT_s) \end{bmatrix} - \int_0^t e^{\Lambda A_m \Lambda^{-1}(t-\tau)}\Lambda b_m \sigma(iT_s+\tau)d\tau.\end{aligned} \tag{4.115}$$

Next we prove that for all $iT_s \leq t'$ one has

$$|\tilde{y}(iT_s)| \leq \varsigma(T_s), \tag{4.116}$$

$$\tilde{z}^\top(iT_s)P_2\tilde{z}(iT_s) \leq \alpha, \tag{4.117}$$

where $\varsigma(T_s)$ and α were defined in (4.79).

We start by noting that, since $\tilde{\xi}(0) = 0$, it is straightforward to show that $|\tilde{y}(0)| \leq \varsigma(T_s)$, $\tilde{z}^\top(0)P_2\tilde{z}(0) \leq \alpha$. Next, for arbitrary $(j+1)T_s \leq t'$, we prove that if

$$|\tilde{y}(jT_s)| \leq \varsigma(T_s), \tag{4.118}$$

$$\tilde{z}^\top(jT_s)P_2\tilde{z}(jT_s) \leq \alpha, \tag{4.119}$$

then the inequalities (4.118)–(4.119) hold for $j+1$ as well, which would imply that the bounds in (4.116)–(4.117) hold for all $iT_s \leq t'$.

To this end, assume that (4.118)–(4.119) hold for j and, in addition, that $(j+1)T_s \leq t'$. Then, it follows from (4.114) that

$$\tilde{\xi}((j+1)T_s) = \chi((j+1)T_s) + \zeta((j+1)T_s),$$

where

$$\chi((j+1)T_s) = \mathrm{e}^{\Lambda A_m \Lambda^{-1} T_s} \begin{bmatrix} \tilde{y}(jT_s) \\ 0 \end{bmatrix} + \int_0^{T_s} \mathrm{e}^{\Lambda A_m \Lambda^{-1}(T_s-\tau)} \Lambda \hat{\sigma}(jT_s) d\tau, \tag{4.120}$$

$$\zeta((j+1)T_s) = \mathrm{e}^{\Lambda A_m \Lambda^{-1} T_s} \begin{bmatrix} 0 \\ \tilde{z}(jT_s) \end{bmatrix} - \int_0^{T_s} \mathrm{e}^{\Lambda A_m \Lambda^{-1}(T_s-\tau)} \Lambda b_m \sigma(jT_s+\tau) d\tau. \tag{4.121}$$

Substituting the adaptive law from (4.88) in (4.120), we have

$$\chi((j+1)T_s) = 0. \tag{4.122}$$

It follows from (4.115) that $\zeta(t)$ is the solution to the following dynamics:

$$\dot{\zeta}(t) = \Lambda A_m \Lambda^{-1} \zeta(t) - \Lambda b_m \sigma(t), \tag{4.123}$$

$$\zeta(jT_s) = \begin{bmatrix} 0 \\ \tilde{z}(jT_s) \end{bmatrix}, \quad t \in [jT_s,\ (j+1)T_s]. \tag{4.124}$$

Consider now the function

$$V(t) = \zeta^\top(t)\Lambda^{-\top} P \Lambda^{-1} \zeta(t)$$

over $t \in [jT_s,\ (j+1)T_s]$. Since Λ is nonsingular and P is positive definite, $\Lambda^{-\top} P \Lambda^{-1}$ is positive definite and, hence, $V(t)$ is a positive-definite function. It follows from Lemma 4.2.1 and the relationship in (4.124) that

$$V(\zeta(jT_s)) = \tilde{z}^\top(jT_s)P_2\tilde{z}(jT_s),$$

which, along with the upper bound in (4.119), leads to

$$V(\zeta(jT_s)) \leq \alpha. \tag{4.125}$$

It follows from (4.123) that over $t \in [jT_s,\ (j+1)T_s]$

$$\begin{aligned} \dot{V}(t) &= \zeta^\top(t)\Lambda^{-\top} P \Lambda^{-1} \Lambda A_m \Lambda^{-1} \zeta(t) + \zeta^\top(t)\Lambda^{-\top} A_m^\top \Lambda^\top \Lambda^{-\top} P^\top \Lambda^{-1} \zeta(t) \\ &\quad - 2\zeta^\top(t)\Lambda^{-\top} P \Lambda^{-1} \Lambda b_m \sigma(t) \\ &= -\zeta^\top(t)\Lambda^{-\top} Q \Lambda^{-1} \zeta(t) - 2\zeta^\top(t)\Lambda^{-\top} P b_m \sigma(t). \end{aligned}$$

Using the upper bound from (4.112), one can derive over $t \in [jT_s,\ (j+1)T_s]$

$$\dot{V}(t) \le -\lambda_{\min}(\Lambda^{-\top} Q \Lambda^{-1}) \|\zeta(t)\|^2 + 2\|\zeta(t)\| \|\Lambda^{-\top} P b_m\| \Delta . \tag{4.126}$$

Notice that for all $t \in [jT_s,\ (j+1)T_s]$, if

$$V(t) > \alpha , \tag{4.127}$$

we have

$$\|\zeta(t)\| > \sqrt{\frac{\alpha}{\lambda_{\max}(\Lambda^{-\top} P \Lambda^{-1})}} = \frac{2\Delta \|\Lambda^{-\top} P b_m\|}{\lambda_{\min}(\Lambda^{-\top} Q \Lambda^{-1})} ,$$

and the upper bound in (4.126) yields

$$\dot{V}(t) < 0 . \tag{4.128}$$

It follows from (4.125), (4.127), and (4.128) that

$$V(t) \le \alpha , \quad \forall\, t \in [jT_s,\ (j+1)T_s] ,$$

and therefore

$$V((j+1)T_s) = \zeta^\top((j+1)T_s)(\Lambda^{-\top} P \Lambda^{-1})\zeta((j+1)T_s) \le \alpha . \tag{4.129}$$

Since

$$\tilde{\xi}((j+1)T_s) = \chi((j+1)T_s) + \zeta((j+1)T_s) , \tag{4.130}$$

the equality in (4.122) and the upper bound in (4.129) lead to the following inequality:

$$\tilde{\xi}^\top((j+1)T_s)(\Lambda^{-\top} P \Lambda^{-1})\tilde{\xi}((j+1)T_s) \le \alpha .$$

Using the result of Lemma 4.2.1, one can derive

$$\tilde{z}^\top((j+1)T_s) P_2 \tilde{z}((j+1)T_s) \le \tilde{\xi}^\top((j+1)T_s)(\Lambda^{-\top} P \Lambda^{-1})\tilde{\xi}((j+1)T_s) \le \alpha ,$$

which implies that the upper bound in (4.119) holds for $j+1$.

Next, it follows from (4.106), (4.122), and (4.130) that

$$\tilde{y}((j+1)T_s) = \mathbf{1}_1^\top \zeta((j+1)T_s) ,$$

and the definition of $\zeta((j+1)T_s)$ in (4.121) leads to the following expression:

$$\begin{aligned} \tilde{y}((j+1)T_s) = {} & \mathbf{1}_1^\top e^{\Lambda A_m \Lambda^{-1} T_s} \begin{bmatrix} 0 \\ \tilde{z}(jT_s) \end{bmatrix} \\ & - \mathbf{1}_1^\top \int_0^{T_s} e^{\Lambda A_m \Lambda^{-1}(T_s - \tau)} \Lambda b_m \sigma(jT_s + \tau) d\tau . \end{aligned}$$

The upper bounds in (4.112) and (4.119) yield the following upper bound:

$$
\begin{aligned}
|\tilde{y}((j+1)T_s)| &\leq \|\eta_2(T_s)\| \, \|\tilde{z}(jT_s)\| + \int_0^{T_s} |\mathbf{1}_1^\top e^{\Lambda A_m \Lambda^{-1}(T_s-\tau)} \Lambda b_m| \, |\sigma(jT_s+\tau)| d\tau \\
&\leq \|\eta_2(T_s)\| \sqrt{\frac{\alpha}{\lambda_{\max}(P_2)}} + \kappa(T_s)\Delta = \varsigma(T_s),
\end{aligned}
$$

where $\eta_2(T_s)$ and $\kappa(T_s)$ were defined in (4.77) and (4.78), and $\varsigma(T_s)$ was defined in (4.79). This confirms the upper bound in (4.118) for $j+1$. Hence, (4.116)–(4.117) hold for all $iT_s \leq t'$.

For all $iT_s + t \leq t'$, where $0 \leq t \leq T_s$, using the expression from (4.113) we can write that

$$
\begin{aligned}
\tilde{y}(iT_s + t) = \mathbf{1}_1^\top e^{\Lambda A_m \Lambda^{-1} t} \tilde{\xi}(iT_s) + \mathbf{1}_1^\top \int_0^t e^{\Lambda A_m \Lambda^{-1}(t-\tau)} \Lambda \hat{\sigma}(iT_s) d\tau \\
- \mathbf{1}_1^\top \int_0^t e^{\Lambda A_m \Lambda^{-1}(t-\tau)} \Lambda b_m \sigma(iT_s+\tau) d\tau \, .
\end{aligned}
$$

The upper bound in (4.112) and the definitions of $\eta_1(t)$, $\eta_2(t)$, $\eta_3(t)$, and $\eta_4(t)$ lead to the following upper bound:

$$
|\tilde{y}(iT_s+t)| \leq |\eta_1(t)| \, |\tilde{y}(iT_s)| + \|\eta_2(t)\| \, \|\tilde{z}(iT_s)\| + \eta_3(t)|\tilde{y}(iT_s)| + \eta_4(t)\Delta \, .
$$

Taking into consideration (4.116)–(4.117), and recalling the definitions of $\beta_1(T_s)$, $\beta_2(T_s)$, $\beta_3(T_s)$, $\beta_4(T_s)$ in (4.81)–(4.82), for all $0 \leq t \leq T_s$ and for arbitrary nonnegative integer i subject to $iT_s + t \leq t'$, we have

$$
|\tilde{y}(iT_s+t)| \leq \beta_1(T_s)\varsigma(T_s) + \beta_2(T_s)\sqrt{\frac{\alpha}{\lambda_{\max}(P_2)}} + \beta_3(T_s)\varsigma(T_s) + \beta_4(T_s)\Delta \, .
$$

Since the right-hand side coincides with the definition of $\gamma_0(T_s)$ in (4.83), for all $t \in [0,\ t']$ we have the bound

$$
|\tilde{y}(t)| \leq \gamma_0(T_s),
$$

which along with the design constraint on T_s introduced in (4.107) yields

$$
\|\tilde{y}_{t'}\|_{\mathcal{L}_\infty} < \bar{\gamma}_0 \, .
$$

This clearly contradicts the statement in (4.110). Therefore, $\|\tilde{y}\|_{\mathcal{L}_\infty} < \bar{\gamma}_0$, which proves (4.108). Further, it follows from Lemma 4.2.4 that

$$
\|e_t\|_{\mathcal{L}_\infty} \leq \frac{\|H_2(s)\|_{\mathcal{L}_1}}{1 - \|G(s)\|_{\mathcal{L}_1} L} \bar{\gamma}_0 ,
$$

which holds uniformly for all $t \geq 0$ and therefore leads to the first upper bound in (4.109).

To prove the second bound in (4.109), we note that from (4.98) and (4.100), it follows that

$$
u(s) = \frac{M(s)C(s)r(s) - M(s)\tilde{\sigma}(s) - C(s)A(s)d(s)}{C(s)A(s) + (1 - C(s))M(s)} ,
$$

which can be used along with the expression of $u_{\rm ref}(s)$ in (4.93) to derive

$$\begin{aligned} u(s)_{\rm ref} - u(s) &= -H_2(s)d_e(s) - \frac{H_3(s)}{C(s)}\tilde{\sigma}(s) \\ &= -H_2(s)d_e(s) - \frac{H_3(s)}{M(s)}\frac{M(s)}{C(s)}\tilde{\sigma}(s). \end{aligned} \tag{4.131}$$

Hence, it follows from (4.103) and (4.131) that

$$\|u_{\rm ref} - u\|_{\mathcal{L}_\infty} \leq L\|H_2(s)\|_{\mathcal{L}_1}\|y_{\rm ref} - y\|_{\mathcal{L}_\infty} + \|H_3(s)/M(s)\|_{\mathcal{L}_1}\|\tilde{y}\|_{\mathcal{L}_\infty},$$

which, along with the bound in (4.108) and the first bound in (4.109), leads to the second bound in (4.109). The proof is complete. □

Thus, the tracking error between $y(t)$ and $y_{\rm ref}(t)$, as well as between $u(t)$ and $u_{\rm ref}(t)$, is uniformly bounded by a constant proportional to T_s. This implies that during the transient phase, one can achieve arbitrary improvement of tracking performance by uniformly reducing T_s.

Remark 4.2.1 Notice that the parameter T_s is the fixed time step in the definition of the adaptive law. The adaptive parameters in $\hat{\sigma}(t) \in \mathbb{R}^n$ take constant values during $[iT_s, (i+1)T_s)$ for every $i = 0, 1, \ldots$. Reducing T_s imposes hardware (CPU) requirements, and Theorem 4.2.1 further implies that the performance limitations are consistent with the hardware limitations. This in turn is consistent with the results in Chapter 2, where improvement of the transient performance was achieved by increasing the adaptation rate in the projection-based adaptive laws.

Remark 4.2.2 We notice that the *ideal* control signal

$$u_{\rm id}(t) = r(t) - \sigma_{\rm ref}(t)$$

is the one that leads to the desired system response

$$y_{\rm id}(s) = M(s)r(s)$$

by canceling the uncertainties exactly. Thus, the reference system in (4.90)–(4.91) has a different response as compared to the *ideal* one. It cancels only the uncertainties within the bandwidth of $C(s)$, which can be selected compatible with the control channel specifications. This is exactly what one can hope to achieve with any feedback in the presence of uncertainties.

Remark 4.2.3 We notice that stability of $H(s)$ is equivalent to stabilization of $A(s)$ by

$$\frac{C(s)}{M(s)(1 - C(s))}. \tag{4.132}$$

Indeed, consider the closed-loop system, comprised of the system $A(s)$ and the negative feedback of (4.132). The closed-loop transfer function is

$$\frac{A(s)}{1 + A(s)\frac{C(s)}{M(s)(1-C(s))}}. \tag{4.133}$$

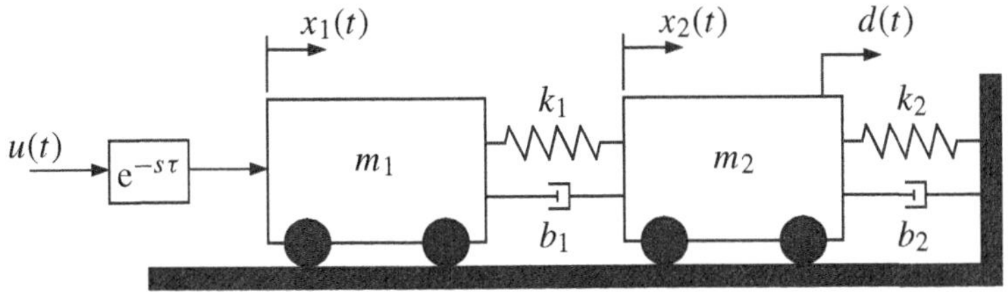

Figure 4.5: The two–cart MSD system.

Incorporating (4.67), one can verify that the denominator of the system in (4.133) is exactly $H_d(s)$. Hence, stability of $H(s)$ is equivalent to the stability of the closed-loop system in (4.133). This implies that the class of systems $A(s)$ that can be stabilized by the $\mathcal{L}_1$ adaptive output feedback controller, given by (4.87), (4.88), and (4.89), is not empty. Moreover, we note that methods from disturbance observer literature can be effectively used for parametrization of $C(s)$ that would achieve stabilization of $H(s)$ for a sufficiently broad class of systems $A(s)$ [155]. Research in this direction is underway.

Remark 4.2.4 We also notice that, while the feedback in (4.132) may stabilize the system in (4.61) for some classes of unknown nonlinearities, it will not ensure uniform transient performance in the presence of unknown $A(s)$. On the contrary, the $\mathcal{L}_1$ adaptive controller ensures *uniform transient performance* for a system's signals, both input and output, independent of the unknown nonlinearity and independent of $A(s)$.

Remark 4.2.5 Finally, it is important to mention that the output predictor of the $\mathcal{L}_1$ adaptive output feedback controller presented in this section can be modified similar to the state predictor of the full-state feedback $\mathcal{L}_1$ controllers introduced in Sections 2.1.6 and 3.3. This modification can be used to tune the frequency response and robustness margins of the closed-loop adaptive system. In general, this modification may require an increase in the adaptation sampling rate.

4.2.4 Simulation Example: Two-Cart Benchmark Problem

The two-cart mass-spring-damper example was originally proposed as a benchmark problem for robust control design. In a paper published in 2006, Fekri, Athans, and Pascoal [52] used a slightly modified version of the original two-cart system to illustrate the design methodology and performance of the Robust Multiple-Model Adaptive Control. Next, we will revisit the two-cart example with the $\mathcal{L}_1$ output feedback adaptive controller presented in this section. The reader will find additional explanations and simulations in [178].

The two-cart system is shown in Figure 4.5. The states $x_1(t)$ and $x_2(t)$ represent the absolute position of the two carts, whose masses are m_1 and m_2, respectively (only $x_2(t)$ is measured), $d(t)$ is a random colored disturbance force acting on the mass m_2, and $u(t)$ is the control force, which acts upon the mass m_1. The disturbance force $d(t)$ is modeled as a first-order (colored) stochastic process generated by driving a low-pass filter with continuous-time white noise $\xi(s)$, with zero-mean and unit intensity, i.e., $\Xi = 1$, as follows:

$$d(s) = \frac{\alpha}{s+\alpha}\xi(s), \quad \alpha > 0.$$

The overall state-space representation is

$$\dot{x}(t) = Ax(t) + Bu(t) + L\xi(t),$$
$$y(t) = Cx(t) + \theta(t),$$

where the state vector is

$$x^\top(t) = \begin{bmatrix} x_1(t) & x_2(t) & \dot{x}_1(t) & \dot{x}_2(t) & d(t) \end{bmatrix}^\top,$$

and

$$A = \begin{bmatrix} 0 & 0 & 1 & 0 & 0 \\ 0 & 0 & 0 & 1 & 0 \\ -\frac{k_1}{m_1} & \frac{k_1}{m_1} & -\frac{b_1}{m_1} & \frac{b_1}{m_1} & 0 \\ \frac{k_1}{m_2} & -\frac{k_1+k_2}{m_2} & \frac{b_2}{m_2} & -\frac{b_1+b_2}{m_2} & \frac{1}{m_2} \\ 0 & 0 & 0 & 0 & -\alpha \end{bmatrix},$$

$$B^\top = \begin{bmatrix} 0 & 0 & \frac{1}{m_1} & 0 & 0 \end{bmatrix}, \quad L^\top = \begin{bmatrix} 0 & 0 & 0 & 0 & \alpha \end{bmatrix},$$
$$C = \begin{bmatrix} 0 & 1 & 0 & 0 & 0 \end{bmatrix}, \tag{4.134}$$

while $\theta(t)$ is additive sensor noise affecting the only measurement, and it is modeled as white noise, independent of $\xi(t)$, and defined by

$$\mathrm{E}\{\theta(t)\} = 0, \quad \mathrm{E}\{\theta(t)\theta(\tau)\} = 10^{-6}\delta(t-\tau).$$

The following parameters in (4.134) are fixed and known:

$$m_1 = m_2 = 1, \; k_2 = 0.15, \; b_1 = b_2 = 0.1, \; \alpha = 0.1,$$

while the spring constant, k_1, is unknown with known upper and lower bounds

$$0.25 \leq k_1 \leq 1.75.$$

In addition to the uncertain spring stiffness, an *unmodeled* time delay τ, whose maximum possible value is 0.05 s, is assumed to be present in the control channel.

The control objective is to design a control law $u(t)$ so that the mass m_2 tracks a reference step signal $r(t)$ following a desired model, while minimizing the effects of the disturbance $d(t)$ and the sensor noise $\theta(t)$.

For application of the $\mathcal{L}_1$ adaptive output feedback controller to the two-cart example, the design procedure described in [169] leads to

$$M(s) = \frac{1}{s^3 + 1.4s^2 + 0.17s + 0.052},$$
$$C(s) = \frac{0.18s + 0.19}{s^5 + 2.8s^4 + 3.3s^2 + 2.0s^2 + 0.66s + 0.19},$$

while the sample time for adaptation is set to $T_s = 1$ ms.

Figures 4.6 to 4.8 show the response of the closed-loop system with the $\mathcal{L}_1$ adaptive output feedback controller to step-reference inputs of different amplitudes and for different

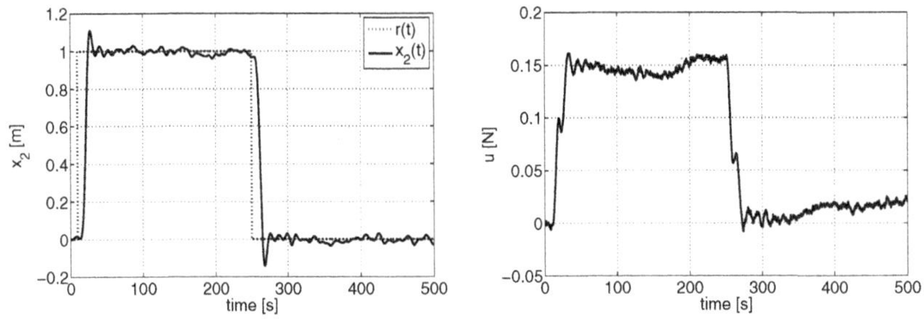

(a) system output $x_2(t)$ (b) control signal $u(t)$

Figure 4.6: Closed-loop response to a step input of 1 m with $k_1 = 0.25$ and $\tau = 0.05$ s.

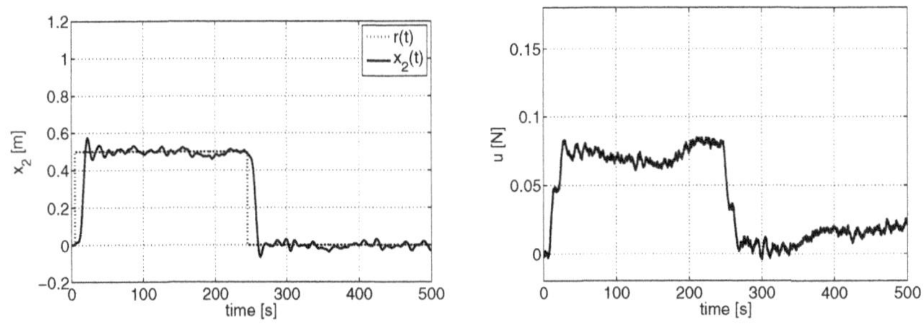

(a) system output $x_2(t)$ (b) control signal $u(t)$

Figure 4.7: Closed-loop response to a step input of 0.5 m with $k_1 = 0.25$ and $\tau = 0.05$ s.

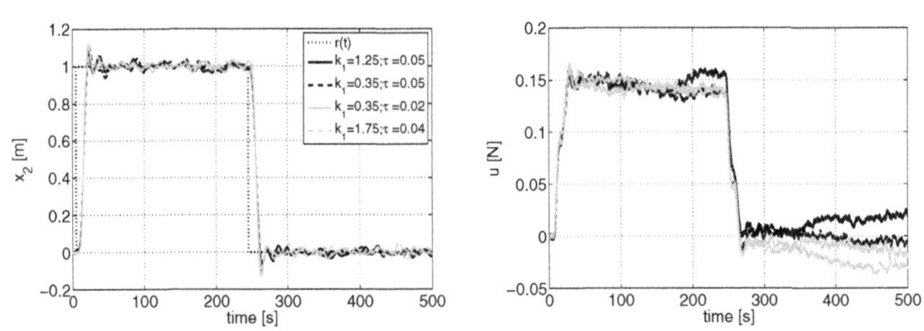

(a) system output $x_2(t)$ (b) control signal $u(t)$

Figure 4.8: Closed-loop response to a step input of 1 m for different values of the unknowns k_1 and τ.

values of the unknown parameters k_1 and τ. As one can see, the $\mathcal{L}_1$ adaptive controller drives the mass m_2 to the desired position in about 15 s for arbitrary values of the unknown parameters while minimizing the effects of both the disturbance and the sensor noise present in the system. Also, for a scaled reference input, the closed-loop system achieves scaled transient response as expected (see Figures 4.6 and 4.7).

Chapter 5

$\mathcal{L}_1$ Adaptive Controller for Time-Varying Reference Systems

Because of growing complexity of engineering systems, the closed-loop system may need to meet different performance specifications at different points of the operational envelope. This implies that the *desired reference system behavior is time varying*. A classical example to support this statement is flight control system design, which relies on gain scheduling of the control parameters across the flight envelope to meet different performance specifications for different flight regimes. This section presents the $\mathcal{L}_1$ adaptive control architecture for time-varying reference systems. The presented solution ensures that in the presence of fast adaptation, a single design of the $\mathcal{L}_1$ adaptive controller leads to uniform performance bounds with respect to a time-varying reference system, without the need for gain scheduling.

5.1 $\mathcal{L}_1$ Adaptive Controller for Linear Time-Varying Systems

In this section, we derive the $\mathcal{L}_1$ adaptive controller for LTV systems with its performance specifications given via another LTV system. We prove that the fast adaptation ability of the $\mathcal{L}_1$ adaptive controller ensures that the signals of the closed-loop system remain close to the corresponding signals of a bounded LTV reference system. We follow the framework of [171], where the methodology was first outlined and was applied to an aerial refueling problem for a racetrack maneuver.

5.1.1 Problem Formulation

Consider the class of systems

$$\begin{aligned} \dot{x}(t) &= A_m(t)x(t) + b(t)\big(\omega u(t) + \theta^\top(t)x(t) + \sigma(t)\big), \quad x(0) = x_0, \\ y(t) &= c^\top x(t), \end{aligned} \tag{5.1}$$

where $x(t) \in \mathbb{R}^n$ is the system state (measured); $A_m(t) \in \mathbb{R}^{n\times n}$ and $b(t) \in \mathbb{R}^n$ are a known time-varying matrix and a known vector, respectively; $c \in \mathbb{R}^n$ is a known constant vector;

$\omega \in \mathbb{R}$ is a constant unknown parameter representing the uncertainty in the system input gain; $\theta(t) \in \mathbb{R}^n$ is the unknown system parameter vector while $\sigma(t) \in \mathbb{R}$ is the disturbance; $y(t) \in \mathbb{R}$ is the system output; and $u(t) \in \mathbb{R}$ is the control input.

The system above verifies the following assumptions.

Assumption 5.1.1 (Uniform Asymptotic Stability of Desired System) The matrix $A_m(t)$ is continuously differentiable and there exist positive constants $\mu_A > 0$, $d_A > 0$, and $\mu_\lambda > 0$, such that for all $t \geq 0$, $\|A_m(t)\|_\infty \leq \mu_A$, $\|\dot{A}_m(t)\|_\infty \leq d_A$, and $\mathrm{Re}[\lambda_i(A_m(t))] \leq -\mu_\lambda$ $\forall\, i = 1, \ldots, n$, where $\lambda_i(A_m(t))$ is a pointwise eigenvalue of $A_m(t)$. Further, for all $t \geq 0$, the equilibrium of the state equation

$$\dot{x} = A_m x(t), \quad x(t_0) = x_0$$

is exponentially stable, and the solution of

$$A_m^\top(t)P(t) + P(t)A_m(t) = -\mathbb{I}$$

satisfies $\|\dot{P}(t)\|_\infty < 1$.

Remark 5.1.1 The existence of d_A in Assumption 5.1.1 is guaranteed by Lemma A.6.2.

Assumption 5.1.2 (Uniform boundedness of $b(t)$ and its derivative) There exist positive constants $\mu_b, d_b > 0$, such that $\|b(t)\| \leq \mu_b$ and $\|\dot{b}(t)\| \leq d_b$.

Assumption 5.1.3 (Strong controllability of $(A(t), b(t))$) The pair $(A(t), b(t))$ is strongly controllable.

Assumption 5.1.4 (Uniform boundedness of unknown parameters) Let

$$\omega \in \Omega = [\omega_l, \omega_u], \quad \theta(t) \in \Theta, \quad |\sigma(t)| \leq \Delta, \quad \forall\, t \geq 0, \tag{5.2}$$

where $0 < \omega_l < \omega_u$ are given known upper and lower bounds, Θ is a known convex compact set, and $\Delta \in \mathbb{R}^+$ is a known (conservative) bound of $\sigma(t)$.

Assumption 5.1.5 (Uniform boundedness of the rate of variation of parameters) We assume that $\theta(t)$ and $\sigma(t)$ are continuously differentiable, and their derivatives are uniformly bounded:

$$\|\dot{\theta}(t)\| \leq d_\theta < \infty, \quad |\dot{\sigma}(t)| \leq d_\sigma < \infty, \quad \forall\, t \geq 0.$$

In this section we present the $\mathcal{L}_1$ adaptive controller, which ensures that the system output $y(t)$ follows a given bounded piecewise-continuous reference signal $r(t) \in \mathbb{R}$ with uniform and quantifiable transient and steady-state performance bounds.

5.1.2 $\mathcal{L}_1$ Adaptive Control Architecture

Definitions and $\mathcal{L}_1$-Norm Sufficient Condition for Stability

Let $\mathcal{H}$ be the input-to-state map of the system

$$\dot{x}(t) = A_m(t)x(t) + b(t)u(t), \quad x(0) = 0.$$

Then, the system in (5.1) can be rewritten as

$$x = \mathcal{H}\left(\omega u + \theta^\top x + \sigma\right) + x_{\text{in}},$$

where $x_{\text{in}}(t)$ is the solution of

$$\dot{x}_{\text{in}}(t) = A_m(t) x_{\text{in}}(t), \quad x_{\text{in}}(0) = x_0 .$$

Notice that from Assumption 5.1.1 and Lemma A.6.2, it follows that $\|x_{\text{in}}\|_{\mathcal{L}_\infty}$ is bounded.

The design of the $\mathcal{L}_1$ adaptive controller proceeds by considering a positive feedback gain $k > 0$ and a strictly proper stable transfer function $D(s)$, which lead, for all $\omega \in \Omega$, to a strictly proper stable

$$C(s) \triangleq \frac{\omega k D(s)}{1 + \omega k D(s)} \tag{5.3}$$

with DC gain $C(0) = 1$. Let $\mathfrak{C}$ denote the input-output map for the transfer function $C(s)$.

For the proofs of stability and performance bounds, the choice of k and $D(s)$ needs to ensure that the following $\mathcal{L}_1$-norm condition holds:

$$\|\mathcal{G}\|_{\mathcal{L}_1} L < 1, \qquad \mathcal{G} \triangleq \mathcal{H}(1 - \mathfrak{C}), \tag{5.4}$$

where

$$L \triangleq \max_{\theta \in \Theta} \|\theta\|_1 , \tag{5.5}$$

with Θ being the compact set defined in (5.2).

The elements of the $\mathcal{L}_1$ adaptive controller are introduced next.

State Predictor

We consider the following state predictor with time-varying $A_m(t)$ and $b(t)$:

$$\begin{aligned} \dot{\hat{x}}(t) &= A_m(t)\hat{x}(t) + b(t)\left(\hat{\omega}(t)u(t) + \hat{\theta}^\top(t)x(t) + \hat{\sigma}(t)\right), \quad \hat{x}(0) = x_0 , \\ \hat{y}(t) &= c^\top \hat{x}(t), \end{aligned} \tag{5.6}$$

where $\hat{x}(t) \in \mathbb{R}^n$ is the state vector of the predictor, while $\hat{\omega}(t), \hat{\sigma}(t) \in \mathbb{R}$ and $\hat{\theta}(t) \in \mathbb{R}^n$ are the adaptive estimates.

Adaptation Laws

The adaptive laws are given by

$$\begin{aligned} \dot{\hat{\omega}}(t) &= \Gamma \text{Proj}\left(\hat{\omega}(t), -\tilde{x}^\top(t)P(t)b(t)u(t)\right), \quad \hat{\omega}(0) = \hat{\omega}_0 , \\ \dot{\hat{\theta}}(t) &= \Gamma \text{Proj}\left(\hat{\theta}(t), -\tilde{x}^\top(t)P(t)b(t)x(t)\right), \quad \hat{\theta}(0) = \hat{\theta}_0 , \\ \dot{\hat{\sigma}}(t) &= \Gamma \text{Proj}\left(\hat{\sigma}(t), -\tilde{x}^\top(t)P(t)b(t)\right), \quad \hat{\sigma}(0) = \hat{\sigma}_0 , \end{aligned} \tag{5.7}$$

where $\tilde{x}(t) \triangleq \hat{x}(t) - x(t)$, $\Gamma \in \mathbb{R}^+$ is the adaptation gain, $\text{Proj}(\cdot,\cdot)$ denotes the projection operator defined in Appendix B, and the symmetric positive definite matrix $P(t) = P^\top(t) >$

0 was defined in Assumption 5.1.1. The projection operator ensures that $\hat{\omega}(t) \in \Omega$, $\hat{\theta}(t) \in \Theta$, $|\hat{\sigma}(t)| \leq \Delta$.

Control Law

The control law is generated as the output of the (feedback) system

$$u(s) = -kD(s)(\hat{\eta}(s) - r_g(s)), \tag{5.8}$$

where $\hat{\eta}(s)$ is the Laplace transform of $\hat{\eta}(t) \triangleq \hat{\omega}(t)u(t) + \hat{\theta}^\top(t)x(t) + \hat{\sigma}(t)$, while $r_g(s)$ is the Laplace transform of

$$r_g(t) \triangleq k_g(t)r(t), \quad k_g(t) \triangleq -1/(c^\top A_m^{-1}(t)b(t)).$$

The $\mathcal{L}_1$ adaptive controller is defined via (5.6), (5.7), and (5.8), subject to the $\mathcal{L}_1$-norm condition in (5.4).

5.1.3 Analysis of $\mathcal{L}_1$ Adaptive Controller

Closed-Loop Reference System

Consider the following closed-loop reference system:

$$\begin{aligned} \dot{x}_{\text{ref}}(t) &= A_m(t)x_{\text{ref}}(t) + b(t)\big(\omega u_{\text{ref}}(t) + \theta^\top(t)x_{\text{ref}}(t) + \sigma(t)\big), \\ u_{\text{ref}}(s) &= \frac{C(s)}{\omega}(r_g(s) - \eta_{\text{ref}}(s)), \\ y_{\text{ref}}(t) &= c^\top x_{\text{ref}}(t), \quad x_{\text{ref}}(0) = x_0, \end{aligned} \tag{5.9}$$

where $x_{\text{ref}}(t) \in \mathbb{R}^n$ is the reference system state vector and $\eta_{\text{ref}}(s)$ is the Laplace transform of $\eta_{\text{ref}}(t) \triangleq \theta^\top(t)x_{\text{ref}}(t) + \sigma(t)$. Notice that this reference system is not implementable as it depends upon the unknown quantities ω, $\theta(t)$, and $\sigma(t)$. Therefore, it is used only for analysis purposes. In particular, it helps to establish stability and performance bounds of the closed-loop adaptive system. The next lemma proves the stability of this closed-loop reference system.

Lemma 5.1.1 If the choice of k and $D(s)$ verifies the $\mathcal{L}_1$-condition in (5.4), then the closed-loop reference system in (5.9) is BIBS stable with respect to $r(t)$ and x_0.

Proof. The closed-loop reference system in (5.9) can be rewritten as

$$x_{\text{ref}} = \mathcal{G}\eta_{\text{ref}} + \mathcal{H}\mathcal{C}r_g + x_{\text{in}}. \tag{5.10}$$

Lemmas A.6.2 and A.7.5 imply

$$\|x_{\text{ref}\tau}\|_{\mathcal{L}_\infty} \leq \|\mathcal{G}\|_{\mathcal{L}_1}\|\eta_{\text{ref}\tau}\|_{\mathcal{L}_\infty} + \|\mathcal{H}\mathcal{C}\|_{\mathcal{L}_1}\|r_g\|_{\mathcal{L}_\infty} + \|x_{\text{in}}\|_{\mathcal{L}_\infty}. \tag{5.11}$$

The definition of $\eta_{\text{ref}}(t)$ along with (5.5) gives

$$\|\eta_{\text{ref}\tau}\|_{\mathcal{L}_\infty} \leq L\|x_{\text{ref}\tau}\|_{\mathcal{L}_\infty} + \|\sigma_\tau\|_{\mathcal{L}_\infty}.$$

Substituting it into (5.11) and solving with respect to $\|x_{\text{ref}\tau}\|_{\mathcal{L}_\infty}$ leads to

$$\|x_{\text{ref}\tau}\|_{\mathcal{L}_\infty} \leq \frac{\|\mathcal{H}\mathcal{C}\|_{\mathcal{L}_1}\|r_g\|_{\mathcal{L}_\infty} + \|\mathcal{G}\|_{\mathcal{L}_1}\Delta + \|x_{\text{in}}\|_{\mathcal{L}_\infty}}{1 - \|\mathcal{G}\|_{\mathcal{L}_1} L}.$$

Since k and $D(s)$ verify the $\mathcal{L}_1$-norm condition in (5.4), $\|x_{\text{ref}}\|_{\mathcal{L}_\infty}$ is uniformly bounded, and hence, the closed-loop system in (5.9) is BIBS stable. □

Transient and Steady-State Performance

The error dynamics between the state predictor and the plant are given by

$$\dot{\tilde{x}}(t) = A_m(t)\tilde{x}(t) + b(t)\left(\tilde{\omega}(t)u(t) + \tilde{\theta}^\top(t)x(t) + \tilde{\sigma}(t)\right), \quad \tilde{x}(0) = 0, \tag{5.12}$$

where $\tilde{\omega}(t) \triangleq \hat{\omega}(t) - \omega$, $\tilde{\theta}(t) \triangleq \hat{\theta}(t) - \theta(t)$, and $\tilde{\sigma}(t) \triangleq \hat{\sigma}(t) - \sigma(t)$.

Lemma 5.1.2 The prediction error $\tilde{x}(t)$ in (5.12) is bounded,

$$\|\tilde{x}\|_{\mathcal{L}_\infty} \leq \sqrt{\frac{\theta_m}{\lambda_{P_{\min}}\Gamma}}, \tag{5.13}$$

where

$$\theta_m \triangleq (\omega_u - \omega_l)^2 + 4\max_{\theta\in\Theta}\|\theta\|^2 + 4\Delta^2 + 4\frac{\lambda_{P_{\max}}}{1-\epsilon_P}\left(\max_{\theta\in\Theta}\|\theta\|d_\theta + \Delta d_\sigma\right), \tag{5.14}$$

and

$$\lambda_{P_{\min}} \triangleq \inf_{\substack{t\in[0,\infty),\\ i=1\ldots n}} \lambda_i(P(t)), \quad \lambda_{P_{\max}} \triangleq \sup_{\substack{t\in[0,\infty),\\ i=1\ldots n}} \lambda_i(P(t)).$$

Proof. Consider the following Lyapunov function candidate:

$$V(\tilde{x}(t),\tilde{\omega}(t),\tilde{\theta}(t),\tilde{\sigma}(t)) = \tilde{x}^\top(t)P(t)\tilde{x}(t) + \frac{1}{\Gamma}(\tilde{\omega}^2(t) + \tilde{\theta}^\top(t)\tilde{\theta}(t) + \tilde{\sigma}^2(t)). \tag{5.15}$$

Using the adaptation laws in (5.7) and Property B.2 of the Proj$(\cdot,\cdot)$ operator, we can derive the following upper bound on the derivative of the Lyapunov function:

$$\begin{aligned}
\dot{V}(t) &= \tilde{x}^\top(t)\big(P(t)A_m(t) + A_m^\top(t)P(t) + \dot{P}(t)\big)\tilde{x}^\top(t) + 2\tilde{x}^\top(t)P(t)b(t)\tilde{\omega}(t)u(t)\\
&\quad + 2\tilde{x}^\top(t)P(t)b(t)\tilde{\theta}^\top(t)x(t) + 2\tilde{x}^\top(t)P(t)b(t)\tilde{\sigma}(t)\\
&\quad + \frac{2}{\Gamma}(\tilde{\omega}(t)\dot{\hat{\omega}}(t) + \tilde{\theta}^\top(t)\dot{\hat{\theta}}(t) + \tilde{\sigma}(t)\dot{\hat{\sigma}}(t)) - \frac{2}{\Gamma}(\tilde{\theta}^\top(t)\dot{\theta}(t) + \tilde{\sigma}(t)\dot{\sigma}(t))\\
&= -\tilde{x}^\top(t)\big(\mathbb{I} - \dot{P}(t)\big)\tilde{x}^\top(t) - \frac{2}{\Gamma}(\tilde{\theta}^\top(t)\dot{\theta}(t) + \tilde{\sigma}(t)\dot{\sigma}(t))\\
&\quad + 2\tilde{\omega}(t)\left(\tilde{x}^\top(t)P(t)b(t)u(t) + \text{Proj}\left(\hat{\omega}(t), -\tilde{x}^\top(t)P(t)b(t)u(t)\right)\right)\\
&\quad + 2\tilde{\theta}^\top(t)\left(\tilde{x}^\top(t)P(t)b(t)x(t) + \text{Proj}\left(\hat{\theta}(t), -\tilde{x}^\top(t)P(t)b(t)x(t)\right)\right)\\
&\quad + 2\tilde{\sigma}(t)\left(\tilde{x}^\top(t)P(t)b(t) + \text{Proj}\left(\hat{\sigma}(t), -\tilde{x}^\top(t)P(t)b(t)\right)\right)\\
&\leq -\tilde{x}^\top(t)(\mathbb{I} - \dot{P}(t))\tilde{x}(t) + \frac{2}{\Gamma}|\tilde{\theta}^\top(t)\dot{\theta}(t) + \tilde{\sigma}(t)\dot{\sigma}(t)|.
\end{aligned}$$

Projection ensures that $\hat{\theta}(t) \in \Theta$, $\hat{\omega}(t) \in \Omega$, and $|\hat{\sigma}(t)| \le \Delta$, for all $t \ge 0$, and therefore

$$\max_{t\ge 0}\left(\frac{1}{\Gamma}\left(\tilde{\theta}^\top(t)\tilde{\theta}(t)+\tilde{\omega}^2(t)+\tilde{\sigma}^2(t)\right)\right) \le \frac{1}{\Gamma}\left(4\max_{\theta\in\Theta}\|\theta\|^2+4\Delta^2+(\omega_u-\omega_l)^2\right)$$

for all $t \ge 0$. If at some $t_1 > 0$ one has $V(t_1) > \theta_m/\Gamma$, then it follows from (5.14) and (5.15) that

$$\tilde{x}^\top(t_1)P(t_1)\tilde{x}(t_1) > 4\frac{\lambda_{P_{\max}}}{\Gamma(1-\epsilon_P)}(\max_{\theta\in\Theta}\|\theta\|d_\theta+\Delta d_\sigma).$$

Then, Lemma A.6.2 implies

$$\tilde{x}^\top(t_1)(\mathbb{I}-\dot{P}(t_1))\tilde{x}(t_1) \ge \frac{1-\epsilon_P}{\lambda_{P_{\max}}}\tilde{x}^\top(t_1)P(t_1)\tilde{x}(t_1) > \frac{4}{\Gamma}(\max_{\theta\in\Theta}\|\theta\|d_\theta+\Delta d_\sigma). \tag{5.16}$$

Notice that

$$\frac{1}{\Gamma}|\tilde{\theta}(t)\dot{\theta}(t)+\tilde{\sigma}(t)\dot{\sigma}(t)| \le \frac{2}{\Gamma}(\max_{\theta\in\Theta}\|\theta\|d_\theta+\Delta d_\sigma), \qquad \forall\, t\ge 0,$$

which, along with the bound in (5.16), leads to

$$\dot{V}(t_1) < 0.$$

Since $\hat{x}(0) = x(0)$, we can verify that

$$V(0) \le \frac{1}{\Gamma}\left(4\max_{\theta\in\Theta}\|\theta\|^2+4\Delta^2+(\omega_u-\omega_l)^2\right) < \frac{\theta_m}{\Gamma}.$$

Therefore, we have

$$V(t) \le \frac{\theta_m}{\Gamma}, \qquad \forall\, t \ge 0.$$

Since $\lambda_{P_{\min}}\|\tilde{x}(t)\|^2 \le \tilde{x}^\top(t)P\tilde{x}(t) \le V(t)$, then

$$\|\tilde{x}(t)\| \le \sqrt{\frac{\theta_m}{\lambda_{P_{\min}}\Gamma}}.$$

The result in (5.13) follows from the fact that this bound holds uniformly for all $t \ge 0$. □

Theorem 5.1.1 Given the system in (5.1) and the $\mathcal{L}_1$ adaptive controller defined via (5.6), (5.7), and (5.8), subject to the $\mathcal{L}_1$-norm condition in (5.4), we have

$$\|x_{\text{ref}}-x\|_{\mathcal{L}_\infty} \le \gamma_1, \tag{5.17}$$

$$\|u_{\text{ref}}-u\|_{\mathcal{L}_\infty} \le \gamma_2, \tag{5.18}$$

where

$$\gamma_1 \triangleq \frac{\|\mathcal{H}\|_{\mathcal{L}_1}\kappa_0}{1-\|\mathcal{G}\|_{\mathcal{L}_1}L}\sqrt{\frac{\theta_m}{\lambda_{P_{\min}}\Gamma}}, \tag{5.19}$$

$$\gamma_2 \triangleq \left\|\frac{C(s)}{\omega}\right\|_{\mathcal{L}_1}L\gamma_1+\frac{\kappa_0}{\omega}\sqrt{\frac{\theta_m}{\lambda_{P_{\min}}\Gamma}}, \tag{5.20}$$

$$\kappa_0 \triangleq \left(\sum_{i=0}^{n-1}\|\mathfrak{C}a_{i+1}\|_{\mathcal{L}_1}\left\|\frac{s^i}{\bar{c}_n s^{n-1}+\cdots+\bar{c}_1}\right\|_{\mathcal{L}_1}+\left\|\frac{C(s)s^n}{\bar{c}_n s^{n-1}+\cdots+\bar{c}_1}\right\|_{\mathcal{L}_1}\right)\|\bar{c}^\top T\|_{\mathcal{L}_\infty},$$

and $\bar{c}_i$, $i = 1, \ldots, n$, are the coefficients of an arbitrary Hurwitz polynomial $p(s) \triangleq \bar{c}_n s^{n-1} + \cdots + \bar{c}_1$, while $T(t)$ and $a_i(t)$ are, respectively, the transformation matrix and the coefficients of the characteristic polynomial for the system given in (5.1), defined according to Lemma A.11.2.

Proof. Let

$$\eta(t) \triangleq \theta^\top(t)x(t) + \sigma(t), \quad \tilde{\eta}(t) \triangleq \tilde{\omega}(t)u(t) + \tilde{\theta}^\top(t)x(t) + \tilde{\sigma}(t).$$

It follows from (5.8) that

$$u(s) = -KD(s)(\omega u(s) + \eta(s) - r_g(s) + \tilde{\eta}(s)),$$

which can be rewritten as

$$u(s) = -\frac{C(s)}{\omega}(\eta(s) + \tilde{\eta}(s) - r_g(s)). \tag{5.21}$$

Then, the system in (5.1) takes the form

$$x = \mathcal{G}\eta + \mathcal{H}\mathcal{C}(r_g - \tilde{\eta}) + x_{\text{in}}.$$

The expression above, together with (5.10), leads to

$$x_{\text{ref}} - x = \mathcal{G}\eta_e + \mathcal{H}\mathcal{C}\tilde{\eta}, \quad \eta_e(t) \triangleq \theta^\top(t)(x_{\text{ref}}(t) - x(t)). \tag{5.22}$$

Lemma A.7.5 gives the following upper bound:

$$\|(x_{\text{ref}} - x)_\tau\|_{\mathcal{L}_\infty} \leq \|\mathcal{G}\|_{\mathcal{L}_1}\|\eta_{e\tau}\|_{\mathcal{L}_\infty} + \|\mathcal{H}\|_{\mathcal{L}_1}\|(\mathcal{C}\tilde{\eta})_\tau\|_{\mathcal{L}_\infty}. \tag{5.23}$$

From the definition of L in (5.5), it follows that $\|\theta^\top(x_{\text{ref}} - x)_\tau\|_{\mathcal{L}_\infty} \leq L\|(x_{\text{ref}} - x)_\tau\|_{\mathcal{L}_\infty}$, and hence $\|\eta_{e\tau}\|_{\mathcal{L}_\infty} \leq L\|(x_{\text{ref}} - x)_\tau\|_{\mathcal{L}_\infty}$. Substituting this back into (5.23), and solving for $\|(x_{\text{ref}} - x)_\tau\|_{\mathcal{L}_\infty}$, one gets

$$\|(x_{\text{ref}} - x)_\tau\|_{\mathcal{L}_\infty} \leq \frac{\|\mathcal{H}\|_{\mathcal{L}_1}}{1 - \|\mathcal{G}\|_{\mathcal{L}_1}L}\|(\mathcal{C}\tilde{\eta})_\tau\|_{\mathcal{L}_\infty}.$$

Applying Lemma A.12.2 to the last term of this bound, we obtain

$$\|(x_{\text{ref}} - x)_\tau\|_{\mathcal{L}_\infty} \leq \frac{\|\mathcal{H}\|_{\mathcal{L}_1}}{1 - \|\mathcal{G}\|_{\mathcal{L}_1}L}\kappa_0\|\tilde{x}_\tau\|_{\mathcal{L}_\infty}.$$

Taking into account the upper bound from Lemma 5.1.2, one gets

$$\|(x_{\text{ref}} - x)_\tau\|_{\mathcal{L}_\infty} \leq \frac{\|\mathcal{H}\|_{\mathcal{L}_1}\kappa_0}{1 - \|\mathcal{G}\|_{\mathcal{L}_1}L}\sqrt{\frac{\theta_m}{\lambda_{P_{\min}}\Gamma}},$$

which holds uniformly for all $t \geq 0$, leading to the bound in (5.17).

To prove the bound in (5.18), we notice that from (5.9) and (5.21) one can derive

$$u_{\text{ref}}(s) - u(s) = -\frac{C(s)}{\omega}(\eta_e(s) - \tilde{\eta}(s)).$$

It follows from Lemma A.7.5 that

$$\|(u_{\text{ref}}-u)_\tau\|_{\mathcal{L}_\infty} \leq \left\|\frac{C(s)}{\omega}\right\|_{\mathcal{L}_1} L\|(x_{\text{ref}}-x)_\tau\|_{\mathcal{L}_\infty} + \frac{1}{\omega}\|(\mathfrak{C}\tilde{\eta})_\tau\|_{\mathcal{L}_\infty}.$$

Using the upper bound on $\|(x_{\text{ref}}-x)_\tau\|_{\mathcal{L}_\infty}$ and applying Lemma A.12.2 to the last term of this bound, we obtain

$$\|(u_{\text{ref}}-u)_\tau\|_{\mathcal{L}_\infty} \leq \left\|\frac{C(s)}{\omega}\right\|_{\mathcal{L}_1} L\gamma_1 + \frac{\kappa_0}{\omega}\sqrt{\frac{\theta_m}{\lambda_{P_{\min}}\Gamma}},$$

which leads to the bound in (5.18) and completes the proof. □

Remark 5.1.2 It follows from the definition of γ_1 and γ_2 in (5.19) and (5.20) that one can achieve arbitrary desired performance bounds for a system's signals, both input and output, simultaneously by increasing the adaptive gain.

5.1.4 Simulation Example

To verify numerically the results proved in this section, we consider the following second-order dynamics:

$$\begin{aligned}\dot{x}(t) &= A_m(t)x(t) + b(t)(\omega u(t) + \theta(t)x(t) + \sigma(t)), \quad x(0) = x_0,\\ y(t) &= c^\top x(t),\end{aligned} \tag{5.24}$$

where

$$A_m(t) = \begin{bmatrix} 0 & 1 \\ -\omega_m^2(t) & -2\zeta\omega_m(t)\end{bmatrix}, \quad b(t) = \begin{bmatrix} 0 \\ \omega_m^2(t)\end{bmatrix}, \quad c = \begin{bmatrix} 1 \\ 0\end{bmatrix}^\top,$$

with $\zeta = 0.7$, and

$$\omega_m(t) = 1 + 0.4\sin\left(\frac{\pi}{40}t\right).$$

In the simulations, we consider the system uncertainties ω and $\theta(t)$,

$$\begin{aligned}\omega_1 &= 0.8, & \theta_1(t) &= [2,\ 1]^\top,\\ \omega_2 &= 1.2, & \theta_2(t) &= [2+\sin(0.3t),\ 0.5\sin(0.5t)]^\top,\end{aligned}$$

and the disturbances $\sigma(t)$,

$$\begin{aligned}\sigma_1(t) &= 1 + 3\cos(0.5t),\\ \sigma_2(t) &= 3 + \sin(0.5t) + 0.5\sin(t),\end{aligned}$$

so that the compact sets can be conservatively chosen as $\Omega = [0.1,\ 3]$, $\Theta = \{\vartheta = [\vartheta_1,\ \vartheta_2]^\top \in \mathbb{R}^2 : \vartheta_i \in [-4,\ 4], \text{ for all } i = 1,2\}$, $\Delta = 50$.

We implement the $\mathcal{L}_1$ adaptive controller according to (5.6), (5.7), and (5.8). In the implementation of the control law we use the filter and feedback gain

$$k = 60, \quad D(s) = \frac{1}{s},$$

and we set the adaptation gain to $\Gamma = 10^5$. The time-varying desired system is given by

$$\dot{x}_{\mathrm{id}}(t) = A_m(t)x_{\mathrm{id}}(t) + b(t)k_g(t)r(t)\,,$$
$$y_{\mathrm{id}}(t) = c^\top x_{\mathrm{id}}(t)\,.$$

First, we verify the assumptions and the $\mathcal{L}_1$-norm upper bound from (5.4). From the problem formulation in (5.24), one can easily see that Assumptions 5.1.2, 5.1.3, 5.1.4, and 5.1.5 are satisfied for the class of uncertainties and disturbances, introduced above. To verify Assumption 5.1.1, we solve the Lyapunov equation $A_m^\top(t)P(t) + P(t)A_m(t) = -\mathbb{I}$ for $P(t)$ and obtain

$$P(t) = \begin{bmatrix} \frac{\zeta}{\omega_m(t)} + \frac{\omega_m(t)}{4\zeta} + \frac{1}{4\zeta\omega_m(t)} & \frac{1}{2\omega_m^2(t)} \\ \frac{1}{2\omega_m^2(t)} & \frac{1}{4\zeta\omega_m(t)}\left(1 + \frac{1}{\omega_m^2(t)}\right) \end{bmatrix},$$

which leads to

$$\dot{P}(t) = \dot{\omega}_m(t)\begin{bmatrix} \frac{1}{4\zeta} - \frac{\zeta}{\omega_m^2(t)} - \frac{1}{4\zeta\omega_m^2(t)} & -\frac{1}{\omega_m^3(t)} \\ -\frac{1}{\omega_m^3(t)} & -\frac{3}{4\zeta\omega_m^4(t)} - \frac{1}{4\zeta\omega_m^2(t)} \end{bmatrix}.$$

Notice that $\dot{\omega}_m(t) = \frac{\pi}{80}\cos(\frac{\pi}{40}t) \leq \frac{\pi}{80}$, which leads to $\max_{t\geq 0}(\|\dot{P}(t)\|_\infty) = 0.55 < 1$. Thus, Assumption 5.1.1 is satisfied. We also compute $L = 8$, using the conservatively chosen Θ, and numerically compute the $\mathcal{L}_1$-norm of the map $\mathcal{G}$ using Definition A.7.6. This results in

$$\|\mathcal{G}\|_{\mathcal{L}_1} L = 0.9 < 1\,.$$

Thus, the $\mathcal{L}_1$-norm condition holds for our choice of control design parameters.

Figure 5.1 shows the simulation results for both ω_1 and ω_2 with $\theta(t) = \theta_1(t)$ and $\sigma(t) = \sigma_1(t)$. Figure 5.2 shows the simulation results for both $\theta_1(t)$ and $\theta_2(t)$ with $\omega = \omega_1$ and $\sigma(t) = \sigma_1(t)$, and Figure 5.3 shows the simulation results for both disturbances $\sigma_1(t)$ and $\sigma_2(t)$ with $\omega = \omega_1$ and $\theta(t) = \theta_1(t)$. From these results one can see that the fast adaptation ability of the $\mathcal{L}_1$ adaptive controller ensures uniform transient performance for different uncertainties and disturbances. We notice that while the system's output remains close to the desired reference signal in the presence of different uncertainties and disturbances, the control signal changes significantly to ensure adequate compensation for the uncertainties and the disturbances.

Next, we test the tracking performance of the closed-loop adaptive system. We set the reference signal to $r(t) = \sin\left(\frac{\pi}{5}t\right)$ and let $\omega = \omega_1$, $\theta(t) = \theta_1(t)$, and $\sigma(t) = \sigma_1(t)$. The simulation results are shown in Figure 5.4. One can see that the closed-loop adaptive system system has satisfactory tracking performance. It compensates for the uncertainties in the system and rejects the disturbance within the bandwidth of the control channel specified via $C(s)$, given in (5.3).

Figure 5.5 shows the simulation results for step-reference signals of different amplitudes. We observe that the system response is close to scaled response, similar to linear systems.

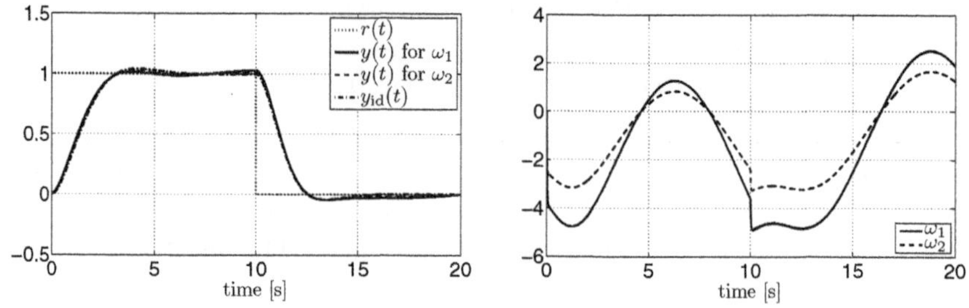

(a) $r(t)$, $y(t)$, and $y_{id}(t)$ (b) Time history of $u(t)$

Figure 5.1: Performance of the $\mathcal{L}_1$ adaptive controller for ω_1 and ω_2 with fixed $\theta_1(t)$ and $\sigma_1(t)$.

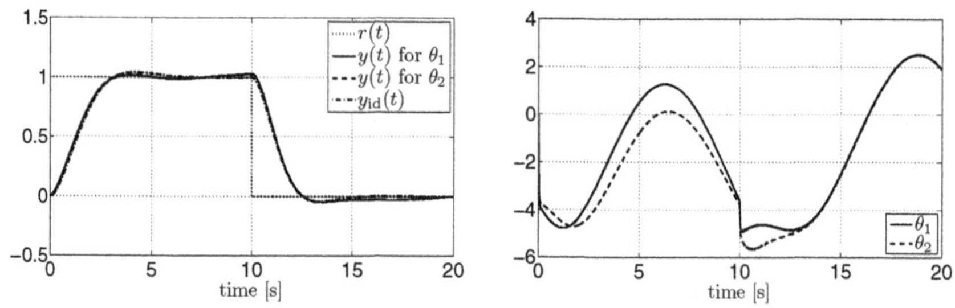

(a) $r(t)$, $y(t)$, and $y_{id}(t)$ (b) Time history of $u(t)$

Figure 5.2: Performance of the $\mathcal{L}_1$ adaptive controller for $\theta_1(t)$ and $\theta_2(t)$ with fixed ω_1 and $\sigma_1(t)$.

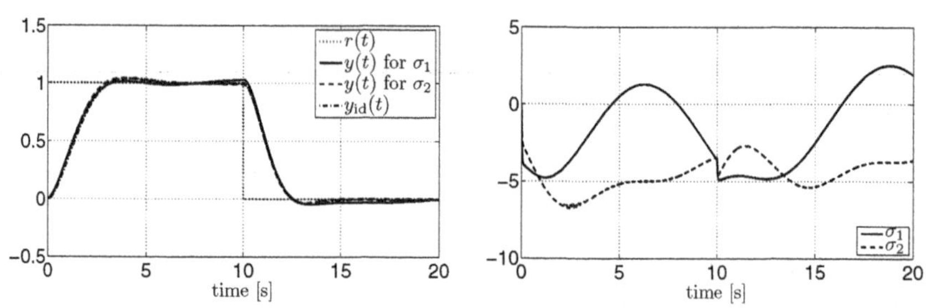

(a) $r(t)$, $y(t)$, and $y_{id}(t)$ (b) Time history of $u(t)$

Figure 5.3: Performance of the $\mathcal{L}_1$ adaptive controller for $\sigma_1(t)$ and $\sigma_2(t)$ with fixed ω_1 and $\theta_1(t)$.

Next, we test the system performance in the presence of nonzero initialization error. We set the initial conditions of the system different from the initial conditions of the state predictor: $x_0 = [0.5,\ 1]^\top$, $\hat{x}_0 = [1.5,\ 0.1]^\top$. The simulation results in Figure 5.6 verify the performance of the $\mathcal{L}_1$ adaptive controller in the presence of nonzero initialization errors.

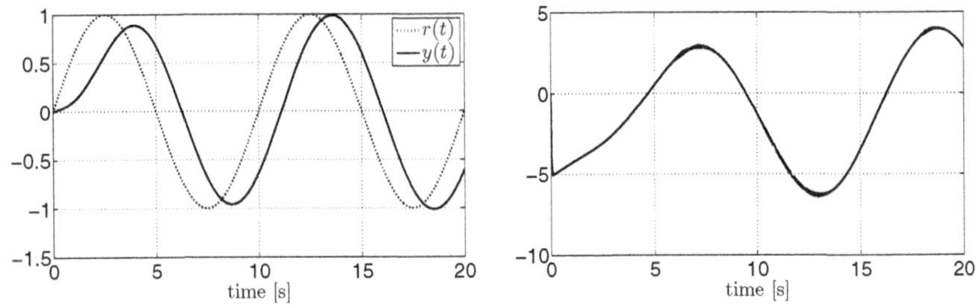

(a) $r(t)$, $y(t)$ (b) Time history of $u(t)$

Figure 5.4: Performance of the $\mathcal{L}_1$ adaptive controller for $r(t) = \sin(\frac{\pi}{5}t)$.

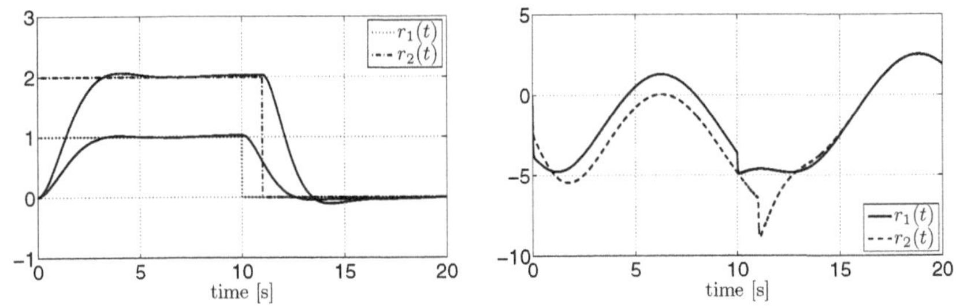

(a) $r(t)$, $y(t)$ (b) Time history of $u(t)$

Figure 5.5: Performance of the $\mathcal{L}_1$ adaptive controller for step-reference signals.

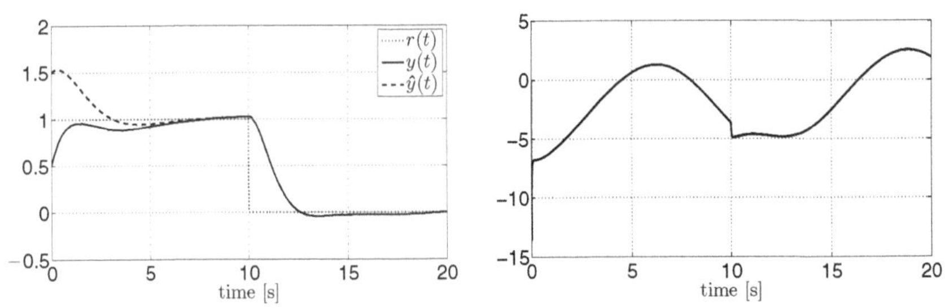

(a) $r(t)$, $y(t)$, and $\hat{y}(t)$ (b) Time history of $u(t)$

Figure 5.6: Performance of the $\mathcal{L}_1$ adaptive controller in the presence of nonzero initialization error.

Finally, we numerically test the robustness of the $\mathcal{L}_1$ adaptive controller to time delays. Figure 5.7 shows the simulation results in the presence of time delay of 20 ms for the uncertainties considered above. One can see that the system has some expected degradation in the performance but remains stable. Moreover, the system output in the presence of time delay remains close to the one in the absence of time delay for both types of the above considered uncertainties.

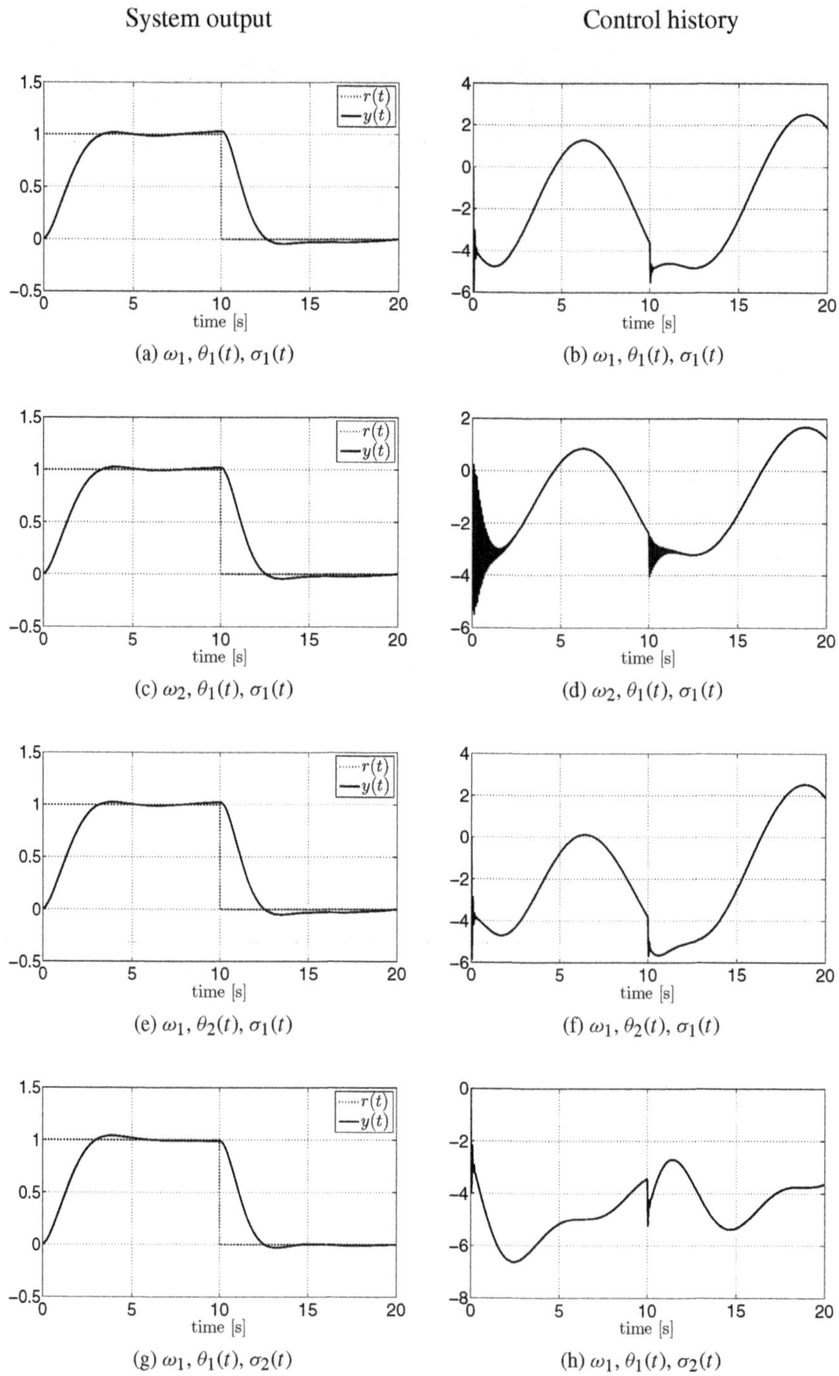

Figure 5.7: Performance of the $\mathcal{L}_1$ adaptive controller with time delay of 20 ms.

It is important to emphasize that in the simulations above there is no retuning of the $\mathcal{L}_1$ adaptive controller from one scenario to another, and the same *constant* control parameters are used for every simulation. The time-varying nature of the desired reference system is reflected in the state predictor, which uses $A_m(t)$ and $b(t)$.

5.2 $\mathcal{L}_1$ Adaptive Controller for Nonlinear Systems in the Presence of Unmodeled Dynamics

This section considers the class of uncertain systems with time- and state-dependent unknown nonlinearities and unmodeled dynamics. The $\mathcal{L}_1$ adaptive controller yields uniform performance bounds with respect to a bounded LTV reference system, which hold semiglobally [91].

5.2.1 Problem Formulation

Consider the following class of systems:

$$\begin{aligned}
\dot{x}(t) &= A_m(t)x(t) + b(t)\big(\mu(t) + f(t,x(t),z(t))\big), \quad x(0) = x_0,\\
\dot{x}_z(t) &= g(t,x_z(t),x(t)), \quad x_z(0) = x_{z0},\\
z(t) &= g_o(t,x_z(t)),\\
y(t) &= c^\top x(t),
\end{aligned} \tag{5.25}$$

where $x(t) \in \mathbb{R}^n$ is the system state; $A_m(t) \in \mathbb{R}^{n\times n}$ and $b(t) \in \mathbb{R}^n$ are a known time-varying matrix and a vector, respectively; $c \in \mathbb{R}^n$ is a known constant vector; $y(t) \in \mathbb{R}$ is the system output; $f : \mathbb{R} \times \mathbb{R}^n \times \mathbb{R}^l \to \mathbb{R}$ is an unknown nonlinear map, which represents system nonlinearities; $x_z(t) \in \mathbb{R}^m$ and $z(t) \in \mathbb{R}^l$ are the state and the output of the unmodeled nonlinear dynamics; $g : \mathbb{R} \times \mathbb{R}^m \times \mathbb{R}^n \to \mathbb{R}^m$, $g_0 : \mathbb{R} \times \mathbb{R}^m \to \mathbb{R}^l$ are unknown nonlinear maps continuous in their arguments; and $\mu(t) \in \mathbb{R}$ is the output of the following system:

$$\mu(s) = F(s)u(s),$$

where $u(t) \in \mathbb{R}$ is the control signal, and $F(s)$ is an unknown BIBO-stable and proper transfer function with known sign of its DC gain. The initial condition x_0 is assumed to be inside an arbitrarily large known set, so that $\|x_0\|_\infty \le \rho_0 < \infty$ with known $\rho_0 > 0$.

Let $X \triangleq [x^\top, z^\top]^\top$, and with a slight abuse of notation let $f(t,X) \triangleq f(t,x,z)$. Similar to the previous section, Assumptions 5.1.1, 5.1.2, and 5.1.3 hold. In addition, we impose the following assumptions.

Assumption 5.2.1 (Uniform boundedness of $f(t,0,0)$) There exists $B > 0$, such that $|f(t,0)| \le B$ holds for all $t \ge 0$.

Assumption 5.2.2 (Semiglobal uniform boundedness of partial derivatives) For arbitrary $\delta > 0$, there exist positive constants $d_{f_x}(\delta) > 0$ and $d_{f_t}(\delta) > 0$ independent of time, such that for all $\|X\|_\infty \le \delta$ the partial derivatives of $f(t,X)$ are piecewise-continuous and bounded,

$$\left\|\frac{\partial f(t,X)}{\partial X}\right\|_1 \le d_{f_x}(\delta), \qquad \left|\frac{\partial f(t,X)}{\partial t}\right| \le d_{f_t}(\delta).$$

Assumption 5.2.3 (Stability of unmodeled dynamics) The x_z-dynamics are BIBO stable with respect to both initial conditions x_{z0} and input $x(t)$, i.e., for arbitrary initial condition x_{z0} there exist $L_1, L_2 > 0$ such that for all $t \geq 0$

$$\|z_t\|_{\mathcal{L}_\infty} \leq L_1 \|x_t\|_{\mathcal{L}_\infty} + L_2 .$$

Assumption 5.2.4 (Partial knowledge of actuator dynamics) There exists $L_F > 0$ verifying $\|F(s)\|_{\mathcal{L}_1} \leq L_F$. Also, we assume that there exist known constants $\omega_l, \omega_u \in \mathbb{R}$ satisfying

$$0 < \omega_l \leq F(0) \leq \omega_u ,$$

where, without loss of generality, we have assumed $F(0) > 0$. Finally, we assume (for design purposes) that we know a set $\mathbb{F}_\Delta$ of all admissible actuator dynamics.

Next, we present the $\mathcal{L}_1$ adaptive controller that ensures that the system output $y(t)$ tracks a given bounded piecewise-continuous reference signal $r(t) \in \mathbb{R}$ with uniform and quantifiable performance bounds.

5.2.2 $\mathcal{L}_1$ Adaptive Control Architecture

Definitions and $\mathcal{L}_1$-Norm Sufficient Condition for Stability

Let $\mathcal{H}$ be the input-to-state map of the system

$$\dot{x}(t) = A_m(t)x(t) + b(t)u(t), \quad x(0) = 0 .$$

Then, the system in (5.25) can be rewritten as

$$x = \mathcal{H}\mu + \mathcal{H}f + x_{\text{in}} ,$$

where f denotes $f(t, x(t), z(t))$, and $x_{\text{in}}(t)$ is the solution of

$$\dot{x}_{\text{in}}(t) = A_m(t)x_{\text{in}}(t), \quad x_{\text{in}}(0) = x_0 .$$

Notice that from Assumption 5.1.1 and Lemma A.6.2 it follows that $\|x_{\text{in}}\|_{\mathcal{L}_\infty}$ is bounded, and $\|x_{\text{in}}\|_{\mathcal{L}_\infty} \leq \rho_{\text{in}}$, where

$$\rho_{\text{in}} \triangleq \max_{\|x_0\|_\infty \leq \rho_0} \|x_{\text{in}}\|_{\mathcal{L}_\infty} .$$

Further, let

$$L_\delta \triangleq \frac{\bar{\delta}(\delta)}{\delta} d_{f_x}(\bar{\delta}(\delta)), \quad \bar{\delta}(\delta) \triangleq \max\{\delta + \bar{\gamma}_1, L_1(\delta + \bar{\gamma}_1) + L_2\}, \tag{5.26}$$

where $d_{f_x}(\cdot)$ was introduced in Assumption 5.2.2, and $\bar{\gamma}_1 > 0$ is an arbitrary positive constant.

Similar to Section 5.1.2, the design of $\mathcal{L}_1$ adaptive controller proceeds by considering a positive feedback gain $k > 0$ and a strictly proper stable transfer function $D(s)$, which lead, for all $F(s) \in \mathbb{F}_\Delta$, to a strictly proper stable

$$C(s) \triangleq \frac{kF(s)D(s)}{1 + kF(s)D(s)}, \tag{5.27}$$

with DC gain $C(0) = 1$. Also, let $\mathfrak{C}$ denote the input-output map for $C(s)$.

For the proofs of stability and performance bounds, the choice of k and $D(s)$ needs to ensure that, for a given ρ_0, there exists $\rho_r > \rho_{\text{in}}$, such that the following $\mathcal{L}_1$-norm condition can be verified:

$$\|\mathcal{G}\|_{\mathcal{L}_1} < \frac{\rho_r - \|\mathcal{H}\mathfrak{C}k_g\|_{\mathcal{L}_1}\|r\|_{\mathcal{L}_\infty} - \rho_{\text{in}}}{L_{\rho_r}\rho_r + B}, \tag{5.28}$$

where $\mathcal{G} \triangleq \mathcal{H}(1-\mathfrak{C})$, and $k_g(t) \triangleq -1/(c^\top A_m^{-1}(t)b(t))$ is the feedforward gain required for tracking the reference signal $r(t)$.

To streamline the subsequent analysis of stability and performance bounds, we introduce the following notation. Similar to the previous section, let $r_g(t) \triangleq k_g(t)r(t)$. Also let $C_u(s) \triangleq C(s)/F(s)$, and let $\mathfrak{C}_u$ be the input-output map for $C_u(s)$. Note that $C_u(s)$ is a strictly proper BIBO-stable transfer function. Next, let $\bar{c}_i$, $i = 1,\ldots,n$, be the coefficients of an arbitrary Hurwitz polynomial $\bar{c}_n s^{n-1} + \cdots + \bar{c}_1$. Let $T(t)$ be the transformation matrix, reducing $(A_m(t), b(t))$ to its controllable canonical form, and let a_i, $i = 1,\ldots,n$, be the coefficients of the characteristic polynomial, as discussed in Lemma A.11.2.

Let

$$\rho \triangleq \rho_r + \bar{\gamma}_1$$

and

$$\gamma_1 \triangleq \frac{\|\mathcal{H}\mathcal{F}\|_{\mathcal{L}_1}\kappa_0}{1 - \|\mathcal{G}\|_{\mathcal{L}_1}L_{\rho_r}}\gamma_0 + \beta, \tag{5.29}$$

where $\mathcal{F}$ is the input-to-state map of the system $F(s)$ and κ_0 is defined as

$$\kappa_0 \triangleq \left(\sum_{i=0}^{n-1}\|\mathfrak{C}_u a_{i+1}\|_{\mathcal{L}_1}\left\|\frac{s^i}{\bar{c}_n s^{n-1} + \cdots + \bar{c}_1}\right\|_{\mathcal{L}_1} + \left\|\frac{C_u(s)s^n}{\bar{c}_n s^{n-1} + \cdots + \bar{c}_1}\right\|_{\mathcal{L}_1}\right)\|\bar{c}^\top T\|_{\mathcal{L}_\infty},$$

while β and γ_0 are arbitrary small positive constants such that $\gamma_1 \le \bar{\gamma}_1$. Moreover, let

$$\rho_u \triangleq \rho_{ur} + \gamma_2,$$

where ρ_{ur} and γ_2 are defined as

$$\rho_{ur} \triangleq \|C_u(s)\|_{\mathcal{L}_1}(\|k_g\|_{\mathcal{L}_\infty}\|r\|_{\mathcal{L}_\infty} + L_{\rho_r}\rho_r + B),$$
$$\gamma_2 \triangleq \|C_u(s)\|_{\mathcal{L}_1}L_{\rho_r}\gamma_1 + \kappa_0\gamma_0. \tag{5.30}$$

Finally, using the conservative knowledge of $F(s)$, let

$$\Delta_1 \triangleq L_\rho L_2 + B + \epsilon, \tag{5.31}$$
$$\Delta_2 \triangleq \max_{F(s)\in\mathbb{F}_\Delta}\|F(s) - (\omega_l + \omega_u)/2\|_{\mathcal{L}_1}\rho_u, \tag{5.32}$$
$$\Delta \triangleq \Delta_1 + \Delta_2, \tag{5.33}$$
$$\rho_{\dot{u}} \triangleq \|ksD(s)\|_{\mathcal{L}_1}(\rho_u\omega_u + L_\rho\rho + \Delta + \|k_g\|_{\mathcal{L}_\infty}\|r\|_{\mathcal{L}_\infty}),$$

where $\epsilon > 0$ is an arbitrarily small positive constant.

Remark 5.2.1 In the following analysis we demonstrate that ρ_r and ρ characterize the positively invariant sets for the state of the closed-loop reference system (yet to be defined) and the state of the closed-loop adaptive system, respectively. We notice that, since $\bar{\gamma}_1$ can be set arbitrarily small, ρ can approximate ρ_r arbitrarily closely.

The elements of the $\mathcal{L}_1$ adaptive controller are introduced next.

State Predictor

The following state predictor is used for derivation of the adaptive laws:

$$\begin{aligned}\dot{\hat{x}}(t) &= A_m(t)\hat{x}(t) + b(t)\left(\hat{\omega}(t)u(t) + \hat{\theta}(t)\|x_t\|_{\mathcal{L}_\infty} + \hat{\sigma}(t)\right), \quad \hat{x}(0) = x_0,\\ y(t) &= c^\top \hat{x}(t),\end{aligned} \tag{5.34}$$

where $\hat{x}(t) \in \mathbb{R}^n$ is the state of the predictor, while $\hat{\omega}(t), \hat{\theta}(t), \hat{\sigma}(t) \in \mathbb{R}$ are the adaptive estimates.

Adaptation Laws

The adaptive laws are defined via the projection operator as follows:

$$\begin{aligned}\dot{\hat{\omega}}(t) &= \Gamma \mathrm{Proj}\left(\hat{\omega}(t), -\tilde{x}^\top(t)P(t)b(t)u(t)\right), \quad \hat{\omega}(0) = \hat{\omega}_0,\\ \dot{\hat{\theta}}(t) &= \Gamma \mathrm{Proj}\left(\hat{\theta}(t), -\tilde{x}^\top(t)P(t)b(t)\|x_t\|_{\mathcal{L}_\infty}\right), \quad \hat{\theta}(0) = \hat{\theta}_0,\\ \dot{\hat{\sigma}}(t) &= \Gamma \mathrm{Proj}\left(\hat{\sigma}(t), -\tilde{x}^\top(t)P(t)b(t)\right), \quad \hat{\sigma}(0) = \hat{\sigma}_0,\end{aligned} \tag{5.35}$$

where $\tilde{x}(t) \triangleq \hat{x}(t) - x(t)$, $\Gamma \in \mathbb{R}^+$ is the adaptation gain, and the symmetric positive definite matrix $P(t) = P^\top(t) > 0$ solves the Lyapunov equation $A_m^\top(t)P(t) + P(t)A_m(t) = -\mathbb{I}$. The projection operator ensures that $\hat{\omega}(t) \in \Omega \triangleq [\omega_l, \omega_u]$, $\hat{\theta}(t) \in \Theta \triangleq [-L_\rho, L_\rho]$ and $|\hat{\sigma}(t)| \le \Delta$.

Control Law

The control law is generated as the output of the following (feedback) system:

$$u(s) = -kD(s)(\hat{\eta}(s) - r_g(s)), \tag{5.36}$$

where $r_g(s)$ and $\hat{\eta}(s)$ are the Laplace transforms of $r_g(t)$ and $\hat{\eta}(t) \triangleq \hat{\omega}(t)u(t) + \hat{\theta}(t)\|x_t\|_{\mathcal{L}_\infty} + \hat{\sigma}(t)$, respectively, while k and $D(s)$ were introduced in (5.27).

The $\mathcal{L}_1$ adaptive controller is defined via (5.34), (5.35), and (5.36), subject to the $\mathcal{L}_1$-norm condition in (5.28).

5.2.3 Analysis of the $\mathcal{L}_1$ Adaptive Controller

Closed-Loop Reference System

Consider the following closed-loop reference system:

$$\begin{aligned}\dot{x}_{\mathrm{ref}}(t) &= A_m(t)x_{\mathrm{ref}}(t) + b(t)\left(\mu_{\mathrm{ref}}(t) + f(t, x_{\mathrm{ref}}(t), z(t))\right), \quad x_{\mathrm{ref}}(0) = x_0,\\ \mu_{\mathrm{ref}}(s) &= F(s)u_{\mathrm{ref}}(s),\\ u_{\mathrm{ref}}(s) &= -C_u(s)(\eta_{\mathrm{ref}}(s) - r_g(s)),\\ y_{\mathrm{ref}}(t) &= c^\top x_{\mathrm{ref}}(t),\end{aligned} \tag{5.37}$$

where $x_{\text{ref}}(t) \in \mathbb{R}^n$ is the reference system state vector, $\eta_{\text{ref}}(s)$ is the Laplace transform of $\eta_{\text{ref}}(t) \triangleq f(t, x_{\text{ref}}(t), z(t))$. Notice that this reference system is not implementable, as it contains the unknowns $f(t, x_{\text{ref}}(t), z(t))$ and $F(s)$. This system is used only for analysis purposes. The next lemma proves the stability of this closed-loop reference system subject to an assumption on $z(t)$, which will be verified later in the proof of stability and performance bounds of the closed-loop adaptive system (Theorem 5.2.1).

Lemma 5.2.1 For the closed-loop reference system given in (5.37), subject to the $\mathcal{L}_1$-norm condition in (5.28), if for some $\tau \geq 0$

$$\|z_\tau\|_{\mathcal{L}_\infty} \leq L_1(\|x_{\text{ref}\tau}\|_{\mathcal{L}_\infty} + \gamma_1) + L_2, \tag{5.38}$$

then the following bounds hold:

$$\|x_{\text{ref}\tau}\|_{\mathcal{L}_\infty} < \rho_r, \quad \|u_{\text{ref}\tau}\|_{\mathcal{L}_\infty} < \rho_{ur}. \tag{5.39}$$

Proof. The response of the closed-loop reference system in (5.37) can be rewritten as

$$x_{\text{ref}} = \mathcal{G}\eta_{\text{ref}} + \mathcal{H}\mathcal{C}k_g r + x_{\text{in}}. \tag{5.40}$$

Lemmas A.6.2 and A.7.5, along with the fact $\|x_{\text{in}}\|_{\mathcal{L}_\infty} \leq \rho_{\text{in}}$, imply that

$$\|x_{\text{ref}\tau}\|_{\mathcal{L}_\infty} \leq \|\mathcal{G}\|_{\mathcal{L}_1}\|\eta_{\text{ref}\tau}\|_{\mathcal{L}_\infty} + \|\mathcal{H}\mathcal{C}k_g\|_{\mathcal{L}_1}\|r\|_{\mathcal{L}_\infty} + \rho_{\text{in}}. \tag{5.41}$$

Next, we use a contradictive argument. Assume that the first bound in (5.39) does not hold. Then, since $\|x_{\text{ref}}(0)\|_\infty = \|x_0\|_\infty \leq \rho_0 < \rho_r$ and $x_{\text{ref}}(t)$ is continuous, there exists a time instant $\tau_1 \in (0, \tau]$, such that

$$\|x_{\text{ref}}(t)\|_\infty < \rho_r, \quad \forall\, t \in [0, \tau_1), \qquad \text{and} \qquad \|x_{\text{ref}}(\tau_1)\|_\infty = \rho_r,$$

which implies that

$$\|x_{\text{ref}\tau_1}\|_{\mathcal{L}_\infty} = \rho_r. \tag{5.42}$$

It follows from (5.38) that $\|z_{\tau_1}\|_{\mathcal{L}_\infty} \leq L_1(\rho_r + \gamma_1) + L_2$, which implies

$$\|X_{\text{ref}\tau_1}\|_{\mathcal{L}_\infty} = \left\| \begin{bmatrix} x_{\text{ref}} \\ z \end{bmatrix}_{\tau_1} \right\|_{\mathcal{L}_\infty} \leq \max\{\rho_r + \gamma_1, L_1(\rho_r + \gamma_1) + L_2\} \leq \bar{\rho}_r(\rho_r).$$

Assumption 5.2.2 further implies that, for all $\|X_{\text{ref}}\|_\infty \leq \bar{\rho}_r(\rho_r)$, the following inequality holds:

$$|f(t, X_{\text{ref}}) - f(t, 0)| \leq d_{f_x}(\bar{\rho}_r(\rho_r))\|X_{\text{ref}}\|_\infty, \quad \forall\, t \in [0, \tau].$$

Further, Assumption 5.2.1 and the definition of L_δ in (5.26) lead to

$$\|\eta_{\text{ref}\tau_1}\|_{\mathcal{L}_\infty} \leq d_{f_x}(\bar{\rho}_r(\rho_r))\bar{\rho}_r(\rho_r) + B = L_{\rho_r}\rho_r + B. \tag{5.43}$$

Thus, from (5.41) we get

$$\|x_{\text{ref}\tau_1}\|_{\mathcal{L}_\infty} \leq \|\mathcal{G}\|_{\mathcal{L}_1}(L_{\rho_r}\rho_r + B) + \|\mathcal{H}\mathcal{C}k_g\|_{\mathcal{L}_1}\|r\|_{\mathcal{L}_\infty} + \rho_{\text{in}}.$$

From the $\mathcal{L}_1$-norm condition in (5.28) one has

$$\|\mathcal{G}\|_{\mathcal{L}_1}(L_{\rho_r}\rho_r + B) + \|\mathcal{H}\mathcal{C}k_g\|_{\mathcal{L}_1}\|r\|_{\mathcal{L}_\infty} + \rho_{\text{in}} < \rho_r\,,$$

which implies $\|x_{\text{ref}\tau_1}\|_{\mathcal{L}_\infty} < \rho_r$. This contradicts (5.42), which proves the first bound in (5.39).

Because this bound is strict and holds uniformly for all $\tau_1 \in (0, \tau]$, one can rewrite (5.43) as a strict inequality,

$$\|\eta_{\text{ref}\tau}\|_{\mathcal{L}_\infty} < L_{\rho_r}\rho_r + B\,.$$

Then, from (5.37) it follows that

$$\begin{aligned}\|u_{\text{ref}\tau}\|_{\mathcal{L}_\infty} &\le \|C_u(s)\|_{\mathcal{L}_1}(\|r_g\|_{\mathcal{L}_\infty} + \|\eta_{\text{ref}\tau}\|_{\mathcal{L}_\infty})\\ &< \|C_u(s)\|_{\mathcal{L}_1}(\|k_g\|_{\mathcal{L}_\infty}\|r\|_{\mathcal{L}_\infty} + L_{\rho_r}\rho_r + B) = \rho_{ur}\,,\end{aligned}$$

which completes the proof. □

Equivalent (Semi-)Linear Time-Varying System

In this section, we transform the original nonlinear system with unmodeled dynamics in (5.25) into an equivalent (semi-)linear time-varying system with unknown time-varying parameters and disturbances. This transformation requires us to impose the following assumptions on the signals of the system: the control signal $u(t)$ is continuous, and moreover the following bounds hold

$$\|x_\tau\|_{\mathcal{L}_\infty} \le \rho\,, \quad \|u_\tau\|_{\mathcal{L}_\infty} \le \rho_u\,, \quad \|\dot{u}_\tau\|_{\mathcal{L}_\infty} \le \rho_{\dot{u}}\,; \tag{5.44}$$

these will be verified later in the proof of Theorem 5.2.1. Next we construct the equivalent LTV system in two steps.

First Equivalent System Consider the system in (5.25). From (5.25) and the first two bounds in (5.44), it follows that $\|\dot{x}_\tau\|_{\mathcal{L}_\infty}$ is bounded for all $\tau \in [0, \infty)$. Thus, Lemma A.9.1 implies that there exist continuous $\theta(t)$ and $\sigma_1(t)$ with (piecewise)-continuous derivative, defined over $t \in [0, \tau]$, such that

$$\begin{aligned}|\theta(t)| &< L_\rho\,, \quad |\dot{\theta}(t)| \le d_\theta\,,\\ |\sigma_1(t)| &< \Delta_1\,, \quad |\dot{\sigma}_1(t)| \le d_{\sigma_1}\,,\end{aligned} \tag{5.45}$$

and

$$f(t, x(t), z(t)) = \theta(t)\|x_t\|_{\mathcal{L}_\infty} + \sigma_1(t)\,,$$

where L_ρ and Δ_1 are as defined in (5.26) and (5.31), while the algorithm for computing $d_\theta > 0$, $d_{\sigma_1} > 0$ is derived in the proof of Lemma A.9.1. Thus, the system in (5.25) can be rewritten over $t \in [0, \tau]$ as

$$\begin{aligned}\dot{x}(t) &= A_m(t)x(t) + b(t)\big(\mu(t) + \theta(t)\|x_t\|_{\mathcal{L}_\infty} + \sigma_1(t)\big)\,, \quad x(0) = x_0\,,\\ y(t) &= c^\top x(t)\,.\end{aligned} \tag{5.46}$$

Second Equivalent System Keeping in mind the assumptions on $u(t)$ and its derivative in (5.44), and using Lemma A.10.1, we can rewrite the signal $\mu(t)$ as

$$\mu(t) = \omega u(t) + \sigma_2(t),$$

where $\omega \in (\omega_l, \omega_u)$ is an unknown constant and $\sigma_2(t)$ is a continuous signal with (piecewise)-continuous derivative, defined over $t \in [0, \tau]$, such that

$$|\sigma_2(t)| \leq \Delta_2, \quad |\dot{\sigma}_2(t)| \leq d_{\sigma_2},$$

with Δ_2 as introduced in (5.32), and $d_{\sigma_2} \triangleq \|F(s) - (\omega_l + \omega_u)/2\|_{\mathcal{L}_1} \rho_{\dot{u}}$. This implies that one can rewrite the system in (5.46) over $t \in [0, \tau]$ as

$$\begin{aligned} \dot{x}(t) &= A_m(t)x(t) + b(t)\big(\omega u(t) + \theta(t)\|x_t\|_{\mathcal{L}_\infty} + \sigma(t)\big), \quad x(0) = x_0, \\ y(t) &= c^\top x(t), \end{aligned} \tag{5.47}$$

where $\sigma(t) \triangleq \sigma_1(t) + \sigma_2(t)$ is an unknown time-varying signal subject to $|\sigma(t)| < \Delta$, with Δ as introduced in (5.33), $|\dot{\sigma}(t)| < d_\sigma$, $d_\sigma \triangleq d_{\sigma_1} + d_{\sigma_2}$, and $\theta(t)$ as introduced in (5.45).

Transient and Steady-State Performance

Using (5.47), one can write the error dynamics over $t \in [0, \tau]$,

$$\dot{\tilde{x}}(t) = A_m(t)\tilde{x}(t) + b(t)\left(\tilde{\omega}(t)u(t) + \tilde{\theta}(t)\|x_t\|_{\mathcal{L}_\infty} + \tilde{\sigma}(t)\right), \quad \tilde{x}(0) = 0, \tag{5.48}$$

where $\tilde{\omega}(t) \triangleq \hat{\omega}(t) - \omega$, $\tilde{\theta}(t) \triangleq \hat{\theta}(t) - \theta(t)$, and $\tilde{\sigma}(t) \triangleq \hat{\sigma}(t) - \sigma(t)$.

Lemma 5.2.2 If

$$\|x_\tau\|_{\mathcal{L}_\infty} \leq \rho, \quad \|u_\tau\|_{\mathcal{L}_\infty} \leq \rho_u, \quad \|\dot{u}_\tau\|_{\mathcal{L}_\infty} \leq \rho_{\dot{u}},$$

then

$$\|\tilde{x}_\tau\|_{\mathcal{L}_\infty} \leq \sqrt{\frac{\theta_m(\rho, \rho_u, \rho_{\dot{u}})}{\lambda_{P_{\min}} \Gamma}}, \tag{5.49}$$

where

$$\theta_m(\rho, \rho_u, \rho_{\dot{u}}) \triangleq (\omega_u - \omega_l)^2 + 4L_\rho^2 + 4\Delta^2 + 4\frac{\lambda_{P_{\max}}}{1 - \epsilon_P}(L_\rho d_\theta + \Delta d_\sigma),$$

and

$$\lambda_{P_{\min}} \triangleq \inf_{\substack{t \in [0,\infty), \\ i = 1\ldots n}} \lambda_i(P(t)), \qquad \lambda_{P_{\max}} \triangleq \sup_{\substack{t \in [0,\infty), \\ i = 1\ldots n}} \lambda_i(P(t)).$$

Proof. Consider the following Lyapunov function candidate:

$$V(\tilde{x}(t), \tilde{\omega}(t), \tilde{\theta}(t), \tilde{\sigma}(t)) = \tilde{x}^\top(t)P(t)\tilde{x}(t) + \frac{1}{\Gamma}(\tilde{\omega}^2(t) + \tilde{\theta}^2(t) + \tilde{\sigma}^2(t)). \tag{5.50}$$

Using the adaptation laws in (5.35) and Property B.2 for the Proj(·,·) operator, we compute the upper bound on the derivative of the Lyapunov function similar to the proof of Lemma 5.1.2:

$$\dot{V}(t) \leq -\tilde{x}^\top(t)(\mathbb{I} - \dot{P}(t))\tilde{x}(t) + \frac{2}{\Gamma}|\tilde{\theta}(t)\dot{\theta}(t) + \tilde{\sigma}(t)\dot{\sigma}(t)|.$$

Let $t_1 \in (0, \tau]$ be the time instant when the first discontinuity of $\dot{\theta}(t)$ or $\dot{\sigma}(t)$ occurs, or $t_1 = \tau$ if there are no discontinuities. Consider the Lyapunov function candidate given in (5.50). Notice that

$$\max_{t\in[0,t_1]} \frac{1}{\Gamma}\left(\tilde{\omega}^2(t) + \tilde{\theta}^2(t) + \tilde{\sigma}^2(t)\right) \leq \frac{(\omega_u - \omega_l)^2 + 4L_\rho^2 + 4\Delta^2}{\Gamma} < \frac{\theta_m(\rho, \rho_u, \rho_{\dot{u}})}{\Gamma}.$$

This inequality, along with the fact that $\tilde{x}(0) = 0$, leads to

$$V(0) < \frac{\theta_m(\rho, \rho_u, \rho_{\dot{u}})}{\Gamma}.$$

If at arbitrary time $t_2 \in [0, t_1]$ we have

$$V(t_2) > \frac{\theta_m(\rho, \rho_u, \rho_{\dot{u}})}{\Gamma},$$

then it follows that

$$\tilde{x}^\top(t_2)P(t_2)\tilde{x}(t_2) > 4\frac{\lambda_{P_{\max}}}{\Gamma(1-\epsilon_P)}(L_\rho d_\theta + \Delta d_\sigma).$$

Further, from Lemma A.6.2, one can write

$$\tilde{x}^\top(t_2)(\mathbb{I} - \dot{P}(t_2))\tilde{x}(t_2) \geq \frac{1-\epsilon_P}{\lambda_{P_{\max}}}\tilde{x}^\top(t_2)P(t_2)\tilde{x}(t_2) > \frac{4}{\Gamma}(L_\rho d_\theta + \Delta d_\sigma). \tag{5.51}$$

Moreover, notice that

$$\frac{1}{\Gamma}|\tilde{\theta}(t)\dot{\theta}(t) + \tilde{\sigma}(t)\dot{\sigma}(t)| \leq \frac{2}{\Gamma}(L_\rho d_\theta + \Delta d_\sigma),$$

which, together with the bound in (5.51), leads to

$$\dot{V}(t_2) < 0.$$

Therefore, we have

$$V(t_2) \leq \frac{\theta_m(\rho, \rho_u, \rho_{\dot{u}})}{\Gamma}.$$

The continuity of $V(t)$ allows for repeating these derivations for all points of discontinuity of $\dot{\theta}(t)$ or $\dot{\sigma}(t)$, which leads to the following uniform bound:

$$V(t) \leq \frac{\theta_m(\rho, \rho_u, \rho_{\dot{u}})}{\Gamma}, \qquad \forall\, t \in [0, \tau].$$

This further implies

$$\|\tilde{x}(t)\| \leq \sqrt{\frac{\theta_m(\rho, \rho_u, \rho_{\dot{u}})}{\lambda_{P_{\min}} \Gamma}}, \qquad \forall\, t \in [0, \tau],$$

which leads to the bound in (5.49). □

Theorem 5.2.1 If the adaptive gain verifies the lower bound

$$\Gamma \geq \frac{\theta_m(\rho, \rho_u, \rho_{\dot{u}})}{\lambda_{P_{\min}} \gamma_0^2}, \tag{5.52}$$

where $\gamma_0 > 0$ is an arbitrary constant introduced in (5.30), then the following bounds hold:

$$\|u_{\text{ref}}\|_{\mathcal{L}_\infty} \leq \rho_u, \tag{5.53}$$

$$\|x_{\text{ref}}\|_{\mathcal{L}_\infty} \leq \rho_r, \tag{5.54}$$

$$\|\tilde{x}\|_{\mathcal{L}_\infty} \leq \gamma_0, \tag{5.55}$$

$$\|x_{\text{ref}} - x\|_{\mathcal{L}_\infty} \leq \gamma_1, \tag{5.56}$$

$$\|u_{\text{ref}} - u\|_{\mathcal{L}_\infty} \leq \gamma_2. \tag{5.57}$$

Proof. We prove (5.56) and (5.57) following a contradicting argument. Assume that (5.56) and (5.57) do not hold. Then, since

$$\|x_{\text{ref}}(0) - x(0)\|_\infty = 0, \quad \|u_{\text{ref}}(0) - u(0)\|_\infty = 0,$$

continuity of $x_{\text{ref}}(t)$, $x(t)$, $u_{\text{ref}}(t)$, $u(t)$ implies that there exists time $\tau > 0$ for which

$$\|x_{\text{ref}}(t) - x(t)\|_\infty < \gamma_1, \qquad \|u_{\text{ref}}(t) - u(t)\|_\infty < \gamma_2, \qquad \forall\, t \in [0, \tau),$$

and

$$\|x_{\text{ref}}(\tau) - x(\tau)\|_\infty = \gamma_1, \quad \text{or} \quad \|u_{\text{ref}}(\tau) - u(\tau)\|_\infty = \gamma_2.$$

This implies that at least one of the following equalities holds:

$$\|(x_{\text{ref}} - x)_\tau\|_{\mathcal{L}_\infty} = \gamma_1, \quad \|(u_{\text{ref}} - u)_\tau\|_{\mathcal{L}_\infty} = \gamma_2. \tag{5.58}$$

The first equality above leads to

$$\|z_\tau\|_{\mathcal{L}_\infty} \leq L_1(\|x_{\text{ref}\tau}\|_{\mathcal{L}_\infty} + \gamma_1) + L_2.$$

Then, since all the conditions of Lemma 5.2.1 hold, the following bounds are valid:

$$\|x_{\text{ref}\tau}\|_{\mathcal{L}_\infty} < \rho_r, \quad \|u_{\text{ref}\tau}\|_{\mathcal{L}_\infty} < \rho_{ur}, \tag{5.59}$$

which in turn leads to

$$\|x_\tau\|_{\mathcal{L}_\infty} \leq \rho_r + \gamma_1 \leq \rho, \quad \|u_\tau\|_{\mathcal{L}_\infty} \leq \rho_{ur} + \gamma_2 = \rho_u. \tag{5.60}$$

Further, consider the control law in (5.36). From the properties of the projection operator, we have

$$\|\hat{\eta}_\tau\|_{\mathcal{L}_\infty} \leq \omega_u \rho_u + L_\rho \rho + \Delta\,,$$

and consequently

$$\|\dot{u}_\tau\| \leq \|ksD(s)\|_{\mathcal{L}_1}(\omega_u \rho_u + L_\rho \rho + \Delta + \|k_g\|_{\mathcal{L}_\infty}\|r\|_{\mathcal{L}_\infty}) = \rho_{\dot{u}}\,. \tag{5.61}$$

Then, since all the conditions of Lemma 5.2.2 hold, selection of Γ following (5.52) gives

$$\|\tilde{x}_\tau\|_{\mathcal{L}_\infty} \leq \gamma_0\,. \tag{5.62}$$

Next, let $\eta(t)$ and $\tilde{\eta}(t)$ be defined as

$$\eta(t) \triangleq \theta(t)\|\,x_t\,\|_{\mathcal{L}_\infty} + \sigma(t)\,, \qquad \tilde{\eta}(t) \triangleq \tilde{\omega}(t)u(t) + \tilde{\theta}(t)\|x_t\|_{\mathcal{L}_\infty} + \tilde{\sigma}(t)\,.$$

From the upper bounds in (5.60) and the bound in (5.61), it follows that, for all $t \in [0,\tau]$, the following equality holds:

$$\eta(t) = f(t, x(t), z(t))\,.$$

Then, notice that, for all $t \in [0,\ \tau]$, we have

$$\hat{\eta}(t) = \mu(t) + \tilde{\eta}(t) + \eta(t)\,,$$

which implies that the control law in (5.36) can be written as

$$u(s) = -C_u(s)(\tilde{\eta}(s) + \eta(s) - r_g(s))\,. \tag{5.63}$$

Further, the system in (5.25) can be rewritten as

$$x = \mathcal{G}\eta - \mathcal{H}\mathcal{C}\tilde{\eta} + \mathcal{H}\mathcal{C}r_g + x_{\text{in}}\,.$$

Recall that in (5.40) the reference system is presented as

$$x_{\text{ref}} = \mathcal{G}\eta_{\text{ref}} + \mathcal{H}\mathcal{C}r_g + x_{\text{in}}\,.$$

The two expressions above yield

$$x_{\text{ref}} - x = \mathcal{G}(\eta_{\text{ref}} - \eta) + \mathcal{H}\mathcal{C}\tilde{\eta}\,. \tag{5.64}$$

Notice now that, from Assumption 5.2.3 and (5.60), it follows that

$$\|z_\tau\|_{\mathcal{L}_\infty} \leq L_1(\rho_r + \gamma_1) + L_2\,,$$

which, along with (5.60), leads to

$$\|X_\tau\|_{\mathcal{L}_\infty} \leq \max\{\rho_r + \gamma_1,\, L_1(\rho_r + \gamma_1) + L_2\} \leq \bar{\rho}_r(\rho_r)\,.$$

Similarly, one can show that

$$\|X_{\text{ref}\tau}\|_{\mathcal{L}_\infty} \leq \bar{\rho}_r(\rho_r)\,.$$

Thus, using Assumption 5.2.2, one can write

$$\|(\eta_{\text{ref}} - \eta)_\tau\|_{\mathcal{L}_\infty} \le d_{f_x}(\bar{\rho}_r(\rho_r))\|(x_{\text{ref}} - x)_\tau\|_{\mathcal{L}_\infty},$$

and since $d_{f_x}(\bar{\rho}_r(\rho_r)) < L_{\rho_r}$, it follows that

$$\|(\eta_{\text{ref}} - \eta)_\tau\|_{\mathcal{L}_\infty} \le L_{\rho_r}\|(x_{\text{ref}} - x)_\tau\|_{\mathcal{L}_\infty}.$$

From (5.64), it follows that

$$\|(x_{\text{ref}} - x)_\tau\|_{\mathcal{L}_\infty} \le \|\mathcal{G}\|_{\mathcal{L}_1} L_{\rho_r}\|(x_{\text{ref}} - x)_\tau\|_{\mathcal{L}_\infty} + \|\mathcal{H}\mathcal{F}\|_{\mathcal{L}_1}\|(\mathcal{C}_u\tilde{\eta})_\tau\|_{\mathcal{L}_\infty}.$$

The $\mathcal{L}_1$-norm condition in (5.28) implies $\|\mathcal{G}\|_{\mathcal{L}_1} L_{\rho_r} < 1$, which allows for the following derivation:

$$\|(x_{\text{ref}} - x)_\tau\|_{\mathcal{L}_\infty} \le \frac{\|\mathcal{H}\mathcal{F}\|_{\mathcal{L}_1}}{1 - \|\mathcal{G}\|_{\mathcal{L}_1} L_{\rho_r}}\|(\mathcal{C}_u\tilde{\eta})_\tau\|_{\mathcal{L}_\infty}.$$

Since $C_u(s)$ is a strictly proper BIBO-stable transfer function, application of Lemma A.40 to the linear time-varying prediction error dynamics in (5.48) yields

$$\|(\mathcal{C}_u\tilde{\eta})_\tau\|_{\mathcal{L}_\infty} \le \kappa_0\|\tilde{x}_\tau\|_{\mathcal{L}_\infty},$$

which implies that

$$\|(x_{\text{ref}} - x)_\tau\|_{\mathcal{L}_\infty} \le \frac{\|\mathcal{H}\mathcal{F}\|_{\mathcal{L}_1}\kappa_0}{1 - \|\mathcal{G}\|_{\mathcal{L}_1} L_{\rho_r}}\|\tilde{x}_\tau\|_{\mathcal{L}_\infty}.$$

This bound, along with (5.62), leads to

$$\|(x_{\text{ref}} - x)_\tau\|_{\mathcal{L}_\infty} \le \frac{\|\mathcal{H}\mathcal{F}\|_{\mathcal{L}_1}\kappa_0}{1 - \|\mathcal{G}\|_{\mathcal{L}_1} L_{\rho_r}}\gamma_0 = \gamma_1 - \beta < \gamma_1. \tag{5.65}$$

Thus, we obtain a contradiction to the first equality in (5.58).

To show that the second equation in (5.58) also cannot hold, consider (5.37) and (5.63), which lead to

$$u_{\text{ref}}(s) - u(s) = -C_u(s)(\eta_{\text{ref}}(s) - \eta(s)) + C_u(s)\tilde{\eta}(s).$$

Using the bounds on $\|(\eta_{\text{ref}} - \eta)_\tau\|_{\mathcal{L}_\infty}$ and $\|(\mathcal{C}_u\tilde{\eta})_\tau\|_{\mathcal{L}_\infty}$, one can write

$$\|(u_{\text{ref}} - u)_\tau\|_{\mathcal{L}_\infty} \le \|C_u(s)\|_{\mathcal{L}_1} L_{\rho_r}\|(x_{\text{ref}} - x)_\tau\|_{\mathcal{L}_\infty} + \kappa_0\|\tilde{x}_\tau\|_{\mathcal{L}_\infty}.$$

Finally, the upper bounds on $\|\tilde{x}_\tau\|_{\mathcal{L}_\infty}$ and $\|(x_{\text{ref}} - x)_\tau\|_{\mathcal{L}_\infty}$ in (5.62) and (5.65), together with the above inequality, lead to

$$\|(u_{\text{ref}} - u)_\tau\|_{\mathcal{L}_\infty} \le \|C_u(s)\|_{\mathcal{L}_1} L_{\rho_r}(\gamma_1 - \beta) + \kappa_0\gamma_0 < \gamma_2, \tag{5.66}$$

which contradicts the second equality in (5.58), which implies that the upper bounds in (5.65) and (5.66) hold uniformly. The upper bound in (5.55) follows from (5.62) directly. The upper bounds in (5.53) and (5.54) follow from (5.59). □

Remark 5.2.2 It follows from (5.52) that one can prescribe an arbitrary desired performance bound γ_0 by increasing the adaptive gain, which further implies, from (5.29) and (5.30),

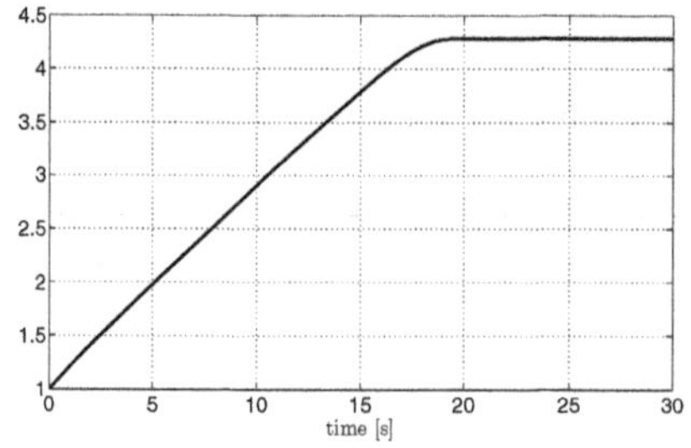

(a) $\omega_m(t)$

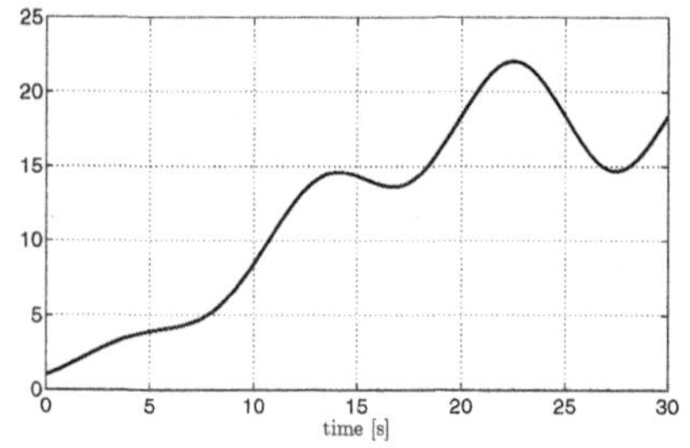

(b) $b_m(t)$

Figure 5.8: Plots of $\omega_m(t)$ and $b_m(t)$.

that one can achieve arbitrarily small γ_1 and γ_2 for the system's signals, both input and output, simultaneously.

5.2.4 Simulation Example

To illustrate the results presented in this chapter, consider the system dynamics in (5.25), with

$$A_m(t) = \begin{bmatrix} 0 & 1 \\ -\omega_m^2(t) & -2\zeta_m\omega_m(t) \end{bmatrix}, \; b(t) = \begin{bmatrix} 0 \\ b_m(t) \end{bmatrix},$$
$$c = \begin{bmatrix} 1 & 0 \end{bmatrix},$$

$$\omega_m(t) = \begin{cases} 1 + 20(1 - \mathrm{e}^{-0.01t}), & t \in [0,\ 15], \\ 21 - 20\mathrm{e}^{-0.15} + 0.5\sin(0.4\mathrm{e}^{-0.15}(t-15)), & t \in (15,\ 15 + \mathrm{e}^{0.15}5\pi/4], \\ 21.5 - 20\mathrm{e}^{-0.15}, & t > 15 + \mathrm{e}^{0.15}5\pi/4, \end{cases}$$

and $\zeta_m = 0.7$, $b_m(t) = \omega_m^2(t) + 0.2\omega_m^2(t)\sin(0.2\pi t)$. The plots of $\omega_m(t)$ and $b_m(t)$ are given in Figure 5.8. One can see that the natural frequency of the system is continuously growing with time and is changing from 1 to ≈ 4.5. Also, $b_m(t)$ is significantly changing within the interval $[1,\ 20]$ s of the simulation time.

The unmodeled dynamics are given by the Lorenz attractor:

$$g(t, x_z, x) = t_s \begin{bmatrix} \sigma_l(x_{z_2} - x_{z_1}) \\ r_l x_{z_1} - x_{z_2} - x_{z_1}x_{z_3} + k_l x_1 \\ x_{z_1}x_{z_2} - b_l x_{z_3} \end{bmatrix},$$
$$g_o(t, x_z) = c_l^\top x_z,$$

where $t_s = 0.1$ is a time-scaling coefficient; $\sigma_l = 10$, $r_l = 28$, $b_l = 8/3$ are the Lorenz system parameters; and $k_l = 50$ and $c_l^\top = [0,\ 1/15,\ 0]$ are the input and the output gains, respectively. To illustrate the behavior of the unmodeled dynamics, we consider it separately from the system in (5.25) and excite it with the step-impulse signal of the form $x(t) = [\mathrm{u}(t) - \mathrm{u}(t-1),\ 0]$, where $\mathrm{u}(t)$ is a unit step function. The response is given in Figure 5.9. One can see that it has a stable limit cycle, biased from the equilibrium. Moreover, the response to the input signal is aggressive and has a relatively high derivative.

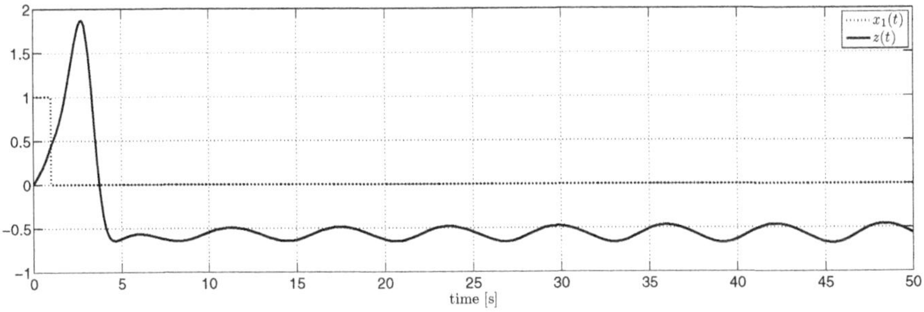

Figure 5.9: Response of the unmodeled dynamics to $x = [\mathrm{u}(t) - \mathrm{u}(t-1),\ 0]^\top$.

To illustrate the performance of $\mathcal{L}_1$ adaptive controller, we consider two scenarios with different unmodeled actuator dynamics and uncertainties:

- Scenario 1: Let

$$F_1(s) = \frac{6400}{s^2 + 112s + 6400}$$

and

$$f(t,x,z) = x_1^2 + \sin(x_1) + z + d(t),$$

where $d(t) = 0.5\sin(0.3\pi t) + 0.3\sin(0.2\pi t)$.

- Scenario 2: Let

$$F_2(s) = \frac{10000}{s^2 + 100s + 10000}$$

and

$$f(t,x,z) = 2x_1^2 + x_1x_2 + z + d(t),$$

where $d(t) = 0.7\sin(0.1\pi t) + 0.1\sin(0.2\pi t)$.

For the implementation of the $\mathcal{L}_1$ adaptive controller, we set

$$k = 25, \qquad D(s) = \frac{100}{s(s^2 + 140s + 100)}.$$

Further, we set the adaptive gain and the projection bounds to be $\Gamma = 10^5$, $\Theta = [-100,\ 100]$, $\Omega = [0.1,\ 10]$, $\Delta = 100$.

The simulation results of the $\mathcal{L}_1$ adaptive controller are shown in Figures 5.10–5.14. Figure 5.10 shows the simulations results for Scenario 1 for a series of step-reference signals, introduced at $t_1 = 0$ s, $t_2 = 10$ s, $t_3 = 15$ s. One can see that $y(t) = x_1(t)$ tracks $r(t)$, and the closed-loop system is behaving close to the time-varying ideal system $x_{\mathrm{id}}(t)$, given by

$$\dot{x}_{\mathrm{id}}(t) = A_m(t)x_{\mathrm{id}}(t) + b(t)k_g(t)r(t),$$
$$y_{\mathrm{id}}(t) = c^\top x_{\mathrm{id}}(t).$$

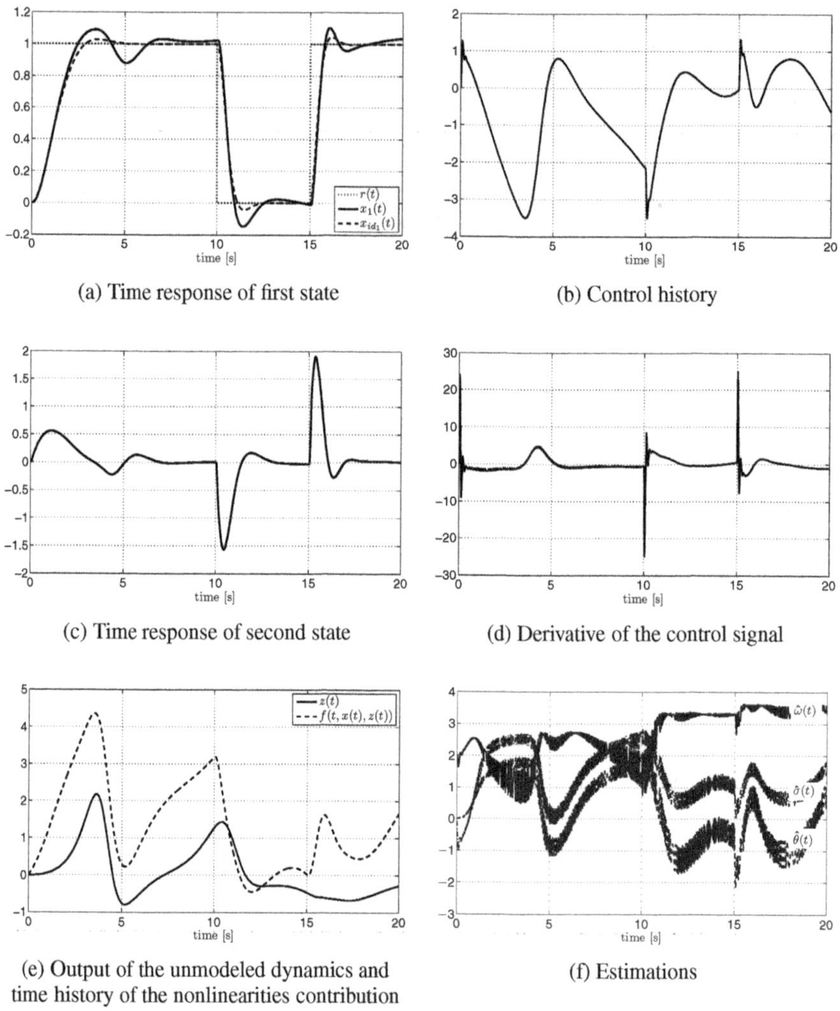

(a) Time response of first state

(b) Control history

(c) Time response of second state

(d) Derivative of the control signal

(e) Output of the unmodeled dynamics and time history of the nonlinearities contribution

(f) Estimations

Figure 5.10: Performance of the $\mathcal{L}_1$ adaptive controller for Scenario 1.

Notice that the desired time constant changes its value from 1 s to $\approx 1/4$ s, and the closed-loop system response reflects this change by decreasing the settling time for step commands at t_2 and t_3.

Figure 5.11 shows the results for Scenario 2 for two series of step signals with different amplitudes $r_1(t) \equiv 1$, $r_2(t) \equiv 0.5$. One can see that the transient for both commands is scaled, similar to linear systems response. Figure 5.12 shows the tracking results for Scenarios 1 and 2 with the reference signal $r(t) = \sin(t)$. One can see that the closed-loop system has tracking performance very close to ideal.

Next we simulate the closed-loop system with nonzero initialization error. We set the initial condition of the system to $x_0 = [0.5,\ 0.5]^\top$ and the initial condition of the state predictor to $\hat{x}_0 = [1.5,\ -0.2]^\top$. We let the initial conditions of unmodeled dynamics be

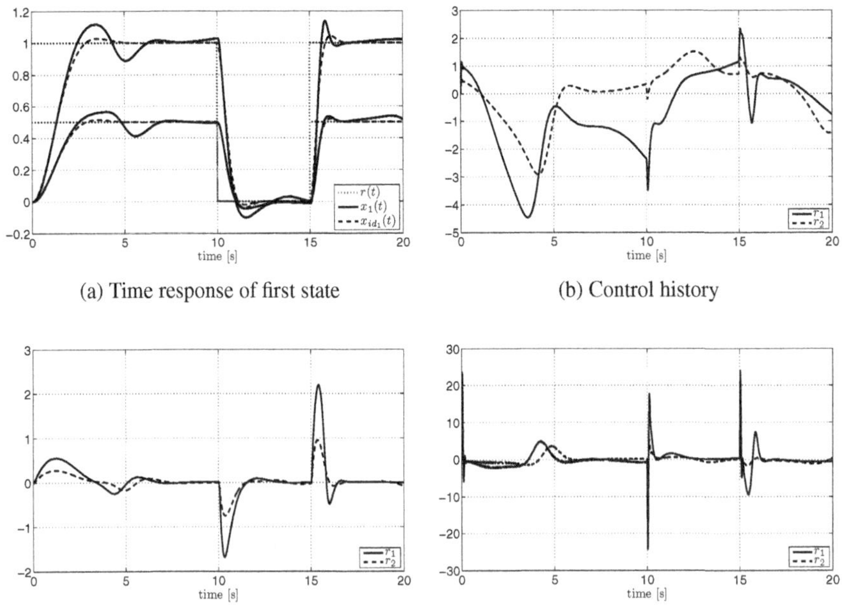

(a) Time response of first state

(b) Control history

(c) Time response of second state

(d) Derivative of the control signal

Figure 5.11: Response of the $\mathcal{L}_1$ adaptive controller for Scenario 2 to the reference commands $r_1(t)$ and $r_2(t)$.

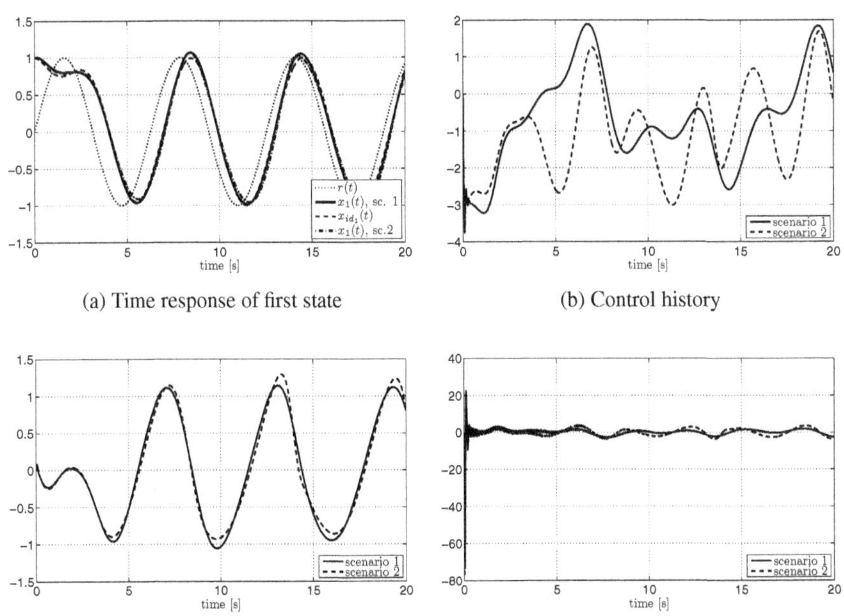

(a) Time response of first state

(b) Control history

(c) Time response of second state

(d) Derivative of the control signal

Figure 5.12: Tracking performance of the $\mathcal{L}_1$ adaptive controller for $r(t) = \sin(t)$ for Scenarios 1 and 2.

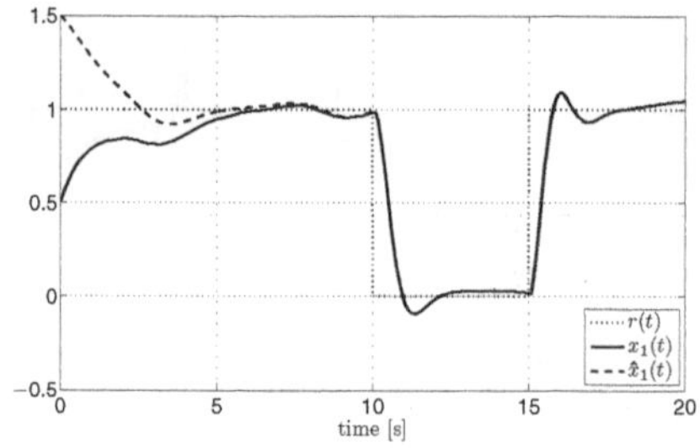

(a) Time response of first state

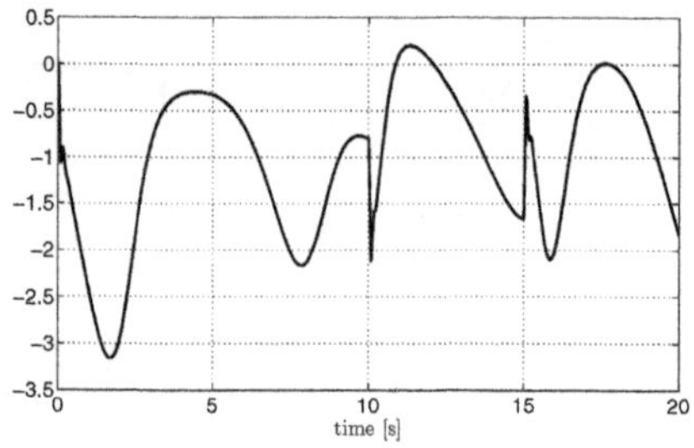

(b) Control history

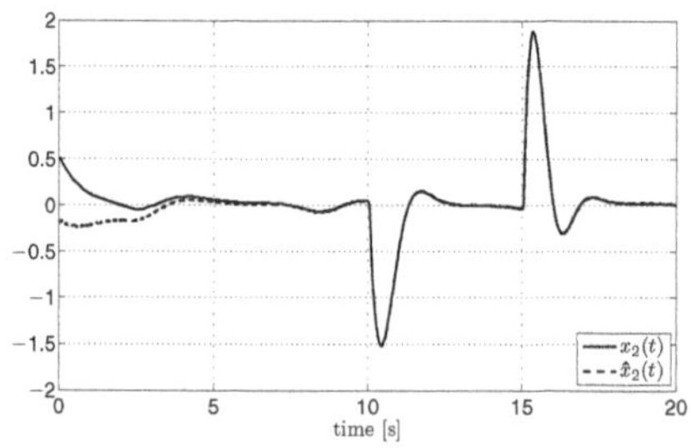

(c) Time response of second state

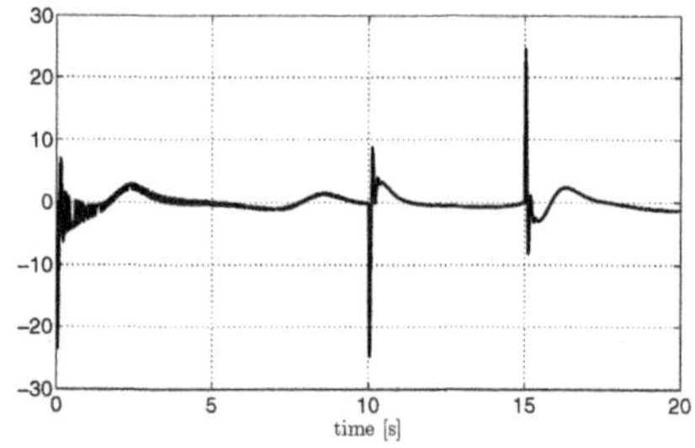

(d) Derivative of the control signal

Figure 5.13: System response in the presence of nonzero initialization error.

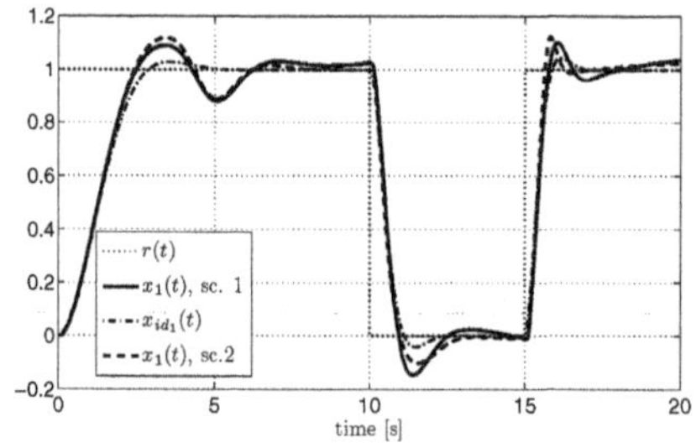

(a) Time response of first state

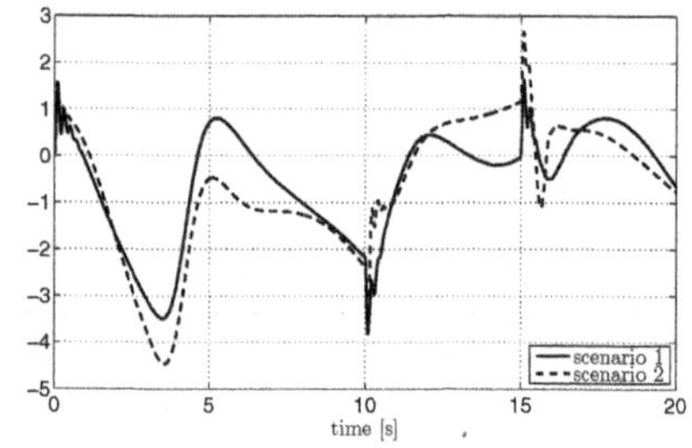

(b) Control history

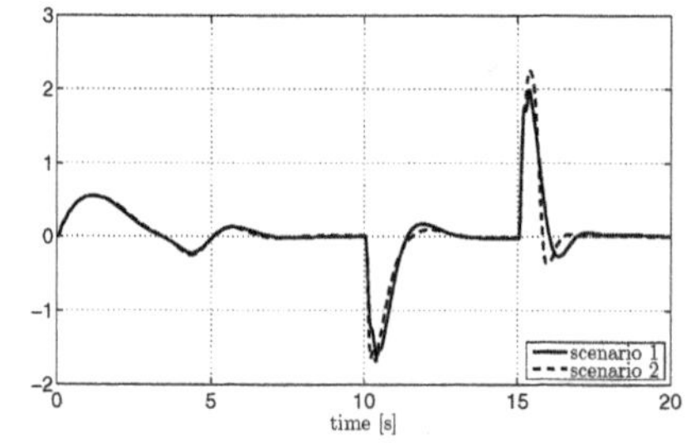

(c) Time response of second state

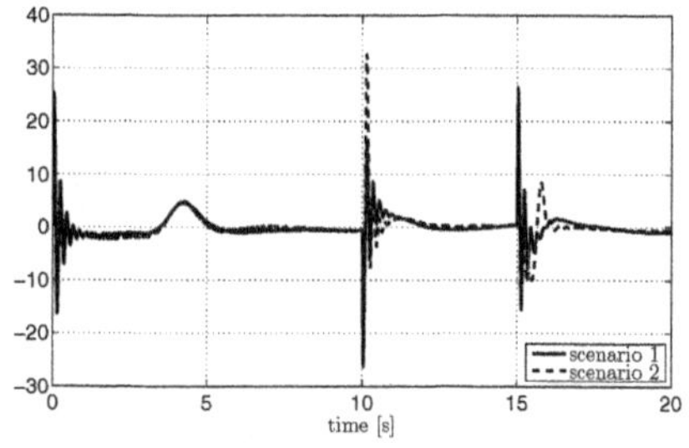

(d) Derivative of the control signal

Figure 5.14: System response in the presence of time delay of 13 ms for Scenarios 1 and 2.

$x_{z0} = [3,\ 7,\ 10]^\top$. The results are shown in Figure 5.13. One can see that the closed-loop system is stable and the state of the predictor rapidly converges to the state of the system. We observe that the nonzero initialization error does not deteriorate significantly the system's transient response.

Further, we verify the robustness of the $\mathcal{L}_1$ adaptive controller to time delay in the control channel. Figure 5.14 shows the simulation results in the presence of time delay of 13 ms for Scenarios 1 and 2. One can see that system has insignificant degradation in the performance and remains stable.

We note that the $\mathcal{L}_1$ adaptive controller guarantees smooth and uniform transient response for time-varying reference systems without any retuning of the controller in the presence of different types of nonlinear uncertainties, unmodeled actuator dynamics, and disturbances.

Chapter 6

Applications, Conclusions, and Open Problems

In this book we attempted to present a unified treatment of the $\mathcal{L}_1$ adaptive control theory with detailed proofs of the main results. The key feature of its architectures is the *guaranteed robustness in the presence of fast adaptation*. We considered a broad class of deterministic systems and presented state-feedback and output-feedback architectures, summarizing the main assumptions and proofs for the *uniform guaranteed performance bounds*. This chapter presents preliminary results on the development and application of $\mathcal{L}_1$ adaptive control architectures to the design of inner-loop flight control systems and then gives a brief description of the results not covered in the book, draws some concluding remarks, and summarizes open problems for future research.

6.1 $\mathcal{L}_1$ Adaptive Control in Flight

Inner-loop adaptive flight control systems may provide the opportunity to improve aircraft performance and reduce pilot compensation in challenging flight envelope conditions or in the event of control surface failures and vehicle damage. However, implementing adaptive control technologies can increase the complexity of the flight control systems beyond the capability of current Verification and Validation (V&V) processes [175]. This fact, combined with the criticality of inner-loop flight control systems, leads to high certification costs and renders difficult the transition of these technologies to military and commercial applications. Programs like NASA's Integrated Resilient Aircraft Control (IRAC) program and Wright-Patterson AFRL's Certification Techniques for Advanced Flight Critical Systems represent an effort to advance the state of the art in adaptive control technology, to analyze the deficiencies of current V&V practices, and to advance airworthiness certification of adaptive flight control systems.

These two programs have significantly contributed to the ongoing efforts in the development, flight verification and validation, and transition of $\mathcal{L}_1$ adaptive control from a theoretical research field into a viable and reliable technology toward improving the robustness and performance of advanced flight control systems. The main goal of implementing an $\mathcal{L}_1$ adaptive controller onboard is to guarantee that an aircraft, suddenly experiencing an adverse flight regime or an unexpected failure, will not "escape" its α-β *wind-tunnel data envelope* (see Figure 6.1), provided that some control redundancy remains. It is important

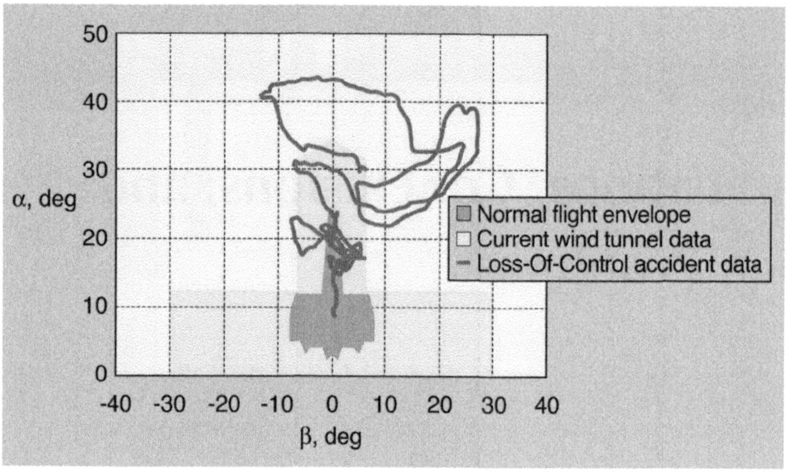

Figure 6.1: Loss-of-control accident data relative to angle of attack and angle of sideslip. This figure appears here with the permission of NASA [53] .

to notice that, outside this wind-tunnel envelope, the aerodynamic models available are usually obtained by *extrapolation* of wind-tunnel test data, implying that these models are highly uncertain. This fact suggests that pilots might not be correctly trained to fly the aircraft in these regimes (or, in the case of unmanned vehicles, the guidance loops might not be properly designed for safe recovery). Moreover, it does not seem reasonable to rely on a flight control system to compensate for the uncertainties in these flight conditions, as aircraft controllability is not even guaranteed in such regimes. In this sense, flight control systems with *only asymptotic guarantees* might *not prevent* the aircraft from entering these adverse flight conditions with unusual attitudes, and therefore inner-loop control architectures ensuring *transient response with desired specifications and guaranteed robustness* appear to be imperative for safe operation of manned (and unmanned) aircraft in the presence of anomalies. In particular, it is important to note that successful recovery from a failure, if possible at all, can be achieved only during the first few seconds after it occurs, in which the airplane is still in a regime with some controllability guarantees (Figure 6.1). Hence, the *guaranteed fast and robust adaptation* of $\mathcal{L}_1$ adaptive control architectures makes this control theory ideally suited for such eventualities. In fact, the $\mathcal{L}_1$ adaptive flight control system has been already shown to be capable of compensating for sudden, unknown, severe failure events, while delivering predictable performance across the flight envelope without resorting to gain scheduling of the control parameters, persistency of excitation, or control reconfiguration (see Sections 6.1.1 and 6.1.2 for details). Also, a *graceful* degradation in performance and handling qualities has been observed as the failures and structural damage impose increasingly severe limitations on the controllability of the aircraft.

The scope of this section is to demonstrate the advantages of $\mathcal{L}_1$ adaptive control as a *verifiable* robust adaptive control architecture with the potentiality of reducing flight control design costs and facilitating the transition of adaptive control into advanced flight control systems.

6.1.1 Flight Validation of $\mathcal{L}_1$ Adaptive Control at Naval Postgraduate School

Recognizing the value of experimental V&V of advanced flight control algorithms, the Naval Postgraduate School (NPS) team has developed the so-called Rapid Flight Test Prototyping System (RFTPS) [47]. The RFTPS consists of a testbed unmanned aerial vehicle (UAV) equipped with a commercial autopilot (AP), an embedded computer running the research algorithms in real time, and a ground control station for flight management and data monitoring and collection. This system facilitates the real-time onboard integration of advanced control algorithms and provides the opportunity to design and conduct comprehensive flight test programs to evaluate the robustness and performance characteristics of these algorithms.

In order to demonstrate the benefits of $\mathcal{L}_1$ adaptive control, the commercial autopilot of the RFTPS was augmented with the $\mathcal{L}_1$ adaptive output feedback architectures presented in Chapter 4. The $\mathcal{L}_1$ augmentation loop is introduced to enhance the angular-rate tracking capabilities of the autopilot across the flight envelope in the event of control surface failures and in the presence of significant environmental disturbances. The inner-loop $\mathcal{L}_1$ adaptive flight control architecture implemented on the RFTPS is represented in Figure 6.2.

In the following sections, we present an overview of the main results from the extensive flight test program conducted by NPS since 2006 in Camp Roberts, CA. The reader will find detailed explanations and further hardware-in-the-loop simulations and flight test results in [3, 46, 82, 94, 110, 117].

Aggressive Path Following

Conventional autopilots are normally designed to provide only guidance loops for waypoint navigation. In order to extend the range of possible applications of (small) UAVs equipped with traditional autopilots, a solution to the problem of three-dimensional (3D) path-following control was presented in [82]. The solution proposed exhibits a multiloop control structure, with (i) an outer-loop path-following control law that relies on a nonlinear control strategy derived at the kinematic level, and (ii) an inner-loop consisting of the commercial autopilot augmented with the $\mathcal{L}_1$ adaptive controller. The overall closed-loop system with the $\mathcal{L}_1$ adaptive augmentation loop is presented in Figure 6.3.

Flight test results comparing the performance of the path-following algorithm with and without $\mathcal{L}_1$ adaptation are shown in Figure 6.4. The flight test data include the two-dimensional (2D) horizontal projection of the commanded and the actual paths, the commanded $r_c(t)$ and the measured $r(t)$ turn rate responses, and the path-tracking errors $y_F(t)$ and $z_F(t)$. The results show that the UAV is able to follow the path, keeping the path-following tracking errors reasonably small during the whole experiment. The plots also demonstrate the improved path-following performance when the $\mathcal{L}_1$ augmentation loop is enabled. One can observe that the nominal outer-loop path-following controller exhibits significant oscillatory behavior, with rate commands going up to 0.35 rad/s and with maximum path-tracking errors around 18 m, while meantime the $\mathcal{L}_1$ augmentation loop is able to improve the angular-rate tracking capabilities of the inner-loop controller, which results in rate commands not exceeding 0.15 rad/s and path-tracking errors below 8 m. Furthermore, it is important to note that the adaptive controller does not introduce any

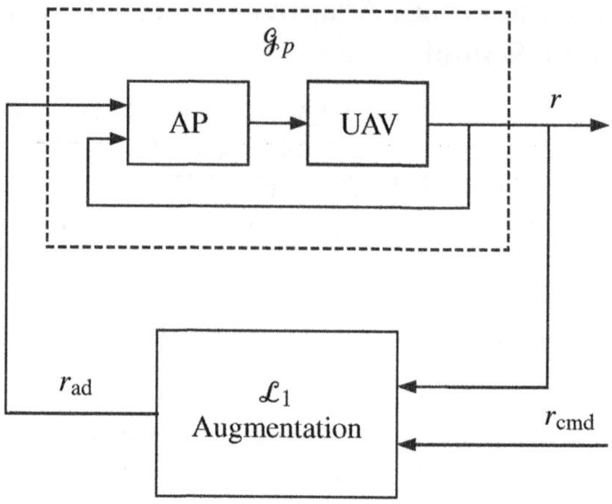

(a) Inner-loop structure with $\mathcal{L}_1$ adaptive augmentation

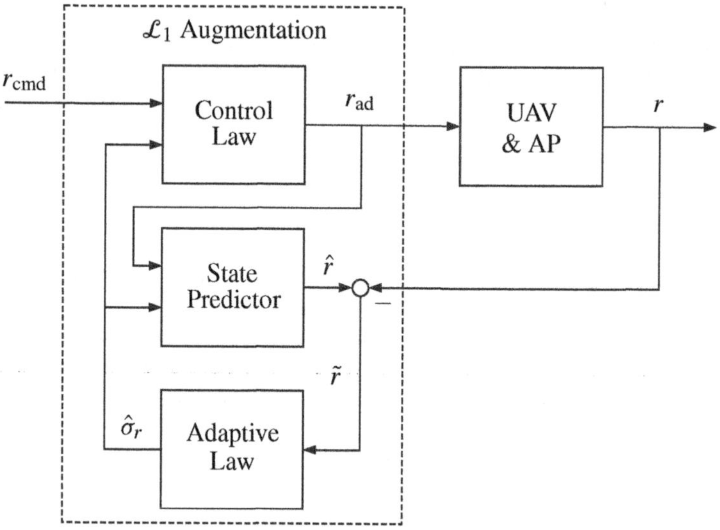

(b) $\mathcal{L}_1$ adaptive controller for turn-rate control

Figure 6.2: Inner-loop $\mathcal{L}_1$ adaptive augmentation loop tested by NPS.

high-frequency content into the commanded turn-rate signal, as can be seen by comparing Figures 6.4(d) and 6.4(c).

Finally, it is important to mention that, in the derivations in [82], the uniform performance bounds that the $\mathcal{L}_1$ adaptive controller guarantees in both transient and steady-state are critical to prove stability of the path-following closed-loop system, which takes into account the dynamics of the UAV with its autopilot.

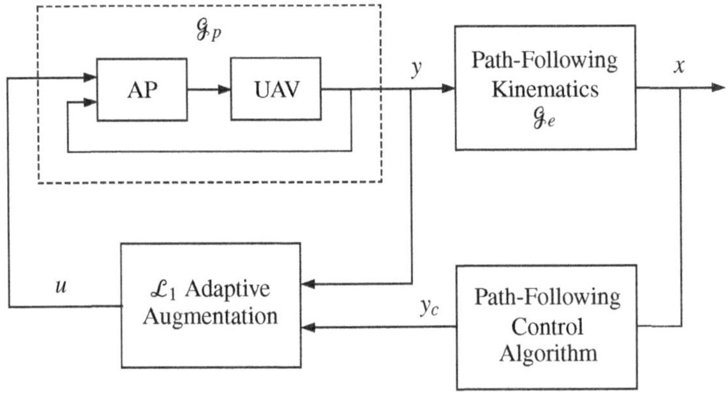

Figure 6.3: Closed-loop path-following system with $\mathcal{L}_1$ adaptive augmentation.

$\mathcal{L}_1$ Adaptation in the Presence of Control Surface Failures

The path-following flight test setup introduced in the previous section was used in [46] to demonstrate that the $\mathcal{L}_1$ augmentation loop provides fast recovery to sudden locked-in-place failures in either one of the ailerons or in the rudder of the RFTPS, while the nominal unaugmented system goes unstable. In these experiments, we took advantage of the capability of the RFTPS to instantaneously deflect and hold a preprogrammed combination of control surfaces at a predefined position without notifying or reconfiguring the nominal autopilot.

While the flight experiments considered failures in the left aileron covering the range from 0 deg to -12 deg (with respect to a trim value -2.34 deg), and rudder failures from 0 deg to 2 deg, in this section we present only an extract from these results. In particular, Figures 6.5 and 6.6 illustrate the performance of the path-following system with two levels of sudden left-aileron locked-in-place failures at -2 deg and -10 deg (with respect to trim value -2.34 deg):

(i) Analysis of the -2 deg case showed that even such a small deflection pushes the UAV away to the right from the desired path (see Figure 6.5(a)), resulting in almost 25 m of lateral miss distance (see Figure 6.5(b)). After the failure is introduced, the UAV converges to a 5-m lateral error boundary in about 20 s.

(ii) The results of the -10 deg left-aileron failure (Figure 6.6) are similar to the ones for the previous case. Naturally, the errors and the control efforts increase due to the increased severity of the failure, and the impaired UAV converges to the 5-m boundary in approximately 30 s.

Moreover, analysis of the entire series of results with left-aileron failures (covering the range from 0 deg to -12 deg) shows a *graceful* and predictable degradation in the path-following performance. It is important to mention that the predictability in the response provided by the $\mathcal{L}_1$ adaptive controller is especially critical for manned aircraft.

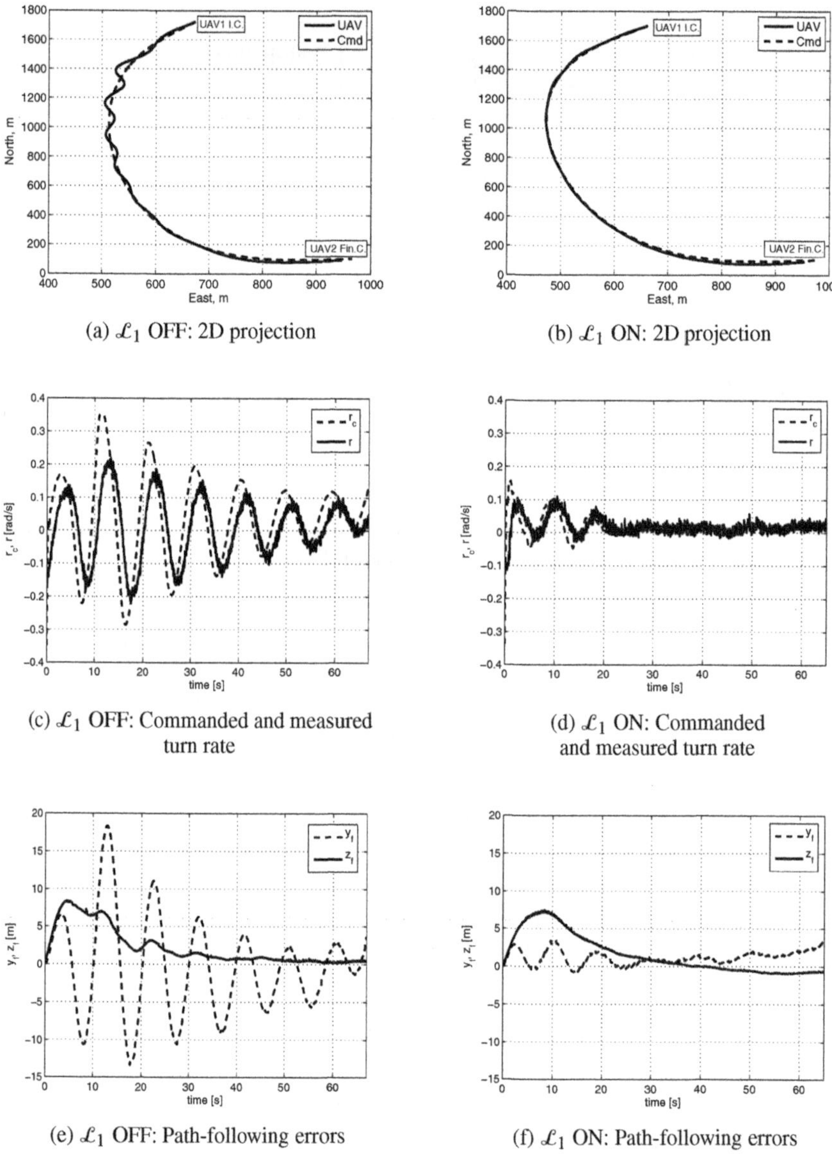

(a) $\mathcal{L}_1$ OFF: 2D projection

(b) $\mathcal{L}_1$ ON: 2D projection

(c) $\mathcal{L}_1$ OFF: Commanded and measured turn rate

(d) $\mathcal{L}_1$ ON: Commanded and measured turn rate

(e) $\mathcal{L}_1$ OFF: Path-following errors

(f) $\mathcal{L}_1$ ON: Path-following errors

Figure 6.4: Fight test. Path-following performance with and without $\mathcal{L}_1$ adaptive augmentation.

Finally, we notice that the $\mathcal{L}_1$ adaptive controller automatically readjusts the control signals in order to stabilize the impaired airplane and uses the remaining control authority to steer the airplane along the path, without resorting to fault detection and isolation methods or reconfiguration of the existing inner-loop control structure. Therefore, integration of $\mathcal{L}_1$ adaptation onboard increases the fault tolerance of the system.

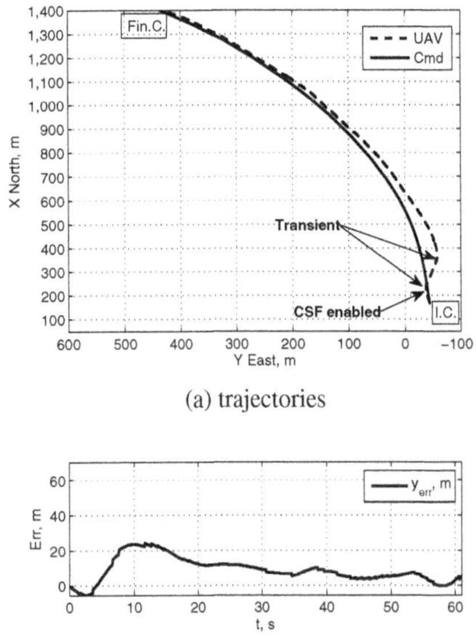

(a) trajectories

(b) lateral error

Figure 6.5: Flight test. 2 deg locked-in-place left-aileron failure.

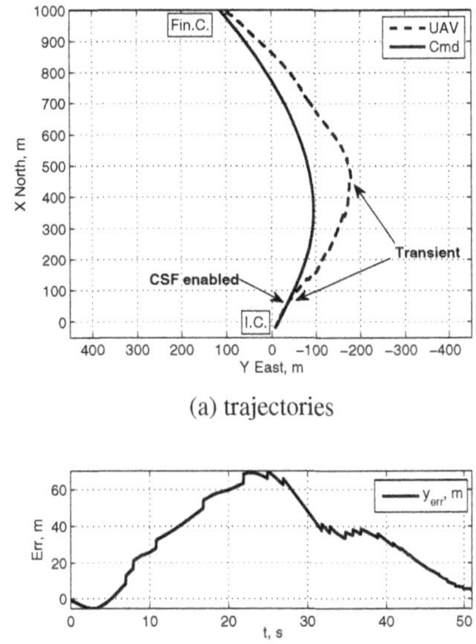

(a) trajectories

(b) lateral error

Figure 6.6: Flight test. 10 deg locked-in-place left-aileron failure.

Rohrs' Example in Flight

Motivated by Rohrs' example, which was analyzed in detail in Section 2.3, the paper in [94] extends the setup introduced by Rohrs in [147] to the flight test environment, in which the first-order nominal plant is replaced by the UAV with its commercial autopilot. In [94], different unmodeled dynamics were considered, ranging from very lightly damped second-order systems (representing, for example, a flexible body mode of an airplane) to control surface failures. Figure 6.7 shows the block diagrams with the implementation of the unmodeled dynamics for two particular cases of second-order transfer functions at the output of the nominal plant.

Initially, a conventional output feedback MRAC algorithm from [75] with properties similar to the adaptive controller used in [147] was used as an augmentation loop. This was done to verify correctness of the flight test set-up—the results obtained in flight should be similar to the ones obtained in [147]. The same scenario was then used to evaluate the performance of the $\mathcal{L}_1$ adaptive augmentation. In [94], the authors also considered the implementation of some of the adaptive law modifications developed to overcome the problem of parameter drift in conventional MRAC. We note that the determination of the phase crossover frequency, necessary to reproduce Rohrs' example, required identification of the frequency response of the nominal plant consisting of the UAV with the autopilot. Next, we present an extract of the results presented in [94], in which the second-order unmodeled dynamics introduced artificially at the output of the nominal plant were characterized by a natural frequency of 1.5 rad/s and a damping ratio of 0.45.

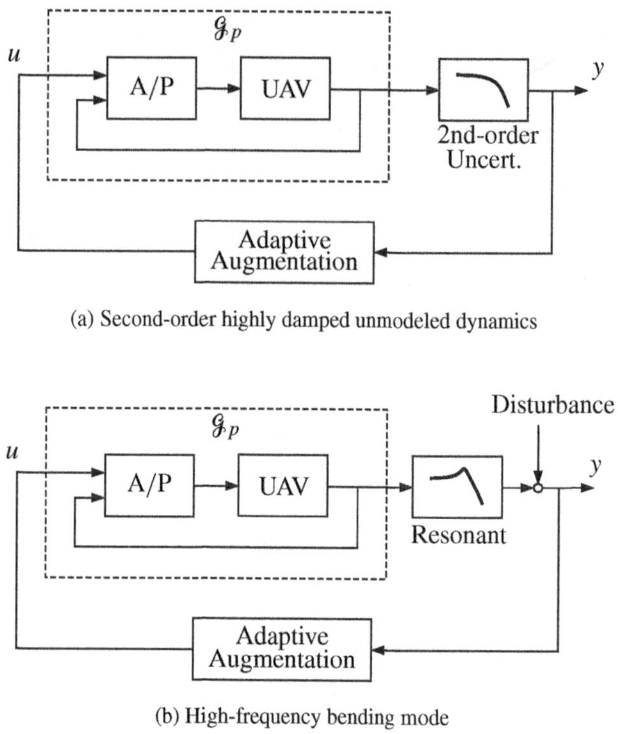

Figure 6.7: System with two cases of unmodeled dynamics and output disturbance.

First, Figure 6.8 shows the response of the closed-loop MRAC adaptive system in the presence of the unmodeled dynamics to a biased sinusoidal reference signal r_{cmd} at the phase-crossover frequency. One can see that the parameters drift slowly, generating command signals r_{ad} larger than $\pm 100\ \frac{\text{deg}}{\text{s}}$, which are sent to the autopilot. It is important to note that for safety reasons, in the actual implementation the autopilot limits the commands received from the adaptive controller to avoid undesirable attitudes that might lead to the loss of the UAV. The same test was performed considering three different modifications of the adaptive laws (σ-modification, e-modification, and projection operator) to verify that, in fact, the closed-loop adaptive system remains stable. Flight test results for the e-modification, which are not included in this book, can be found in [94]. The results obtained show that under appropriate (trial-and-error-based) tuning, the stability of the closed-loop system can be preserved, although the resulting performance is in general poor, and, similar to pure MRAC architectures (without adaptive law modifications), the transient response characteristics of the system are highly unpredictable.

The same experiment was conducted for the $\mathcal{L}_1$ adaptive augmentation; see Figure 6.9. The system maintains stability during the whole flight, and the control signal r_{ad} remains inside reasonable bounds during the experiment. As one would expect, since the frequency of the reference signal is well beyond the bandwidth of the low-pass filter in the control law, the $\mathcal{L}_1$ adaptive controller is not able to recover desired performance of

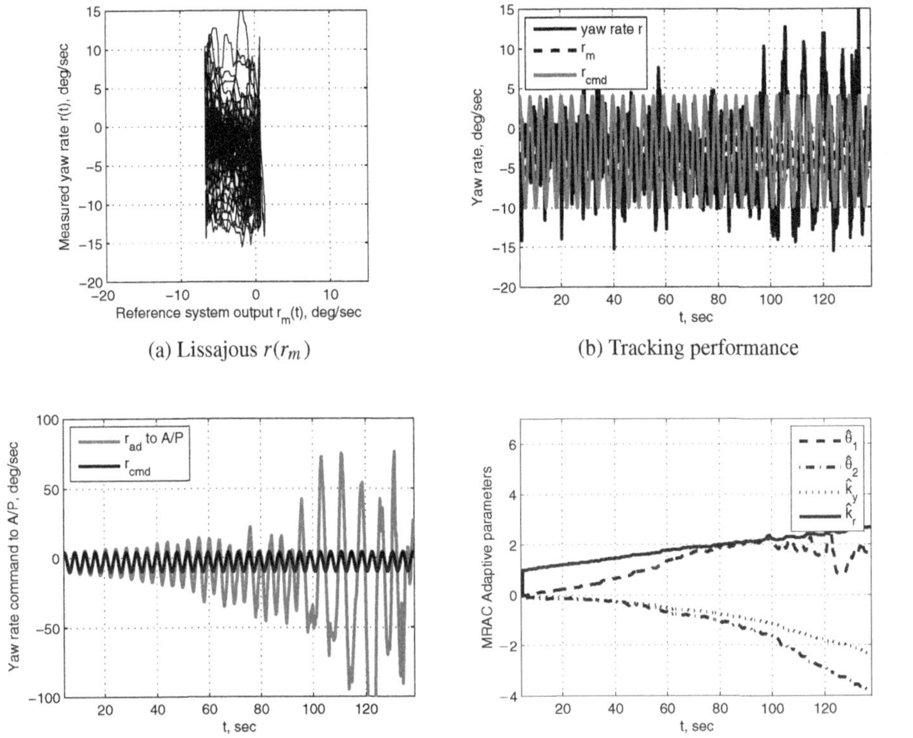

(a) Lissajous $r(r_m)$

(b) Tracking performance

(c) Adaptive command to AP

(d) MRAC parameters

Figure 6.8: MRAC augmentation. Closed-loop response in the presence of second-order unmodeled dynamics to a biased sinusoidal reference signal at the phase-crossover frequency.

the closed-loop adaptive system. The response with $\mathcal{L}_1$ adaptive controller is consistent during the entire flight and does not exhibit undesirable characteristics like bursting. The setup above was also used to illustrate the degradation of performance as the frequency of the reference signal increases beyond the bandwidth of the low-pass filter. To this end, the closed-loop system with the second-order unmodeled dynamics and the $\mathcal{L}_1$ controller implemented on board was driven with a set of biased sinusoidal reference signals at different frequencies. Figure 6.10 shows the results of these experiments. It can be seen that the output r of the closed-loop adaptive system is able to track the output of the reference system r_{m} for reference signals at low frequencies ($\omega = 0.3\ \frac{\mathrm{rad}}{\mathrm{s}}$), and, as the frequency of the reference signal increases, the performance degrades slowly and progressively. This *graceful* degradation in the performance of the system is consistent with the theoretical claims of the $\mathcal{L}_1$ adaptive control theory, which predict the response of the closed-loop adaptive system and ensure graceful degradation of it outside the bandwidth of the design.

Finally, a combined experiment is presented in Figure 6.11. In this experiment, the second-order unmodeled dynamics were first injected to the nominal plant (unaugmented autopilot), and then the adaptive algorithms were enabled, first the MRAC adaptive al-

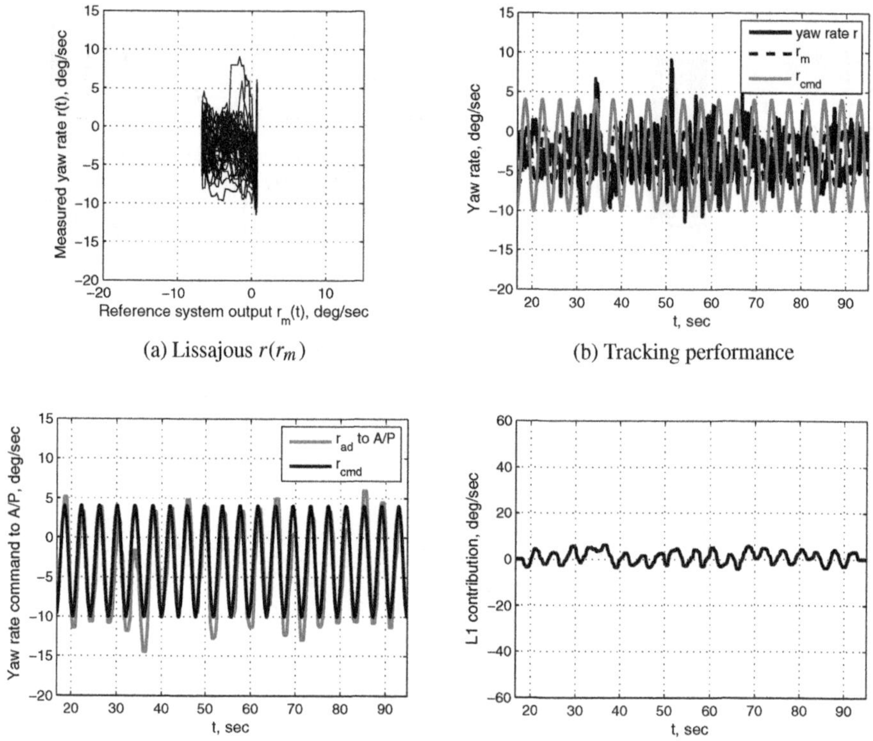

(a) Lissajous $r(r_m)$

(b) Tracking performance

(c) Adaptive command to AP

(d) $\mathcal{L}_1$ contribution

Figure 6.9: $\mathcal{L}_1$ augmentation. Closed-loop response in the presence of second-order unmodeled dynamics to a biased sinusoidal reference signal at the phase-crossover frequency.

gorithm and then the $\mathcal{L}_1$ adaptive controller. The figure shows that with the unmodeled dynamics injected, the closed-loop MRAC system becomes unstable, and the control command from the adaptive algorithm eventually hits the saturation limits of the autopilot. Then, at $t = 56$ s, the adaptive augmentation loop was switched from MRAC to $\mathcal{L}_1$ control, and the UAV recovered stability in around 1.5 s, which confirms the theoretical claims of the $\mathcal{L}_1$ adaptive control architectures regarding their fast adaptation with guaranteed robustness.

Implementation Details

Because the performance bounds and the stability margins of the $\mathcal{L}_1$ adaptive controller can be systematically improved by increasing the adaptation gain, it is critical to ensure that the implementation of the *fast estimation* scheme does not lead to numerical instabilities and that the onboard CPU has enough computational power to robustly execute the fast integration. Therefore, in this section we summarize implementation details of the $\mathcal{L}_1$ controller in the RFTPS, providing some intuitive guidelines in the required hardware specifications.

We start by noting that the entire control system implemented onboard the small

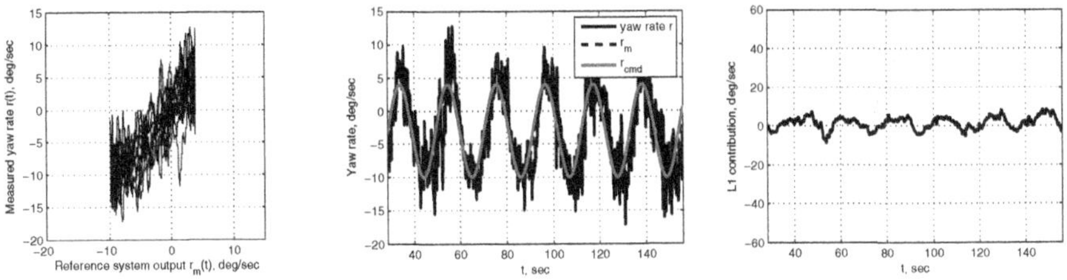

(a) Lissajous $r(r_m)$ (b) Tracking performance (c) $\mathcal{L}_1$ contribution

Figure 6.10: $\mathcal{L}_1$ augmentation. Closed-loop response in the presence of second-order unmodeled dynamics to biased sinusoidal reference signal at $\omega = 0.3\ \frac{\text{rad}}{\text{s}}$.

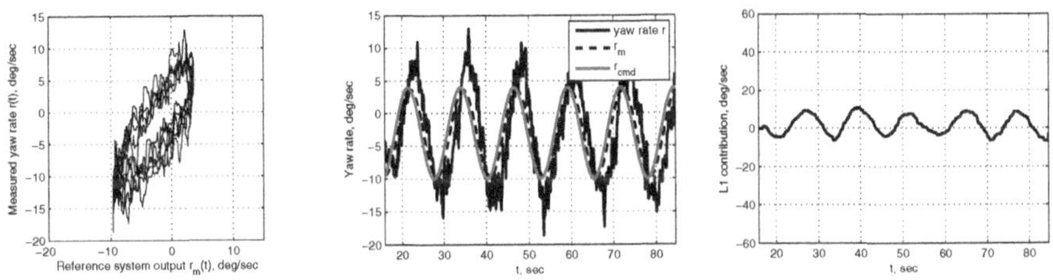

(d) Lissajous $r(r_m)$ (e) Tracking performance (f) $\mathcal{L}_1$ contribution

Figure 6.10: $\mathcal{L}_1$ augmentation. Closed-loop response in the presence of second-order unmodeled dynamics to biased sinusoidal reference signal at $\omega = 0.5\ \frac{\text{rad}}{\text{s}}$.

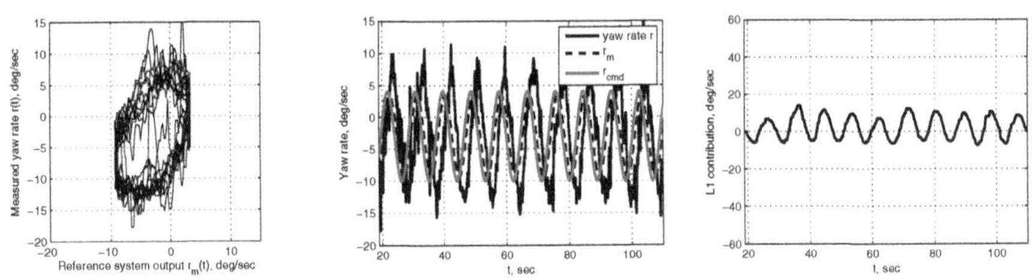

(g) Lissajous $r(r_m)$ (h) Tracking performance (i) $\mathcal{L}_1$ contribution

Figure 6.10: $\mathcal{L}_1$ augmentation. Closed-loop response in the presence of second-order unmodeled dynamics to biased sinusoidal reference signal at $\omega = 0.7\ \frac{\text{rad}}{\text{s}}$.

UAV is a multirate algorithm consisting of two primary subsystems: (i) the hardware interfacing modules executing at the rate allowed by the sensors or actuators, and (ii) the control subsystem that utilizes sensor data to produce a control signal to be sent back to the actuators.

In most of the cases, the hardware interfacing subsystem is executed at a significantly lower rate (10–100 Hz) than the control algorithm (100–1000 Hz), therefore demanding

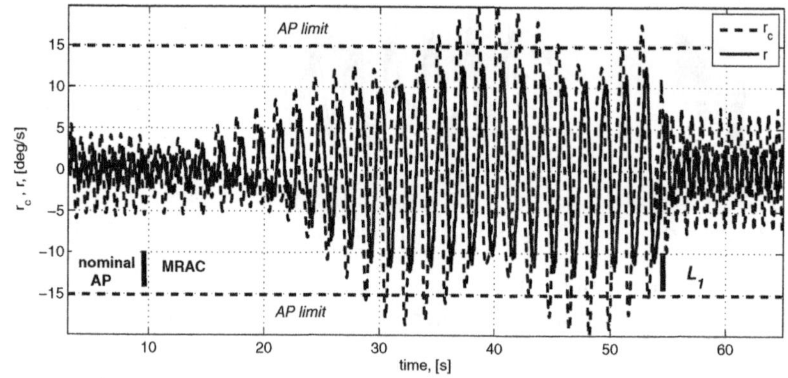

Figure 6.11: Combined experiment: switching from MRAC to $\mathcal{L}_1$ adaptive controller in the presence of second-order unmodeled dynamics.

insignificant CPU power. This provides a suitable framework for the implementation of $\mathcal{L}_1$ adaptive control architectures, as the computational power can be effectively used for fast adaptation. Furthermore, as soon as the *base sampling time* (fastest execution rate) of the multirate system is chosen to satisfy the performance requirements of the $\mathcal{L}_1$ adaptive controller (see equations (4.31) and (4.107)), and the multirate transitions are matched, the task execution time (TET) of the entire code (including the hardware interfacing loop) can be precisely predicted. Finally, knowing the overhead required by the real-time operating system (5–10% of TET), the performance (CPU frequency, bus speed, memory) of the required processing board can be calculated. More details on the real-time scheduling of custom developed algorithms can be found in the technical documentation to the xPC Target.[1]

Figure 6.2(b) shows implementation of the $\mathcal{L}_1$ adaptive output feedback controller for a single turn rate control channel (SISO). The solution looks rather trivial with an integrator as the most computationally demanding elementary block (see equations in Sections 4.1.2 and 4.2.2). Being a part of the control subsystem, the $\mathcal{L}_1$ controller execution is scheduled at the highest available rate. Clearly, the "price" associated with the implementation of the $\mathcal{L}_1$ adaptive controller lies primarily in the computational power required to accommodate the high adaptation rate ($\Gamma = 30000$ in the current NPS setup), resulting in significant stiffness of the underlying differential equation. From a numerical point of view, this translates into the boundedness of the integration error during the entire execution time. While offline, one can use, for example, the advances of MATLAB/Simulink that allow for iterative or multistep integration algorithms. However, in almost every real-time implementation that uses discretized algorithms running at a fixed sampling time, the conflict between numerical accuracy and fixed sampling interval might result in the loss of precision during integration [40, 41]. When a dynamically adjustable integration step is not available, this may lead to the cessation of execution and system failure. Therefore, finding the optimal

[1] xPC Target: Perform Real-Time Rapid Prototyping and Hardware-in-the-Loop Simulation Using PC Hardware. http://www.mathworks.com/products/xpctarget/

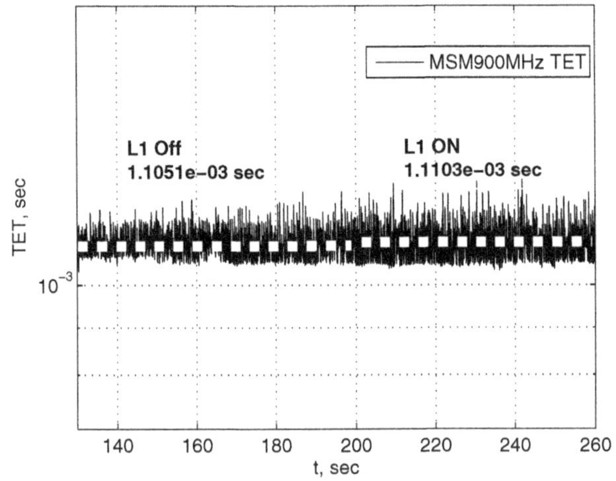

Figure 6.12: Hardware in the loop. TET: negligible increase in CPU load.

trade-off between the complexity of the integration algorithm, a feasible (fastest available) sampling time, and the numerical stability is of paramount importance.

The complexity of numerical real-time implementation can be resolved via the development of accurate and stable numerical integration algorithms and by utilizing the latest advancements of the automatic code generation. Besides verifying and generating a highly optimized executable code targeted for almost all existing CPU architectures, these tools provide powerful profiling methods that suggest possible bottlenecks in the code execution. Based on the detailed reports provided by these tools, the correct numerical algorithms, the sampling time, and the appropriate scheduling can be chosen to enable optimal multirate systems execution, providing full utilization of the available CPU power.

As an illustration of the feasibility of the above-described system implementation, we have analyzed the computational load of the algorithms. Figure 6.12 shows the computational power required for implementation of the outer-loop path-following controller in real time with and without $\mathcal{L}_1$ adaptive augmentation. The parameter chosen to represent the computational load is the average TET of the entire code (including hardware interfacing and control) during one sampling interval (10 ms). The control code with the standard ODE3 Bogacki–Shampine solver was implemented onboard an MSM900BEV[2] industrial PC104 computer using the xPC/RTW Target development environment. This figure highlights two important points. First, the average TET ($\approx$ 1 ms) is an order of magnitude less than the base sampling time of the real-time code (10 ms), which implies that the sampling time of the code implementation was chosen quite conservatively and could be reduced to improve the closed-loop performance. Second, the difference in the CPU load when the $\mathcal{L}_1$ adaptive controller is enabled or disabled is negligible (an additional 0.052% with respect to the nominal controller), which supports the ease of implementation in almost any platform. With the current pace of evolution of new processors and advances in automatic

[2]Advanced Digital Logic (ADL) | PC104+ : AMD Geode LX900 CPU 500MHz, MSM900BEV, http://www.adl-usa.com/products/cpu/index.php

code generation, validation, and verification tools, the resources available for embedded $\mathcal{L}_1$ adaptive control implementation are practically unlimited.

6.1.2 $\mathcal{L}_1$ Adaptive Control Design for the NASA AirSTAR Flight Test Vehicle

The research control law developed for the GTM aircraft has as its primary objective achieving tracking for a variety of tasks with guaranteed stability and robustness in the presence of uncertain dynamics, such as changes due to rapidly varying flight conditions during standard maneuvers, and unexpected failures. All of this must be achieved while providing Level I [39] handling qualities under nominal as well as adverse flight conditions. In particular, one essential objective for safe flight under adverse conditions is for the aircraft never to leave the extended α-β flight envelope; once outside the boundary and in uncontrollable space, no guarantees for recovery can be made (see Figure 6.1). Consequently, the adaptive controller should learn fast enough to keep the aircraft within the extended flight envelope. This implies that the control law action in the initial 2 to 3 seconds after initiation of an adverse condition is the key to safe flight.

The $\mathcal{L}_1$ control system used for this application is a three-axes angle of attack (α), roll rate (p)-sideslip angle (β) command augmentation system, and is based on the theory developed in Section 3.3, which compensates for both *matched* and *unmatched* dynamic uncertainties. For inner-loop flight control system design, the effects of slow outer-loop variables (e.g., airspeed, pitch angle, bank angle) may appear as unmatched uncertainties in the dynamics of the fast inner-loop variables we are trying to control (e.g., angle of attack, sideslip angle, roll rate). Also, unmodeled nonlinearities, cross coupling between channels, and dynamic asymmetries may introduce unmatched uncertainties in the inner-loop system dynamics. If the design of the inner-loop flight control system does not account for these uncertainties, their effect in the inner-loop dynamics will require continuous compensation by the pilot, thereby increasing the pilot's workload. Therefore, automatic compensation for the undesirable effects of these unmatched uncertainties on the output of the system is important to achieve desired performance, reduce pilot's workload, and improve the aircraft's handling qualities.

It is important to note that the $\mathcal{L}_1$ adaptive flight control system provides a systematic framework for adaptive controller design that allows for explicit enforcement of MIL-Standard requirements [1] and significantly reduces the tuning effort required to achieve desired closed-loop performance, which in turn reduces the design cycle time and development costs. In particular, the design of the $\mathcal{L}_1$ adaptive flight control system for the GTM is based on the linearized dynamics of the aircraft at an (equivalent) airspeed of 80 knots and at an altitude of 1000 ft. Since the airplane is Level I at this flight condition, the nominal desired dynamics of the (linear) state predictor were chosen to be similar to those of the actual airplane; only some additional damping was added to both longitudinal and directional dynamics, while the lateral dynamics were set to be slightly faster than the original ones in order to satisfy performance specifications. The state predictor was scheduled to specify different performance requirements at special flight regimes (high-speed regimes and high-α regimes). In order to improve the handling qualities of the airplane, a linear prefilter was added to the adaptive flight control system to ensure desired decoupling properties as well as desired command tracking performance. Overdamped second-order low-pass filters with unity DC gain were used in all the control channels, while their bandwidths were set to ensure (at least) a time-delay margin of 130 ms and a gain margin of 6 dB. Finally, the adaptation sampling time was set to $\frac{1}{600}$ s, which corresponds to the maximum integration step allowed in the AirSTAR flight control computer. We notice that the same control parameters for the prefilter, the low-pass filters, and the adaptation rate were used across the entire flight envelope with no scheduling or reconfiguration. Further details about the design of the $\mathcal{L}_1$ adaptive controller for the GTM can be found in [177].

This section presents the preliminary results of a piloted-simulation evaluation on the GTM aircraft high-fidelity simulator, which includes full nonlinear, asymmetric aerodynamics, actuator dynamics, sensor dynamics including nonlinearities, noise, biases, and scaling factors, and other nonlinear elements typical of these simulators. This piloted-simulation evaluation was part of development and profile planning of flight test tasks, and the tasks were flown with no training or repeatability and thus are considered to be a preliminary evaluation. At the time of this evaluation, the GTM aircraft had three flight control modes. *Mode 1* was the revisionary stick-to-surface control (the aircraft is Level I in this configuration); *Mode 2* was referred to as the *baseline control law* and was an α-command, p-β stability augmentation system; *Mode 3* was the *research control law* with the $\mathcal{L}_1$ adaptive α, p-β command control. The baseline α-command was an LQR-PI based and was designed to an operational limit of $\alpha \leq 10$ deg. It served as the operational baseline feedback control law for the GTM. Hence, comparison between the $\mathcal{L}_1$ controller and the baseline α-command could be performed only for $\alpha \leq 10$ deg; comparisons for higher α range were performed with the stick-to-surface control law (Mode 1).

The main results and conclusions of this evaluation are presented in [59]. Flight test results obtained during the AirSTAR deployments in March and June 2010 can be found in [60]. The reader can also find in [37] a study on the handling qualities and possible adverse pilot interactions in the GTM equipped with the $\mathcal{L}_1$ adaptive flight control system.

Angle of Attack Captures in the Presence of Static Stability Degradation

The task is trim at $V = 80$ knots end air speed (KEAS) and $Alt = 1200$ ft and then capture $\alpha = 8$ deg within 1 s, hold for 2 s, with α-desired ± 1 deg, α-adequate ± 2 deg. This task was repeated for various levels of static stability expressed as $\Delta C_{m\alpha}$ and ranging from 0 to 100%, i.e., from nominal stable aircraft to neutral static longitudinal stability. The change in $\Delta C_{m\alpha}$ is achieved by using both inboard elevator sections, scheduled with α, to produce a destabilizing effect. These two elevator sections also become unavailable to the control law. In a sense, it is a double fault—a destabilized aircraft and a reduction in control power in the affected axis.

This longitudinal task was evaluated for the $\mathcal{L}_1$ adaptive control law and the baseline, both of which are α-command response type. The performance of both control laws is provided in Figure 6.13. For nominal GTM aircraft, performance of both control laws is very similar, as illustrated in Figures 6.13(a) and 6.13(b), and solid Level I flying qualities (FQ) according to pilot comments. However, as the static stability is decreased by 50% (Figures 6.13(c) and 6.13(d)), the performance of the baseline controller degrades to high Level II (Cooper–Harper rating (CHR) 4), while the $\mathcal{L}_1$ adaptive controller remains solid Level I FQ [39]. With $\Delta C_{m\alpha} = 75\%$ (Figures 6.13(e) and 6.13(f)), the $\mathcal{L}_1$ adaptive controller remains predictable and Level I, while the baseline performance degrades to achieving only adequate performance (Level II, CHR 5). At the point of neutral static stability (Figures 6.13(g) and 6.13(h)), the $\mathcal{L}_1$ adaptive controller is still described as predictable but does experience some oscillations, and its performance is reduced to high Level II (CHR 4), while the baseline controller is described as pilot-induced oscillation prone and

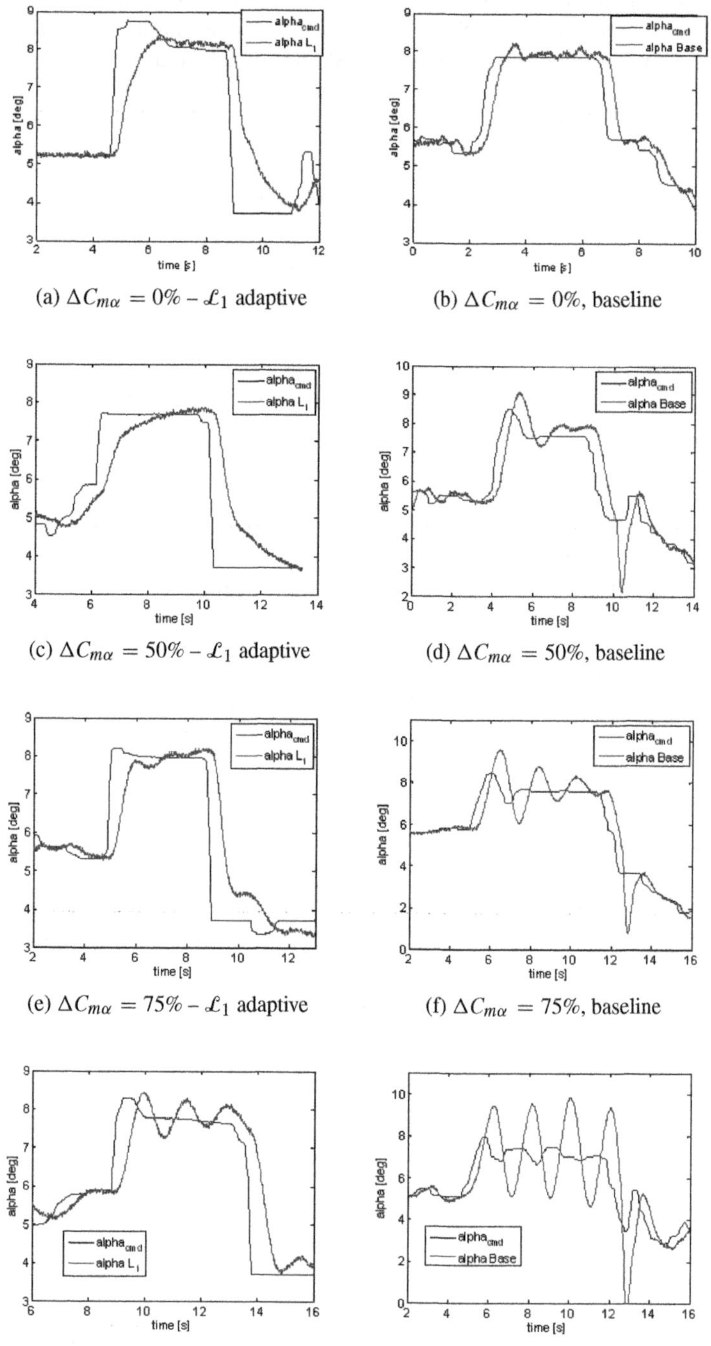

(a) $\Delta C_{m\alpha} = 0\% - \mathcal{L}_1$ adaptive

(b) $\Delta C_{m\alpha} = 0\%$, baseline

(c) $\Delta C_{m\alpha} = 50\% - \mathcal{L}_1$ adaptive

(d) $\Delta C_{m\alpha} = 50\%$, baseline

(e) $\Delta C_{m\alpha} = 75\% - \mathcal{L}_1$ adaptive

(f) $\Delta C_{m\alpha} = 75\%$, baseline

(g) $\Delta C_{m\alpha} = 100\% - \mathcal{L}_1$ adaptive

(h) $\Delta C_{m\alpha} = 100\%$, baseline

Figure 6.13: Angle-of-attack capture task with variable static stability.

FQ are reduced to Level 3 bordering on uncontrollable CHR 10. We would like to emphasize that the performance of the $\mathcal{L}_1$ adaptive control law was found predictable by the pilot for all levels of static stability.

High Angle of Attack Captures

One of the several research objectives for the AirSTAR facility is to identify high angle-of-attack dynamics and verify these against obtained wind-tunnel and CFD data. In order to do so, the GTM aircraft flight must be able to safely fly at the very edges of the attainable flight envelope. Part of the scheduled flight test is the high α envelope expansion. In addition, the GTM exhibits highly nonlinear *pitch break* phenomena for $12 \leq \alpha \leq 18$ deg. In other words, if in open loop the aircraft reaches $\alpha = 12$ deg, it will pitch up and could be recovered only once it reaches $\alpha = 18$ deg. Thus the envelope expansion is flown in the revisionary stick-to-surface control mode with 2 deg-α increments from $\alpha = 18$ deg to $\alpha = 28$ deg. On the other hand, the $\mathcal{L}_1$ adaptive controller is expected to perform α capture task for entire poststall region starting at $\alpha = 12$ deg.

The aerodynamics in the poststall region are nonlinear and increasingly asymmetric with increased α. For $12 \leq \alpha \leq 20$ deg the aerodynamics are expected to be asymmetric, the roll damping (C_{lp}) is expected to be low, and nose roll-off is also expected. Beyond $\alpha = 28$ deg, in addition to aerodynamic asymmetry and low roll damping, a pronounced nose-slice due $C_{n\beta}$ is expected. The α capture performance for the $\mathcal{L}_1$ adaptive controller is illustrated in Figure 6.14. The task is, starting in trim at $V = 80$ KEAS, to capture the indicated α at the rate of 3 $\frac{\text{deg}}{\text{s}}$, hold for 4 s, with α-desired ± 1 deg, α-adequate ± 2 deg. Note that the desired and the adequate criteria are the same as for the α capture in the linear region; additionally, holding for 4 s would expose any control law to instability in this "pitch break" region. Due to the nature of expected dynamic behavior in the high α region, β and ϕ are additional variables of interest plotted in Figure 6.14. The $\mathcal{L}_1$ adaptive controller performance was judged as close to nominal α capture task. For $\alpha = 20$ deg case (Figure 6.14(b)), the approach was at high pitch rate with $V_{\min}$ getting into the 30s (KEAS); this created some nascent small oscillations as α approached 20 deg. The $\alpha = 26$ deg case was performed less rapidly and the slight oscillations are no longer present (Figure 6.14(c)). For comparison purposes, the same α case for the revisionary stick-to-surface mode is shown in Figure 6.14(d). Note the oscillatory nature of α as it follows the ramping α_{cmd}; also note the bank and sideslip angle excursions, which illustrate the expected roll-off (ϕ) and nose-slice (β) dynamics. Figure 6.15 provides another way of looking at the high α capture task by illustrating the coupling between these variables. Ideally, an α excursion would be completely decoupled from β and would produce a straight vertical line on the plot. The $\mathcal{L}_1$ adaptive controller produces $|\beta| \leq 1$ deg excursions, as shown in Figure 6.15(a). On the other hand, the revisionary stick-to-surface control law shows significantly greater degree of coupling between the α-β axis, as illustrated in Figure 6.15(b).

From Figure 6.15, and recalling the α-β flight envelope in Figure 6.1, it is evident that the $\mathcal{L}_1$ adaptive controller keeps the airplane inside the *normal flight envelope* during any of the high-α maneuvers, thus ensuring controllability of the airplane during the whole task. The same cannot be said about the revisionary stick-to-surface control law, for which the airplane experiences high-α/high-β excursions.

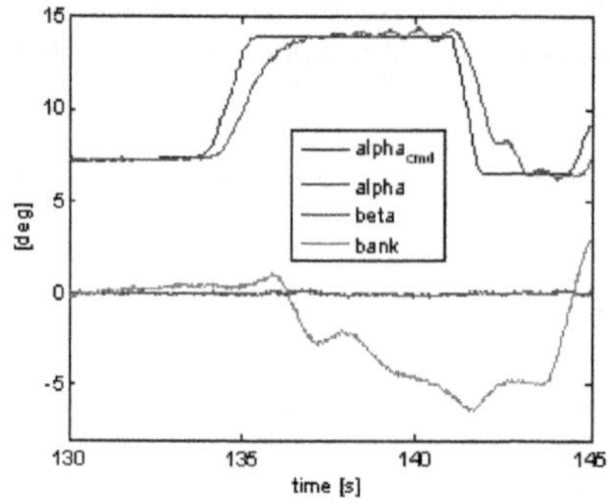

(a) $\alpha_{\text{cmd}} = 14$ deg – $\mathcal{L}_1$ adaptive

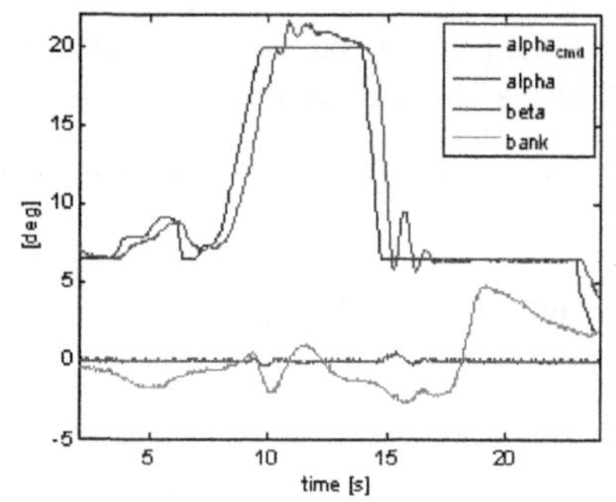

(b) $\alpha_{\text{cmd}} = 20$ deg – $\mathcal{L}_1$ adaptive

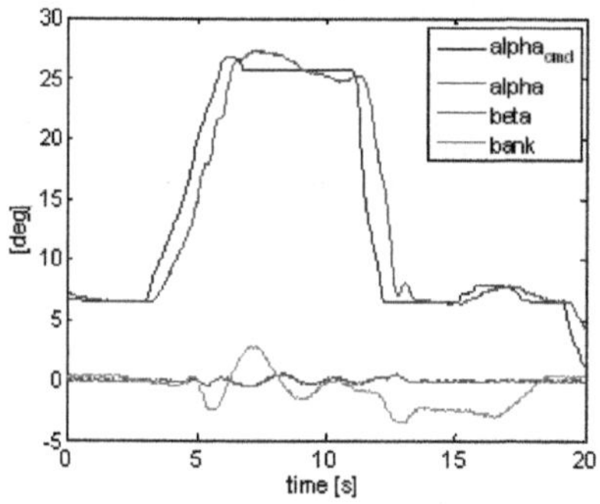

(c) $\alpha_{\text{cmd}} = 26$ deg – $\mathcal{L}_1$ adaptive

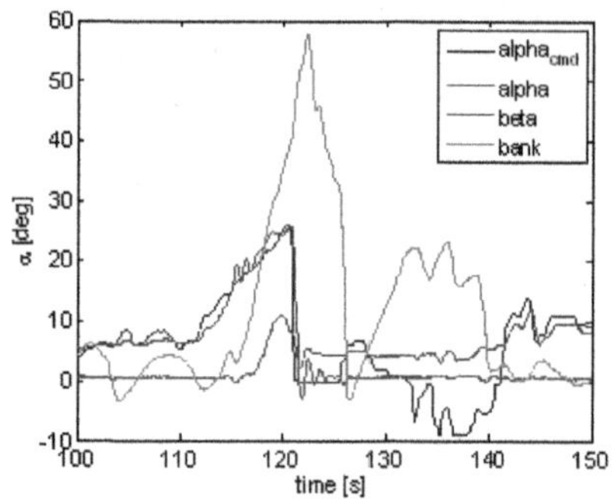

(d) $\alpha_{\text{cmd}} = 26$ deg, stick-to-surface

Figure 6.14: High α capture task.

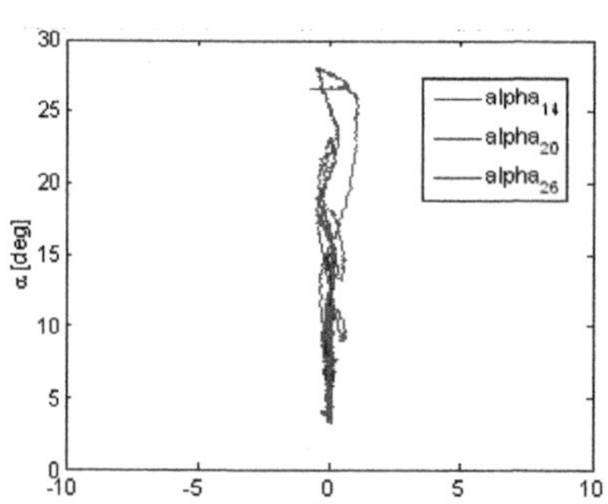

(a) $\alpha_{\text{cmd}} = 14/20/26$ deg
$\mathcal{L}_1$ adaptive

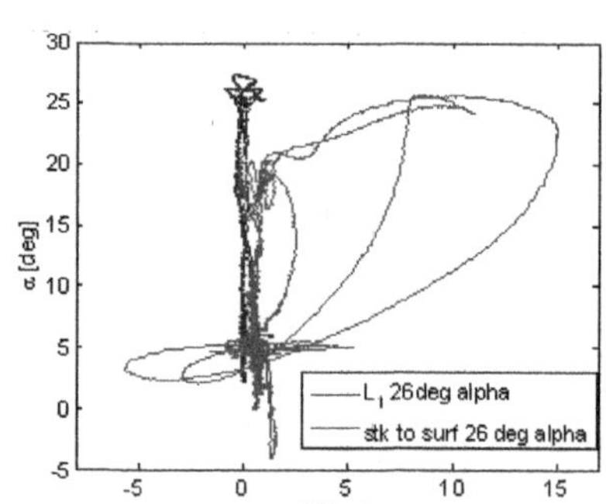

(b) $\alpha_{\text{cmd}} = 26$ deg
Stick-to-surface vs. $\mathcal{L}_1$ adaptive

Figure 6.15: α-β excursions for high-α capture task.

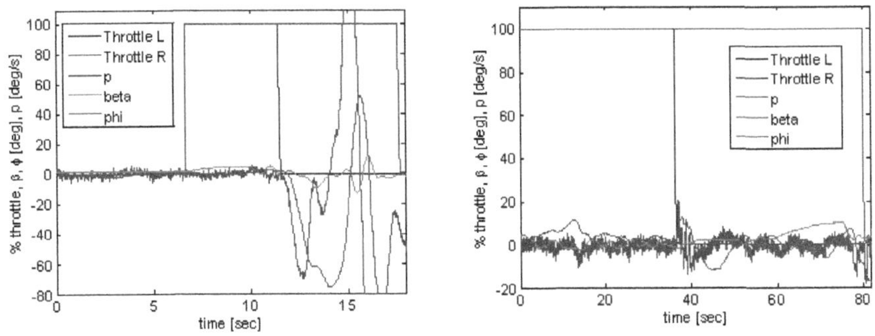

(a) Stick-to-surface (b) $\mathcal{L}_1$ adaptive control

Figure 6.16: Asymmetric engine failure.

Sudden Asymmetric Engine Failure

This maneuver was performed unrehearsed once for each control law—reversion stick-to-surface mode and $\mathcal{L}_1$ adaptive controller. The results of this maneuver are used for qualitative comparison between the two control law responses with pilot in the loop and no training. The task starts the airplane climbing at $\approx$ 30 deg attitude and throttles at full power, then at some time the left throttle is reduced from 100% to 0% thrust in less than 0.5 s, as illustrated in Figure 6.16. This is primarily a lateral-directional task since the sudden change in thrust induces a rolling moment affecting p, ϕ and a side force affecting β. These variables are coplotted with throttle activity in Figure 6.16. The loss of control for the revisionary stick-to-surface control law is evident from Figure 6.16(a). From Figure 6.16(b) it is also evident that this sudden asymmetric thrust is a nonevent for the $\mathcal{L}_1$ adaptive controller, especially from the stability perspective.

We would like to emphasize that the $\mathcal{L}_1$ adaptive controller was *not* redesigned or retuned in all these scenarios and that a *single set* of control parameters was used for all the piloted tasks and throughout the whole flight envelope. As stated earlier, only the reference model (state predictor) is scheduled in order to specify different performance requirements at different flight regimes.

6.1.3 Other Applications

The $\mathcal{L}_1$ adaptive controller presented in Section 3.3 and used for the development of the $\mathcal{L}_1$ adaptive flight control system on GTM has also been validated for the X-48B aircraft [106] as an augmentation of a dynamic inversion baseline controller, and for the X-29 aircraft [63] as an augmentation of an LQR-PI baseline. The same $\mathcal{L}_1$ adaptive architecture has been applied to the longitudinal control of a flexible fixed-wing aircraft [143]. Also, the $\mathcal{L}_1$ adaptive output feedback adaptive augmentation architecture that is being flight tested at NPS was used in [124] for the control of indoor autonomous quadrotors and a fixed-wing aerobatic aircraft. The output feedback architecture presented in Section 4.2 has been applied in a MIMO setting to the ascent control of a generic flexible crew launch vehicle [88]. In [64], the authors design a high-bandwidth inner-loop controller to provide attitude and velocity stabilization of an autonomous small-scale rotorcraft in the presence of wind disturbances. Also, the $\mathcal{L}_1$ adaptive control architecture presented in Section 2.2 has been used on a vision-based tracking and motion estimation system as an augmentation loop for the control of a gimbaled pan-tilt camera onboard a UAV [110].

The $\mathcal{L}_1$ adaptive controller has been applied to other areas outside aerospace applications. Reference [111] explored application of an integrated estimator and $\mathcal{L}_1$ adaptive controller for pressure control in well-drilling systems. In [51], the authors explored the application of the $\mathcal{L}_1$ adaptive controller to compensate for undesired hysteresis and constitutive nonlinearities present in smart-material-based transducers. Also, the application of $\mathcal{L}_1$ adaptive control in a nuclear power plant to improve recovery of the system from unexpected faults and emergency situations was studied in [78].

6.2 Key Features, Extensions, and Open Problems

6.2.1 Main Features of the $\mathcal{L}_1$ Adaptive Control Theory

The main features of the $\mathcal{L}_1$ adaptive control theory, proved in theory and consistently verified in experiments, can thus be summarized as follows:

- Guaranteed robustness in the presence of fast adaptation;
- Separation (decoupling) between adaptation and robustness;
- Guaranteed transient response, *without* resorting to
  persistency of excitation type assumptions,
  high-gain feedback, or
  gain-scheduling of the controller parameters;
- Guaranteed (bounded-away-from-zero) time-delay margin;
- Uniform *scaled transient response* dependent on changes in
  initial conditions,
  unknown parameters, and
  reference inputs.

With these features the architectures of the $\mathcal{L}_1$ adaptive control theory reduce the performance limitations to hardware limitations and provide a suitable framework for development of theoretically justified tools for V&V of feedback systems. The next section summarizes the open problems in this direction.

6.2.2 Extensions Not Covered in the Book

The $\mathcal{L}_1$ adaptive control theory has been developed for a broader class of systems, the presentation of which is outside the limits of this book. Some of the important extensions are presented in [107, 116, 179, 183, 188].

Robustness features of the $\mathcal{L}_1$ adaptive controller have also been verified using the framework for robustness analysis of nonlinear systems in the gap metric [109]. By an appropriate extension of a classical result [55], the $\mathcal{L}_1$ adaptive controller has been proved to have guaranteed robust stability margin. The computation of the robust stability margin in the gap metric confirmed that in the absence of the low-pass filter one loses the robustness guarantees of this feedback structure (similar to Theorem 2.2.4). The computational tractability of this method leads, in turn, to explicit derivation of the margins for a wide class of system uncertainties such as time delay and multiplicative unmodeled dynamics.

The results in [116] summarize the extension to nonaffine-in-control systems. Following Lemma A.8.1, the linear parametrization of the control-dependent nonlinearity leads to time-dependent control input gain $\omega(t)$. This further implies that the low-pass filter, used for definition of the control signal, can no longer be effectively employed. An appropriate extension is developed, which considers an LTV system for definition of the control signal, and its analysis is pursued by elaborating the tools used in the proofs of Chapter 5 for LTV reference systems. These new mathematical developments also supported the extension to nonlinear systems in the presence of input hysteresis [188].

Reference [107] analyzes the performance bounds of the $\mathcal{L}_1$ adaptive controller in the presence of input saturation. Following the approach in [85], appropriate modification of the state predictor is considered to remove the effect of the control deficiency from the adaptation process. The uniform performance bounds are computed with respect to a bounded reference system, which can be designed to meet the performance specifications for the given input constraints.

The results in [183] extended the $\mathcal{L}_1$ adaptive controller to decentralized setup by considering large-scale interconnected nonlinear systems in the presence of unmodeled dynamics. The decentralized local $\mathcal{L}_1$ adaptive controllers compensate for the effect of unmodeled dynamics and nonlinearities on the system output without having access to the states or outputs of other subsystems.

Reference [162] analyzed the performance of the $\mathcal{L}_1$ adaptive controller in the presence of input quantization. The resulting performance bounds are shown to be decoupled into two terms, one of which is the standard term that one has in the absence of quantization and can be systematically improved by increasing the *rate of adaptation*, while the other term can be reduced by improving the *quality of the quantizer*. This *decoupled* nature of the performance bounds allows for independent design of the quantizers and can broaden the application domain of quantized control.

Parallel to this, reference [173] considered implementation of the $\mathcal{L}_1$ adaptive controller over real-time networks using event triggering. Event-triggering schedules the data transmission dependent upon errors exceeding certain threshold. Similar to [162], with the proposed event-triggering schemes and with the $\mathcal{L}_1$ adaptive controller in the feedback loop, the performance bounds of the networked system are decoupled into two terms, one of which is the standard term that one has in the absence of networking and event-triggering and can be systematically improved by increasing the *rate of adaptation*, while the other term can be reduced by increasing the *data transmission frequency*. This further implies that the performance limitations of the $\mathcal{L}_1$ adaptive closed-loop systems are consistent with the hardware limitations.

6.2.3 Open Problems

The main challenge of the $\mathcal{L}_1$ adaptive control theory is the *optimal design* of the bandwidth-limited filter. Because the filter defines the trade-off between performance and robustness, its *optimal design* is a problem of constrained optimization. Moreover, the $\mathcal{L}_1$ norm condition for minimization of the performance bounds renders the optimization problem nonconvex and hence more challenging. The design of the filter involves consideration of its order and parametrization. While we have provided partial design guidelines in Section 2.6 for the full state-feedback architecture, the problem is still largely open and hard to address.

The design problem is especially challenging in output feedback due to the stability condition for the transfer function $H(s)$ defined in (4.5) and (4.65) as

$$H(s) \triangleq \frac{A(s)M(s)}{C(s)A(s)+(1-C(s))M(s)}.$$

As shown in Chapter 4, the definition of $H(s)$ involves the uncertain plant $A(s)$, and therefore the uncertainty is not decoupled in the $\mathcal{L}_1$-norm condition as it is in the state-feedback solution. Hence, stability of $H(s)$ adds an additional constraint to the choice of the filter and the desired reference system.

In [169], tools from robust control were invoked to address the problem for the case of reference systems of higher order, which do not verify the SPR property for their input-output transfer functions. In this case, partial results have been obtained by resorting the analysis to Kharitonov's theorem from robust control [169]. We note that the disturbance observer literature has methods on offer for parametrization of $C(s)$ that would achieve stabilization of $H(s)$ for a sufficiently broad class of systems $A(s)$ [155]. The current research is focused on appropriate extension of these methods to capture non-minimum phase systems among $A(s)$.

Special attention is deserved in the case of MIMO systems in the presence of unmatched uncertainties, analyzed in Sections 3.2 and 3.3. The sufficient condition for stability and performance, given in terms of $\mathcal{L}_1$ norm bound, involves a restriction on the rate of variation of uncertainties in both matched and unmatched channels, in addition to the desired performance specifications, given by A_m and B_m matrices. From (3.71) and (3.126) it follows that the cross-coupling, expressed in $G_{um}(s)$, will directly affect the choice of A_m and B_m and also the invariant set where the solutions lie. While the sufficient conditions are intuitive, their complete analytical investigation is largely an open area of research. In particular, in the case of the design of the $\mathcal{L}_1$ flight control system for the GTM, which was discussed in Section 6.1.2, the selection of A_m and B_m followed from (MIL-standard [1]) performance requirements at different flight regimes, while the (second-order) low-pass filters in the control laws were designed to guarantee stability of the closed-loop adaptive system and achieve a satisfactory level of robustness. Currently there exists no general methodology that one could formalize for this purpose, and we expect this analysis to be done on case-by-case basis dependent upon the nature of the application.

An important direction for future research is the extension of the time-delay margin proof to more complex classes of systems. The proof in Section 2.2.5 is pursued for LTI (open-loop) systems with time-varying disturbance. The lower bound for the time-delay margin is provided via an LTI system, given in (2.80). Notice that this system depends upon the original LTI system, given by $\bar{H}(s)$. Obviously, if the original system were not LTI, then the explicit computation of the time-delay margin could not be reduced to an expression similar to the one in (2.80). More elaborate tools are needed from nonlinear systems theory to address the problem for more complex classes of systems.

While a complete answer to the above described problems may not be obtained in the near future, $\mathcal{L}_1$ adaptive control, with its *key feature of decoupling adaptation from robustness*, has already facilitated new opportunities in the area of networked systems and event-driven adaptation [3,162,173]. We expect rapid developments in these directions over the next couple of years.

Appendix A

Systems Theory

In this chapter we provide a brief review of some basic facts from stability theory and robust control, which are used throughout the book.

A.1 Vector and Matrix Norms

The norm of a vector or a matrix is a real valued function $\|\cdot\|$, defined on the vector or the matrix space, verifying the following properties for all vectors or matrices u, v, and $\lambda \in \mathbb{R}$:

- $\|u\| > 0$ if $u \neq 0$, and $\|u\| = 0$ if and only if $u = 0$;
- $\|u + v\| \leq \|u\| + \|v\|$;
- $\|\lambda u\| = |\lambda| \|u\|$.

Obviously, the norm is not uniquely defined. Below we introduce the most frequently used norms.

A.1.1 Vector Norms

1. The **1-norm** of a vector $u = [u_1, \ldots, u_m]^\top \in \mathbb{R}^m$ is defined as

$$\|u\|_1 \triangleq \sum_{i=1}^{m} |u_i| .$$

2. The **2-norm** of a vector $u = [u_1, \ldots, u_m]^\top \in \mathbb{R}^m$ is defined as

$$\|u\|_2 \triangleq \sqrt{u^\top u} .$$

3. The ***p*-norm** of a vector $u = [u_1, \ldots, u_m]^\top \in \mathbb{R}^m$ for $1 \leq p < \infty$ is defined as

$$\|u\|_p \triangleq \left(\sum_{i=1}^{m} |u_i|^p \right)^{1/p} .$$

4. The **∞-norm** of a vector $u = [u_1, \dots, u_m]^\top \in \mathbb{R}^m$ is defined as
$$\|u\|_\infty \triangleq \max_{1 \le i \le m} |u_i|.$$

Throughout the book, when the type of norm is not explicitly specified, the 2-norm is assumed.

A.1.2 Induced Norms of Matrices

The matrix $A \in \mathbb{R}^{n \times m}$ can be viewed as an operator that maps $\mathbb{R}^m$, the space of m-dimensional vectors, into $\mathbb{R}^n$, the space of n-dimensional vectors. The operator norm, or the induced p-norm, of a matrix is defined as
$$\|A\|_p \triangleq \sup_{x \ne 0} \frac{\|Ax\|_p}{\|x\|_p} = \sup_{\|x\|_p = 1} \|Ax\|_p.$$

The proof of the last equality is straightforward. Indeed, if $x \ne 0$, then
$$\frac{\|Ax\|_p}{\|x\|_p} = \left\| A \frac{x}{\|x\|_p} \right\|_p.$$

Taking the sup of both sides proves the last equality above. This definition leads to the following matrix norms:

1. The induced **1-norm** of the matrix $A \in \mathbb{R}^{n \times m}$ is defined as
$$\|A\|_1 \triangleq \max_{1 \le j \le m} \sum_{i=1}^{n} |a_{ij}| \text{ (column sum)}.$$

2. The induced **2-norm** of the matrix $A \in \mathbb{R}^{n \times m}$ is defined as
$$\|A\|_2 \triangleq \sqrt{\lambda_{\max}(A^\top A)},$$
where $\lambda_{\max}(\cdot)$ denotes the maximum eigenvalue.

3. The induced **∞-norm** of the matrix $A \in \mathbb{R}^{n \times m}$ is defined as
$$\|A\|_\infty \triangleq \max_{1 \le i \le n} \sum_{j=1}^{m} |a_{ij}| \text{ (row sum)}.$$

Throughout the book, and similar to vector norms, when the type of norm is not explicitly specified, the 2-norm is assumed. The following properties are straightforward to verify for all the norms:

1. $\|A\| = \|A^\top\|$.

2. For arbitrary vector x and arbitrary matrix A with appropriate dimensions, the following inequality holds:
$$\|Ax\| \le \|A\| \|x\|.$$

3. All the norms are equivalent, i.e., if $\|\cdot\|_p$ and $\|\cdot\|_q$ are two different norms, then, for arbitrary vector or matrix X, there exist two constants $c_1 > 0$ and $c_2 > 0$ such that
$$c_1 \|X\|_p \le \|X\|_q \le c_2 \|X\|_p.$$

A.2 Symmetric and Positive Definite Matrices

We recall definitions of symmetric and positive definite matrices plus one key property often used in this book.

Definition A.2.1 The matrix $M \in \mathbb{R}^{n\times n}$ is symmetric if $M = M^\top$.

Definition A.2.2 The matrix $M \in \mathbb{R}^{n\times n}$ is

- positive definite if, for an arbitrary nonzero vector $x \in \mathbb{R}^n$, the following inequality holds:
$$x^\top Mx > 0 \qquad \text{and} \qquad x^\top Mx = 0 \quad \text{only for} \quad x = 0;$$
- positive semidefinite if, for arbitrary vector $x \in \mathbb{R}^n$,
$$x^\top Mx \geq 0;$$
- negative definite if, for arbitrary nonzero vector $x \in \mathbb{R}^n$,
$$x^\top Mx < 0 \qquad \text{and} \qquad x^\top Mx = 0 \quad \text{only for} \quad x = 0;$$
- negative semidefinite if, for arbitrary vector $x \in \mathbb{R}^n$,
$$x^\top Mx \leq 0.$$

For arbitrary vector $x \in \mathbb{R}^n$ and arbitrary positive definite symmetric matrix $M \in \mathbb{R}^{n\times n}$, the following inequalities hold:

$$\lambda_{\min}(M)\|x\|^2 \leq x^\top Mx \leq \lambda_{\max}(M)\|x\|^2,$$

where $\lambda_{\min}(M)$ and $\lambda_{\max}(M)$ denote, respectively, the minimum and maximum eigenvalues of M.

Definition A.2.3 The matrix A is Hurwitz if all its eigenvalues have negative real part:

$$\Re(\lambda_i) < 0, \quad \det|\lambda_i \mathbb{I} - A| = 0, \quad i = 1,\ldots,n.$$

Lemma A.2.1 The matrix A is Hurwitz if and only if given an arbitrary positive definite symmetric matrix Q, there exists a positive definite symmetric matrix P solving the **algebraic Lyapunov equation**

$$A^\top P + PA = -Q.$$

A.3 $\mathcal{L}$-spaces and $\mathcal{L}$-norms

Using the definition for the vector norms, we define norms of functions $f : [0,+\infty) \to \mathbb{R}^n$ as follows:

- **$\mathcal{L}_1$-norm and $\mathcal{L}_1$-space:** The space of piecewise-continuous integrable functions with bounded $\mathcal{L}_1$-norm

$$\|f\|_{\mathcal{L}_1} \triangleq \int_0^\infty \|f(\tau)\| d\tau < \infty$$

is denoted $\mathcal{L}_1^n$, where any of the vector norms can be used for $\|f(\tau)\|$.

- **$\mathcal{L}_p$-norm and $\mathcal{L}_p$-space:** The space of piecewise-continuous integrable functions with bounded $\mathcal{L}_p$-norm

$$\|f\|_{\mathcal{L}_p} \triangleq \left(\int_0^\infty \|f(\tau)\|^p d\tau \right)^{1/p} < \infty$$

is denoted $\mathcal{L}_p^n$. As above, any of the vector norms can be used for $\|f(\tau)\|$. However, for the $\mathcal{L}_2$-space it is conventional to use the 2-norm of the vector.

- **$\mathcal{L}_\infty$-norm:** The space of piecewise-continuous bounded functions with $\mathcal{L}_\infty$-norm

$$\|f\|_{\mathcal{L}_\infty} \triangleq \max_{1 \le i \le n} \left\{ \sup_{\tau \ge 0} |f_i(\tau)| \right\} < \infty$$

is denoted $\mathcal{L}_\infty^n$.

Notice that the requirement for the $\mathcal{L}$-norm to be finite restricts the class of functions that can belong to the $\mathcal{L}_p^n$ space. So as not to be restricted by this, we consider the **extended space $\mathcal{L}_e^n$**, defined as the space of functions

$$\mathcal{L}_e^n \triangleq \{ f(t) \mid f_\tau(t) \in \mathcal{L}^n, \ \forall \tau \in [0, \infty) \},$$

where $f_\tau(t)$ is the truncation of the function $f(t)$ defined by

$$f_\tau(t) = \begin{cases} f(t), & 0 \le t \le \tau, \\ 0, & t > \tau. \end{cases}$$

Thus, every function that does not have finite escape time belongs to the extended space $\mathcal{L}_e^n$. Notice that any of the $\mathcal{L}_p$-norms can be used in the definition of the extended space.

The definitions above can also be extended for functions defined on $[t_0, \infty)$, $t_0 > 0$. The $\mathcal{L}$-norm of the function $f : [t_0, \infty) \to \mathbb{R}^n$ and its truncated $\mathcal{L}$-norm are given as the $\mathcal{L}$-norms of $f_{[t_0,\infty)}$ and $f_{[t_0,\tau]}$, respectively, where

$$f_{[t_0,\infty)}(t) = \begin{cases} 0, & 0 \le t < t_0, \\ f(t), & t_0 \le t, \end{cases} \qquad f_{[t_0,\tau]}(t) = \begin{cases} 0, & 0 \le t < t_0, \\ f(t), & t_0 \le t \le \tau, \\ 0, & t > \tau. \end{cases}$$

In this book we omit the index t_0 from the notation of the norms $\|f_{[t_0,\infty)}\|_{\mathcal{L}}$, $\|f_{[t_0,\tau]}\|_{\mathcal{L}}$ and simply write $\|f\|_{\mathcal{L}}$, $\|f_\tau\|_{\mathcal{L}}$ if it is clear from the context which norm has been used.

It is worth mentioning that for function norms the principle of equivalence does not hold (some of the norms can be finite, while some can be unbounded, i.e., not defined).

Example A.3.1 Consider the piecewise-continuous function

$$f(t) = \begin{cases} 1/\sqrt{t}, & 0 < t \le 1, \\ 0, & t > 1. \end{cases}$$

It has bounded $\mathcal{L}_1$-norm:

$$\|f\|_{\mathcal{L}_1} = \int_0^1 \frac{1}{\sqrt{t}} dt = 2.$$

Its $\mathcal{L}_\infty$-norm does not exist, since $\|f\|_{\mathcal{L}_\infty} = \sup_{t\ge 0} |f(t)| = \infty$, and its $\mathcal{L}_2$-norm is unbounded because the integral of $1/t$ is divergent. Thus, $f(t) \in \mathcal{L}_1$, but $f(t) \notin \mathcal{L}_2 \cup \mathcal{L}_\infty$. ∎

Example A.3.2 Next, consider the continuous function

$$f(t) = \frac{1}{1+t}.$$

It has bounded $\mathcal{L}_\infty$-norm and bounded $\mathcal{L}_2$-norm:

$$\|f\|_{\mathcal{L}_\infty} = \sup_{t\ge 0} \left| \frac{1}{1+t} \right| = 1, \quad \|f\|_{\mathcal{L}_2} = \left(\int_0^\infty \frac{1}{(1+t)^2} dt \right)^{1/2} = 1.$$

Its $\mathcal{L}_1$-norm does not exist, since

$$\|f\|_{\mathcal{L}_1} = \int_0^\infty \frac{1}{1+t} dt = \lim_{t\to\infty} \ln(1+t) = \infty.$$

Thus $f(t) \in \mathcal{L}_2 \cap \mathcal{L}_\infty$, but $f(t) \notin \mathcal{L}_1$. ∎

A.4 Impulse Response of Linear Time-Invariant Systems

A SISO LTI system has the following state-space representation:

$$\begin{aligned} \dot{x}(t) &= Ax(t) + bu(t), \quad x(0) = x_0, \\ y(t) &= c^\top x(t), \end{aligned} \tag{A.1}$$

where $x \in \mathbb{R}^n$ is the *state vector*, $u \in \mathbb{R}$ is the *input*, $y \in \mathbb{R}$ is the *output* of the system, and x_0 is the *initial condition*. Further, $A \in \mathbb{R}^{n\times n}$ is the *state matrix*, and $b, c \in \mathbb{R}^n$ are vectors of appropriate dimensions. In the frequency domain, the system is defined by means of its transfer function $y(s) = G(s)u(s)$, where $G(s) \triangleq c^\top (s\mathbb{I} - A)^{-1} b$. Notice that in this representation $x_0 = 0$. Obviously, given $y(s) = G(s)u(s)$, the state-space realization in (A.1) is not unique.

Letting $u(s) = 1$, which corresponds to the Dirac-delta impulse function,

$$u(t) = \begin{cases} \infty & t = 0, \\ 0 & t \ne 0, \end{cases} \quad \text{and} \quad \int_{-\infty}^{\infty} u(t) dt = 1,$$

we have $y(s) = G(s)$. The inverse Laplace transform of the transfer function $G(s)$, given by $g(t) = \mathfrak{L}^{-1}(G(s))$, is called the **impulse response** of the system.

For MIMO systems, the state-space representation is given by

$$\begin{aligned} \dot{x}(t) &= Ax(t) + Bu(t), \quad x(0) = x_0, \\ y(t) &= Cx(t), \end{aligned} \tag{A.2}$$

where $x \in \mathbb{R}^n$, $u \in \mathbb{R}^m$, and $y \in \mathbb{R}^l$ are the state, the input, and the output of the system, while $A \in \mathbb{R}^{n \times n}$, $B \in \mathbb{R}^{n \times m}$, and $C \in \mathbb{R}^{l \times n}$ are the corresponding matrices.

The **transfer matrix** of the system in (A.2) is given by

$$G(s) \triangleq C(s\mathbb{I} - A)^{-1}B\,.$$

Let $g_j : \mathbb{R} \to \mathbb{R}^l$ be the response of the system to the unit Dirac-delta function applied at the jth input with all the initial conditions set to zero. The matrix $g(t) \in \mathbb{R}^{l \times m}$ with its entries $g_j(t)$ is called the **impulse response matrix** of the system and can be computed as the inverse Laplace transform of the transfer matrix:

$$g(t) = \mathfrak{L}^{-1}(G(s))\,.$$

A.5 Impulse Response of Linear Time-Varying Systems

When the system matrices in (A.2) depend upon time, i.e.,

$$\begin{aligned} \dot{x}(t) &= A(t)x(t) + B(t)u(t), \quad x(t_0) = x_0, \\ y(t) &= C(t)x(t), \end{aligned} \tag{A.3}$$

then the linear system is time varying. For an LTV system, the impulse response is defined via its state transition matrix.

Definition A.5.1 The **state transition matrix $\Phi(t,t_0)$** of the LTV system in (A.3) is the solution of the following linear homogeneous matrix equation:

$$\frac{\partial \Phi(t,t_0)}{\partial t} = A(t)\Phi(t,t_0), \quad \Phi(t_0,t_0) = \mathbb{I}\,. \tag{A.4}$$

When $u(t) \equiv 0$, the state trajectory of (A.3) and the state transition matrix are related as follows:

$$x(t) = \Phi(t,t_0)x(t_0)\,.$$

The impulse response of the LTV system in (A.3) is computed according to the following equation:

$$g(t,t_0) = C(t)\Phi(t,t_0)B(t_0), \quad t \geq t_0\,. \tag{A.5}$$

Notice that in the case of LTI systems, when $A(t) = A$, $B(t) = B$, $C(t) = C$ in (A.3) are independent of time, assuming $t_0 = 0$, the solution of (A.4) for $\Phi(t,0) = \Phi(t)$ in frequency domain can be written as

$$\Phi(s) = (s\mathbb{I} - A)^{-1}\,.$$

Then, (A.5) in the frequency domain takes the form

$$\mathfrak{L}(g(t)) = C\Phi(s)B = C(s\mathbb{I} - A)^{-1}B = G(s)\,,$$

which recovers the result for LTI systems.

A.6 Lyapunov Stability

In this section we recall several fundamental results of Lyapunov stability theory. We begin by considering the nonlinear autonomous system of the form

$$\dot{x}(t) = f(x(t)), \quad x(0) = x_0, \tag{A.6}$$

where $x(t) \in \mathbb{R}$ is the system state, and $f(x) : \mathbb{R}^n \to \mathbb{R}^n$ is a locally Lipschitz nonlinearity.

A.6.1 Autonomous Systems

We assume that the system (A.6) has an equilibrium at $x = 0$, i.e., $f(0) = 0$.

Definition A.6.1 The equilibrium point at the origin is

stable if for all $\varepsilon > 0$ there exists $\delta = \delta(\varepsilon) > 0$, such that, if $\|x_0\| < \delta$, then $\|x(t)\| < \varepsilon$ for all $t \geq 0$;

(locally) asymptotically stable if it is stable and δ can be chosen such that, if $\|x_0\| < \delta$, then $\lim_{t\to\infty} \|x(t)\| = 0$;

globally asymptotically stable if it is stable and for all $x_0 \in \mathbb{R}^n$, $\lim_{t\to\infty} \|x(t)\| = 0$;

unstable if it is not stable.

The following result, known as Lyapunov's direct method, gives sufficient conditions for stability. For this result we consider an open set $\mathcal{D} \subset \mathbb{R}^n$, which contains the equilibrium $x = 0$, and a continuously differentiable function $V : \mathcal{D} \to \mathbb{R}$ with its derivative along the trajectories given by $\dot{V}(x(t)) = \frac{\partial V(x)}{\partial x} f(x(t))$.

Definition A.6.2 A function $V : \mathcal{D} \to \mathbb{R}$ with $V(0) = 0$ is called

positive definite if $V(x) > 0, \ x \in \mathcal{D} \setminus \{0\}$;

positive semidefinite if $V(x) \geq 0, \ x \in \mathcal{D} \setminus \{0\}$;

negative definite if $V(x) < 0, \ x \in \mathcal{D} \setminus \{0\}$;

negative semidefinite if $V(x) \leq 0, \ x \in \mathcal{D} \setminus \{0\}$.

Theorem A.6.1 Consider the system (A.6) and assume that there exists a continuously differentiable positive definite function $V : \mathcal{D} \to \mathbb{R}$, such that

$$V(0) = 0, \quad V(x) > 0 \text{ for } x \in \mathcal{D} \setminus \{0\}.$$

Then, the equilibrium $x = 0$ is (locally) stable if

$$\dot{V}(x(t)) \leq 0, \quad x \in \mathcal{D},$$

and the equilibrium $x = 0$ is (locally) asymptotically stable if

$$\dot{V}(x(t)) < 0, \quad x \in \mathcal{D} \setminus \{0\}.$$

Additionally, if $\mathcal{D} = \mathbb{R}^n$, and $V(x)$ is radially unbounded, i.e., $\lim_{\|x\|\to\infty} V(x) = \infty$, then these results hold globally.

The proof of this theorem can be found in [86]. If $V(x)$ verifies the properties above, it is called the **Lyapunov function** for the system.

A.6.2 Time-Varying Systems

Next we consider nonautonomous systems of the type

$$\dot{x}(t) = f(t, x(t)), \quad x(t_0) = x_0, \tag{A.7}$$

with $f(t,x) : [t_0, \infty) \times \mathbb{R}^n \to \mathbb{R}^n$ being piecewise-continuous in t and locally Lipschitz in x. We assume that $f(t,0) \equiv 0$, so that the system has an equilibrium at the origin. A key feature of the nonautonomous systems is the *uniformity* of its convergence properties.

Definition A.6.3 The equilibrium point $x = 0$ is

stable if for all $\varepsilon > 0$ there exists $\delta = \delta(\varepsilon, t_0) > 0$, such that if $\|x_0\| < \delta$, then $\|x(t)\| < \varepsilon$ for all $t \geq t_0 \geq 0$;

(locally) asymptotically stable if it is stable and δ can be chosen such that, if $\|x_0\| < \delta$, then $\lim_{t\to\infty} \|x(t)\| = 0$;

globally asymptotically stable if it is stable and for all $x_0 \in \mathbb{R}^n$, $\lim_{t\to\infty} \|x(t)\| = 0$;

unstable if it is not stable.

The stability is said to hold uniformly if $\delta = \delta(\varepsilon) > 0$ is independent of t_0.

Definition A.6.4 The equilibrium point $x = 0$ of (A.7) is (locally) exponentially stable if there exist positive constants $c > 0$, $a > 0$, $\lambda > 0$, such that

$$\|x(t)\| \leq a\|x_0\| e^{-\lambda(t-t_0)}, \quad \forall\ \|x_0\| \leq c.$$

It is globally exponentially stable if this bound holds for arbitrary initial condition x_0.

Example A.6.1 (see [86]) To highlight the significance of *uniformity*, consider the following first-order system:

$$\dot{x}(t) = (6t\sin t - 2t)x(t), \quad x(t_0) = x_0.$$

Its solution is given by

$$\begin{aligned} x(t) &= x(t_0)\exp\left[\int_{t_0}^{t} (6\tau\sin\tau - 2\tau)d\tau\right] \\ &= x(t_0)\exp[6\sin t - 6t\cos t - t^2 - 6\sin t_0 + 6t_0\cos t_0 + t_0^2]. \end{aligned}$$

For arbitrary t_0, the term $-t^2$ will dominate, which implies that

$$\exists\ c(t_0) \in \mathbb{R} \quad \text{such that} \quad |x(t)| < |x(t_0)|c(t_0), \quad \forall\, t \geq t_0.$$

For arbitrary $\varepsilon > 0$, the choice of $\delta = \varepsilon / c(t_0)$ shows that the origin is stable. However, assuming that $t_0 = 2n\pi$, $n = 0, 1, 2, \ldots$, we get

$$x(t_0 + \pi) = x(t_0)\exp[(4n+1)(6-\pi)\pi],$$

which implies that for $x(t_0) \neq 0$

$$\lim_{n\to\infty} \frac{x(t_0+\pi)}{x(t_0)} = \infty.$$

Thus, given $\varepsilon > 0$ there is no δ independent of t_0 that would verify the stability definition uniformly in t. ∎

In the time-varying case $V : \mathbb{R} \times \mathcal{D} \to \mathbb{R}$ may also be time dependent. Thus, its derivative along the trajectories of the system is given by $\dot{V}(t,x(t)) = \frac{\partial V(t,x)}{\partial t} + \frac{\partial V(t,x)}{\partial x} f(t,x(t))$.

Theorem A.6.2 Consider the system (A.7), and assume that there exists a continuously differentiable function $V(t,x)$ such that

$$W_1(x) \le V(t,x) \le W_2(x), \quad \forall\, x \in \mathcal{D},$$

where $W_1(x)$, $W_2(x)$ are continuous and positive definite functions on $\mathcal{D}$. If

- $\dot{V}(t,x(t)) \le 0$, for all $x \in \mathcal{D}$, then the equilibrium is uniformly stable;
- $\dot{V}(t,x(t)) \le -W_3(x)$, for all $x \in \mathcal{D}$, where $W_3(x)$ is a continuous and positive definite function on $\mathcal{D}$, then the equilibrium is uniformly asymptotically stable.

Additionally, if $\mathcal{D} = \mathbb{R}^n$, and $W_1(x)$ is radially unbounded, then these results hold globally.

The proof of this theorem can be found in [86].

Example A.6.2 Consider the second-order system

$$\begin{aligned} \dot{x}_1(t) &= x_2(t), \\ \dot{x}_2(t) &= -x_1(t) - \phi(t)x_2(t), \end{aligned} \tag{A.8}$$

where $\phi(t)$ is a positive definite continuously differentiable function with bounded derivative. Consider the following Lyapunov function candidate:

$$V(t,x) = \frac{1}{2}(x_1^2 + x_2^2).$$

Its derivative along the trajectories of the system is given by

$$\dot{V}(t,x(t)) = x_1(t)x_2(t) - x_2(t)x_1(t) - \phi(t)x_2^2(t) = -\phi(t)x_2^2(t) \le 0.$$

This allows for concluding uniform stability of the origin. However, using Theorem A.6.2, one cannot conclude asymptotic stability. ■

Next we show the result known as Barbalat's lemma, which in some cases can help to conclude stronger stability properties.

Lemma A.6.1 (Barbalat's lemma) Let $f : \mathbb{R} \to \mathbb{R}$ be a uniformly continuous function on $[0, \infty)$. Assume that $\lim_{t\to\infty} \int_0^t f(\tau)d\tau$ exists. Then

$$\lim_{t\to\infty} f(t) = 0.$$

The proof of this lemma can be found in [86].

Corollary A.6.1 If a scalar function $V(t,x)$ satisfies the conditions

- $V(t,x)$ is lower bounded,
- $\dot{V}(t,x)$ is negative semidefinite, and
- $\dot{V}(t,x)$ is uniformly continuous in time,

then $\dot{V}(t,x(t)) \to 0$ as $t \to \infty$.

Recall Example A.6.2. Notice that the second derivative

$$\ddot{V}(t,x(t)) = -2\phi(t)\dot{x}_2(t) - \dot{\phi}(t)x_2^2(t)$$

is bounded. This implies that $\dot{V}(t,x(t))$ is uniformly continuous, and thus according to Corollary A.6.1,

$$\lim_{t\to\infty} \dot{V}(t,x(t)) = 0,$$

which leads to

$$\lim_{t\to\infty} x_2(t) = 0.$$

From (A.8) one can see that $\dot{x}_2(t)$ is uniformly continuous, and application of Barbalat's lemma gives us

$$\lim_{t\to\infty} \dot{x}_2(t) = 0,$$

which along with (A.8) implies that

$$\lim_{t\to\infty} x_1(t) = 0.$$

Therefore, the origin is asymptotically stable.

Lemma A.6.2 Suppose that for the linear state equation

$$\dot{x}(t) = A(t)x(t), \quad x(0) = x_0,$$

with continuously differentiable $A(t) \in \mathbb{R}^{n\times n}$ there exist positive constants μ_A, μ_λ, such that for all $t \geq 0$, $\|A(t)\|_\infty \leq \mu_A$, and at each time t, the eigenvalues of $A(t)$ (pointwise eigenvalue) satisfy $\Re[\lambda(t)] \leq -\mu_\lambda$. Then, there exists a positive constant ζ such that, if the time derivative of $A(t)$ satisfies $\|\dot{A}(t)\|_\infty \leq \zeta$ for all $t \geq 0$, the equilibrium of the state equation is exponentially stable and $\|\dot{P}(t)\|_\infty < 1$, where $P(t)$ is the solution of

$$A^\top(t)P(t) + P(t)A(t) = -\mathbb{I}.$$

The proof is similar to the proof of Theorem 8.7 in [149].

A.7 $\mathcal{L}$-Stability

Next we consider the stability of input-output models of dynamical systems and refer to a system as

$$y = \mathcal{G}u,$$

where $\mathcal{G}$ denotes the map from the input $u(t) \in \mathbb{R}^m$ to the output $y(t) \in \mathbb{R}^l$. We introduce two special spaces of functions, which will be used in the forthcoming analysis.

Definition A.7.1 A continuous function $\alpha : [0, \infty) \to [0, \infty)$ is said to belong to class $\mathcal{K}$, if it is strictly increasing and $\alpha(0) = 0$. It is said to belong to class $\mathcal{K}_\infty$, if in addition $\lim_{x\to\infty} \alpha(x) = \infty$.

Definition A.7.2 A continuous function $\alpha : [0, \infty) \times [0, \infty) \to [0, \infty)$ is said to belong to class $\mathcal{KL}$, if $\beta(x) = \alpha(t_0, x) \in \mathcal{K}$ for each fixed t_0, and for each fixed x_0 it is a decreasing function in t, i.e., $\lim_{t\to\infty} \alpha(t, x_0) = 0$.

Definition A.7.3 The map $\mathcal{G} : \mathcal{L}_e^m \to \mathcal{L}_e^l$ is $\mathcal{L}$-stable if there exist a class $\mathcal{K}$ function $\alpha(\cdot)$ and a nonnegative constant b such that

$$\|(\mathcal{G}u)_\tau\|_{\mathcal{L}} \leq \alpha(\|u_\tau\|_{\mathcal{L}}) + b$$

for all $u(t) \in \mathcal{L}_e^m$ and $\tau \in [0, \infty)$. It is finite-gain $\mathcal{L}$-stable if there exist nonnegative constants a, b such that

$$\|(\mathcal{G}u)_\tau\|_{\mathcal{L}} \leq a\|u_\tau\|_{\mathcal{L}} + b$$

for all $u(t) \in \mathcal{L}_e^m$ and $\tau \in [0, \infty)$.

If Definition A.7.3 holds for the $\mathcal{L}_\infty$-norm of the signals, the system is called **BIBO stable**.

Consider the particular case in which the input-output map $\mathcal{G}$ is given by the dynamics

$$\begin{aligned} \dot{x}(t) &= f(t, x(t), u(t)), \quad x(t_0) = x_0, \\ y(t) &= g(t, x(t), u(t)), \end{aligned}$$

where $x(t) \in \mathbb{R}^n$, $y(t) \in \mathbb{R}^l$ are the state and the output of the system, respectively; $u(t) \in \mathbb{R}^m$ is the input of the system; $f(t, x, u) : [t_0, \infty) \times \mathbb{R}^n \times \mathbb{R}^m \to \mathbb{R}^n$ is a function satisfying the sufficient conditions of existence and uniqueness of solution; and $g(t, x, u) : [t_0, \infty) \times \mathbb{R}^n \times \mathbb{R}^m \to \mathbb{R}^l$ is a given function. Let $\mathcal{H}$ denote the map from the control input to the state of the system: $x = \mathcal{H}u$. Then if Definition A.7.3 holds for the map $\mathcal{H}$ with $\mathcal{L}_\infty$-norm of the signals, the system is called **BIBS stable**.

Next we provide necessary and sufficient conditions for BIBO stability of LTI and LTV systems.

A.7.1 BIBO Stability of LTI Systems

Definition A.7.4 For a given m-input and l-output LTI system $G(s)$ with impulse response $g(t) \in \mathbb{R}^{l\times m}$, its $\mathcal{L}_1$ norm is defined as

$$\|g\|_{\mathcal{L}_1} \triangleq \max_{i=1,\ldots,l} \left(\sum_{j=1}^{m} \|g_{ij}\|_{\mathcal{L}_1} \right).$$

For the purpose of simplifying the notation in relatively complex derivations, involving cascades of several systems, in the book $\|G(s)\|_{\mathcal{L}_1}$ is used instead of $\|g\|_{\mathcal{L}_1}$.

Lemma A.7.1 Assume that $g(t) \in \mathcal{L}_1$, i.e., $\|g\|_{\mathcal{L}_1} < \infty$. Then for arbitrary $u(t) \in \mathcal{L}_{\infty e}$, we have

$$\|y_\tau\|_{\mathcal{L}_\infty} \leq \|g\|_{\mathcal{L}_1} \|u_\tau\|_{\mathcal{L}_\infty},$$

and $y(t) \in \mathcal{L}_{\infty e}$.

Proof. Let $y_i(t)$ be the ith element of $y(t)$ and $u_j(t)$ be the jth element of $u(t)$. Then, for arbitrary $t \in [t_0, \tau]$, we have

$$y_i(t) = \int_{t_0}^{t} \left(\sum_{j=1}^{m} g_{ij}(t-\xi)u_j(\xi) \right) d\xi .$$

This leads to the following upper bound

$$\begin{aligned}
|y_i(t)| &\leq \int_{t_0}^{t} \left(\sum_{j=1}^{m} |g_{ij}(t-\xi)||u_j(\xi)| \right) d\xi \\
&\leq \max_{j=1,\dots,m} \left(\sup_{t_0 \leq \xi \leq t} |u_j(\xi)| \right) \int_{t_0}^{t} \sum_{j=1}^{m} |g_{ij}(t-\xi)| d\xi \\
&= \max_{j=1,\dots,m} \left(\sup_{t_0 \leq \xi \leq t} |u_j(\xi)| \right) \sum_{j=1}^{m} \int_{t_0}^{t} |g_{ij}(\xi)| d\xi \\
&\leq \|u_t\|_{\mathcal{L}_\infty} \sum_{j=1}^{m} \|g_{ij}\|_{\mathcal{L}_1} , \quad \forall\, t \in [t_0, \tau] .
\end{aligned}$$

Hence, it follows that

$$\|y_\tau\|_{\mathcal{L}_\infty} = \max_{i=1,\dots,l} \|y_{i\tau}\|_{\mathcal{L}_\infty} \leq \|u_\tau\|_{\mathcal{L}_\infty} \max_{i=1,\dots,l} \left(\sum_{j=1}^{m} \|g_{ij}\|_{\mathcal{L}_1} \right) = \|g\|_{\mathcal{L}_1} \|u_\tau\|_{\mathcal{L}_\infty} ,$$

which completes the proof. □

Lemma A.7.2 A continuous-time LTI (proper) system $y(s) = G(s)u(s)$ with impulse response matrix $g(t)$ is BIBO stable if and only if its $\mathcal{L}_1$-norm is bounded, i.e., $\|g\|_{\mathcal{L}_1} < \infty$, or equivalently $g(t) \in \mathcal{L}_1$.

Proof. *Sufficiency* immediately follows from Lemma A.7.1. To show *necessity*, we will prove that if the $\mathcal{L}_1$-norm of $g(t)$ is not bounded, then there exists at least one bounded input that will force the output $y(t)$ to diverge. If $\|g(t)\|_{\mathcal{L}_1} = \infty$, then from Definition A.7.4 it follows that there exists at least one entry $\{ij\}$ in the impulse response matrix such that

$$\int_{0}^{\infty} |g_{ij}(\sigma)| d\sigma = \infty .$$

For every t let the jth element of the vector $u(t-\sigma)$ be

$$u_j(t-\sigma) = \begin{cases} +1 & \text{if } g_{ij}(\sigma) \geq 0, \\ -1 & \text{if } g_{ij}(\sigma) < 0, \end{cases}$$

and let other elements of the vector $u(t-\sigma)$ be zero. Then $g_i(\sigma)u(t-\sigma) = |g_{ij}(\tau)|$, where $g_i(\sigma)$ denotes the ith row of the matrix $g(\sigma)$. This implies that the ith element of the output is given by

$$y_i(t) = \int_0^t g_i(\sigma)u(t-\sigma)d\sigma = \int_0^t |g_{ij}(\sigma)|d\sigma\,.$$

Thus, we have $\lim_{t\to\infty} y_i(t) = \infty$, which implies $\lim_{t\to\infty} \|y\|_{\mathcal{L}_\infty} = \infty$ and contradicts the assumption on system's stability. □

Remark A.7.1 Notice that for a BIBO-stable LTI system with impulse response matrix $g(t)$, if its input $u(t)$ is uniformly bounded, i.e., $u(t) \in \mathcal{L}_\infty$, then one has

$$\|y\|_{\mathcal{L}_\infty} \leq \|g\|_{\mathcal{L}_1}\|u\|_{\mathcal{L}_\infty}\,.$$

Lemma A.7.3 The LTI system

$$\dot{x}(t) = Ax(t) + Bu(t)$$

is finite-gain $\mathcal{L}$-stable if and only if A is Hurwitz.

A proof of this lemma can be found in [86].

Lemma A.7.4 For a cascaded system $G(s) = G_2(s)G_1(s)$, where $G_1(s)$ and $G_2(s)$ are stable proper systems, we have

$$\|G(s)\|_{\mathcal{L}_1} \leq \|G_2(s)\|_{\mathcal{L}_1}\|G_1(s)\|_{\mathcal{L}_1}\,.$$

Proof. Let $y_1(s) = G_1(s)u_1(s)$, $y_2(s) = G_2(s)u_2(s)$, and $u_2(t) \equiv y_1(t)$. Further, let $u_1(t) \in \mathcal{L}_\infty$. From Lemma A.7.1 it follows that

$$\|y_1\|_{\mathcal{L}_\infty} \leq \|G_1(s)\|_{\mathcal{L}_1}\|u_1\|_{\mathcal{L}_\infty}\,,$$
$$\|y_2\|_{\mathcal{L}_\infty} \leq \|G_2(s)\|_{\mathcal{L}_1}\|u_2\|_{\mathcal{L}_\infty}\,.$$

Since $G(s) = G_2(s)G_1(s)$, then $y_2(s) = G(s)u_1(s) = G_2(s)G_1(s)u_1(s) = G_2(s)y_1(s)$. Hence we have

$$\|y_2\|_{\mathcal{L}_\infty} \leq \|G_2(s)\|_{\mathcal{L}_1}\|y_1\|_{\mathcal{L}_\infty} \leq \|G_2(s)\|_{\mathcal{L}_1}\|G_1(s)\|_{\mathcal{L}_1}\|u_1\|_{\mathcal{L}_\infty}\,. \tag{A.9}$$

On the other hand, from BIBO stability of $G(s)$ it follows that

$$\|y_2\|_{\mathcal{L}_\infty} \leq \|G(s)\|_{\mathcal{L}_1}\|u_1\|_{\mathcal{L}_\infty}\,.$$

Next we show that $\|G(s)\|_{\mathcal{L}_1}$ is the least upper bound of $\|y_2\|_{\mathcal{L}_\infty}$. This can be done by contradiction. Without loss of generality, let $\|u_1\|_{\mathcal{L}_\infty} \leq 1$, and assume that there exists a lower upper bound η, such that $\|y_2\|_{\mathcal{L}_\infty} \leq \eta < \|G(s)\|_{\mathcal{L}_1}$. This implies that

$$\sup_{t\geq 0}\|y_2(t)\|_\infty \leq \eta < \|G(s)\|_{\mathcal{L}_1}\,.$$

Then, there exist $t_1 > 0$ and index k such that

$$\sum_{j=1}^{m}\int_0^{t_1}\left|g_{kj}(t_1-\sigma)\right|d\sigma > \eta\,,$$

where $g(t)$ is the impulse response matrix for the system $G(s)$. We can choose the control signal as

$$u_1(\sigma) = \begin{cases} \left[\mathrm{sgn}(g_{k1}(t_1-\sigma)), \dots, \mathrm{sgn}(g_{km}(t_1-\sigma))\right]^\top, & \sigma \in [0, t_1], \\ 0, \quad \sigma > t_1. \end{cases}$$

Notice that for this control signal $\|u_1\|_{\mathcal{L}_\infty} \le 1$. Then we have

$$(y_2)_k(t_1) = \sum_{j=1}^{m} \int_0^{t_1} g_{kj}(t_1-\sigma)(u_1)_j(\sigma)d\sigma = \sum_{j=1}^{m} \int_0^{t_1} \left|g_{kj}(t_1-\sigma)\right| d\sigma > \eta.$$

This implies $\|y_2\|_{\mathcal{L}_\infty} > \eta$, which contradicts the fact that η is an upper bound for $\|y_2\|_{\mathcal{L}_\infty}$. Hence $\|G(s)\|_{\mathcal{L}_1}$ is the least upper bound for $\|y_2\|_{\mathcal{L}_\infty}$. This fact, along with (A.9), completes the proof. □

A.7.2 BIBO Stability for LTV Systems

Consider the LTV system

$$\begin{aligned} \dot{x}(t) &= A(t)x(t) + B(t)u(t), \quad x(t_0) = x_0, \\ y(t) &= C(t)x(t), \end{aligned} \tag{A.10}$$

where $x \in \mathbb{R}^n$ is the system state and $A(t) \in \mathbb{R}^{n\times n}$, $B(t) \in \mathbb{R}^{n\times m}$, $C(t) \in \mathbb{R}^{l\times n}$ are piecewise-continuous in time. Let $\mathcal{G}$ be the input-output map of this system for given t_0 and zero initial condition $x_0 = 0$.

Definition A.7.5 The system in (A.10) is uniformly BIBO stable if there exists a positive constant $a > 0$, such that for arbitrary t_0 and arbitrary bounded input signal $u(t)$ the corresponding response for $x_0 = 0$ verifies

$$\sup_{t\ge t_0} \|y(t)\|_\infty \le a \sup_{t\ge t_0} \|u(t)\|_\infty.$$

Definition A.7.6 The $\mathcal{L}_1$-norm of the m-input l-output LTV system in (A.3) is defined as

$$\|g\|_{\mathcal{L}_1} \triangleq \max_{1\le i\le n} \left(\sum_{j=1}^{l} \|g_{ij}\|_{\mathcal{L}_1} \right),$$

where

$$\|g_{ij}\|_{\mathcal{L}_1} \triangleq \sup_{t\ge\tau, \tau\in\mathbb{R}^+} \int_\tau^t |g_{ij}(t,\sigma)| d\sigma,$$

with $g_{ij}(t,\sigma)$ being the $\{ij\}$th entry of the impulse response matrix.

We will use $\|\mathcal{G}\|_{\mathcal{L}_1} \triangleq \|g\|_{\mathcal{L}_1}$ to denote the $\mathcal{L}_1$-norm of the input-output map $\mathcal{G}$ of the system with impulse response matrix $g(t,t_0)$.

Lemma A.7.5 Consider the system in (A.10) with zero initial condition $x_0 = 0$. Suppose it has a uniformly asymptotically stable equilibrium at the origin, and there exist positive constants $b, c > 0$ such that for all $t \geq 0$

$$\|B(t)\| \leq b, \quad \|C(t)\| \leq c.$$

Then the system is also uniformly BIBO stable. Further, if $u(t) \in \mathcal{L}_\infty$, then

$$\|y\|_{\mathcal{L}_\infty} \leq \|\mathcal{G}\|_{\mathcal{L}_1} \|u\|_{\mathcal{L}_\infty}.$$

The proof of this lemma can be found in [149].

Lemma A.7.6 For a cascaded system $\mathcal{G} = \mathcal{G}_2\mathcal{G}_1$, where $\mathcal{G}_1$ and $\mathcal{G}_2$ are uniformly BIBO-stable systems, we have

$$\|\mathcal{G}\|_{\mathcal{L}_1} \leq \|\mathcal{G}_2\|_{\mathcal{L}_1} \|\mathcal{G}_1\|_{\mathcal{L}_1}.$$

Proof. Let $y_1 = \mathcal{G}_1 u_1$, $y_2 = \mathcal{G}_2 u_2$, and $u_2(t) \equiv y_1(t)$. Further, let $u_1(t) \in \mathcal{L}_\infty$. From Lemma A.7.5 it follows that

$$\|y_1\|_{\mathcal{L}_\infty} \leq \|\mathcal{G}_1\|_{\mathcal{L}_1} \|u_1\|_{\mathcal{L}_\infty},$$
$$\|y_2\|_{\mathcal{L}_\infty} \leq \|\mathcal{G}_2\|_{\mathcal{L}_1} \|u_2\|_{\mathcal{L}_\infty}.$$

Since $\mathcal{G} = \mathcal{G}_2\mathcal{G}_1$, then $y_2 = \mathcal{G}u_1 = \mathcal{G}_2\mathcal{G}_1 u_1 = \mathcal{G}_2 y_1$. Hence we have

$$\|y_2\|_{\mathcal{L}_\infty} \leq \|\mathcal{G}_2\|_{\mathcal{L}_1} \|y_1\|_{\mathcal{L}_\infty} \leq \|\mathcal{G}_2\|_{\mathcal{L}_1} \|\mathcal{G}_1\|_{\mathcal{L}_1} \|u_1\|_{\mathcal{L}_\infty}. \tag{A.11}$$

On the other hand, from uniform BIBO stability of $\mathcal{G}$ it follows that

$$\|y_2\|_{\mathcal{L}_\infty} \leq \|\mathcal{G}\|_{\mathcal{L}_1} \|u_1\|_{\mathcal{L}_\infty}.$$

Next we show that $\|\mathcal{G}\|_{\mathcal{L}_1}$ is the least upper bound of $\|y_2\|_{\mathcal{L}_\infty}$. This can be done by contradiction. Without loss of generality, let $\|u_1\|_{\mathcal{L}_\infty} \leq 1$, and assume that there exists a lower upper bound η, such that $\|y_2\|_{\mathcal{L}_\infty} \leq \eta < \|\mathcal{G}\|_{\mathcal{L}_1}$. This implies that

$$\sup_{t \geq 0} \|y_2(t)\|_\infty \leq \eta < \|\mathcal{G}\|_{\mathcal{L}_1}.$$

Then, there exist t_0 and t_1, $t_1 > t_0$, and index k such that

$$\sum_{j=1}^{m} \int_{t_0}^{t_1} \left|g_{kj}(t_1,\sigma)\right| d\sigma > \eta,$$

where $g(t,t_0)$ is the impulse response matrix for the system $\mathcal{G}$. We can choose the control signal as

$$u_1(\sigma) = \begin{cases} \left[\operatorname{sgn}(g_{k1}(t_1,\sigma)), \ldots, \operatorname{sgn}(g_{km}(t_1,\sigma))\right]^\top, & \sigma \in [t_0,\ t_1], \\ 0, \quad \sigma > t_1. \end{cases}$$

Notice that for this control signal $\|u_1\|_{\mathcal{L}_\infty} \leq 1$. Then we have

$$(y_2)_k(t_1) = \sum_{j=1}^{m} \int_{t_0}^{t_1} g_{kj}(t_1,\sigma)(u_1)_j(\sigma) d\sigma = \sum_{j=1}^{m} \int_{t_0}^{t_1} \left|g_{kj}(t_1,\sigma)\right| d\sigma > \eta.$$

This implies $\|y_2\|_{\mathcal{L}_\infty} > \eta$, which contradicts the fact that η is an upper bound for $\|y_2\|_{\mathcal{L}_\infty}$. Hence $\|\mathcal{G}\|_{\mathcal{L}_1}$ is the least upper bound for $\|y_2\|_{\mathcal{L}_\infty}$. This fact, along with (A.11), completes the proof. □

A.8 Linear Parametrization of Nonlinear Systems

Consider the nonlinear map $f(t,x) : [0, \infty) \times \mathbb{R}^n \to \mathbb{R}$, subject to the following assumptions.

Assumption A.8.1 (Uniform boundedness of $f(t,0)$) There exists $B > 0$, such that $|f(t,0)| \leq B$ for all $t \geq 0$.

Assumption A.8.2 (Semiglobal uniform boundedness of partial derivatives) The map $f(t,x)$ is continuous in its arguments, and moreover, for arbitrary $\delta > 0$, there exist $d_{f_t}(\delta) > 0$ and $d_{f_x}(\delta) > 0$ independent of time, such that for all $\|x\|_\infty \leq \delta$ the partial derivatives of $f(t,x)$ with respect to t and x are piecewise continuous and bounded,

$$\left\| \frac{\partial f(t,x)}{\partial x} \right\|_1 \leq d_{f_x}(\delta), \quad \left| \frac{\partial f(t,x)}{\partial t} \right| \leq d_{f_t}(\delta).$$

The next lemma proves that, subject to Assumptions A.8.1 and A.8.2, on any finite time interval the nonlinear function $f(t,x(t))$ can be linearly parameterized in two time-varying parameters using $\|x(t)\|_\infty$ as a regressor.

Lemma A.8.1 (see [30]) Let $x(t)$ be a continuous and (piecewise)-differentiable function of t for $t \geq 0$. If $\|x_\tau\|_{\mathcal{L}_\infty} \leq \rho$ and $\|\dot{x}_\tau\|_{\mathcal{L}_\infty} \leq d_x$ for $\tau \geq 0$, where ρ and d_x are some positive constants, then there exist continuous $\theta(t)$ and $\sigma(t)$ with (piecewise)-continuous derivative, such that for all $t \in [0, \tau]$

$$f(t,x(t)) = \theta(t)\|x(t)\|_\infty + \sigma(t), \tag{A.12}$$

where

$$|\theta(t)| < \theta_\rho, \quad |\dot{\theta}(t)| < d_\theta,$$
$$|\sigma(t)| < \sigma_b, \quad |\dot{\sigma}(t)| < d_\sigma,$$

with $\theta_\rho \triangleq d_{f_x}(\rho)$, $\sigma_b \triangleq B + \epsilon$, in which $\epsilon > 0$ is an arbitrary constant, and d_θ, d_σ are computable bounds.

Proof. The semiglobal uniform boundedness of the partial derivatives of $f(t,x)$ in Assumption A.8.2 implies that for arbitrary $\|x\|_\infty \leq \rho$

$$|f(t,x) - f(t,0)| \leq d_{f_x}(\rho)\|x\|_\infty. \tag{A.13}$$

Next, from Assumption A.8.1, it follows that if $\|x(0)\|_\infty \leq \rho$, the following bound holds:

$$|f(0,x(0))| \leq d_{f_x}(\rho)\|x(0)\|_\infty + B < d_{f_x}(\rho)\|x(0)\|_\infty + B + \epsilon,$$

where $\epsilon > 0$ is an arbitrary constant. This implies that there exist $\theta(0)$ and $\sigma(0)$ such that

$$|\theta(0)| < \theta_\rho, \quad |\sigma(0)| < \sigma_b, \tag{A.14}$$

and

$$f(0,x(0)) = \theta(0)\|x(0)\|_\infty + \sigma(0). \tag{A.15}$$

Notice that the choice of $\theta(0)$ and $\sigma(0)$ is not unique. For the sake of proof we select arbitrary $\theta(0)$, $\sigma(0)$ that verify (A.14) and (A.15).

We construct the trajectories of $\theta(t)$ and $\sigma(t)$ according to the dynamics

$$\begin{bmatrix} \dot{\theta}(t) \\ \dot{\sigma}(t) \end{bmatrix} = A_\eta^{-1}(t) \begin{bmatrix} \frac{df(t,x(t))}{dt} - \theta(t)\frac{d\|x(t)\|_\infty}{dt} \\ 0 \end{bmatrix}, \tag{A.16}$$

where

$$A_\eta(t) = \begin{bmatrix} \|x(t)\|_\infty & 1 \\ -(\sigma_b - |\sigma(t)|) & \theta_\rho - |\theta(t)| \end{bmatrix}, \tag{A.17}$$

with any initial values satisfying (A.15). The determinant of A_η is

$$\det(A_\eta(t)) = \|x(t)\|_\infty(\theta_\rho - |\theta(t)|) + \sigma_b - |\sigma(t)|. \tag{A.18}$$

If

$$|\theta(t)| < \theta_\rho, \quad |\sigma(t)| < \sigma_b, \tag{A.19}$$

then it follows from (A.18) that $\det(A_\eta(t)) \neq 0$ for all $t \in [0,\bar{\tau})$, where $\bar{\tau} > 0$ is an arbitrary constant or ∞. Hence, it follows from (A.16), (A.17) that

$$\frac{df(t,x(t))}{dt} = \frac{d\Big(\theta(t)\|x(t)\|_\infty + \sigma(t)\Big)}{dt}, \tag{A.20}$$

$$\frac{\dot{\sigma}(t)}{\sigma_b - |\sigma(t)|} = \frac{\dot{\theta}(t)}{\theta_\rho - |\theta(t)|} \tag{A.21}$$

for all $t \in [0,\bar{\tau})$. Using the selected initial condition from (A.15), we can integrate to obtain

$$f(t,x(t)) = \theta(t)\|x(t)\|_\infty + \sigma(t), \tag{A.22}$$

$$\int_0^{\bar{\tau}_-} \frac{\dot{\sigma}(t)}{\sigma_b - |\sigma(t)|} dt = \int_0^{\bar{\tau}_-} \frac{\dot{\theta}(t)}{\theta_\rho - |\theta(t)|} dt, \tag{A.23}$$

where $\int_0^{\bar{\tau}_-}(\cdot)dt \triangleq \lim_{\xi\to\bar{\tau}_-}\int_0^{\xi}(\cdot)dt$.

Next we compute the left-hand side of (A.23) assuming that $|\sigma(t)| < \sigma_b$:

$$\begin{aligned} \int_0^{\bar{\tau}_-} \frac{\dot{\sigma}(t)}{\sigma_b - |\sigma(t)|} dt &= \int_0^{\bar{\tau}_-} \frac{1}{\sigma_b - |\sigma(t)|} \frac{d(\operatorname{sgn}(\sigma(t))|\sigma(t)|)}{dt} dt \\ &= \lim_{t\to\bar{\tau}} \Big(\operatorname{sgn}(\sigma(t))\ln(\sigma_b - |\sigma(t)|)\Big) \\ &\quad - \operatorname{sgn}(\sigma(0))\ln(\sigma_b - |\sigma(0)|). \end{aligned}$$

Notice that $\operatorname{sgn}(\sigma(t))$ is not differentiable when $\sigma(t)$ crosses zero. However, the set of points $\{t_i\}$, where $\sigma(t_i) = 0$, is a countable set with Lebesgue measure zero. This allows us to exclude these points while taking the integral, and thus we can ensure $d(\text{sgn}(\sigma(t))|\sigma(t)|) = \text{sgn}(\sigma(t))d(|\sigma(t)|)$. Similarly, assuming that $|\theta(t)| < \theta_\rho$, the right-hand side of (A.23) is given by

$$\int_0^{\bar{\tau}_-} \frac{\dot{\theta}(t)}{\theta_\rho - |\theta(t)|} dt = \lim_{t\to\bar{\tau}} \left(\text{sgn}(\theta(t))\ln(\theta_\rho - |\theta(t)|)\right) \\ - \text{sgn}\,(\theta(0))\ln(\theta_\rho - |\theta(0)|)\,.$$

Using these arguments, we rewrite (A.23):

$$\lim_{t\to\bar{\tau}} \left(\text{sgn}(\sigma(t))\ln(\sigma_b - |\sigma(t)|)\right) - \text{sgn}(\sigma(0))\ln(\sigma_b - |\sigma(0)|) \\ = \lim_{t\to\bar{\tau}} \left(\text{sgn}(\theta(t))\ln(\theta_\rho - |\theta(t)|)\right) - \text{sgn}(\theta(0))\ln(\theta_\rho - |\theta(0)|)\,. \tag{A.24}$$

In what follows, we prove (A.19) by contradiction. If (A.19) is not true, since $\theta(t)$ and $\sigma(t)$ are continuous, it follows from (A.14) that there exists $\bar{\tau} \in [0,\ \tau]$ such that either

$$\text{(i)} \lim_{t\to\bar{\tau}} |\theta(t)| = \theta_\rho \text{ or} \tag{A.25}$$

$$\text{(ii)} \lim_{t\to\bar{\tau}} |\sigma(t)| = \sigma_b\,, \tag{A.26}$$

while $|\theta(t)| < \theta_\rho,\ |\sigma(t)| < \sigma_b$ for all $t \in [0,\ \bar{\tau})$.

(i) In this case we have

$$|\lim_{t\to\bar{\tau}} \left(\text{sgn}(\theta(t))\ln(\theta_\rho - |\theta(t)|)\right)| = \infty\,.$$

Since it is obvious that $\text{sgn}(\sigma(0))\ln(\sigma_b - |\sigma(0)|)$ and $\text{sgn}(\theta(0))\ln(\theta_\rho - |\theta(0)|)$ are bounded, it follows from (A.24) that

$$|\lim_{t\to\bar{\tau}} \left(\text{sgn}(\sigma(t))\ln(\sigma_b - |\sigma(t)|)\right)| = \infty\,,$$

and hence

$$\lim_{t\to\bar{\tau}} |\sigma(t)| = \sigma_b\,. \tag{A.27}$$

Thus, from (A.22) we have

$$\lim_{t\to\bar{\tau}} f(t,x(t)) = \lim_{t\to\bar{\tau}} \left(\theta(t)\|x(t)\|_\infty + \sigma(t)\right),$$

which along with (A.25) and (A.27) implies that

$$|\lim_{t\to\bar{\tau}} f(t,x(t))| = |f(\bar{\tau},x(\bar{\tau}))| = \theta_\rho\|x(\bar{\tau})\|_\infty + \sigma_b\,. \tag{A.28}$$

From (A.13) and Assumption A.8.1, it follows that

$$|f(\bar{\tau},x(\bar{\tau}))| \le d_{f_x}(\rho)\|x(\bar{\tau})\|_\infty + B = \theta_\rho\|x(\bar{\tau})\|_\infty + \sigma_b - \epsilon\,,$$

which contradicts (A.28), and therefore (A.25) is not true.

(ii) Following the same steps as above, one can derive a contradicting argument to (A.26).

Since (A.25), (A.26) are not true, then the relationships in (A.19) hold. Equality (A.12) follows from (A.19) and (A.22) directly.

Further, if $\|\dot{x}_\tau\|_{\mathcal{L}_\infty}$ is bounded, then in light of Assumption A.8.2, $\frac{df(t,x(t))}{dt}$ and $\frac{d\|x(t)\|_\infty}{dt}$ are bounded, although the derivative $\frac{d\|x(t)\|_\infty}{dt}$ may not be continuous. Since $\theta(t)$ is bounded, then $\frac{df(t,x(t))}{dt} - \theta(t)\frac{d\|x(t)\|_\infty}{dt}$ is bounded. From (A.19), it follows that $\det A_\eta(t) \neq 0$, and therefore we conclude from (A.16) that $\dot{\theta}(t)$ and $\dot{\sigma}(t)$ are bounded. This concludes the proof. □

Consider the following system dynamics:

$$\begin{aligned} \dot{x}(t) &= A_m x(t) + b\big(\omega u(t) + f(t,x(t))\big), \quad x(0) = x_0, \\ y(t) &= c^\top x(t), \end{aligned} \tag{A.29}$$

where $x(t) \in \mathbb{R}^n$ is the system state; $u(t) \in \mathbb{R}$ is the given bounded system input; $y(t) \in \mathbb{R}$ is the system output; $A_m \in \mathbb{R}^{n\times n}$ is a known Hurwitz matrix; $b,\ c \in \mathbb{R}^n$ are known vectors; $\omega \in \mathbb{R}$ is an unknown parameter; and $f : \mathbb{R}\times\mathbb{R}^n \to \mathbb{R}$ is an unknown nonlinear function. Subject to Assumptions A.8.1 and A.8.2, the nonlinear system in (A.29) can be rewritten over $t \in [0,\ \tau]$ for arbitrary $\tau \geq 0$ as a semilinear time-varying system with bounded parameters that have bounded derivatives:

$$\begin{aligned} \dot{x}(t) &= A_m x(t) + b\big(\omega u(t) + \theta(t)\|x(t)\|_\infty + \sigma(t)\big), \quad x(0) = x_0, \\ y(t) &= c^\top x(t). \end{aligned}$$

A.9 Linear Time-Varying Representation of Systems with Linear Unmodeled Dynamics

Consider the following dynamics:

$$\begin{aligned} \dot{x}_z(t) &= g(t, x_z(t), x(t)), \quad x_z(0) = x_0, \\ z(t) &= g_0(t, x_z(t)), \end{aligned}$$

where $x(t) \in \mathbb{R}^n$ is a bounded differentiable signal with bounded derivative, the functions $g : \mathbb{R}\times\mathbb{R}^m\times\mathbb{R}^n \to \mathbb{R}^m$, $g_0 : \mathbb{R}\times\mathbb{R}^m \to \mathbb{R}^l$ are unknown nonlinear maps continuous in their arguments, and $x_z(t) \in \mathbb{R}^m$ and $z(t) \in \mathbb{R}^l$ represent the state and the output of the system.

Further, consider the map $f(t,x,z) : \mathbb{R}\times\mathbb{R}^n\times\mathbb{R}^l \to \mathbb{R}$ on the time interval $t \in [0,\ \tau)$ for arbitrary $\tau \geq 0$. Let $X \triangleq [x^\top,\ z^\top]^\top$, and with a slight abuse of language let $f(t,X) \triangleq f(t,x,z)$, $i = 1,2$. The function $f(\cdot)$ verifies the following assumptions.

Assumption A.9.1 (Uniform boundedness of $f(t,0)$) There exists $B > 0$, such that $|f(t,0)| \leq B$ holds for all $t \geq 0$.

Assumption A.9.2 (Semiglobal uniform boundedness of partial derivatives) The map $f(t,x)$ is continuous in its arguments, and for arbitrary $\delta > 0$, there exist $d_{f_t}(\delta) > 0$ and

$d_{f_x}(\delta) > 0$ independent of time, such that for all $\|X\|_\infty \leq \delta$ the partial derivatives of $f(t,X)$ with respect to t and X are piecewise-continuous and bounded:

$$\left\| \frac{\partial f(t,X)}{\partial X} \right\|_1 \leq d_{f_x}(\delta), \quad \left| \frac{\partial f(t,X)}{\partial t} \right| \leq d_{f_t}(\delta).$$

Assumption A.9.3 (Stability of unmodeled dynamics) The z-dynamics are BIBO stable; i.e., there exist $L_1 > 0$ and $L_2 > 0$ such that for all $t \geq 0$

$$\|z_t\|_{\mathcal{L}_\infty} \leq L_1 \|x_t\|_{\mathcal{L}_\infty} + L_2 .$$

Let $\bar{\delta} \triangleq \max\{\delta + \gamma, L_1(\delta+\gamma) + L_2\}$ for an arbitrary $\gamma > 0$, and define $L_\delta \triangleq \frac{\bar{\delta}}{\delta} d_{f_x}(\bar{\delta})$.

The next lemma proves that, subject to Assumptions A.9.1, A.9.2, and A.9.3, on any finite time interval the nonlinear function $f(t,x(t),z(t))$ can be linearly parameterized in two time-varying parameters using $\|x_t\|_{\mathcal{L}_\infty}$ as a regressor.

Lemma A.9.1 (see [31]) If $\|x_\tau\|_{\mathcal{L}_\infty} \leq \rho$ and $\|\dot{x}_\tau\|_{\mathcal{L}_\infty} \leq d_x$ for $\tau \geq 0$, where ρ and d_x are some positive constants, then there exist continuous and (piecewise)-differentiable $\theta(t) \in \mathbb{R}$ and $\sigma(t) \in \mathbb{R}$ with bounded derivatives, such that for all $t \in [0,\tau]$

$$f(t,X(t)) = \theta(t)\|x_t\|_{\mathcal{L}_\infty} + \sigma(t),$$

and

$$|\theta(t)| < \theta_\rho, \quad |\dot{\theta}(t)| < d_\theta,$$
$$|\sigma(t)| < \sigma_b, \quad |\dot{\sigma}(t)| < d_\sigma,$$

with $\theta_\rho \triangleq L_\rho$, $\sigma_b \triangleq L_\rho L_2 + B + \epsilon$, in which $\epsilon > 0$ is an arbitrary constant, and d_θ, d_σ are computable bounds.

Proof. From Assumption A.9.3, it follows that for all $t \in [0,\ \tau)$, if $\|x(t)\|_\infty \leq \rho$, then

$$\|z_t\|_{\mathcal{L}_\infty} \leq L_1 \|x_t\|_{\mathcal{L}_\infty} + L_2 < L_1(\rho+\gamma) + L_2,$$

where $\gamma > 0$ is an arbitrary constant. Let $\bar{\rho} \triangleq \max\{\rho+\gamma, L_1(\rho+\gamma)+L_2\}$. Then

$$\|X(t)\|_\infty \leq \max\{\|x_t\|_{\mathcal{L}_\infty},\ L_1\|x_t\|_{\mathcal{L}_\infty} + L_2\} < \bar{\rho}.$$

The uniform boundedness of the partial derivatives of $f(t,X)$ in Assumption A.9.2 implies that for arbitrary $\|X\|_\infty \leq \bar{\rho}$

$$|f(t,X) - f(t,0)| \leq d_{f_x}(\bar{\rho})\|X\|_\infty .$$

Further, from Assumptions A.9.1 and the definition of L_δ, it follows that

$$|f(t,X(t))| \leq d_{f_x}(\bar{\rho})\|X(t)\|_\infty + B < L_\rho\|x_t\|_{\mathcal{L}_\infty} + L_\rho L_2 + B + \epsilon, \tag{A.30}$$

where $\epsilon > 0$ is an arbitrary constant.

Next, we follow the same steps as in the proof of Lemma A.8.1, replacing the $\|x(t)\|_\infty$ norm with $\|x_t\|_{\mathcal{L}_\infty}$ norm and keeping in mind the difference in the definition of σ_b. □

The result of Lemma A.9.1 can be extended to the vector case, when $f(t,x,z):\mathbb{R}\times\mathbb{R}^n\times\mathbb{R}^l\to\mathbb{R}^m$. If Assumptions A.9.1, A.9.2 hold for each component of $f(t,x,z)$ with the same upper bound, then they can be rewritten as

$$\|f(t,0)\|_\infty \le B\,,\qquad \left\|\frac{\partial f(t,X)}{\partial X}\right\|_\infty \le d_{f_x}(\delta)\,,\quad \text{and}\quad \left\|\frac{\partial f(t,X)}{\partial t}\right\|_\infty \le d_{f_t}(\delta)\,,$$

where the first and the third norms are vector ∞-norm, and the second is a matrix-induced ∞-norm. This leads to the result stated in the following lemma.

Lemma A.9.2 If $\|x_\tau\|_{\mathcal{L}_\infty}\le\rho$ and $\|\dot{x}_\tau\|_{\mathcal{L}_\infty}\le d_x$ for $\tau\ge 0$, where ρ and d_x are some positive constants, then there exist differentiable $\theta(t)\in\mathbb{R}^m$ and $\sigma(t)\in\mathbb{R}^m$ with bounded derivatives, such that for all $t\in[0,\tau]$

$$f(t,X(t))=\theta(t)\|x_t\|_{\mathcal{L}_\infty}+\sigma(t)\,,$$

where

$$\begin{aligned}\|\theta(t)\|_\infty &< \theta_\rho\,, & \|\dot{\theta}(t)\|_\infty &< d_\theta\,,\\ \|\sigma(t)\|_\infty &< \sigma_b\,, & \|\dot{\sigma}(t)\|_\infty &< d_\sigma\,,\end{aligned}$$

with $\theta_\rho\triangleq L_\rho$, $\sigma_b\triangleq L_\rho L_2+B+\epsilon$, in which $\epsilon>0$ is an arbitrary constant, and d_θ, d_σ are computable bounds.

Consider the following system dynamics:

$$\begin{aligned}\dot{x}(t)&=A_m x(t)+b\big(\omega u(t)+f(t,x(t),z(t))\big)\,,\quad x(0)=x_0\,,\\ y(t)&=c^\top x(t)\,,\\ \dot{x}_z(t)&=g(t,x_z(t),x(t))\,,\quad x_z(0)=x_0\,,\\ z(t)&=g_0(t,x_z(t))\,,\end{aligned}\tag{A.31}$$

where $x(t)\in\mathbb{R}^n$ is the system state; $u(t)\in\mathbb{R}$ is the given bounded system input; $y(t)\in\mathbb{R}$ is the system output; $A_m\in\mathbb{R}^{n\times n}$ is a known Hurwitz matrix; b, $c\in\mathbb{R}^n$ are known vectors; $\omega\in\mathbb{R}$ is an unknown parameter; and $f:\mathbb{R}\times\mathbb{R}^n\times\mathbb{R}^l\to\mathbb{R}$, $g:\mathbb{R}\times\mathbb{R}^m\times\mathbb{R}^n\to\mathbb{R}^m$, $g_0:\mathbb{R}\times\mathbb{R}^m\to\mathbb{R}^l$ are unknown nonlinear maps. Then, subject to Assumptions A.9.1–A.9.3, the nonlinear system in (A.31) can be rewritten over $t\in[0,\tau]$ for arbitrary $\tau\ge 0$ as a semilinear time-varying system with bounded parameters that have bounded derivatives:

$$\begin{aligned}\dot{x}(t)&=A_m x(t)+b\big(\omega u(t)+\theta(t)\|x_t\|_{\mathcal{L}_\infty}+\sigma(t)\big)\,,\quad x(0)=x_0\,,\\ y(t)&=c^\top x(t)\,.\end{aligned}$$

A.10 Linear Time-Varying Representation of Systems with Linear Unmodeled Actuator Dynamics

Consider the following system given by its transfer function:

$$\mu(s)=F(s)u(s)\,,\tag{A.32}$$

where $\mu(s)$, $u(s)\in\mathbb{R}$ are the Laplace transforms of the system output and input, respectively, and $F(s)$ is an unknown BIBO-stable transfer function with known DC gain and known upper bound for its $\mathcal{L}_1$-norm.

Assumption A.10.1 There exists $L_F > 0$ verifying $\|F(s)\|_{\mathcal{L}_1} \leq L_F$. Also, we assume that there exist known constants $\omega_l,\ \omega_u \in \mathbb{R}$, verifying

$$0 < \omega_l \leq F(0) \leq \omega_u\,,$$

where, without loss of generality, we have assumed $F(0) > 0$.

The next lemma proves that the dynamical relationship between $u(t)$ and $\mu(t)$ in (A.32) can be equivalently replaced by an algebraic relationship, linear in its structure.

Lemma A.10.1 (see [28]) Consider the system in (A.32). If for some $\tau > 0$

$$\|u_\tau\|_{\mathcal{L}_\infty} \leq \rho_u\,, \quad \|\dot{u}_\tau\|_{\mathcal{L}_\infty} \leq d_u\,,$$

then there exist ω and differentiable $\sigma(t)$ over $t \in [0, \tau]$, such that

$$\mu(t) = \omega u(t) + \sigma(t)\,,$$

where

$$\omega \in (\omega_l, \omega_u), \quad |\sigma(t)| \leq \sigma_b\,, \quad |\dot{\sigma}(t)| \leq d_\sigma\,,$$

with $\sigma_b \triangleq \|F(s) - (\omega_l + \omega_u)/2\|_{\mathcal{L}_1}\rho_u$, and $d_\sigma \triangleq \|F(s) - (\omega_l + \omega_u)/2\|_{\mathcal{L}_1} d_u$.

Proof. Let

$$\omega \triangleq \frac{\omega_l + \omega_u}{2}\,, \quad \sigma(t) \triangleq \mu(t) - \omega u(t)\,.$$

The second equation leads to

$$\mu(t) = \omega u(t) + \sigma(t)\,,$$

and further

$$\sigma(s) = (F(s) - \omega)u(s)\,, \quad s\sigma(s) = (F(s) - \omega)su(s)\,.$$

Lemma A.7.1 implies

$$\|\sigma_\tau\|_{\mathcal{L}_\infty} \leq \left\| F(s) - \frac{\omega_l + \omega_u}{2} \right\|_{\mathcal{L}_1} \|u_\tau\|_{\mathcal{L}_\infty} \leq \left\| F(s) - \frac{\omega_l + \omega_u}{2} \right\|_{\mathcal{L}_1} \rho_u$$

and

$$\|\dot{\sigma}_\tau\|_{\mathcal{L}_\infty} \leq \left\| F(s) - \frac{\omega_l + \omega_u}{2} \right\|_{\mathcal{L}_1} d_u\,. \qquad \square$$

Consider the system

$$\begin{aligned} \dot{x}(t) &= A_m x(t) + b\mu(t)\,, \quad x(0) = x_0\,, \\ y(t) &= c^\top x(t)\,, \\ \mu(s) &= F(s)u(s)\,, \end{aligned} \tag{A.33}$$

where $x(t) \in \mathbb{R}^n$ is the state of the system; $A_m \in \mathbb{R}^{n\times n}$ and $b \in \mathbb{R}^n$ are a known Hurwitz matrix and a constant vector, respectively; $c \in \mathbb{R}^n$ is a known constant vector; $\mu(t) \in \mathbb{R}$ is the actuator output; $y(t) \in \mathbb{R}$ is the system output; $u(t) \in \mathbb{R}$ is the control input; and $F(s)$ is an unknown BIBO-stable transfer function.

Using Lemma A.10.1, the system (A.33) with multiplicative uncertainty can be rewritten as the following equivalent LTI system with unknown constant system input gain and additive disturbance $\sigma(t)$, which is bounded and has bounded derivative:

$$\begin{aligned}\dot{x}(t) &= A_m x(t) + b\big(\omega u(t) + \sigma(t)\big), \quad x(0) = x_0, \\ y(t) &= c^\top x(t).\end{aligned}$$

A.11 Properties of Controllable Systems

A.11.1 Linear Time-Invariant Systems

Consider an LTI system given by

$$x(s) = (s\mathbb{I} - A)^{-1} b u(s), \tag{A.34}$$

where $x(s)$, $u(s)$ are the Laplace transforms of the system state and the system input, $A \in \mathbb{R}^{n\times n}$, $b \in \mathbb{R}^n$, and assume that

$$(s\mathbb{I} - A)^{-1} b = \frac{N(s)}{D(s)}, \tag{A.35}$$

where $D(s) = \det(s\mathbb{I} - A)$, and $N(s)$ is a $n \times 1$ vector with its ith element being a polynomial function

$$N_i(s) = \sum_{j=1}^{n} N_{ij} s^{j-1}. \tag{A.36}$$

Lemma A.11.1 If $(A,\, b)$ is controllable, then the matrix N of entries N_{ij} is full rank.

Proof. Controllability of $(A,\, b)$ implies that given an initial condition $x(0) = 0$, and arbitrary t_1 and x_{t_1}, there exists $u(\tau)$, $\tau \in [0,\, t_1]$, such that $x(t_1) = x_{t_1}$. If N is not full rank, then there exists a $\mu \in \mathbb{R}^n$, $\mu \neq 0$, such that $\mu^\top N(s) = 0$. Thus, for $x(0) = 0$ one has

$$\mu^\top x(s) = \mu^\top \frac{N(s)}{D(s)} u(s) = 0, \quad \forall\, u(s),$$

which implies that, in particular, $x(t) \neq \mu$ for any t. This contradicts the fact that $x(t_1) = x_{t_1}$ can be any point in $\mathbb{R}^n$. Thus, N must be full rank. □

Corollary A.11.1 If the pair $(A,\, b)$ in (A.34) is controllable, then there exists $c_o \in \mathbb{R}^n$, such that $c_o^\top N(s)/D(s)$ has relative degree one, i.e., $\deg(D(s)) - \deg(c_o^\top N(s)) = 1$, and $N(s)$ has all its zeros in the left half plane.

Proof. It follows from (A.35) that for arbitrary vector $c_o \in \mathbb{R}^n$

$$c_o^\top (s\mathbb{I} - A)^{-1} b = \frac{c_o^\top N [s^{n-1} \ \cdots \ 1]^\top}{D(s)},$$

where $N \in \mathbb{R}^{n\times n}$ is the matrix with its ith row jth column entry N_{ij} introduced in (A.36). Since (A, b) is controllable, it follows from Lemma A.11.1 that N is full rank. Consider an arbitrary vector $\bar{c} \in \mathbb{R}^n$ such that $\bar{c}^\top [s^{n-1} \ \cdots \ 1]^\top$ is a stable $(n-1)$-order polynomial, and let $c_o = (N^{-1})^\top \bar{c}$. Then

$$c_o^\top (s\mathbb{I} - A)^{-1} b = \frac{\bar{c}^\top [s^{n-1} \ \cdots \ 1]^\top}{D(s)}$$

has relative degree 1 with all its zeros in the left half plane. □

A.11.2 Linear Time-Varying Systems

Consider the following time-varying system dynamics:

$$\dot{x}(t) = A(t)x(t) + b(t)u(t), \quad x(t_0) = x_0, \tag{A.37}$$

where $x(t) \in \mathbb{R}^n$ is the state of the system, $u(t) \in \mathbb{R}$ is the input of the system, and $A(t) \in \mathbb{R}^{n\times n}$, $b(t) \in \mathbb{R}^n$ are piecewise-continuous in time.

Definition A.11.1 (see [158]) The system in (A.37) is uniformly controllable if the controllability matrix

$$Q_c(t) = [p_0(t), \ p_1(t), \ldots, \ p_{n-1}(t)] \in \mathbb{R}^{n\times n},$$

where $p_{k+1}(t) = -A(t)p_k(t) + \dot{p}_k(t)$ with $p_0(t) = b(t)$ is nonsingular for all $t \geq t_0$. The representation is strongly controllable if the controllability matrix $Q_c(t)$ is strongly nonsingular, i.e., there exists a constant $q > 0$, such that

$$|\det(Q_c(t))| \geq q, \quad \forall\, t \geq t_0.$$

Lemma A.11.2 (see [156, 157, 167]) Consider the single-input time-varying system in (A.37). There exists a nonsingular continuously differentiable transformation $T(t)$ reducing the system to its controllable (phase-variable) canonical form given by

$$\dot{\bar{x}}(t) = \bar{A}(t)\bar{x}(t) + \bar{b}u(t),$$

where $\bar{A}(t) = (T(t)A(t) + \dot{T}(t))T^{-1}(t)$, $\bar{b} = T(t)b(t)$, with

$$\bar{A}(t) = \begin{bmatrix} 0 & 1 & 0 & \cdots & 0 \\ 0 & 0 & 1 & \cdots & 0 \\ \vdots & \vdots & \vdots & \ddots & \vdots \\ 0 & 0 & 0 & \cdots & 1 \\ -a_1(t) & -a_2(t) & -a_3(t) & \cdots & -a_n(t) \end{bmatrix}, \quad \bar{b} = \begin{bmatrix} 0 \\ 0 \\ \vdots \\ 0 \\ 1 \end{bmatrix},$$

if and only if $(A(t), b(t))$ is uniformly controllable. Moreover, if $A(t)$ and $b(t)$ are uniformly bounded and smooth, and if $(A(t), b(t))$ is strongly controllable, then the transformation $T(t)$ is uniformly bounded, and the resulting entries $a_i(t)$ in $\bar{A}(t)$ (coefficients of the characteristic polynomial) are also uniformly bounded.

A.12 Special Case of State-to-Input Stability

Consider the following dynamics:

$$\dot{x}(t) = Ax(t) + bu(t), \quad x(0) = 0, \tag{A.38}$$

where $x(t) \in \mathbb{R}^n$ is the system state vector, $u(t) \in \mathbb{R}$ is the control input (bounded and piecewise-continuous), $A \in \mathbb{R}^{n\times n}$, and $b \in \mathbb{R}^n$. Let $G(s) \triangleq (s\mathbb{I} - A)^{-1}b$ so that

$$x(s) = G(s)u(s).$$

The solution for $x(t)$ from (A.38) can be written explicitly:

$$x(t) = \int_0^t e^{A(t-\tau)}bu(\tau)d\tau. \tag{A.39}$$

In Section A.7 we defined the notion of BIBO stability for linear systems. Notice that the output of the system in (A.38) is the state $x(t)$ itself. Therefore, the bound on the norm of the system state can be written in the following form:

$$\|x_\tau\|_{\mathcal{L}_\infty} \le \gamma \|u_\tau\|_{\mathcal{L}_\infty}, \quad \gamma = \|G(s)\|_{\mathcal{L}_1}, \quad \forall\, \tau \in [0,\, \infty).$$

Thus, for a BIBO-stable linear system, it is always possible to upper bound the norm of the output by a function of the norm of the input. Looking at (A.39) one may ask the opposite question, namely, whether it is possible to find an upper bound on the system input in terms of its output without invoking the derivatives. While in general the answer to this question is negative [112], we prove that a similar upper bound can be derived for the low-pass-filtered input.

A.12.1 Linear Time-Invariant Systems

Lemma A.12.1 Let the pair $(A,\, b)$ in (A.38) be controllable. Further, let $F(s)$ be an arbitrary strictly proper BIBO-stable transfer function. Then, there exists a proper and stable $G_1(s)$, given by

$$G_1(s) \triangleq \frac{F(s)}{c_o^\top G(s)} c_o^\top,$$

where $c_O \in \mathbb{R}^n$, and $c_o^\top G(s)$ is a minimum phase transfer function with relative degree one, such that

$$F(s)u(s) = G_1(s)x(s).$$

Proof. From Corollary A.11.1, it follows that there exists $c_o \in \mathbb{R}^n$, such that $c_o^\top G(s)$ has relative degree one, and $c_o^\top G(s)$ has all its zeros in the left half plane. Hence, we can write

$$F(s)u(s) = \frac{F(s)}{c_o^\top G(s)} c_o^\top G(s)u(s) = G_1(s)x(s),$$

where the properness of $G_1(s)$ is ensured by the fact that $F(s)$ is strictly proper, while stability follows immediately from its definition. □

Letting $\mu(s) = F(s)u(s)$, it follows from Lemma A.7.1 that

$$\|\mu\|_{\mathcal{L}_\infty} \leq \|G_1(s)\|_{\mathcal{L}_1} \|x\|_{\mathcal{L}_\infty}.$$

A.12.2 Linear Time-Varying Systems

Consider the system

$$\dot{x}(t) = A(t)x(t) + b(t)u(t), \quad x(0) = x_0, \tag{A.40}$$

where $x(t) \in \mathbb{R}^n$ is the system state, $u(t) \in \mathbb{R}$ is the input, and $A(t) \in \mathbb{R}^{n\times n}$, $b(t) \in \mathbb{R}^n$ are piecewise-continuous in time.

The next lemma shows that, for arbitrary strictly proper BIBO-stable transfer function $F(s)$, and under appropriate assumptions on the system in (A.40), the output of the system $F(s)$ to the input $u(t)$ can be upper bounded in terms of the output $x(t)$.

Lemma A.12.2 Let the system in (A.40) be strongly controllable, and let $A(t)$ and $b(t)$ be uniformly bounded and smooth. Further, let $\mathcal{F}$ be the input-output map of $F(s)$. Then for arbitrary $\tau \geq 0$ we have

$$\|(\mathcal{F}u)_\tau\|_{\mathcal{L}_\infty} \leq \gamma \|x_\tau\|_{\mathcal{L}_\infty},$$

where

$$\gamma \triangleq \left(\sum_{i=0}^{n-1} \|\mathcal{F}a_{i+1}\|_{\mathcal{L}_1} \left\| \frac{s^i}{\bar{c}_n s^{n-1} + \cdots + \bar{c}_1} \right\|_{\mathcal{L}_1} + \left\| \frac{F(s)s^n}{\bar{c}_n s^{n-1} + \cdots + \bar{c}_1} \right\|_{\mathcal{L}_1} \right) \|\bar{c}^\top T\|_{\mathcal{L}_\infty},$$

and $\bar{c}_i$, $i = 1, 2, \ldots, n$, are the coefficients of an arbitrary Hurwitz polynomial $p(s) \triangleq \bar{c}_n s^{n-1} + \cdots + \bar{c}_1$, while $T(t)$ and $a_i(t)$ are the transformation matrix and the coefficients of the characteristic polynomial for the system given in (A.40), defined according to Lemma A.11.2.

Proof. Lemma A.11.2 implies that for all $t \in [0, \tau]$ there exists a uniformly bounded transformation $T(t)$, such that $\bar{x}(t) \triangleq T(t)x(t)$ transforms the dynamics in (A.40) to its controllable canonical form:

$$\dot{\bar{x}}(t) = \bar{A}(t)\bar{x}(t) + \bar{b}u(t),$$

where $\bar{A}(t)$ and $\bar{b}$ are given by Lemma A.11.2. Then, the relationship between $\varphi(t) \triangleq \bar{x}_1(t)$ and $u(t)$ is described by the following ODE:

$$\frac{d^n\varphi(t)}{dt^n} + a_n(t)\frac{d^{n-1}\varphi(t)}{dt^{n-1}} + \cdots + a_1(t)\varphi(t) = u(t),$$

where the time-varying coefficients $a_i(t)$ are defined in Lemma A.11.2. Applying the map $\mathcal{F}$ to the both sides of this equation, taking the $\mathcal{L}_\infty$-norms, and using the triangular norm inequality, we obtain the following bound:

$$\|(\mathcal{F}u)_\tau\|_{\mathcal{L}_\infty} \leq \left\| \left(\mathcal{F}\varphi^{(n)}\right)_\tau \right\|_{\mathcal{L}_\infty} + \left\| \left(\mathcal{F}a_n\varphi^{(n-1)} + \cdots + \mathcal{F}a_1\varphi\right)_\tau \right\|_{\mathcal{L}_\infty}. \tag{A.41}$$

Let

$$z(t) \triangleq \bar{c}^\top \bar{x}(t),$$

where $\bar{c} \triangleq [\bar{c}_1, \ \dots, \ \bar{c}_n]^\top$. Then, it follows that

$$\varphi^{(i)}(s) = \frac{s^i}{\bar{c}_n s^{n-1} + \cdots + \bar{c}_1} z(s),$$

which implies

$$F(s)\varphi^{(n)}(s) = \frac{F(s)s^n}{\bar{c}_n s^{n-1} + \cdots + \bar{c}_1} z(s).$$

Thus, (A.41) leads to

$$\|(\mathcal{F}u)_\tau\|_{\mathcal{L}_\infty} \leq \left(\left\| \frac{F(s)s^n}{\bar{c}_n s^{n-1} + \cdots + \bar{c}_1} \right\|_{\mathcal{L}_1} \right.$$
$$\left. + \sum_{i=0}^{n-1} \|\mathcal{F} a_{i+1}\|_{\mathcal{L}_1} \left\| \frac{s^i}{\bar{c}_n s^{n-1} + \cdots + \bar{c}_1} \right\|_{\mathcal{L}_1} \right) \|z_\tau\|_{\mathcal{L}_\infty}.$$

Noticing that $z(t) = \bar{c}^\top T(t)x(t)$, one can write $\|z_\tau\|_{\mathcal{L}_\infty} \leq \|\bar{c}^\top T\|_{\mathcal{L}_\infty} \|x_\tau\|_{\mathcal{L}_\infty}$. This leads to

$$\|(\mathcal{F}u)_\tau\|_{\mathcal{L}_\infty} \leq \gamma \|x_\tau\|_{\mathcal{L}_\infty},$$

which completes the proof. □

Appendix B

Projection Operator for Adaptation Laws

Projection-based adaptation laws are used quite often to prevent parameter drift in adaptation schemes. In this appendix we introduce some definitions and facts from convex analysis and present some properties of the projection operator used throughout the book.

Definition B.1 (see [24]) $\Omega \subseteq \mathbb{R}^n$ is a **convex set** if for all $x, y \in \Omega$ the following holds:

$$\lambda x + (1-\lambda)y \in \Omega, \quad \forall \lambda \in [0, 1].$$

The illustrations of convex and nonconvex sets are shown in Figure B.1.

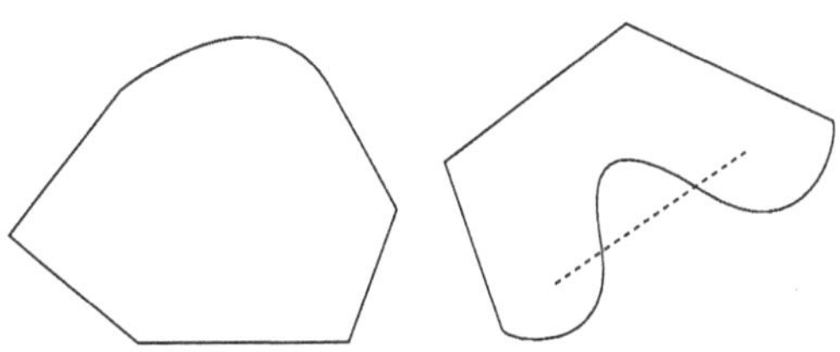

(a) Convex set (b) Nonconvex set

Figure B.1: Illustration of convex and nonconvex sets.

Definition B.2 $f : \mathbb{R}^n \to \mathbb{R}$ is a convex function, if for all $x, y \in \mathbb{R}^n$ the following holds:

$$f(\lambda x + (1-\lambda)y) \leq \lambda f(x) + (1-\lambda)f(y), \quad \forall \lambda \in [0\ 1].$$

A sketch of a convex function is presented in Figure B.2.

Lemma B.1 Let $f : \mathbb{R}^n \to \mathbb{R}$ be a convex function. Then for arbitrary constant δ, the set $\Omega_\delta \triangleq \{\theta \in \mathbb{R}^n | f(\theta) \leq \delta\}$ is convex. The set Ω_δ is called a **sublevel set**.

The proof of this lemma can be found in [24].

Lemma B.2 Let $f : \mathbb{R}^n \to \mathbb{R}$ be a continuously differentiable convex function. Choose a constant δ and consider the convex set $\Omega_\delta \triangleq \{\theta \in \mathbb{R}^n | f(\theta) \leq \delta\}$. Let $\theta, \theta^* \in \Omega_\delta$ and

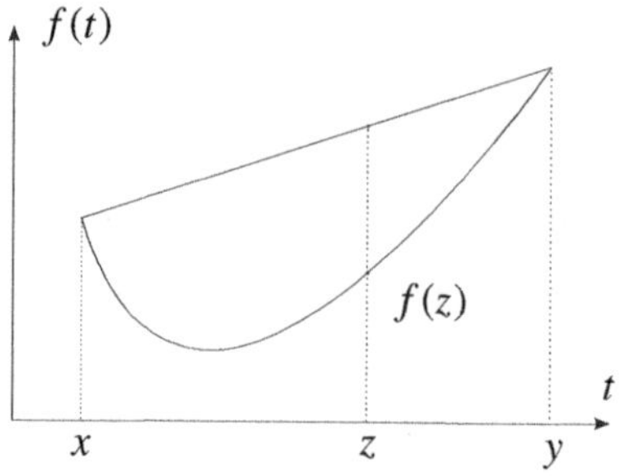

Figure B.2: Illustration of convex function.

$f(\theta^*) < \delta$ and $f(\theta) = \delta$ (i.e., θ^* is not on the boundary of Ω_δ, while θ is on the boundary of Ω_δ). Then, the following inequality takes place:

$$(\theta^* - \theta)^\top \nabla f(\theta) \leq 0,$$

where $\nabla f(\theta)$ is the gradient vector of $f(\cdot)$ evaluated at θ.

The proof of this lemma immediately follows from [145, Theorem 25.1]. Next, we introduce the definition of the projection operator.

Definition B.3 (see [144]) Consider a convex compact set with a smooth boundary given by

$$\Omega_c \triangleq \{\theta \in \mathbb{R}^n \mid f(\theta) \leq c\}, \quad 0 \leq c \leq 1,$$

where $f : \mathbb{R}^n \to \mathbb{R}$ is the following smooth convex function:

$$f(\theta) \triangleq \frac{(\epsilon_\theta + 1)\theta^\top \theta - \theta_{\max}^2}{\epsilon_\theta \theta_{\max}^2},$$

with $\theta_{\max}$ being the norm bound imposed on the vector θ, and $\epsilon_\theta > 0$ is the projection tolerance bound of our choice. The projection operator is defined as

$$\mathrm{Proj}(\theta, y) \triangleq \begin{cases} y & \text{if } f(\theta) < 0, \\ y & \text{if } f(\theta) \geq 0 \text{ and } \nabla f^\top y \leq 0, \\ y - \frac{\nabla f}{\|\nabla f\|}\left\langle \frac{\nabla f}{\|\nabla f\|}, y \right\rangle f(\theta) & \text{if } f(\theta) \geq 0 \text{ and } \nabla f^\top y > 0. \end{cases}$$

Property B.1 (see [144]) The projection operator $\mathrm{Proj}(\theta, y)$ does not alter y if θ belongs to the set $\Omega_0 \triangleq \{\theta \in \mathbb{R}^n \mid f(\theta) \leq 0\}$. In the set $\{\theta \in \mathbb{R}^n \mid 0 \leq f(\theta) \leq 1\}$, if $\nabla f^\top y > 0$, the $\mathrm{Proj}(\theta, y)$ operator subtracts a vector normal to the boundary $\bar{\Omega}_{f(\theta)} = \{\bar{\theta} \in \mathbb{R}^n \mid f(\bar{\theta}) = f(\theta)\}$, so that we get a smooth transformation from the original vector field y to an inward or tangent vector field for Ω_1.

Property B.2 (see [144]) Given the vectors $y \in \mathbb{R}^n$, $\theta^* \in \Omega_0 \subset \Omega_1 \subset \mathbb{R}^n$, and $\theta \in \Omega_1$, we have

$$(\theta - \theta^*)^\top (\mathrm{Proj}(\theta, y) - y) \leq 0. \tag{B.1}$$

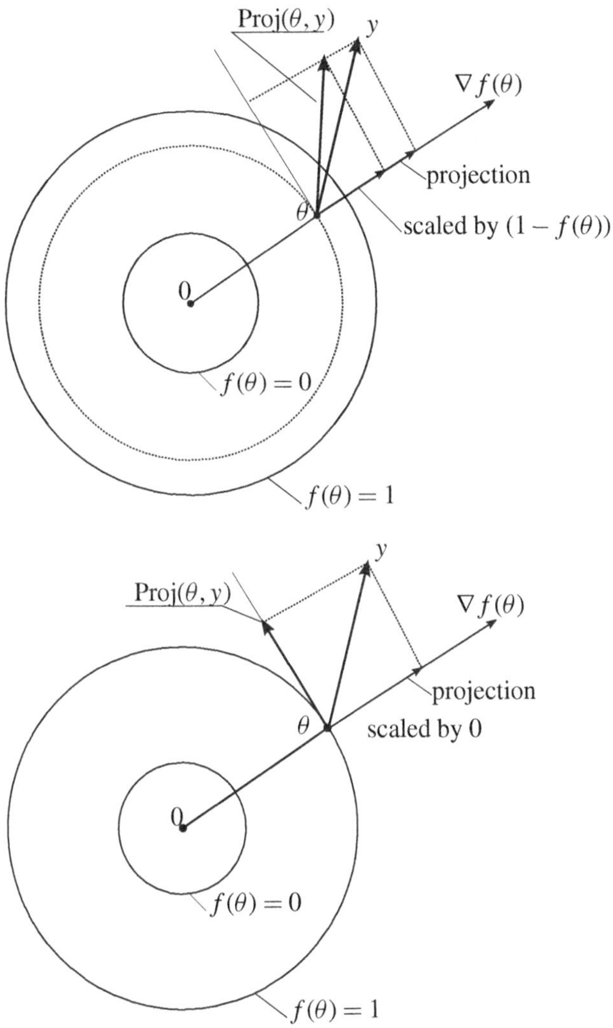

Figure B.3: Illustration of the projection operator.

Indeed,

$$(\theta^* - \theta)^\top (y - \mathrm{Proj}(\theta, y))$$
$$= \begin{cases} 0 & \text{if } f(\theta) < 0, \\ 0 & \text{if } f(\theta) \geq 0 \text{ and } \nabla f^T y \leq 0, \\ \dfrac{\underbrace{(\theta^* - \theta)^\top \nabla f}_{\leq 0} \; \underbrace{\nabla f^T y}_{\geq 0} \; \underbrace{f(\theta)}_{\geq 0}}{\|\nabla f\|^2} & \text{if } f(\theta) \geq 0 \text{ and } \nabla f^\top y > 0. \end{cases}$$

Changing the signs on the left side, one gets (B.1). An illustration of the projection operator is shown in Figure B.3.

Example B.1 In order to illustrate the use of the *projection operator* in the adaptation laws, we consider the system in Section 1.2.1 with the same *direct* model reference adaptive controller. The only difference is that we replace the adaptive law in (1.7) by the following projection-based adaptation law:

$$\dot{\hat{k}}_x(t) \quad = \quad \Gamma \mathrm{Proj}(\hat{k}_x(t), -x(t)e^\top(t)Pb), \quad \hat{k}_x(0) = \hat{k}_{x0}. \tag{B.2}$$

Since the structure of the control law and the definition of the reference model do not change, the tracking error dynamics can still be written as

$$\dot{e}(t) = A_m e(t) + b\tilde{k}_x^\top(t)x(t), \quad e(0) = 0.$$

If we consider the same Lyapunov function candidate as in (1.8), the adaptation law in (B.2) leads to

$$\dot{V}(t) = -e^\top(t)Qe(t) + 2\tilde{k}_x^\top(t)\left(\mathrm{Proj}(\hat{k}_x(t), -x(t)e^\top(t)Pb) + x(t)e^\top(t)Pb\right).$$

Then, Property B.2 implies that

$$\tilde{k}_x^\top(t)\left(\mathrm{Proj}(\hat{k}_x(t), -x(t)e^\top(t)Pb) + x(t)e^\top(t)Pb\right) \le 0,$$

which yields

$$\dot{V}(t) \le -e^\top(t)Qe(t) \le 0.$$

From Barbalat's lemma one concludes that $e(t) \to 0$ as $t \to \infty$. The advantage of using projection-type adaptation is that one ensures boundedness of the adaptive parameters by definition. This property is exploited in the analysis of the $\mathcal{L}_1$ adaptive control architectures. ∎

Remark B.1 Since the MRAC architecture and the predictor-based MRAC architecture lead to the same error dynamics from the same initial conditions, one can also employ the projection-based adaptation law in predictor-based MRAC and refer to Barbalat's lemma to conclude asymptotic stability.

Appendix C

Basic Facts on Linear Matrix Inequalities

C.1 Linear Matrix Inequalities and Convex Optimization

In this section, we give some properties of LMIs. An LMI is an inequality of the form

$$F(x) = F_0 + \sum_{i=1}^{m} x_i F_i < 0, \tag{C.1}$$

where $x \in \mathbb{R}^m$ is the decision variable and $F_i \in \mathbb{R}^{n \times n}$, $i = 0, 1, \ldots, m$, are constant symmetric matrices. By $F(x) < 0$ we mean that $F(x)$ is negative definite.

From the definition of $F(x)$, it follows that it is an affine function of the elements of x. An important property of LMIs is that the set $\{x \,|\, F(x) < 0\}$ is convex, that is, the LMI in (C.1) forms a convex constraint on x.

Lemma C.1.1 (see [23]) A matrix A is positive (or negative) definite if and only if $T^\top A T$ is positive (or negative) definite for arbitrary nonsingular matrix T.

Lemma C.1.2 (Schur complement lemma [23]) The nonlinear inequalities

$$R(x) < 0, \quad Q(x) - S(x)R(x)^{-1}S^\top(x) < 0,$$

where $Q(x) = Q^\top(x)$, $R(x) = R^\top(x)$ and $S(x)$ depend affinely on x, are equivalent to the LMI

$$\begin{bmatrix} Q(x) & S(x) \\ S^\top(x) & R(x) \end{bmatrix} < 0.$$

We consider the following two problems [23]:

LMI Problem Given an LMI in (C.1), the corresponding LMI problem is to find x_f that verifies $F(x_f) < 0$, or to prove that the LMI is infeasible. Further, by solving the LMI $F(x) < 0$, we mean solving the corresponding LMI problem.

Eigenvalue Problem The eigenvalue problem is the minimization of the maximum eigenvalue of a matrix that depends affinely on a variable, subject to an LMI constraint (or proof that the constraint is infeasible), i.e., minimize λ subject to $A(x, \lambda) < 0$, where $A(x, \lambda)$ is affine in (x, λ).

C.2 LMIs for Computation of $\mathcal{L}_1$-Norm of LTI Systems

Consider the following LTI system:

$$\begin{aligned} \dot{x}(t) &= Ax(t) + bu(t), \quad x(0) = x_0, \\ y(t) &= Cx(t), \end{aligned} \tag{C.2}$$

where $A \in \mathbb{R}^{n\times n}$, $C \in \mathbb{R}^{l\times n}$ are given matrices; $b \in \mathbb{R}^n$ is a given vector; $x(t) \in \mathbb{R}^n$ is the state; $y(t) \in \mathbb{R}^l$ is the system output; and $u(t) \in \mathbb{R}$ is the bounded exogenous input. Let $g(t)$ be the impulse response for this system. The following theorem provides a conservative upper bound for the $\mathcal{L}_1$-norm of the system.

Theorem C.2.1 (An upper bound on the $\mathcal{L}_1$-norm [2, 133]) If there exists a symmetric positive definite matrix $P_\alpha \in \mathbb{R}^{n\times n}$ solving the LMI

$$AP_\alpha + P_\alpha A^\top + \alpha P_\alpha + \frac{1}{\alpha} bb^\top \leq 0 \tag{C.3}$$

for some $\alpha > 0$, then

$$\|g\|_{\mathcal{L}_1} \leq \|CP_\alpha C^\top\|_2 .$$

Remark C.2.1 From the Schur complement lemma, it follows that the inequality in (C.3) can be written as

$$\begin{bmatrix} AP_\alpha + P_\alpha A^\top + \alpha P_\alpha & b \\ b^\top & -\alpha \end{bmatrix} \leq 0. \tag{C.4}$$

Remark C.2.2 (Bound on feasible solution for α) The inequality in (C.4) implies that a conservative bound on the feasible solution for α is given by

$$\alpha \in (0, -2\Re(\lambda_{\max}(A))],$$

where $\Re(\lambda_{\max}(A))$ is the maximum real part of the eigenvalues of the matrix A.

Based on the result of Theorem C.2.1 we define the least upper bound on the $\mathcal{L}_1$-norm of the system.

Definition C.2.1 The least upper bound of the $\mathcal{L}_1$-norm over all possible α is called **$*$-norm** [2],

$$\|g\|_* \triangleq \inf_\alpha \|CP_\alpha C^\top\|_2 . \tag{C.5}$$

Remark C.2.3 An important question is, how tight is the $*$-norm for approximation of the $\mathcal{L}_1$-norm? In [168], a family of systems was constructed that illustrates that this approximation can be very poor. It is, nonetheless, quite possible for the $*$-norm upper bound to be useful. It is known, for example, that the simplex algorithm for solving linear programming problems has very poor worst-case computational complexity, but this algorithm has proved to be effective on real-world problems. This example suggests that the engineering experience, and not the theoretical worst-case behavior, will be the definitive test of the $*$-norm approach. In [2], the authors suggest that for many systems the $*$-norm is a quite tight upper bound on the $\mathcal{L}_1$-norm.

C.3 LMIs for Stability Analysis of Systems with Time Delay

Consider the following LTI system in the presence of time delay:

$$\begin{aligned} \dot{x}(t) &= Ax(t) + A_d x(t-\tau), \quad t > 0, \\ x(t) &= \phi(t), \quad t \in [-\tau, 0], \end{aligned} \tag{C.6}$$

where A and $A_d \in \mathbb{R}^{n\times n}$ are the matrices of appropriate dimension, $\phi(t)$ is the given initial condition, and $\tau > 0$ denotes the time delay. The next theorem gives a sufficient condition for stability of this system dependent upon the delay.

Theorem C.3.1 (see [44]) The system (C.6) is asymptotically stable for $\tau \in [0, \bar{\tau}]$ for some $\bar{\tau} > 0$ if there exist $P > 0, P_1 > 0$, and $P_2 > 0$ of appropriate dimensions, satisfying

$$\begin{bmatrix} E & \bar{\tau}PA^\top & \bar{\tau}PA_d^\top \\ \bar{\tau}AP & -\bar{\tau}P_1 & \mathbf{0} \\ \bar{\tau}A_dP & \mathbf{0} & -\bar{\tau}P_2 \end{bmatrix} < 0, \tag{C.7}$$

where $E \triangleq P(A+A_d)^\top + (A+A_d)P + \bar{\tau}A_d(P_1+P_2)A_d^\top$.

Letting $P = P_1 = P_2$, we obtain the following (conservative) result.

Lemma C.3.1 If the LMI

$$\begin{bmatrix} P(A+A_d)^\top + (A+A_d)P & PA^\top & PA_d^\top & A_dP \\ AP & -\frac{1}{\bar{\eta}}P & \mathbf{0} & \mathbf{0} \\ A_dP & \mathbf{0} & -\frac{1}{\bar{\eta}}P & \mathbf{0} \\ PA_d^\top & \mathbf{0} & \mathbf{0} & -\frac{1}{2\bar{\eta}}P \end{bmatrix} \leq 0 \tag{C.8}$$

has a positive definite solution for P, then the system (C.6) is stable for arbitrary $\tau \in [0, \bar{\eta}]$.

The proof follows from the Schur complement (Lemma C.1.2), applied to the inequalities in (C.7).

C.4 LMIs in the Presence of Uncertain Parameters

We finally recall two well-known results from [23], which help to obtain a finite number of LMIs when the uncertain parameter in the original LMI lies in a convex polytope.

Lemma C.4.1 (Vertexization of uncertain LMIs) Let Θ be a convex hull and let Θ_0 be the set of its vertices with finite number of elements. Then, the set

$$\left\{ x \in \mathbb{R}^m \,:\, F(x,\theta) = F_0(\theta) + \sum_{i=1}^{m} x_i F_i(\theta) < 0 \;\forall\, \theta \in \Theta \right\}$$

is nonempty if and only if the set

$$\left\{ x \in \mathbb{R}^m \,:\, F(x,\theta) = F_0(\theta) + \sum_{i=1}^{m} x_i F_i(\theta) < 0 \;\forall\, \theta \in \Theta_0 \right\}$$

is nonempty, provided that $F_i(\theta)$ affinely depend on $\theta \in \Theta$ for each $i = 0, \ldots, m$.

Proof. For the *only-if direction*, the proof immediately follows from the fact that $\Theta_0 \subset \Theta$. For the *if direction*, it follows from the definition of a convex hull and the convexity of $F(x,\theta)$ in θ, given by

$$F(x,\theta) = F\left(x, \sum_i \rho_i \theta_i\right) < \sum_i \rho_i F(x,\theta_i) < 0,$$

which holds for every fixed $x \in \mathbb{R}^m$ and completes the proof. □

Corollary C.4.1 (Polytopic uncertain system) Assume that the system matrices A and A_d in (C.6) are not precisely known but belong to a polytope, such that they can be represented as a convex combination of the vertices of the polytope:

$$\begin{bmatrix} A & A_d \end{bmatrix} = \sum_{j=1}^{n_v} \rho_j \begin{bmatrix} A^{(j)} & A_d^{(j)} \end{bmatrix}, \tag{C.9}$$

where $A^{(j)}$, $A_d^{(j)}$ are the vertices of the polytope, $\rho_j \in [0,1]$ for each index j, $\sum_{j=1}^{n_v} \rho_j = 1$, with n_v being the number of vertices. Then, for arbitrary A and A_d from the polytope (C.9), there exists a feasible solution $P > 0$ for the LMI in (C.8) if and only if the LMI

$$\begin{bmatrix} E^{(j)} & P(A^{(j)})^\top & P(A_d^{(j)})^\top & A_d^{(j)}P \\ A^{(j)}P & -\frac{1}{\bar{\tau}}P & \mathbf{0} & \mathbf{0} \\ A_d^{(j)}P & \mathbf{0} & -\frac{1}{\bar{\tau}}P & \mathbf{0} \\ P(A_d^{(j)})^\top & \mathbf{0} & \mathbf{0} & -\frac{1}{2\bar{\tau}}P \end{bmatrix} < 0$$

has a feasible solution $P > 0$ for all $j = 1, \ldots, n_v$, where $E^{(j)} \triangleq P(A^{(j)} + A_d^{(j)})^\top + (A^{(j)} + A_d^{(j)})P$.

Bibliography

[1] *MIL-STD-1797B. Flying Qualities of Piloted Aircraft*, US Department of Defense Military Specification, February 2006.

[2] JOHN L. ABEDOR, KRSIHAN NAGPAL, AND KAMESHWAR POOLLA, *A linear matrix inequality approach to peak-to-peak gain minimization*, International Journal of Robust and Nonlinear Control, 6 (1996), pp. 899–927.

[3] A. PEDRO AGUIAR, ISAAC KAMINER, REZA GHABCHELOO, ANTÓNIO M. PASCOAL, ENRIC XARGAY, NAIRA HOVAKIMYAN, CHENGYU CAO, AND VLADIMIR DOBROKHODOV, *Time-coordinated path following of multiple UAVs over time-varying networks using $\mathcal{L}_1$ adaptation*, in AIAA Guidance, Navigation and Control Conference, Honolulu, HI, August 2008. AIAA-2008-7131.

[4] BADER ALOLIWI AND HASSAN K. KHALIL, *Adaptive output feedback regulation of a class of nonlinear systems: Convergence and robustness*, IEEE Transactions on Automatic Control, 42 (1997), pp. 1714–1716.

[5] BRIAN D. O. ANDERSON, *Failures of adaptive control theory and their resolution*, Communications in Information and Systems, 5 (2005), pp. 1–20.

[6] BRIAN D. O. ANDERSON, THOMAS BRINSMEAD, DANIEL LIBERZON, AND A. STEPHEN MORSE, *Multiple model adaptive control with safe switching*, International Journal of Adaptive Control and Signal Processing, 15 (2001), pp. 445–470.

[7] ALDAYR D. ARAUJO AND SAHJENDRA N. SINGH, *Variable structure adaptive control of wing-rock motion of slender delta wings*, AIAA Journal of Guidance, Control and Dynamics, 21 (1998), pp. 251–256.

[8] MURAT ARCAK, MARIA SERON, JULIO BRASLAVSKY, AND PETAR V. KOKOTOVIĆ, *Robustification of backstepping against input unmodeled dynamics*, IEEE Transactions on Automatic Control, 45 (2000), pp. 1358–1363.

[9] GÜRDAL ARSLAN AND TAMER BA ̆SAR, *Disturbance attenuating controller design for strict-feedback systems with structurally unknown dynamics*, Automatica J. IFAC, 37 (2001), pp. 1175–1188.

[10] MARCO A. ARTEAGA AND YU TANG, *Adaptive control of robots with an improved transient performance*, IEEE Transactions on Automatic Control, 47 (2002), pp. 1198–1202.

[11] DENIS ARZELIER, ELENA N. GRYAZINA, DIMITRI PEAUCELLE, AND BORIS T. POLYAK, *Mixed LMI/randomized methods for static output feedback control design*, Tech. Report 99535, LAAS-CNRS, Toulouse, October 2009.

[12] KARL J. ÅSTRÖM, *Interactions between excitation and unmodeled dynamics in adaptive control*, in IEEE Conference on Decision and Control, Las Vegas, NV, December 1984, pp. 1276–1281.

[13] KARL J. ÅSTRÖM, *Adaptive feedback control*, Proceedings of the IEEE, 75 (1987), pp. 185–217.

[14] KARL J. ÅSTRÖM, *Adaptive control around 1960*, in IEEE Conference on Decision and Control, New Orleans, LA, January 1995, pp. 2784–2789.

[15] KARL J. ÅSTRÖM AND TORSTEN BOHLIN, *Numerical identification of linear dynamic systems from normal operating records*, in Theory of Self-Adaptive Control Systems, Percival H. Hammond, ed., Plenum Press, New York, 1966, pp. 96–111.

[16] KARL J. ÅSTRÖM AND BJÖRN WITTENMARK, *On self-tuning regulators*, Automatica J. IFAC, 9 (1973), pp. 185–199.

[17] KARL J. ÅSTRÖM AND BJÖRN WITTENMARK, *Adaptive control*, Addison-Wesley, Longman, Boston, MA, 1994.

[18] KARL J. ÅSTRÖM AND BJÖRN WITTENMARK, *Adaptive Control*, 2nd ed., Dover, New York, 2008.

[19] RANDAL W. BEARD, NATHAN B. KNOEBEL, CHENGYU CAO, NAIRA HOVAKIMYAN, AND JOSHUA S. MATTHEWS, *An $\mathcal{L}_1$ adaptive pitch controller for miniature air vehicles*, in AIAA Guidance, Navigation and Control Conference, Keystone, CO, August 2006. AIAA-2006–6777.

[20] RICHARD E. BELLMAN, *Adaptive Control Processes—A Guided Tour*, Princeton University Press, Princeton, NJ, 1961.

[21] DIMITRIS BERTSIMAS AND SANTOSH VEMPALA, *Solving convex programs by random walks*, Journal of the ACM, 51 (2004), pp. 540–556.

[22] BORIS BOSKOVICH AND R. KAUFMAN, *Evaluation of the Honeywell first generation adaptive autopilot and its application to F-94, F-101, X-15 and X-20 vehicles*, Journal of Aircraft, July-August, 1966.

[23] STEPHEN BOYD, LAURENT EL GHAOUI, ERIC FERON, AND VENKATARAMANAN BALAKRISHNAN, *Linear Matrix Inequalities in System and Control Theory*, Studies in Applied Mathematics 15, SIAM, Philadelphia, PA, 1994.

[24] STEPHEN BOYD AND LIEVEN VANDENBERGHE, *Convex Optimization*, Cambridge University Press, Cambridge, UK, 2004.

[25] RICHARD L. BUTCHART AND BARRY SHACKCLOTH, *Synthesis of model reference adaptive systems by Lyapunov's second method*, in IFAC Symposium on the Theory of Self-Adaptive Control Systems, Teddington, UK, September 1965, pp. 145–152.

[26] ANTHONY J. CALISE, YOONGHYUN SHIN, AND MATTHEW D. JOHNSON, *A comparision study of classical and neural network based adaptive control of wing rock*, in AIAA Guidance, Navigation and Control Conference, Providence, RI, August 2004. AIAA-2004-5320.

[27] CHENGYU CAO AND NAIRA HOVAKIMYAN, *Design and analysis of a novel $\mathcal{L}_1$ adaptive control architecture, Part I: Control signal and asymptotic stability*, in American Control Conference, Minneapolis, MN, June 2006, pp. 3397–3402.

[28] CHENGYU CAO AND NAIRA HOVAKIMYAN, *$\mathcal{L}_1$ adaptive controller for systems in the presence of unmodelled actuator dynamics*, in IEEE Conference on Decision and Control, New Orleans, LA, December 2007, pp. 891–896.

[29] CHENGYU CAO AND NAIRA HOVAKIMYAN, *Design and analysis of a novel $\mathcal{L}_1$ adaptive control architecture with guaranteed transient performance*, IEEE Transactions on Automatic Control, 53 (2008), pp. 586–591.

[30] CHENGYU CAO AND NAIRA HOVAKIMYAN, *$\mathcal{L}_1$ adaptive controller for a class of systems with unknown nonlinearities: Part I*, in American Control Conference, Seattle, WA, June 2008, pp. 4093–4098.

[31] CHENGYU CAO AND NAIRA HOVAKIMYAN, *$\mathcal{L}_1$ adaptive controller for nonlinear systems in the presence of unmodelled dynamics: Part II*, in American Control Conference, Seattle, WA, June 2008, pp. 4099–4104.

[32] CHENGYU CAO AND NAIRA HOVAKIMYAN, *$\mathcal{L}_1$ adaptive output feedback controller for systems of unknown dimension*, IEEE Transactions on Automatic Control, 53 (2008), pp. 815–821.

[33] CHENGYU CAO AND NAIRA HOVAKIMYAN, *$\mathcal{L}_1$ adaptive output-feedback controller for non-stricly-positive-real reference systems: Missile longitudinal autopilot design*, AIAA Journal of Guidance, Control, and Dynamics, 32 (2009), pp. 717–726.

[34] CHENGYU CAO AND NAIRA HOVAKIMYAN, *Stability margins of $\mathcal{L}_1$ adaptive control architecture*, IEEE Transactions on Automatic Control, 55 (2010), pp. 480–487.

[35] CHENGYU CAO, NAIRA HOVAKIMYAN, ISAAC KAMINER, VIJAY V. PATEL, AND VLADIMIR DOBROKHODOV, *Stabilization of cascaded systems via $\mathcal{L}_1$ adaptive controller with application to a UAV path following problem and flight test results*, in American Control Conference, New York, July 2007, pp. 1787–1792.

[36] CHENGYU CAO, NAIRA HOVAKIMYAN, AND EUGENE LAVRETSKY, *Application of $\mathcal{L}_1$ adaptive controller to wing rock*, in AIAA Guidance, Navigation and Control Conference, Keystone, CO, August 2006. AIAA-2006-6426.

[37] RONALD CHOE, ENRIC XARGAY, NAIRA HOVAKIMYAN, CHENGYU CAO, AND IRENE M. GREGORY, *$\mathcal{L}_1$ adaptive control under anomaly: Flying qualities and adverse pilot interaction*, in AIAA Guidance, Navigation and Control Conference, Toronto, Canada, August 2010.

[38] MAURICE CLERC AND JAMES KENNEDY, *The particle swarm-explosion, stability, and convergence in a multidimensional complex space*, IEEE Transactions on Evolutionary Computation, 6 (2002), pp. 58–73.

[39] GEORGE E. COOPER AND ROBERT P. HARPER, JR., *The Use of Pilot Rating in the Evaluation of Aircraft Handling Qualities*, Technical Note D–5153, NASA, April 1969.

[40] GERMUND DAHLQUIST, *Convergence and stability in the numerical integration of ordinary differential equations*, Mathematica Scandinavica, 4 (1956), pp. 33–53.

[41] GERMUND DAHLQUIST, *A special stability problem for linear multistep methods*, BIT Numerical Mathematics, 3 (1963), pp. 27–43.

[42] ANIRUDDHA DATTA AND MING-TZU HO, *On modifying model reference adaptive control schemes for performance improvement*, IEEE Transactions on Automatic Control, 39 (1994), pp. 1977–1980.

[43] ANIRUDDHA DATTA AND PETROS A. IOANNOU, *Performance analysis and improvement in model reference adaptive control*, IEEE Transactions on Automatic Control, 39 (1994), pp. 2370–2387.

[44] CARLOS E. DE SOUZA AND XI LI, *Delay-dependent robust $\mathcal{H}_\infty$ control of uncertain linear state-delayed systems*, Automatica J. IFAC, 35 (1999), pp. 1313–1321.

[45] ZHENGTAO DING, *Adaptive control of triangular systems with nonlinear parameterization*, IEEE Transactions on Automatic Control, 44 (2001), pp. 1963–1968.

[46] VLADIMIR DOBROKHODOV, IOANNIS KITSIOS, ISAAC KAMINER, KEVIN D. JONES,
ENRIC XARGAY, NAIRA HOVAKIMYAN, CHENGYU CAO, MARIANO I. LIZÁRRAGA, AND IRENE M. GREGORY, *Flight validation of metrics driven $\mathcal{L}_1$ adaptive control*, in AIAA Guidance, Navigation and Control Conference, Honolulu, HI, August 2008. AIAA-2008-6987.

[47] VLADIMIR DOBROKHODOV, OLEG YAKIMENKO, KEVIN D. JONES, ISAAC KAMINER, EUGENE BOURAKOV, IOANNIS KITSIOS, AND MARIANO LIZARRAGA, *New generation of rapid flight test prototyping system for small unmanned air vehicles*, in Proceedings of AIAA Modelling and Simulation Technologies Conference, Hilton Head Island, SC, August 2007. AIAA-2007-6567.

[48] BO EGARDT, *Stability of Adaptive Controllers*, Springer-Verlag, Berlin, 1979.

[49] BO EGARDT, *Stability of Adaptive Controllers*, Lecture Notes in Control and Information Sciences 20, Springer-Verlag, New York, 1979.

[50] KENAN EZAL, ZIGANG PAN, AND PETAR V. KOKOTOVIĆ, *Locally optimal and robust backstepping design*, IEEE Transactions on Automatic Control, 45 (2000), pp. 260–271.

[51] Xiang Fan and Ralph C. Smith, *Model-based $\mathcal{L}_1$ adaptive control of hysteresis in smart materials*, in IEEE Conference on Decision and Control, Cancun, Mexico, December 2008, pp. 3251–3256.

[52] Sajjad Fekri, Michael Athans, and António M. Pascoal, *Issues, progress and new results in robust adaptive control*, International Journal of Adaptive Control and Signal Processing, 20 (2006), pp. 519–579.

[53] John V. Foster, Kevin Cunningham, Charles M. Fremaux, Gautam H. Shah, Eric C. Stewart, Robert A. Rivers, James E. Wilborn, and William Gato, *Dynamics Modeling and Simulation of Large Transport Airplanes in Upset Conditions*, American Institute of Aeronautics and Astronautics, NASA, Langley Research Center, Hampton, VA, 2005. AIAA-2005-5933.

[54] Oleg N. Gasparyan, *Linear and Nonlinear Multivariable Feedback Control: A Classical Approach*, John Wiley & Sons, New York, 2008.

[55] Tryphon T. Georgiou and Malcolm C. Smith, *Robustness analysis of nonlinear feedback systems: An input-output approach*, IEEE Transactions on Automatic Control, 42 (1997), pp. 1200–1221.

[56] Graham C. Goodwin, Peter J. Ramadge, and Peter E. Caines, *Discrete-time multivariable adaptive control*, IEEE Transactions on Automatic Control, AC-25 (1980), pp. 449–456.

[57] Graham C. Goodwin and Kwai S. Sin, *Adaptive Filtering Prediction and Control*, Prentice-Hall, Englewood Cliffs, NJ, 1984.

[58] Irene M. Gregory, Cehngyu Cao, Vijay V. Patel, and Naira Hovakimyan, *Adaptive control laws for flexible semi-span wind tunnel model of high-aspect ratio flying wing*, in AIAA Guidance, Navigation and Control Conference, Hilton Head, SC, August 2007. AIAA-2007-6525.

[59] Irene M. Gregory, Chengyu Cao, Enric Xargay, Naira Hovakimyan, and Xiaotian Zou, *$\mathcal{L}_1$ adaptive control design for NASA AirSTAR flight test vehicle*, in AIAA Guidance, Navigation and Control Conference, Chicago, IL, August 2009. AIAA-2009-5738.

[60] Irene M. Gregory, E. Xargay, C. Cao, and N. Hovakimyan, *Flight Test of $\mathcal{L}_1$ Adaptive Control on the NASA AirSTAR Flight Test Vehicle*, in AIAA Guidance, Navigation and Control Conference, Toronto, Canada, 2010.

[61] Philip C. Gregory, *Air research and development command plans and programs*, in Proceedings of the Self-Adaptive Flight Control Symposium, Philip C. Gregory, ed., Wright-Patterson Air Force Base, Ohio, 1959, pp. 8–15.

[62] Philip C. Gregory, *Proceedings of Self-Adaptive Flight Control Symposium*, Wright-Patterson Air Force Base, Fairborn, OH, 1959.

[63] BRIAN J. GRIFFIN, JOHN J. BURKEN, ENRIC XARGAY, AND NAIRA HOVAKIMYAN, *$\mathcal{L}_1$ adaptive control augmentation system with application to the X–29 lateral/directional dynamics: A MIMO approach*, in AIAA Guidance, Navigation and Control Conference, Toronto, Canada, August 2010.

[64] BRUNO J. GUERREIRO, CARLOS SILVESTRE, RITA CUNHA, CHENGYU CAO, AND NAIRA HOVAKIMYAN, *$\mathcal{L}_1$ adaptive control for autonomous rotorcraft*, in American Control Conference, St. Louis, MO, June 2009, pp. 3250–3255.

[65] GIORGIO GUGLIERI AND FULVIA QUAGLIOTTI, *Analytical and experimental analysis of wing-rock*, Nonlinear Dynamics, 24 (2001), pp. 129–146.

[66] JOÃO P. HESPANHA, DANIEL LIBERZON, AND A. STEPHEN MORSE, *Overcoming the limitations of adaptive control by means of logic-based switching*, Systems & Control Letters, 49 (2003), pp. 49–65.

[67] RICHARD HINDMAN, CHENGYU CAO, AND NAIRA HOVAKIMYAN, *Designing a high performance, stable $\mathcal{L}_1$ adaptive output feedback controller*, in AIAA Guidance, Navigation and Control Conference, Hilton Head, SC, August 2007. AIAA-2007-6644.

[68] RICHARD HOTZEL AND LAURENT KARSENTI, *Adaptive tracking strategy for a class of nonlinear systems*, IEEE Transactions on Automatic Control, 43 (1998), pp. 1272–1279.

[69] NAIRA HOVAKIMYAN, BONG-JUN YANG, AND ANTHONY J. CALISE, *Adaptive output feedback control methodology applicable to non-minimum phase nonlinear systems*, Automatica J. IFAC, 42 (2006), pp. 513–522.

[70] FAYÇAL IKHOUANE AND MIROSLAV KRSTIĆ, *Robustness of the tuning functions adaptive backstepping design for linear systems*, IEEE Transactions on Automatic Control, 43 (1998), pp. 431–437.

[71] FAYÇAL IKHOUANE, ABDERRAHMAN RABEH, AND FOUAD GIRI, *Transient performance analysis in robust nonlinear adaptive control*, Systems & Control Letters, 31 (1997), pp. 21–31.

[72] PETROS A. IOANNOU AND PETAR V. KOKOTOVIĆ, *An asymptotic error analysis of identifiers and adaptive observers in the presence of parasitics*, IEEE Transactions on Automatic Control, 27 (1982), pp. 921–927.

[73] PETROS A. IOANNOU AND PETAR V. KOKOTOVIĆ, *Adaptive Systems with Reduced Models*, Springer-Verlag, New York, 1983.

[74] PETROS A. IOANNOU AND PETAR V. KOKOTOVIĆ, *Robust redesign of adaptive control*, IEEE Transactions on Automatic Control, 29 (1984), pp. 202–211.

[75] PETROS A. IOANNOU AND JING SUN, *Robust Adaptive Control*, Prentice-Hall, Upper Saddle River, NJ, 1996.

[76] MRDJAN JANKOVIC, *Adaptive nonlinear output feedback tracking with a partial high-gain observer and backstepping*, IEEE Transactions on Automatic Control, 42 (1997), pp. 106–113.

[77] ZHONG-PING JIANG AND DAVID J. HILL, *A robust adaptive backstepping scheme for nonlinear systems with unmodeled dynamics*, IEEE Transactions on Automatic Control, 44 (1999), pp. 1705–1711.

[78] XIN JIN, ASOK RAY, AND ROBERT M. EDWARDS, *Integrated robust and resilient control of nuclear power plants for operational safety and high performance*, IEEE Transactions on Nuclear Science (2010).

[79] THOMAS L. JORDAN, WILLIAM M. LANGFORD, AND JEFFREY S. HILL, *Airborne subscale transport aircraft research testbed-aircraft model development*, in AIAA Guidance, Navigation and Control Conference, San Francisco, CA, August 2005. AIAA-2005-6432.

[80] CLAES G. KÄLLSTRÖM, KARL J. ÅSTRÖM, N. E. THORELL, J. ERIKSSON, AND L. STEN, *Adaptive autopilots for tankers*, Automatica J. IFAC, 15 (1979), pp. 241–254.

[81] RUDOLF KALMAN, *Design of self-optimizing control systems*, ASME Transactions, 80 (1958), pp. 468–478.

[82] ISAAC KAMINER, ANTÓNIO PASCOAL, ENRIC XARGAY, NAIRA HOVAKIMYAN, CHENGYU CAO, AND VLADIMIR DOBROKHODOV, *Path following for unmanned aerial vehicles using $\mathcal{L}_1$ adaptive augmentation of commercial autopilots*, AIAA Journal of Guidance, Control and Dynamics, 33 (2010), pp. 550–564.

[83] ISAAC KAMINER, OLEG A. YAKIMENKO, VLADIMIR DOBROKHODOV, ANTÓNIO M PASCOAL, NAIRA HOVAKIMYAN, VIJAY V. PATEL, CHENGYU CAO, AND AMANDA YOUNG, *Coordinated path following for time-critical missions of multiple UAVs via $\mathcal{L}_1$ adaptive output feedback controllers*, in AIAA Guidance, Navigation and Control Conference, Hilton Head, SC, August 2007. AIAA-2007-6409.

[84] IOANNIS KANELLAKOPOULOS, PETAR V. KOKOTOVIĆ, AND A. STEPHEN MORSE, *Adaptive output-feedback control of systems with output nonlinearities*, IEEE Transactions on Automatic Control, 37 (1992), pp. 1666–1682.

[85] STEINGRÍMUR PÁLL KÁRASON AND ANURADHA M. ANNASWAMY, *Adaptive control in the presence of input constraints*, IEEE Transactions on Automatic Control, 39 (1994), pp. 2325–2330.

[86] HASSAN K. KHALIL, *Nonlinear Systems*, Prentice-Hall, Englewood Cliffs, NJ, 2002.

[87] PRAMOD P. KHARGONEKAR AND ASHOK TIKKUL, *Randomized algorithms for robust control analysis have polynomial complexity*, in IEEE Conference on Decision and Control, Kobe, Japan, December 1996, pp. 3470–3475.

[88] Evgeny Kharisov, Irene M. Gregory, Chengyu Cao, and Naira Hovakimyan, *$\mathcal{L}_1$ adaptive control for flexible space launch vehicle and proposed plan for flight validation*, in AIAA Guidance, Navigation and Control Conference, Honolulu, HI, August 2008. AIAA-2008-7128.

[89] Evgeny Kharisov and Naira Hovakimyan, *Application of $\mathcal{L}_1$ adaptive controller to wing rock*, in AIAA Infotech@Aerospace, Atlanta, GA, April 2010.

[90] Evgeny Kharisov, Naira Hovakimyan, and Karl J. Åström, *Comparison of several adaptive controllers according to their robustness metrics*, in AIAA Guidance, Navigation and Control Conference, Toronto, Canada, August 2010.

[91] Evgeny Kharisov, Naira Hovakimyan, Jiang Wang, and Chengyu Cao, *$\mathcal{L}_1$ adaptive controller for time-varying reference systems in the presence of unmodeled dynamics*, in American Control Conference, Baltimore, MD, June–July 2010, pp. 886–891.

[92] Kwang-Ki Kim and Naira Hovakimyan, *Development of verification and validation approaches for $\mathcal{L}_1$ adaptive control: Multi-criteria optimization for filter design*, in AIAA Guidance, Navigation and Control Conference, Toronto, Canada, August 2010.

[93] Tae-Hyoung Kim, Ichiro Maruta, and Toshiharu Sugie, *Robust PID controller tuning based on the constrained particle swarm optimization*, Automatica J. IFAC, 44 (2008), pp. 1104–1110.

[94] Ioannis Kitsios, Vladimir Dobrokhodov, Isaac Kaminer, Kevin D. Jones, Enric Xargay, Naira Hovakimyan, Chengyu Cao, Mariano I. Lizárraga, Irene M. Gregory, Nhan T. Nguyen, and Kalmanje S. Krishnakumar, *Experimental validation of a metrics driven $\mathcal{L}_1$ adaptive control in the presence of generalized unmodeled dynamics*, in AIAA Guidance, Navigation and Control Conference, Chicago, IL, August 2009. AIAA-2009-6188.

[95] Wolf Kohn, Zelda B. Zabinsky, and Vladimir Brayman, *Optimization of algorithmic parameters using a meta-control approach*, Journal of Global Optimization, 34 (2006), pp. 293–316.

[96] Gerhard Kresselmeier and Kumpati S. Narendra, *Stable model reference adaptive control in the presence of bounded disturbances*, IEEE Transactions on Automatic Control, 27 (1982), pp. 1169–1175.

[97] Prashanth Krishnamurthy and Farshad Khorrami, *A high-gain scaling technique for adaptive output feedback control of feedforward systems*, IEEE Transactions on Automatic Control, 49 (2004), pp. 2286–2292.

[98] Prashanth Krishnamurthy, Farshad Khorrami, and Ramu Sharat Chandra, *Global high-gain-based observer and backstepping controller for generalized output-feedback canonical form*, IEEE Transactions on Automatic Control, 48 (2003), pp. 2277–2283.

[99] PRASHANTH KRISHNAMURTHY, FARSHAD KHORRAMI, AND ZHONG-PING JIANG, *Global output feedback tracking for nonlinear systems in generalized output-feedback canonical form*, IEEE Transactions on Automatic Control, 47 (2002), pp. 814–819.

[100] MIROSLAV KRSTIĆ, IOANNIS KANELLAKOPOULOS, AND PETAR V. KOKOTOVIĆ, *Nonlinear and Adaptive Control Design*, John Wiley & Sons, New York, 1995.

[101] MIROSLAV KRSTIĆ, PETAR V. KOKOTOVIĆ, AND IOANNIS KANELLAKOPOULOS, *Transient-performance improvement with a new class of adaptive controllers*, Systems & Control Letters, 21 (1993), pp. 451–461.

[102] P. R. KUMAR AND PRAVIN P. VARAIYA, *Stochastic systems: Estimation, identification, and adaptive control*, Prentice-Hall, Englewood Cliffs, NJ, 1986.

[103] YOAN D. LANDAU, *Adaptive Control: The Model Reference Approach*, Control & Systems Theory, Marcel Dekker, New York, 1979.

[104] YU LEI, CHENGYU CAO, EUGENE M. CLIFF, NAIRA HOVAKIMYAN, AND ANDREW J. KURDILA, *Design of an $\mathcal{L}_1$ adaptive controller for air-breathing hypersonic vehicle model in the presence of unmodeled dynamics*, in AIAA Guidance, Navigation and Control Conference, Hilton Head, SC, August 2007. AIAA-2006-6527.

[105] YU LEI, CHENGYU CAO, EUGENE M. CLIFF, NAIRA HOVAKIMYAN, ANDREW J. KURDILA, AND KEVIN A. WISE, *$\mathcal{L}_1$ adaptive controller for air-breathing hypersonic vehicle with flexible body dynamics*, in American Control Conference, St. Louis, MO, June 2009, pp. 3166–3171.

[106] TYLER LEMAN, ENRIC XARGAY, GEIR DULLERUD, AND NAIRA HOVAKIMYAN, *$\mathcal{L}_1$ adaptive control augmentation system for the X-48B aircraft*, in AIAA Guidance, Navigation and Control Conference, Chicago, IL, August 2009. AIAA-2009-5619.

[107] DAPENG LI, NAIRA HOVAKIMYAN, AND CHENGYU CAO, *$\mathcal{L}_1$ adaptive controller in the presence of input saturation*, in AIAA Guidance, Navigation and Control Conference, Chicago, IL, August 2009. AIAA-2009-6064.

[108] DAPENG LI, NAIRA HOVAKIMYAN, CHENGYU CAO, AND KEVIN A. WISE, *Filter design for feedback-loop trade-off of $\mathcal{L}_1$ adaptive controller: a linear matrix inequality approach*, in AIAA Guidance, Navigation and Control Conference, Honolulu, HI, August 2008. AIAA-2008-6280.

[109] DAPENG LI, NAIRA HOVAKIMYAN, AND TRYPHON GEORGIOU, *Robustness of $\mathcal{L}_1$ adaptive controllers in the gap metric*, in American Control Conference, Baltimore, MD, June–July 2010, pp. 3247–3252.

[110] ZHIYUAN LI, VLADIMIR DOBROKHODOV, ENRIC XARGAY, NAIRA HOVAKIMYAN, AND ISAAC KAMINER, *Development and implementation of $\mathcal{L}_1$ Gimbal tracking loop onboard of small UAV*, in AIAA Guidance, Navigation and Control Conference, Chicago, IL, August 2009. AIAA-2009-5681.

[111] ZHIYUAN LI, NAIRA HOVAKIMYAN, CHENGYU CAO, AND GLENN-OLE KAASA, *Integrated estimator and $\mathcal{L}_1$ adaptive controller for well drilling systems*, in American Control Conference, St. Louis, MO, June 2009, pp. 1958–1963.

[112] DANIEL LIBERZON, A. STEPHEN MORSE, AND EDUARDO D. SONTAG, *Output-input stability and minimum-phase nonlinear systems*, IEEE Transactions on Automatic Control, 47 (2002), pp. 422–436.

[113] J. LINDAHL, J. MCGUIRE, AND M. REED, *Advanced Flight Vehicle Self-Adaptive Flight Control System*, Wright Air Development (1964).

[114] LENNART LJUNG, *Analysis of recursive stochastic algorithms*, IEEE Transactions on Automatic Control, AC–22 (1977), pp. 551–575.

[115] LENNART LJUNG, *System Identification—Theory for the User*, Prentice-Hall, Englewood Cliffs, NJ, 1987.

[116] JIE LUO, CHENGYU CAO, AND NAIRA HOVAKIMYAN, *$\mathcal{L}_1$ adaptive controller for a class of systems with unknown nonlinearities*, in American Control Conference, Baltimore, MD, June–July 2010, pp. 1659–1664.

[117] LILI MA, CHENGYU CAO, NAIRA HOVAKIMYAN, VLADIMIR DOBROKHODOV, AND ISAAC KAMINER, *Adaptive vision-based guidance law with guaranteed performance bounds for tracking a ground target with time-varying velocity*, in AIAA Guidance, Navigation and Control Conference, Honolulu, HI, August 2008. AIAA-2008-7445.

[118] RICCARDO MARINO AND PATRIZIO TOMEI, *Global adaptive observers for nonlinear systems via filtered transformations*, IEEE Transactions on Automatic Control, 37 (1992), pp. 1239–1245.

[119] RICCARDO MARINO AND PATRIZIO TOMEI, *Global adaptive output-feedback control of nonlinear systems, part I: Linear parameterization*, IEEE Transactions on Automatic Control, 38 (1993), pp. 17–32.

[120] RICCARDO MARINO AND PATRIZIO TOMEI, *Global adaptive output feedback control of nonlinear systems, part II: Nonlinear parameterization*, IEEE Transactions on Automatic Control, 38 (1993), pp. 33–48.

[121] RICCARDO MARINO AND PATRIZIO TOMEI, *Nonlinear Control Design: Geometric, Adaptive, & Robust*, Information and System Sciences, Prentice-Hall, Upper Saddle River, NJ, 1995.

[122] RICCARDO MARINO AND PATRIZIO TOMEI, *An adaptive output feedback control for a class of nonlinear systems with time-varying parameters*, IEEE Transactions on Automatic Control, 44 (1999), pp. 2190–2194.

[123] CHRISTOPHER I. MARRISON AND ROBERT F. STENGEL, *The use of random search and genetic algorithms to optimize stochastic robustness functions*, in American Control Conference, Baltimore, MD, July 1994.

[124] BUDDY MICHINI AND JONATHAN HOW, *$\mathcal{L}_1$ adaptive control for indoor autonomous vehicles: Design process and flight testing*, in AIAA Guidance, Navigation and Control Conference, Chicago, IL, August 2009. AIAA-2009-5754.

[125] DANIEL E. MILLER AND EDWARD J. DAVISON, *An adaptive controller which provides an arbitrarily good transient and steady-state response*, IEEE Transactions on Automatic Control, 36 (1991), pp. 168–81.

[126] ELI MISHKIN AND LUDWIG BRAUN, *Adaptive Control Systems*, McGraw-Hill, New York, 1961.

[127] A. STEPHEN MORSE, *Global stability of parameter-adaptive control systems*, IEEE Transactions on Automatic Control, 25 (1980), pp. 433–439.

[128] A. STEPHEN MORSE, *Supervisory control of families of linear set-point controllers–part 1: Exact matching*, IEEE Transactions on Automatic Control, 41 (1996), pp. 1413–1431.

[129] A. STEPHEN MORSE, *Supervisory control of families of linear set-point controllers–part 2: Robustness*, IEEE Transactions on Automatic Control, 42 (1997), pp. 1500–1515.

[130] KUMPATI S. NARENDRA AND ANURADHA M. ANNASWAMY, *A new adaptive law for robust adaptation without persistent excitation*, IEEE Transactions on Automatic Control, 32 (1987), pp. 134–145.

[131] KUMPATI S. NARENDRA AND JEYENDRAN BALAKRISHNAN, *Improving transient response of adaptive control systems using multiple models and switching*, IEEE Transactions on Automatic Control, 39 (1994), pp. 1861–1866.

[132] KUMPATI S. NARENDRA, YUAN-HAO LIN, AND LENA S. VALAVANI, *Stable adaptive controller design, part II: Proof of stability*, IEEE Transactions on Automatic Control, 25 (1980), pp. 440–448.

[133] SERGEY A. NAZIN, BORIS T. POLYAK, AND MIKHAIL V. TOPUNOV, *Rejection of bounded exogenous disturbances by the method of invariant ellipsoids*, Automation and Remote Control, 68 (2007), pp. 467–486.

[134] VLADIMIR O. NIKIFOROV AND KONSTANTIN V. VORONOV, *Nonlinear adaptive controller with integral action*, IEEE Transactions on Automatic Control, 46 (2001), pp. 2035–2037.

[135] RAÚL ORDÓÑEZ AND KEVIN M. PASSINO, *Adaptive control for a class of nonlinear systems with a time-varying structure*, IEEE Transactions on Automatic Control, 46 (2001), pp. 152–155.

[136] ROMEO ORTEGA, *On Morse's new adaptive controller: parameter convergence and transient performance*, IEEE Transactions on Automatic Control, 38 (1993), pp. 1191–1202.

[137] ZIGANG PAN AND TAMER BAŞAR, *Adaptive controller design for tracking and disturbance attenuation in parametric strict-feedback nonlinear systems*, IEEE Transactions on Automatic Control, 43 (1998), pp. 1066–1083.

[138] ZIGANG PAN, KENAN EZAL, ARTHUR J. KRENER, AND PETAR V. KOKOTOVIĆ, *Backstepping design with local optimality matching*, IEEE Transactions on Automatic Control, 46 (2001), pp. 1014–1027.

[139] PATRICK C. PARKS, *Liapunov redesign of model reference adaptive control systems*, IEEE Transactions on Automatic Control, 11 (1966), pp. 362–367.

[140] VIJAY V. PATEL, CHENGYU CAO, NAIRA HOVAKIMYAN, KEVIN A. WISE, AND EUGENE LAVRETSKY, *$\mathcal{L}_1$ adaptive controller for tailless unstable aircraft*, in American Control Conference, New York, July 2007, pp. 5272–5277.

[141] VIJAY V. PATEL, CHENGYU CAO, NAIRA HOVAKIMYAN, KEVIN A. WISE, AND EUGENE LAVRETSKY, *$\mathcal{L}_1$ adaptive controller for tailless unstable aircraft in the presence of unknown actuator failures*, in AIAA Guidance, Navigation and Control Conference, Hilton Head, SC, August 2007. AIAA-2006-6314.

[142] BENJAMIN B. PETERSON AND KUMPATI S. NARENDRA, *Bounded error adaptive control*, IEEE Transactions on Automatic Control, 27 (1982), pp. 1161–1168.

[143] IRENE A. PIACENZA, ENRIC XARGAY, FULVIA QUAGLIOTTI, GIULIO AVANZINI, AND NAIRA HOVAKIMYAN, *$\mathcal{L}_1$ adaptive control for flexible fixed-wing aircraft: Preliminary results*, in AIAA Guidance, Navigation and Control Conference, Toronto, Canada, August 2010.

[144] JEAN-BAPTISTE POMET AND LAURENT PRALY, *Adaptive nonlinear regulation: estimation from the Lyapunov equation*, IEEE Transactions on Automatic Control, 37 (1992), pp. 729–740.

[145] TYRRELL R. ROCKAFELLAR, *Convex Analysis*, Princeton University Press, Princeton, NJ, 1997.

[146] CHARLES E. ROHRS, LENA S. VALAVANI, MICHAEL ATHANS, AND GÜNTER STEIN, *Robustness of adaptive control algorithms in the presence of unmodeled dynamics*, in IEEE Conference on Decision and Control, vol. 1, Orlando, FL, December 1982, pp. 3–11.

[147] CHARLES E. ROHRS, LENA S. VALAVANI, MICHAEL ATHANS, AND GÜNTER STEIN, *Robustness of continuous-time adaptive control algorithms in the presence of unmodeled dynamics*, IEEE Transactions on Automatic Control, 30 (1985), pp. 881–889.

[148] REUVEN Y. RUBINSTEIN, *Monte-Carlo Optimization, Simulation and Sensitivity of Queueing Networks*, John Wiley, New York, 2003.

[149] WILSON J. RUGH, *Linear System Theory*, Prentice-Hall, Upper Saddle River, NJ, 1996.

[150] SHANKAR SASTRY, *Nonlinear Systems: Analysis, Stability, and Control*, Interdisciplinary Applied Mathematics, Springer-Verlag, New York, 1999.

[151] SHANKAR SASTRY AND MARC BODSON, *Adaptive Control: Stability, Convergence and Robustness*, Prentice-Hall, Englewood Cliffs, NJ, 1989.

[152] JOHANNES J. SCHNEIDER AND SCOTT KIRKPATRICK, *Stochastic Optimization*, Springer, New York, 2006.

[153] HUGO O. SCHUCK, *Honeywell's history and philosophy in the adaptive control field*, in Proceedings of the Self-Adaptive Flight Control Symposium, Philip C. Gregory, ed., Wright-Patterson Air Force Base, Fairborn, OH, 1959, pp. 123–145.

[154] KAI SEDLACZEK AND PETER EBERHARD, *Using augmented lagrangian particle swarm optimization for constrained problems in engineering*, Structural and Multidisciplinary Optimization, 32 (2006), pp. 277–286.

[155] HYUNGBO SHIM AND NAM H. JO, *An almost necessary and sufficient condition for robust stability of closed-loop systems with disturbance observer*, Automatica J. IFAC, 45 (2009), pp. 296–299.

[156] LEONARD M. SILVERMAN, *Transformation of time-variable systems to canonical (phase-variable) form*, IEEE Transactions on Automatic Control, 11 (1966), pp. 300–303.

[157] LEONARD M. SILVERMAN AND BRIAN D. O. ANDERSON, *Controllability, observability and stability of linear systems*, SIAM Journal on Control, 6 (1968), pp. 121–130.

[158] LEONARD M. SILVERMAN AND H. E. MEADOWS, *Controllability and observability in time-variable linear systems*, SIAM Journal on Control, 5 (1967), pp. 64–73.

[159] JEAN-JACQUES E. SLOTINE AND WEIPING LI, *Applied Nonlinear Control*, Prentice-Hall, Englewood Cliffs, NJ, 1991.

[160] JAMES C. SPALL, *Introduction to Stochastic Search and Optimization: Estimation, Simulation and Control*, John Wiley, Hoboken, NJ, 2003.

[161] MARC STEINBERG, *Historical overview of research in reconfigurable flight control*, Proceedings of the IMechE Part G: J. Aerospace Engineering, 219 (2005), pp. 263–275.

[162] HUI SUN, NAIRA HOVAKIMYAN, AND TAMER BAŞAR, *$\mathcal{L}_1$ adaptive controller for systems with input quantization*, in American Control Conference, Baltimore, MD, June–July 2010, pp. 253–258.

[163] JING SUN, *A modified model reference adaptive control scheme for improved transient performance*, IEEE Transactions on Automatic Control, 38 (1993), pp. 1255–1259.

[164] LAWRENCE W. TAYLOR, JR., AND ELMOR J. ADKINS, *Adaptive control and the X-15*, in Princeton University Conference on Aircraft Flying Qualities, Princeton, NJ, June 1965.

[165] ROBERTO TEMPO, ER-WEI BAI, AND FABRIZIO DABBENE, *Probabilistic robustness analysis: Explicit bounds for the minimum number of samples*, System & Control Letters, 30 (1997), pp. 237–242.

[166] ROBERTO TEMPO, GIUSEPPE CALAFIORE, AND FABRIZIO DABBENE, *Randomized Algorithms for Analysis and Control of Uncertain Systems*, Springer-Verlag, London, 2005.

[167] KOSTAS S. TSAKALIS AND PETROS A. IOANNOU, *Linear Time-Varing Systems: Control and Adaptation*, Prentice-Hall, Englewood Cliffs, NJ, 1993.

[168] SALIGRAMA R. VENKATESH AND MUNTHER A. DAHLEH, *Does star norm capture ℓ_1 norm?*, in American Control Conference, Vol. 1, Seattle, WA, June 1995, pp. 944–945.

[169] JIANG WANG, CHENGYU CAO, NAIRA HOVAKIMYAN, RICHARD HINDMAN, AND D. BRETT RIDGELY, *$\mathcal{L}_1$ adaptive controller for a missile longitudinal autopilot design*, in AIAA Guidance, Navigation and Control Conference, Honolulu, HI, August 2008. AIAA-2008-6282.

[170] JIANG WANG, CHENGYU CAO, VIJAY V. PATEL, NAIRA HOVAKIMYAN, AND EUGENE LAVRETSKY, *$\mathcal{L}_1$ adaptive neural network controller for autonomous aerial refueling with guaranteed transient performance*, in AIAA Guidance, Navigation and Control Conference, Keystone, CO, August 2006. AIAA-2006-6206.

[171] JIANG WANG, NAIRA HOVAKIMYAN, AND CHENGYU CAO, *$\mathcal{L}_1$ adaptive augmentation of gain-scheduled controller for racetrack maneuver in aerial refueling*, in AIAA Guidance, Navigation and Control Conference, Chicago, IL, August 2009. AIAA-2009-5739.

[172] JIANG WANG, VIJAY V. PATEL, CHENGYU CAO, NAIRA HOVAKIMYAN, AND EUGENE LAVRETSKY, *$\mathcal{L}_1$ adaptive neural network controller for autonomous aerial refueling in the presence of unknown actuator failures*, in AIAA Guidance, Navigation and Control Conference, Hilton Head, SC, August 2007. AIAA-2006-6313.

[173] XIAOFENG WANG AND NAIRA HOVAKIMYAN, *$\mathcal{L}_1$ adaptive control of event-triggered networked systems*, in American Control Conference, Baltimore, MD, June–July 2010, pp. 2458–2463.

[174] H. PHILIP WHITAKER, *Masschussets Institute of Technology presentation*, in Proceedings of the Self-Adaptive Flight Control Symposium, Philip C. Gregory, ed., Wright-Patterson Air Force Base, Fairborn, OH, 1959, pp. 50–78.

[175] KEVIN A. WISE, EUGENE LAVRETSKY, AND NAIRA HOVAKIMYAN, *Adaptive control in flight: Theory, application, and open problems*, in American Control Conference, Minneapolis, MN, June 2006, pp. 5966–5971.

[176] KEVIN A. WISE, EUGENE LAVRETSKY, JEFFREY ZIMMERMAN, JAMES FRANCIS, DAVE DIXON, AND BRIAN WHITEHEAD, *Adaptive flight control of a sensor guided munition*, in AIAA Guidance, Navigation and Control Conference, San Francisco, CA, August 2005. AIAA-2005-6385.

[177] ENRIC XARGAY, VLADIMIR DOBROKHODOV, ISAAC KAMINER, NAIRA HOVAKIMYAN, CHENGYU CAO, IRENE M. GREGORY, AND ROMAN B. STATNIKOV, *$\mathcal{L}_1$ adaptive flight control system: Systematic design and V&V of control metrics*, in AIAA Guidance, Navigation and Control Conference, Toronto, Canada, August 2010.

[178] ENRIC XARGAY, NAIRA HOVAKIMYAN, AND CHENGYU CAO, *Benchmark problems of adaptive control revisited by $\mathcal{L}_1$ adaptive control*, in Mediterranean Conference on Control and Automation, Thessaloniki, Greece, June 2009, pp. 31–36.

[179] ENRIC XARGAY, NAIRA HOVAKIMYAN, AND CHENGYU CAO, *$\mathcal{L}_1$ adaptive output feedback controller for nonlinear systems in the presence of unmodeled dynamics*, in American Control Conference, St. Louis, MO, June 2009, pp. 5091–5096.

[180] ENRIC XARGAY, NAIRA HOVAKIMYAN, AND CHENGYU CAO, *$\mathcal{L}_1$ adaptive controller for multi-input multi-output systems in the presence of nonlinear unmatched uncertainties*, in American Control Conference, Baltimore, MD, June–July 2010, pp. 875–879.

[181] BIN YAO AND MASAYOSHI TOMIZUKA, *Adaptive robust control of SISO nonlinear systems in a semi-strict feedback form*, Automatica J. IFAC, 33 (1997), pp. 893–900.

[182] B. ERIK YDSTIE, *Transient performance and robustness of direct adaptive control*, IEEE Transactions on Automatic Control, 37 (1992), pp. 1091–1105.

[183] SUNG-JIN YOO, NAIRA HOVAKIMYAN, AND CHENGYU CAO, *Decentralized $\mathcal{L}_1$ adaptive control for large-scale systems with unknown time-varying interaction parameters*, in American Control Conference, Baltimore, MD, June–July 2010, pp. 5590–5595.

[184] ZELDA B. ZABINSKY, *Stochastic Adaptive Search for Global Optimization*, Kluwer Academic Publishers, Boston, 2003.

[185] ZHUQUAN ZANG AND ROBERT R. BITMEAD, *Transient bounds for adaptive control systems*, in IEEE Conference on Decision and Control, Vol. 5, Honolulu, HI, December 1990, pp. 2724–2729.

[186] YOUPING ZHANG, BARIŞ FIDAN, AND PETROS A. IOANNOU, *Backstepping control of linear time-varying systems with known and unknown parameters*, IEEE Transactions on Automatic Control, 48 (2003), pp. 1908–1925.

[187] YOUPING ZHANG AND PETROS A. IOANNOU, *A new linear adaptive controller: Design, analysis and performance*, IEEE Transactions on Automatic Control, 45 (2000), pp. 883–897.

[188] XIAOTIAN ZOU, CHENGYU CAO, AND NAIRA HOVAKIMYAN, *$\mathcal{L}_1$ adaptive controller for systems with hysteresis uncertainties*, in American Control Conference, Baltimore, MD, June–July 2010, pp. 6662–6667.

Index

$\mathcal{L}_1$ design system, 25
$\mathcal{L}_1$ filter design, 26, 113, 189
$\mathcal{L}_1$ adaptive augmentation, 243, 259

Adaptation law modifications
    σ-modification, 2, 248
    e-modification, 2, 248
    projection operator, 248
Adaptation sampling time, 164
AirSTAR
    GTM, 161, 254
    Mobile Operations Station, xix
    piloted evaluations, 254
    T1 research aircraft, xix
    T2 research aircraft, xix
Applications
    AirSTAR, 161, 254
    crew launch vehicle, 192, 259
    flexible aircraft, 259
    nuclear power plant, 261
    rotorcraft, 259
    SIG Rascal, 243
    smart materials, 261
    vision-based control, 260
    well drilling, 261
    X-29, 259
    X-48B, 161, 259

Backstepping, xii, 2, 123
Barbalat's lemma, 273
Base sampling time, 252, 253
Bursting, 78

Certainty equivalence, 2
Control reconfiguration, 242
Control signal saturation, 262
Controllability
    strong, 288
    uniform, 288
Convex
    function, 293
    set, 293
Crossover frequency
    gain, 9
    phase, 77

Decentralized control, 262
Disturbance rejection, 33

Event triggering, 262

Fault detection and isolation, 246
Feedforward prefilter, 147, 163
Final Value Theorem, 23
Flight control systems, 241
Flight envelope, 241, 254
Function
    $\mathcal{KL}$ class, 275
    $\mathcal{K}$ class, 274
    $\mathcal{K}_\infty$ class, 274
    Dirac-delta, 269
    Lyapunov, 271
    positive definite, 271
    truncated, 268

Gain scheduling, 4, 211, 242, 261
Gap metric, 261
Guidance system, 242

Handling qualities, 254
    Cooper–Harper rating, 254, 257
High-fidelity simulator, 255
High-gain feedback, 2, 3, 26, 261

Impulse response, 270
    matrix, 270
Input quantization, 262
Invariant set

positively, 83, 98

Limit cycle, 92
Linear matrix inequality, 114, 297
Locked-in-place failure, 245
Loss of control, 242, 259
Lyapunov equation, 267

Margin
gain, 9, 59
phase, 9
time-delay, 12, 51
Matching conditions, 142, 161
Matrix
Hurwitz, 267
positive definite, 267
state transition, 270
transfer, 270
MIL-Standard requirements, 254
Model Reference Adaptive Control
direct, 1, 4
indirect, 1
neural-network based, 90
state-predictor based, 6
Model-following control, 8

Naval Postgraduate School
flight tests, xviii, 243
hardware-in-the-loop, 243
SIG Rascal 110, xviii
Noise sensitivity, 33
Norm, 265
$*$-norm, 116, 298
$\mathcal{L}_1$-norm for LTI systems, 275
$\mathcal{L}_1$-norm for LTV systems, 278
function, 268
induced matrix, 266
truncated, 268
vector, 265
Nyquist criterion, 8

Optimization problem
constrained, 114
convex, 116
generalized eigenvalue, 117, 297
nonconvex, 114

Parameter drift, 2, 77, 78
Path-following control, 243
Performance optimization, 114, 116
Persistency of excitation, 1, 3, 242, 261
Piecewise-constant adaptation law, 165, 192, 198
Pilot-induced oscillations, 255
Pitch break, 257
Polytopic uncertain system, 299, 300
Projection operator, 18, 294

Recursive design methods, 123, 142
Reference system
LTI, 17
LTV, 211
non-SPR, 192
strictly positive real, 179
Robust Multiple-Model Adaptive Control, 207
Rohrs' example, 1, 77
in flight, 247

Scaled response, 261
Self-oscillating adaptive controller, xi
Self-tuning regulator, xi
Semi-linear system, 85, 101, 127, 149
Space
$\mathcal{L}$-space, 268
extended $\mathcal{L}$-space, 268
Stability
asymptotic, 271, 272
BIBO, 275
uniform, 278
BIBS, 275
exponential, 272
Lyapunov, 271, 272
uniform, 272
State predictor, 6
modification, 22, 174, 207
Stochastic optimization, 115
adaptive random search, 115
meta-control methodologies, 116
particle swarm optimization, 115
randomized algorithms, 115
Strict-feedback system, 3, 123
Supervisory control, 2

Task execution time, 252
Trajectory initialization error, 44

Transmission zeros, 144, 162
Two-cart benchmark problem, 207

Uncertainties
  actuator, 68, 96, 223
  internal dynamics, 96
  matched, 17
  system input gain, 35, 123, 144, 162
  unmatched, 123
Uniform performance, 14, 24
  scaled response, 19, 24, 29
  transient specification, 25

Verification and Validation, 241, 261

Wing rock, 91

X-15, 1

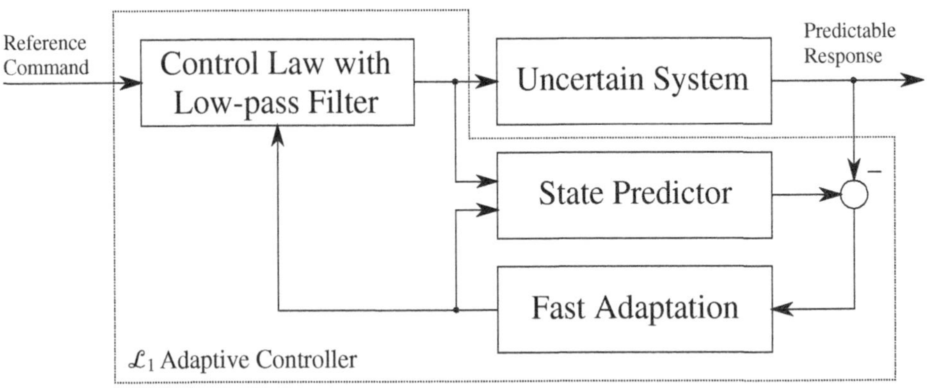

Reference
Command
Control Law with
Low-pass Filter
Uncertain System
Predictable
Response
State Predictor
−
Fast Adaptation
$\mathcal{L}_1$ Adaptive Controller